ELEMENTS OF
# PHYSICAL METALLURGY

ELEMENTS OF

# PHYSICAL METALLURGY

**ALBERT G. GUY**

*and*

**JOHN J. HREN**

*Department of Materials Science and Engineering*
*University of Florida*

**Third Edition**

**ADDISON-WESLEY PUBLISHING COMPANY**
Reading, Massachusetts
Menlo Park, California • London • Don Mills, Ontario

This book is in the
ADDISON-WESLEY SERIES IN METALLURGY AND MATERIALS

*Consulting Editor*
Morris Cohen

ISBN 0-201-02633-3
FGHIJKLMN-HA-8987654321

# PREFACE

Metallurgy is the broad field of the science and technology of metals. It is conveniently divided into two areas: physical metallurgy, which constitutes the subject of this book, and process metallurgy. *Process metallurgy*, sometimes called chemical metallurgy, deals with the extraction of metals from their ores and with the refining, alloying, and initial stage in production (such as a massive ingot of steel). *Physcial metallurgy* is concerned primarily with the products of process metallurgy and their mechanical, physical, and chemical properties. Useful combinations of properties in a given alloy are achieved by control of chemical composition and of "metallurgical structure", which includes such aspects as grain size, constituent particles, and perfection of crystalline structure. Operations such as mechanical working and heat treatment are employed to produce the desired metallurgical structure.

This book, which treats the fundamentals of physical metallurgy, is intended for students in engineering and in science as well as for those whose major interest lies in metallurgy. Chapters 1 through 4 build a background for understanding mechanical properties. The treatment begins with some essential concepts of crystal structure, including the bonding of atoms and the useful quantitative procedure of stereographic projection. The elastic behavior of metals is studied from the crystallographic viewpoint as well as by use of continuum mechanics. The elastic properties of dislocations, the basic imperfections in a crystalline structure, are explained in introductory sections of Chapter 3. This chapter gives a broad coverage of the nature and behavior of vacancies, dislocations, and crystal boundaries. Chapter 4 builds on the previous subject matter in presenting the essential phenomena of plastic deformation and fracture. A section on tests of plastic properties summarizes the test procedures for tensile, hardness, impact, endurance, rupture, and creep tests of alloys.

Utilizing this background on the crystal structure and mechanical behavior of metals, Chapters 5 through 8 present the concepts and tools employed in thermal and mechanical treatments of metallic structures. Chapter 5 describes the study of metallic structures by microscopy, diffraction methods, electron-beam microprobe, and the techniques of quantitative metallography. Chapter 6 shows how thermodynamic concepts can aid in understanding the nature and behavior of the vapor, liquid, and solid phases in metallic systems. The next topic—phase diagrams—covers actual alloy systems, such as iron-carbon, copper-nickel, and aluminum-silicon. Gas-metal systems and isothermal sections of three-component diagrams are also studied. The relation of structure to the properties of alloys is illustrated by commercial alloys such as steels, brasses, and bronzes. Nonequilibrium aspects of phase relations in metal systems are also discussed, including the technique of zone melting. This section of the book concludes with a treatment of diffusion in metals and alloys. The basic laws of diffusion are derived and then applied to such topics as self-diffusion, oxidation, carburizing, and the Kirkendall effect. The influence on diffusion of short-circuit paths, temperature gradients, and electric fields is also described.

The final section of the book—Chapters 9 through 12—treats primarily the important topic of reactions in metals and alloys. A description of the types of phase transformations is followed by important examples of liquid → solid and solid $\alpha$ → solid $\beta$ transformations. Quantitative aspects of recrystallization are explained by use of isothermal-reaction curves. The concepts of phase transformations are applied in Chapter 10 to several processes for strengthening metals and alloys which include cold working, grain refinement, and solid-solution strengthening. The heat treatment of steel and the precipitation hardening of aluminum alloys are discussed in some detail. Chapter 11 considers the chemical and electrochemical reactions that cause deterioration of metals. The treatment focuses on the mechanisms of corrosion and explores the corresponding approaches to control of corrosion. The final chapter treats the electron-theory of metals in sufficient detail to serve as a basis for a study of electrical, magnetic, thermal, and optical properties of metals and alloys. Special attention is given to topics such as superconductivity, thermocouples, magnetic alloys, and prediction of alloy phases.

Although the major aim of our book is to give the student an understanding of the basic principles of physical metallurgy, the study of practical applications forms an integral part of this main purpose. In part, to combine theory and practice, we treat current practical achievements in physical metallurgy as examples of fundamental principles. In addition, quantitative problems are given at the end of most chapters. These exercises have been chosen for their value in illustrating the use of metallurgical principles in answering practical questions concerning the behavior of metals and alloys. Numerous tables concisely presenting representative data on commercial alloys give the student an overall picture of the uses of these alloys. However, for detailed information on properties and processes the student is referred to standard reference works such as the *Metals Handbook*. The index is arranged for easy location of information on a given alloy; the information may sometimes be distributed among several chapters to illustrate several fundamental principles.

Many scientists have contributed through their achievements in the past and present to the subject matter of our book, and we as authors can claim responsibility only for the way of presenting and treating the topics. Even here much of the credit must go to others, especially to Morris Cohen, the consulting editor, whose comments influenced literally every page. Although it is not possible to acknowledge individually the assistance of all the persons who generously supplied information, illustrations, and suggestions, their help provided valuable material for our book. The list of experts who read parts of the manuscript and whose critical suggestions are gratefully acknowledged, includes W. A. Anderson, B. D. Cullity, S. A. Duran, M. G. Fontana, S. M. Gehl, P. F. George, R. E. Grace, S. D. Harkness, J. R. Low, L. Martin, J. B. Newkirk, R. W. Newman, F. H. Wilson, and R. A. Wood.

Many prominent faculty members of metallurgy departments throughout the country made suggestions for our work, based on classroom use of the earlier editions. These courtesies which resulted in many improvements are sincerely appreciated.

The illustrations made available by individuals and companies are credited in the text. We express a especial gratitidue for illustrations which were prepared specifically for this use. Professor Shotaro Morozumi kindly contributed to several of the illustrative models. The equilibrium diagrams are, with a few exceptions, adpated from those given in the compilation *Constitution of Binary Alloys* and its two supplements. Many rough sketches were skillfully converted into finished and effective illustrations by the Addison-Wesley Art Department.

*Gainesville, Florida*                                          A.G.G.
*August 1973*                                                    J.J.H.

# CONTENTS

## Chapter 7   Phase Diagrams

## Chapter 8   Diffusion in Metals

# CHAPTER 1

# CRYSTAL STRUCTURE

Most metals and alloys are in the solid, *crystalline* state when they are used in technology and science. Therefore we begin our study of physical metallurgy with a study of the regular atomic arrangements known as crystal structures. Gaseous, liquid, and noncrystalline-solid aggregations of metal atoms will be considered in later chapters. Since interatomic forces among individual atoms are the basic cause of the various types of aggregation, including the crystalline state, we now consider some elementary principles of electronic theory that underlie the several types of atomic bonding.

## PRINCIPLES OF ATOMIC THEORY

The tiny particles that compose an atomic system behave differently from the massive objects which are a part of our daily experience. A familiar example of this difference is the contrast between the violent Brownian motion in a liquid on the microscopic level and the engineering concept of the same liquid at rest. The counterpart of this difference on the atomic level is described in *Heisenberg's uncertainty principle*, an important restriction on the observation of the properties of any object. This principle defines the minimum, inevitable disturbance of the object that occurs in the process of observing it. A useful statement of the principle is

$$\Delta x \times \Delta p \approx h. \tag{1.1}$$

That is, the product of $\Delta x$, the uncertainty inherent in measuring the position of an object, and $\Delta p$, the uncertainty in momentum, cannot simultaneously be less than (about) Planck's constant, $h$, which has the value $6.62 \times 10^{-34}$ J·s.* These uncertainties are completely negligible in ordinary observations, such as the description of a marble rolling down an inclined plane. They are of vital importance, however, in adequately treating phenomena on an atomic scale. A quantitative illustration is provided by Problem 1, which considers how a single electron might be observed with a minimum disturbance of its motion.

Since, unlike a marble, a moving electron cannot be said to be at a certain point at a given instant, there is an advantage in using a representation of the electron that embodies the inevitable uncertainty in its position. A convenient representation of this kind is based on the de Broglie equation,

$$\lambda = \frac{h}{p}, \tag{1.2}$$

which associates a wavelength $\lambda$ with a particle having a momentum $p$. Thus an electron with a velocity of $0.727 \times 10^7$ m/s would be characterized by the wave-

---

* J · s is the abbreviation for joule · second in SI units. This system of units is used wherever feasible in this book. A brief description of SI units and a list of conversion factors are given in the Appendix.

length*

$$\lambda = \frac{h}{mv} = \frac{6.62 \times 10^{-34}}{(9.11 \times 10^{-31})(0.727 \times 10^7)} = 1.00 \times 10^{-10} \text{ m.}$$

Comparison of Eqs. (1.1) and (1.2) shows that the de Broglie wavelength $\lambda$ is equal to the minimum uncertainty in the position of a particle having the momentum $p$ [see also Problem 1(d)]. Thus a representation of the electron in terms of $\lambda$ nicely meets the requirements of the uncertainty principle. This method is illustrated below by several examples.

Another concept of fundamental importance in atomic theory is *quantization of energy.* A definite packet of energy, one *quantum* of energy, is involved in a unit atomic process. For example, if energy is liberated in a given reaction, the energy is radiated as quanta of a definite magnitude, $E$. The magnitude of $E$ is determined by the unit atomic process. The corresponding frequency, $v$, of this radiant energy is related to $E$ by the *Einstein equation*

$$E = hv, \tag{1.3}$$

where $E$ is the energy in joules contained in one quantum and $v$ is the frequency in hertz (cycles per second). This equation gives the relationship between the frequency of the radiation (green light or x-rays, for example) and the magnitude of the energy packets, the quanta, that make up the radiation. The frequency of green light having a wavelength $\lambda = 0.5 \times 10^{-6}$ m is

$$v = \frac{c}{\lambda} = \frac{3 \times 10^8}{0.5 \times 10^{-6}}$$

$$= 6 \times 10^{14} \text{ hertz,}$$

where $c$ is the velocity of light. A quantum of this radiation therefore is equal to

$$E = hv = (6.62 \times 10^{-34})(6 \times 10^{14})$$

$$= 3.97 \times 10^{-19} \text{ J} = 2.48 \text{ eV.}\dagger$$

Since the wavelength of x-rays is about 5000 times shorter than that of green light, the energy of an x-ray quantum is about 5000 times greater than the energy of a quantum of green light.

Equation (1.3) applies not only to the release of energy but also to the *absorption* of energy. Thus, an atomic process involving the absorption of energy will occur only if adequately large quanta are supplied by a means such as high-

---

* For simplicity the approximation $p = mv$ is used throughout this chapter, although it is quantitatively correct only when $v$ is much less than the speed of light.

† The electron volt (eV) is a conveniently small unit compared with the joule, since 1 eV = $1.60 \times 10^{-19}$ J. By definition, one eV is the energy gained by an electron in falling through a potential difference of one volt.

frequency radiation. Quanta smaller than the critical magnitude are incapable of causing the process to occur. Problem 3 considers a typical phenomenon produced because of this difference in the magnitude of quanta.

## ENERGY LEVELS IN FREE ATOMS

Although the uncertainty principle shows that the *positions* of electrons in an atom are not sharply defined, both theory and experiment give precise *energy levels* for these electrons. Consequently modern theory describes the state of an atom primarily in terms of the energies of its electrons and is relatively unconcerned with the "physical picture" of the atom. To illustrate this approach we choose the simple atomic system represented by the hydrogen atom, which consists of a nucleus with a single positive charge $e$, and an electron with an equal negative charge. The electrical potential energy $W$ of this system depends only on the distance $r$ separating the two charges:

$$W = -\frac{1}{4\pi\varepsilon_0}\frac{e^2}{r}, \qquad (1.4)$$

where $\varepsilon_0$ is the permittivity of free space, $8.85 \times 10^{-12}$ farad/meter. Here we adopt the usual convention that the potential energy is zero when the charges are separated by an infinite distance (see Problem 4).

Employing this knowledge of the potential energy, we will show below how the wave representation of the electron can be used to calculate the values of total energy, $E_n$, that are possible for this system. This demonstration will be more meaningful, however, if the same result is first obtained using the Bohr theory. Although the theory has serious limitations, it has the advantage of mathematical simplicity and therefore continues to find a number of applications such as the present one.

**Bohr theory.** Essentially, the Bohr theory determines the kinetic energy $K$ of the electron and then finds the total energy from the relation $E = K + W$. However, this procedure requires two arbitrary assumptions about the behavior of electrons: (1) electrons can have certain definite energy values, and (2) these values are determined by the condition that the angular momentum of an electron must be an integral multiple of $h/2\pi$. That is,

$$mvr = n\left(\frac{h}{2\pi}\right), \qquad (1.5)$$

where $n$ can have the values 1, 2, 3, . . .

This theory pictures the electron as moving with the velocity $v$ in an orbit of radius $r$. Since the kinetic energy of the electron is $\frac{1}{2}mv^2$, it can be shown, as in Problem 5, that

$$E = K + W = -\frac{1}{8\pi\varepsilon_0}\frac{e^2}{r}. \qquad (1.6)$$

When the "quantizing condition" of Eq. (1.5) is applied, the energy values of the atomic system are found to be

$$E_n = -\left(\frac{me^4}{8\varepsilon_0^2 h^2}\right)\frac{1}{n^2}.$$
(1.7)

That is, the electron in the simple atomic system being considered can have only certain energy values. Ordinarily the electron is in the lowest energy state, which is given by $n = 1$:

$$E_1 = -\frac{me^4}{8\varepsilon_0^2 h^2} = -13.6\,\text{eV}.$$

If energy is supplied to the system, for example in the form of heat, the electron cannot increase its energy by any arbitrary amount. Rather, it must occupy one of the states given by Eq. (1.7). Thus, it might change to the state for which $n = 2$, $E_2 = -3.4\,\text{eV}$, by absorbing the necessary 10.2 eV of energy from its surroundings. The atom would then be in an *excited state* since it would have more energy than the lowest energy state, $E_1$. Such excited states are unstable and tend to change to a state of lower energy with the release of a quantum of energy. In this case the change $E_2 \rightarrow E_1$ would release a 10.2 eV quantum of energy in the form of ultraviolet light of wavelength 0.1216 μm (see Problem 6).

**Wave mechanics.** If this simple atomic system is analyzed in terms of de Broglie waves, the same result can be obtained without the use of arbitrary assumptions. Since the general treatment of wave mechanics involves advanced mathematics, a simplified approach is used here to allow emphasis on the essential concepts. When wave motion is involved in a physical system, the characteristics of the system determine the principal waveforms that can exist. The influence of the length of an organ pipe on the musical tones that it produces is a familiar example of this principle. In this case two equivalent methods can describe these tones: (1) by explicitly describing the tones in question (particular sinusoidal waves), or (2) by stating the features of the organ pipe that determine what tones are produced. The second method is a general description covering a variety of special cases.

Schroedinger recognized that de Broglie waves were similar to the first method and that it was desirable to recast this concept into the second, general form. An example of the type of description that Schroedinger developed is the equation

$$\frac{d^2\psi}{dx^2} + k^2\psi = 0.$$
(1.8)

Trial will show that a solution of this equation is the *wave function* $\psi = A \sin kx$, which is simply a sine wave whose amplitude is $A$. This wave function can be expressed in terms of the de Broglie wavelength $\lambda$, as $\psi = A \sin(2\pi x/\lambda)$, since the term $2\pi x/\lambda$ correctly expresses the condition that $\psi$ should describe one wavelength (from $\sin 0$ to $\sin 2\pi$) as $x$ goes from zero to $\lambda$. Furthermore, it follows from Eq. (1.2) that $\lambda$ can be written in terms of the total energy $E$ and the potential energy

$W$ since, as shown in Problem 7,

$$\lambda = \frac{h}{p} = \frac{h}{\sqrt{2m(E - W)}}. \tag{1.9}$$

The wave function then becomes

$$\psi = A \sin \frac{\sqrt{8\pi^2 m(E - W)}x}{h}. \tag{1.10}$$

Comparison of this expression for $\psi$ with the original solution of Eq. (1.8) shows that

$$k = \frac{\sqrt{8\pi^2 m(E - W)}}{h} \tag{1.11}$$

and

$$k^2 = \frac{8\pi^2 m}{h^2}(E - W). \tag{1.12}$$

Substitution of this value for $k^2$ in Eq. (1.8) gives a form of the famous *Schroedinger equation**

$$\frac{d^2\psi}{dx^2} + \frac{8\pi^2 m}{h^2}(E - W)\psi = 0. \tag{1.13}$$

In using the Schroedinger equation to determine the values of $E_n$ for the hydrogen atom, the first step is to substitute for $W$ the actual expression for the potential energy, $-(1/4\pi\varepsilon_0)(e^2/r)$, Eq. (1.4). The equation can then be solved, and only certain values of the wave function, $\psi_1, \psi_2, \ldots, \psi_n$, are found to satisfy the equation. Furthermore, the equation is satisfied only if for each of these wave functions there is a corresponding, definite value of the energy $E$. Thus, values $E_1, E_2, \ldots, E_n$ identical with those given by the Bohr theory, Eq. (1.7), are obtained as a natural consequence of the wave representation of the electron. Also, the wave functions can be used to picture the atom in terms of a calculated probability of finding the electron at a given distance from the nucleus. Figure 1.1(a) is a sketch of the diffuse electron density calculated for an unexcited hydrogen atom. The corresponding illustration for the more complex magnesium atom, Fig. 1.1(b), shows a high electron density near the center where the ten core electrons are concentrated and a lower density for the two outer, valence electrons.

This simplified treatment of wave mechanics has focused attention on the principal quantum number $n$, which plays a dominant role in determining the energies of an atomic system. However, a more rigorous treatment reveals that four quantum numbers (principal, secondary, magnetic, and spin) are needed for

---

* The corresponding equation for a three-dimensional problem is obtained by replacing $d^2\psi/dx^2$ by the sum $\partial^2\psi/\partial x^2 + \partial^2\psi/\partial y^2 + \partial^2\psi/\partial z^2$. For convenience, this sum can be represented by the symbol $\nabla^2\psi$, so that the Schroedinger equation is often written as $\nabla^2\psi + (8\pi^2 m/h^2)(E - W)\psi = 0$.

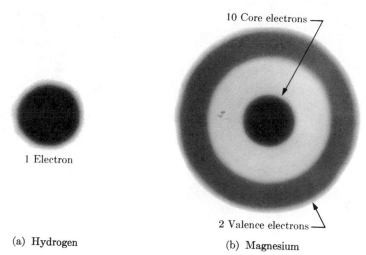

10 Core electrons

2 Valence electrons

1 Electron

(a)  Hydrogen                    (b)  Magnesium

**Fig. 1.1**  Schematic illustration of the electron densities in free atoms.

the most general description of electronic "states" and energies. This mathematical analysis also demonstrates that only a limited number of electronic states are possible for a given value of each quantum number.  Some of these essential features of modern atomic theory are illustrated in the following discussion.

**Electronic structures of the elements.**  The characteristic properties of an element such as magnesium or iron are determined by the energies of the various electrons in its atoms.  These energies in turn are dependent on the quantum numbers that enter into the solution of the appropriate Schroedinger wave equation.  Therefore the electron energies are conveniently designated by sets of *quantum numbers*. In general, electrons with different *principal quantum numbers* (1, 2, 3, etc.) differ greatly in energy; those that are different only in their *secondary quantum numbers* (s, p, d, f, etc.) usually differ only slightly in energy.  The relative values of the four different electron energy levels in a magnesium atom, Fig. 1.2(a), illustrate these rules.

Because of the differences in energy, the electrons in a given atom tend to occupy the lowest possible quantum states.  For example, the single electron of the hydrogen atom is in the 1s state, since this state has the lowest energy.  Also, the two electrons in the helium atom occupy this state,* and the electron con-

---

* These two electrons do not have identical quantum numbers.  In this case the *spin quantum numbers* are different (one is $+\frac{1}{2}$ and the other is $-\frac{1}{2}$).  If all the quantum numbers are taken into account, it can be shown that (1) the maximum number of electrons that can exist in the $n$th principal quantum state is $2n^2$, (2) the maximum number of electrons that can exist in a secondary quantum state is $2(2l - 1)$, where $l$-values of 1, 2, 3, and 4 correspond to secondary quantum numbers s, p, d, and f, respectively.

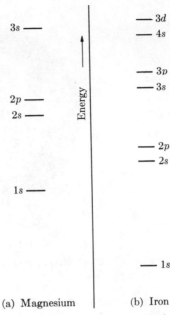

Fig. 1.2 Schematic illustration of the relative electron energies in two free atoms.

figuration of helium can be represented as $(1s)^2$, where the 2 designates the number of electrons in the $1s$ state. The *Pauli exclusion principle* requires that a given quantum state (specified by four quantum numbers) be occupied by only one electron. Consequently, no more than two electrons can be accommodated in this first energy level. Therefore in a lithium atom, which has three electrons, one of the electrons is in the next higher energy state, $2s$. The electron configuration of the lithium atom is thus $(1s)^2(2s)^1$. With further increases in the number of electrons, the electron configurations of the atoms change in the following regular manner:

| Atom | Atomic number | Electron configuration |
|---|---|---|
| Beryllium | 4 | $(1s)^2 (2s)^2$ |
| Boron | 5 | $(1s)^2 (2s)^2(2p)^1$ |
| Carbon | 6 | $(1s)^2 (2s)^2(2p)^2$ |
| Nitrogen | 7 | $(1s)^2 (2s)^2(2p)^3$ |
| Oxygen | 8 | $(1s)^2 (2s)^2(2p)^4$ |
| Fluorine | 9 | $(1s)^2 (2s)^2(2p)^5$ |
| Neon | 10 | $(1s)^2 (2s)^2(2p)^6$ |
| Sodium | 11 | $(1s)^2 (2s)^2(2p)^6 (3s)^1$ |
| Magnesium | 12 | $(1s)^2 (2s)^2(2p)^6 (3s)^2$ |

This regularity in the sequence of principal and secondary quantum numbers is interrupted by several series of *transition* elements. In these elements, after one or two electrons have entered an outer orbit (the 4s orbit in the instance shown here), the additional electrons enter an inner orbit (3d in this example). Thus, the first series of transition elements develops as follows.

| Atom | Atomic number | Electron configuration | | | |
|------|---------------|------------------------|---|---|---|
| Argon | 18 | $(1s)^2$ | $(2s)^2(2p)^6$ | $(3s)^2(3p)^6$ | |
| Potassium | 19 | $(1s)^2$ | $(2s)^2(2p)^6$ | $(3s)^2(3p)^6$ | $(4s)^1$ |
| Calcium | 20 | $(1s)^2$ | $(2s)^2(2p)^6$ | $(3s)^2(3p)^6$ | $(4s)^2$ |
| Scandium | 21 | $(1s)^2$ | $(2s)^2(2p)^6$ | $(3s)^2(3p)^6(3d)^1$ | $(4s)^2$ |
| Titanium | 22 | $(1s)^2$ | $(2s)^2(2p)^6$ | $(3s)^2(3p)^6(3d)^2$ | $(4s)^2$ |
| Vanadium | 23 | $(1s)^2$ | $(2s)^2(2p)^6$ | $(3s)^2(3p)^6(3d)^3$ | $(4s)^2$ |
| Chromium | 24 | $(1s)^2$ | $(2s)^2(2p)^6$ | $(3s)^2(3p)^6(3d)^5$ | $(4s)^1$ |
| Manganese | 25 | $(1s)^2$ | $(2s)^2(2p)^6$ | $(3s)^2(3p)^6(3d)^5$ | $(4s)^2$ |
| Iron | 26 | $(1s)^2$ | $(2s)^2(2p)^6$ | $(3s)^2(3p)^6(3d)^6$ | $(4s)^2$ |
| Cobalt | 27 | $(1s)^2$ | $(2s)^2(2p)^6$ | $(3s)^2(3p)^6(3d)^7$ | $(4s)^2$ |
| Nickel | 28 | $(1s)^2$ | $(2s)^2(2p)^6$ | $(3s)^2(3p)^6(3d)^8$ | $(4s)^2$ |
| Copper | 29 | $(1s)^2$ | $(2s)^2(2p)^6$ | $(3s)^2(3p)^6(3d)^{10}$ | $(4s)^1$ |

(Atomic numbers 21 through 28 are bracketed as *transition elements*.)

The reason for this behavior is that the energy of the 4s electrons is less than but close to that of the 3d electrons for this group of transition elements. The relation of these energies is shown in Fig. 1.2(b) for iron. This circumstance causes not only the variable valences of these elements, but also the pronounced property changes on alloying, since the electron energy states are easily changed by the effect of atomic interactions in the alloys.

**Periodic table.**  The approximate regularity of the development of electron configurations in the various elements leads to a similar approximate regularity in many of the properties of the elements. Figure 1.3 illustrates this behavior with respect to the distances between neighboring atoms in solid metals. The fact that the alkali metals lithium, sodium, potassium, rubidium, and cesium have larger interatomic distances than their neighboring elements in the periodic table is a direct result of the possession by these metals of a single s-type electron. The remaining electrons are in stable levels closer to the nucleus and screen the valence electron from the positive charge of the nucleus. Thus the attractive force is weak, and consequently the orbit of the valence electron is large. Undoubtedly element 87, francium, also will be found to have a maximum value of interatomic distance.

Since useful predictions of this kind can be made on the basis of similarities in electronic structure, many tables highlighting these similarities have been devised. A convenient version is given in Fig. 1.4. Elements belonging to the

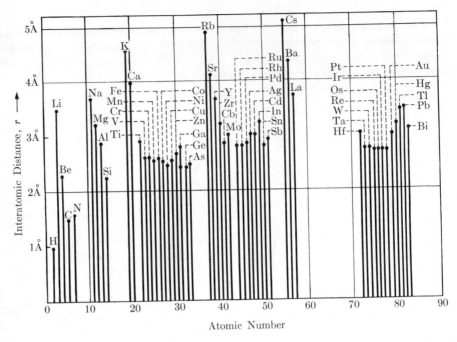

**Fig. 1.3** The periodicity of the distance $r$ between the centers of neighboring atoms in solid metals. One angstrom unit, Å, is $10^{-10}$ meters. (After F. Seitz.)

same group can often adequately replace one another in alloys. Although sulfur is the element commonly used to increase the machinability of steel, both selenium and tellurium have been successfully employed for this purpose in stainless steels and in nonferrous alloys. Tungsten, which increases the resistance to softening of steels at high temperatures, can be replaced by molybdenum or, less satisfactorily, by chromium. The number of such relationships in the periodic table is extremely large, and for this reason the periodic table is constantly used in development and research problems.

## BONDING WITHIN CRYSTAL STRUCTURES

Metals are *aggregates* of atoms, usually arranged in a crystal structure. Metallic properties, therefore, depend not only on the nature of the constituent atoms, but also on the manner in which the atoms are assembled.

The essential, crystalline nature of metals is seldom apparent in final products, such as girders, automobile fenders, or door knobs, but the properties of the individual crystals in a metal are responsible for its ultimate engineering usefulness and they strongly influence the entire processing of the metals. The photographs

| IA | IIA | IIIB | IVB | VB | VIB | VIIB | VIII | | | IB | IIB | IIIA | IVA | VA | VIA | VIIA | 0 |
|---|---|---|---|---|---|---|---|---|---|---|---|---|---|---|---|---|---|
| 1<br>H<br>1.0080 | | | | | | | | | | | | | | | | | 2<br>He<br>4.0026 |
| 3<br>Li<br>6.939 | 4<br>Be<br>9.012 | | | | | | | | | | | 5<br>B<br>10.811 | 6<br>C<br>12.011 | 7<br>N<br>14.007 | 8<br>O<br>15.9994 | 9<br>F<br>18.998 | 10<br>Ne<br>20.183 |
| 11<br>Na<br>22.990 | 12<br>Mg<br>24.312 | | | | | | | | | | | 13<br>Al<br>26.98 | 14<br>Si<br>28.086 | 15<br>P<br>30.96 | 16<br>S<br>32.064 | 17<br>Cl<br>35.453 | 18<br>Ar<br>39.95 |
| 19<br>K<br>39.102 | 20<br>Ca<br>40.08 | 21<br>Sc<br>44.96 | 22<br>Ti<br>47.90 | 23<br>V<br>50.94 | 24<br>Cr<br>52.04 | 25<br>Mn<br>54.94 | 26<br>Fe<br>55.85 | 27<br>Co<br>58.93 | 28<br>Ni<br>58.71 | 29<br>Cu<br>63.55 | 30<br>Zn<br>65.37 | 31<br>Ga<br>69.72 | 32<br>Ge<br>72.59 | 33<br>As<br>74.92 | 34<br>Se<br>78.96 | 35<br>Br<br>79.90 | 36<br>Kr<br>83.80 |
| 37<br>Rb<br>85.47 | 38<br>Sr<br>87.62 | 39<br>Y<br>88.91 | 40<br>Zr<br>91.22 | 41<br>Nb<br>92.91 | 42<br>Mo<br>95.94 | 43<br>Tc<br>97 | 44<br>Ru<br>101.07 | 45<br>Rh<br>102.91 | 46<br>Pd<br>106.4 | 47<br>Ag<br>107.87 | 48<br>Cd<br>112.40 | 49<br>In<br>114.82 | 50<br>Sn<br>118.69 | 51<br>Sb<br>121.75 | 52<br>Te<br>127.60 | 53<br>I<br>126.90 | 54<br>Xe<br>131.30 |
| 55<br>Cs<br>132.91 | 56<br>Ba<br>137.34 | 57–71<br>La<br>series* | 72<br>Hf<br>178.49 | 73<br>Ta<br>180.95 | 74<br>W<br>183.85 | 75<br>Re<br>186.2 | 76<br>Os<br>190.2 | 77<br>Ir<br>192.2 | 78<br>Pt<br>195.1 | 79<br>Au<br>196.97 | 80<br>Hg<br>200.59 | 81<br>Tl<br>204.37 | 82<br>Pb<br>207.19 | 83<br>Bi<br>208.98 | 84<br>Po<br>209 | 85<br>At<br>210 | 86<br>Rn<br>222 |
| 87<br>Fr<br>223 | 88<br>Ra<br>226 | 89–<br>Ac<br>series† | | | | | | | | | | | | | | | |

Transition elements (shaded)

VIII

| * | 58<br>Ce<br>140.12 | 59<br>Pr<br>140.91 | 60<br>Nd<br>144.24 | 61<br>Pm<br>145 | 62<br>Sm<br>150.35 | 63<br>Eu<br>151.96 | 64<br>Gd<br>157.25 | 65<br>Tb<br>158.92 | 66<br>Dy<br>162.50 | 67<br>Ho<br>164.93 | 68<br>Er<br>167.26 | 69<br>Tm<br>168.93 | 70<br>Yb<br>173.04 | 71<br>Lu<br>174.97 |
|---|---|---|---|---|---|---|---|---|---|---|---|---|---|---|
| † | 90<br>Th<br>232.04 | 91<br>Pa<br>231 | 92<br>U<br>238.03 | 93<br>Np<br>237 | 94<br>Pu<br>244 | 95<br>Am<br>243 | 96<br>Cm<br>247 | 97<br>Bk<br>247 | 98<br>Cf<br>251 | 99<br>Es<br>254 | 100<br>Fm<br>257 | 101<br>Md<br>256 | 102<br>No<br>254 | 103<br>Lw<br>257 |

**Fig. 1.4** A periodic table of the elements.

(a) Crystals of magnesium metal.
(Courtesy Dow Chemical Company.)

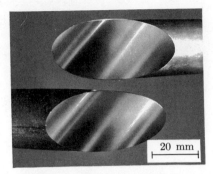

(b) Cleavage of a crystal of zinc on a
crystal plane.

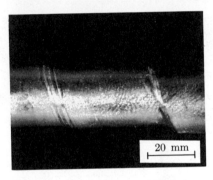

(c) Deformation of a crystal of zinc on
parallel crystal planes. (Courtesy
E. R. Parker.)

(d) Evidence of crystal planes in grains of
a copper-silicon alloy. (Courtesy James
Franck Institute, University of Chicago.)

**Fig. 1.5**  Evidences of the crystalline nature of metals.

in Fig. 1.5 show a few of the striking effects that arise because of the crystalline
nature of metals.  External "crystal" faces similar to those seen on mineralogical
specimens of quartz, galena, etc., also develop on metals under suitable conditions.
The attractive crystals of magnesium in Fig. 1.5(a) were grown slowly by deposition
from vapor in a commercial sublimation process.  Even when the crystal planes
cannot be seen on the surface of a metal crystal they can sometimes be clearly
revealed, as in Fig. 1.5(b), by cleavage along an important crystal plane.  This zinc
crystal was cooled in liquid nitrogen and was then easily cleaved by the pressure
of a knife blade.  Action of similar crystal planes can be seen in the "slip" process
that occurs when metals are deformed, Fig. 1.5(c).

The three previous illustrations involved *single crystals*; a more subtle example
of crystalline effects in metals is given for an ordinary *polycrystalline* alloy in

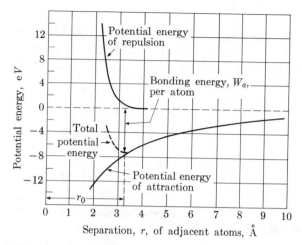

**Fig. 1.6** Schematic illustration of typical bonding of atoms in a crystal as a function of interatomic spacing.

Fig. 1.5(d). The large grain that occupies most of this photomicrograph* was originally a single crystal. The parallel sets of bands were produced when a second, darker constituent formed on certain crystal planes of the original crystal. The reason for the existence of several sets of equivalent crystal planes in typical metal crystals will be explained in a later section.

**Interatomic bonding.** The forces that bond the atoms of a metal or alloy in a crystal structure are predominantly electrical (rather than magnetic or gravitational). In addition to electrostatic attraction, which is the basis of ionic bonding, other types of electrical forces exist and lead to covalent bonding, van der Waals bonding, resonance stabilization, and exchange energy. Each of these bonding mechanisms exists in a given metallic crystal, but the relative contribution varies from one substance to another. Although the bonding in a typical alloy is too complex to be analyzed quantitatively even by modern quantum mechanics, we can picture schematically the nature of the attraction (and repulsion) that exist in a stable crystal structure.

Consider the hypothetical situation in which the atoms (or ions) in a given crystal structure retain their relative positions except that the interatomic spacing is increased by a large factor, Fig. 1.6. If the atoms were separated an infinite

---

* Photographs of metal structures may be at less than natural size (Fig. 1.5(b) and (c), for example), but *photomicrographs,* such as Fig. 1.5(d), are photographed through a microscope and are invariably at high magnifications. In either case, the magnification can be conveniently indicated by a "reference length" on the photograph. The numerical value of the magnification is given by the ratio (actual length)/(indicated length). For example, the magnification of Fig. 1.5(d) is: $16.5 \text{ mm}/50 \text{ }\mu\text{m} = 16.5/0.050 = 330 \times$ or 330 diameters of magnification.

distance, the interatomic forces would be zero and the potential energy can be set equal to its reference value, zero. As the atoms begin to approach one another, attractive forces first become important. The potential energy, $W$, is the negative of the work that could be done by these forces, $F'$, between infinity and some interatomic spacing, $r$,

$$W = -\int_\infty^r F' \, dr \qquad (1.14)$$

The potential energy continues to decrease to lower (and therefore more stable) values as the distance of separation decreases toward the actual interatomic separation in the crystal $r_0$. Near $r_0$, however, repulsion of the ion cores (filled electron shells) of the atoms becomes significant and leads to a sharp increase in potential energy of repulsion. Consequently, the total potential energy reaches a minimum, $W_a$, at $r_0$, the equilibrium atomic spacing. Problem 10 considers an approximate calculation of $W_a$ for a pair of ions in a crystal of sodium chloride.

**Types of bonding.** The bonding mechanism in a given metallic material is some combination of the four major types, Fig. 1.7: metallic, covalent, ionic, and van der Waals. Although a monovalent metal such as sodium is characterized by almost pure metallic bonding, tetravalent tin has appreciable covalent character and even exists in a form (gray tin) having few metallic properties. The intermetallic compound $Mg_3Sb_2$ is strongly ionic in nature. We now consider briefly the essential nature of each of the four major types of bonding.

*Metallic Bonding.* This type of bonding results when each of the atoms of the metal or alloy contributes its valence electrons to the formation of an "electron cloud," which is shared by the entire solid metal. Figure 1.7 contains a sketch of the metal ions and the electron cloud. The conduction of electricity and the principal conduction of heat are produced by the free movement of these electrons through the metal. Since the negative electron cloud surrounds each of the positive ions that make up the orderly three-dimensional crystal structure, strong electrical attraction holds the metal together. A characteristic of metallic bonding is the fact that every positive ion is equivalent. Thus, the metal can behave in a ductile fashion under stress because a group of positive ions break their bonding at one location, slip to a new position, and re-establish their bonds. This aspect of the plastic deformation of metals is considered in detail in Chapters 3 and 4.

Ideally, a symmetrical ion is produced when a valence electron is removed from the metal atom. As a result of this ion symmetry, metals tend to form highly symmetrical, close-packed crystal structures. Figure 1.8(a) shows a representative portion of the crystal structure of copper with the copper ions drawn as spheres in contact and with the electron cloud omitted. Since a crystal structure is made up of a unit that is repeated, this basic group of ions or atoms, the *unit cell*, is of principal interest and is shown in Fig. 1.8(b). In Fig. 1.8(c) the unit cell is represented in the ordinary diagrammatic fashion with the positions of the centers of the ions

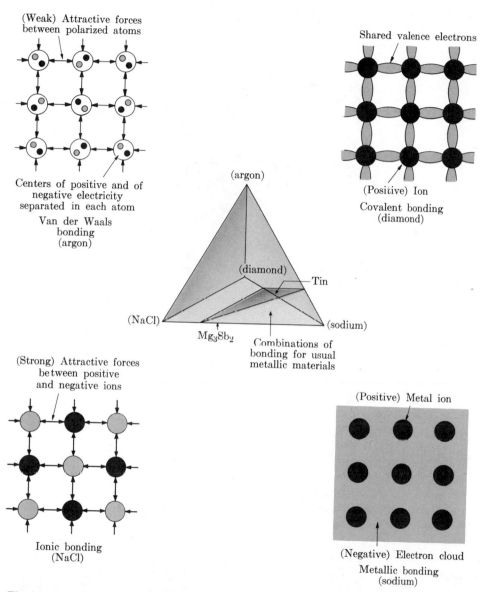

(Weak) Attractive forces between polarized atoms

Centers of positive and of negative electricity separated in each atom

Van der Waals bonding (argon)

Shared valence electrons

(Positive) Ion

Covalent bonding (diamond)

(argon)

(diamond) — Tin

(NaCl)

Mg₃Sb₂    Combinations of bonding for usual metallic materials

(sodium)

(Strong) Attractive forces between positive and negative ions

Ionic bonding (NaCl)

(Positive) Metal ion

(Negative) Electron cloud

Metallic bonding (sodium)

**Fig. 1.7**  Diagram showing the four major types of bonding and the combinations of bonding existing in typical metallic materials.

(a) A portion of the crystal lattice of copper,
showing the copper ions as spheres in contact.

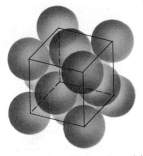

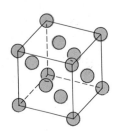

(b) A unit cell taken from (a).      (c) The unit cell of (b) drawn in the
                                     usual manner, with the centers of
                                     the ions indicated by small spheres.

**Fig. 1.8** Schematic representations of the highly symmetrical, close-packed crystal structure
of copper.

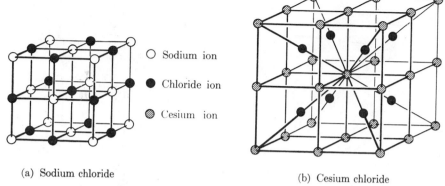

O   Sodium ion

●   Chloride ion

◎   Cesium ion

(a) Sodium chloride

(b) Cesium chloride

**Fig. 1.9** The crystal structures of typical ionic solids.

indicated by small spheres.  Not all metals have structures as simple as copper, and the appearance of more complex, less symmetrical structures is evidence that metallic characteristics are being replaced by those of other solid types.

*Van der Waals Bonding.*  The rare gases and molecules like methane, $CH_4$, which have no valence electrons available for crystalline bonding, obtain a weak attractive force for this purpose as a result of polarization of their electric charges.  Polarization is separation of the centers of positive and negative charges in an electrically neutral atom or molecule as it is brought close to its neighbors.  Its neighbors also become polarized.  A schematic picture of such polarization is given in Fig. 1.7.  The resulting weak electrical attraction between neighboring atoms or molecules is the van der Waals force.  It can overcome the disrupting effect of thermal motion of the atoms and molecules only at low temperatures; therefore the so-called *molecular crystals* produced by this type of bonding are weak and many of them melt at temperatures far below 0°C.

*Ionic Bonding.*  The elementary picture of electrostatic attraction between *one* sodium ion and *one* chlorine ion as a result of the transfer of the valence electron of the sodium atom to the chlorine atom must be slightly modified for an understanding of ionic bonding in solids.  Figure 1.7 contains a schematic illustration of ionic bonding showing that this type of bonding is caused by electrical attraction among alternately placed positive and negative ions.  A portion of the crystal structure of sodium chloride is shown in Fig. 1.9(a).  Here each ion has as nearest neighbors six ions of opposite charge.  Another type of crystal structure characteristic of ionic solids is that of cesium chloride, Fig. 1.9(b).

No electronic conduction of the kind found in metals is possible in ionic crystals, but weak ionic conduction occurs as a result of the motion of the individual ions.  When subjected to stresses, ionic crystals tend to cleave (break) along certain planes of atoms rather than to deform in a ductile fashion as metals do.

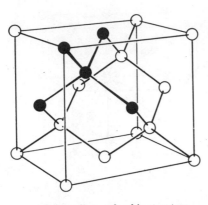

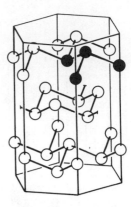

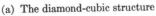

(a)  The diamond-cubic structure

(b)  The arsenic-type structure

**Fig. 1.10**  The crystal structures of typical covalent solids.

*Covalent or Homopolar Bonding.*  Many elements that have three or more valence electrons are bound in their crystal structures by forces arising from the sharing of electrons. The nature of this covalent bonding is shown schematically in Fig. 1.7. To complete the octet of electrons needed for atomic stability, electrons must be shared with $8 - N$ (8 minus $N$) neighboring atoms, where $N$ is the number of valence electrons in the given element.   Diamond is a typical example of the *covalent* or *homopolar crystal* produced by this type of bonding, and Fig. 1.10(a) shows that each carbon atom has four nearest neighbors, corresponding to the $8 - N$ *rule* applied to its four valence electrons.  Bismuth has five electrons in its outer shell and requires three additional electrons to achieve a stable configuration. Figure 1.10(b) shows how the crystal structure, which is of the arsenic type, permits a given atom in this case to share valence electrons with three nearest neighbors. High hardness and low electrical conductivity are general characteristics of solids of this type.

## QUANTITATIVE DESCRIPTION OF CRYSTAL STRUCTURES

When two or more kinds of atoms combine to form a crystalline material, several factors may determine the structure of the crystal. The interatomic bonds, especially if covalent, can impose geometrical requirements on neighboring atoms.  The relative size of two different atoms in a compound can determine a type of packing that promotes attractive forces and avoids repulsive forces.  In some instances the structure must also satisfy the valence requirements of the constituent atoms. In any event, a certain atomic arrangement extends over long distances (perhaps a million atoms aligned in a given direction) and is characterized by a variation of properties with direction in the crystal (anisotropy).

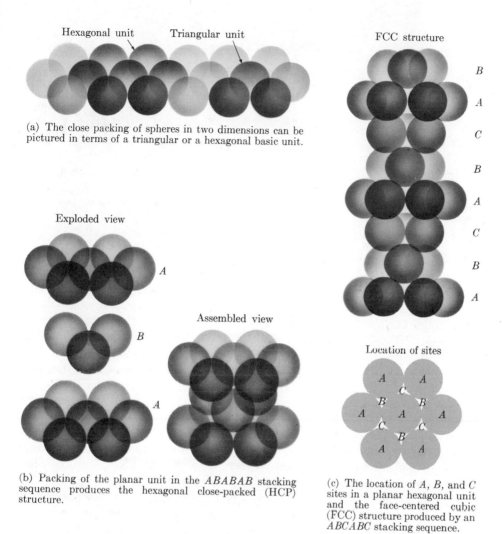

Hexagonal unit    Triangular unit

FCC structure

B

A

C

B

A

C

B

A

(a) The close packing of spheres in two dimensions can be pictured in terms of a triangular or a hexagonal basic unit.

Exploded view

A

B

A

Assembled view

Location of sites

(b) Packing of the planar unit in the *ABABAB* stacking sequence produces the hexagonal close-packed (HCP) structure.

(c) The location of *A*, *B*, and *C* sites in a planar hexagonal unit and the face-centered cubic (FCC) structure produced by an *ABCABC* stacking sequence.

**Fig. 1.11**  Illustration of the two types of closest packing of spheres in three dimensions, the hexagonal close-packed (HCP) and the face-centered cubic (FCC).

**Packing of spheres.**  A number of the common metals have their crystal structures determined in an especially simple manner. To a first approximation, each atom acts like a hard sphere and packs together with other atoms to form an arrangement of closest possible packing. The representation of the crystal structure of copper in Fig. 1.8(a) and (b) is an example of this concept. The metallic type of bonding provided by the cloud of valence electrons of copper adjusts itself easily to such a configuration of maximum close-packing of spheres.

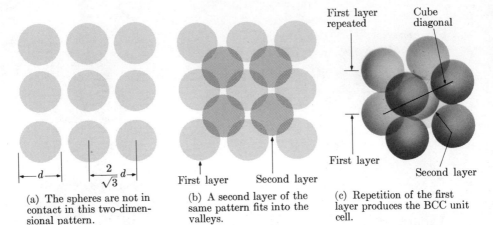

(a) The spheres are not in contact in this two-dimensional pattern.

(b) A second layer of the same pattern fits into the valleys.

(c) Repetition of the first layer produces the BCC unit cell.

**Fig. 1.12** The body-centered cubic (BCC) structure, an example of non-close packing of spheres.

One can easily observe the two possible types of closest packing by simply agitating a box partially filled with identical hard spheres; ping-pong balls or marbles, for example. In two dimensions the only close-packed arrangement is that shown in Fig. 1.11(a), which can be represented by a triangular or a hexagonal unit of structure. Consideration of the central sphere in the hexagonal unit shows that each sphere in this planar array touches six neighboring spheres; that is, it has six nearest neighbors. Three-dimensional close-packed arrangements can be obtained by stacking the planar arrays in two different sequences. The hexagonal close-packed structure, Fig. 1.11(b), consists in placing a second planar array (B) on an initial planar array (A) in such a position that the spheres of array B occupy "valleys" in array A. A third layer is in a position similar to array A, etc., so that a close-packed spatial structure with *stacking sequence ABABAB* ... is obtained.

Closer examination of the "valleys" in a planar hexagonal array (A), Fig. 1.11(c), shows that an adjacent array can be stacked either in valley positions B or in valley positions C. If the hexagonal layer *above* A occupies the B positions and the hexagonal layer *below* A occupies the C positions, the result is the stacking sequence *ABCABC* .... This sequence forms a spatial arrangement of spheres identical to the face-centered cubic (FCC) structure previously illustrated for copper in Fig. 1.8. The hexagonal-type planes lie obliquely in the *FCC* representation. In both the *HCP* and *FCC* structures the number of nearest neighbors is 12; in other words, the *coordination number* of a given atom is 12. Also, in both structures the fraction of the total space that is actually filled by the hard spheres (the packing fraction) is 0.74, the largest value possible (see Problem 11). Later we will consider the nature of the remaining (interstitial) spaces.

A common metallic structure, body-centered cubic (BCC), represented by iron at room temperature, exemplifies a less dense packing of spheres. The two-dimensional pattern in this case, Fig. 1.12(a), is a square array of spheres *not* in contact. A second layer of the same pattern fits into the valleys of the first layer, Fig. 1.12(b). Repetition of the first layer, Fig. 1.12(c), then produces the BCC arrangement. The spheres are in contact only in the directions of the cube diagonals; therefore, each atom has eight nearest neighbors (the coordination number is eight). The packing fraction is 0.68.

**Space lattices.** The crystal structures already discussed, and also the packing of spheres, illustrate the essential characteristic of the crystalline state; that is, regularity in the arrangement of atoms. We now wish to consider the most general types of crystal structure that can be formed by an aggregate of atoms, including atoms of more than one kind. Although the number of possible crystal structures is limitless, the science of crystallography has developed a convenient system of classifying these structures, starting from a few basic concepts. The concepts needed for effective work with the crystal structures found in metals are considered here.

When discussing crystal structure, one usually assumes that the structure continues to infinity in all directions. In terms of the conventional crystal (or grain) of iron one hundredth of an inch in size, this may appear to be an absurd assumption, but when it is realized that there are $10^{18}$ iron atoms in such a grain the approximation to infinity seems much closer. The fundamental definition of regularity of distribution of atoms in space is that of a *space lattice. A distribution of points (or atoms) in three dimensions is said to form a space lattice if every point has identical surroundings.* What a boring, repetitious place an infinite space lattice would be for a hypothetical flea-like creature! As he hopped from one point to the next he would see exactly the same endless vista.

A portion of a general space lattice is shown in Fig. 1.13(a). Because of the regularity in distribution of the points that compose a lattice, the essential geometry can be described by three *lattice vectors,* **a, b,** and **c** (Fig. 1.13(b)). These vectors describe the *unit cell* that is outlined in the corner of the space lattice, Fig. 1.13(a). The geometry of a space lattice is completely specified by the lattice constants (vector lengths) $a$, $b$, and $c$, and the interaxial angles $\alpha$, $\beta$, and $\gamma$. Only fourteen possible arrangements of points can satisfy the definition of a space lattice, and fewer than half of these are important in metallic structures. Figure 1.14 shows the unit cells of six of the more common space lattices found in metals and alloys.

Although a space lattice is fundamentally a distribution of points in space, it is convenient to connect these points by lines (axes) like those shown in Fig. 1.13(a) and to describe the space lattice in terms of the geometric figure formed by these lines. For example, three different distributions of points are conveniently described by a *cubic* network of axes, that is, by three equal axes at right angles to each other. The *simple cubic* space lattice, having points lying only at the intersections of the network of axes, is not an important space lattice in metals. The *body-centered cubic* and the *face-centered cubic* space lattices (previously illustrated by hard-sphere models) are common in metals and are shown in Fig. 1.14. These lattices

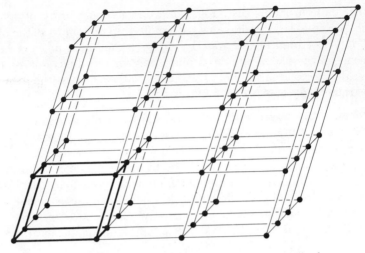

(a) A portion of a space lattice, with a unit cell outlined.

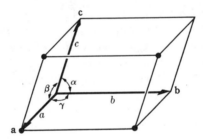

(b) Description of the unit cell in terms of the lattice vectors **a, b, c**. The lattice constants $a$, $b$, and $c$, and the interaxial angles $\alpha$, $\beta$, and $\gamma$ are shown. (From *Elements of X-ray Diffraction* by B. D. Cullity.)

**Fig. 1.13**  Schematic illustration of the nature of a space lattice.

have points at the cube center and at the centers of the cube faces, respectively, as well as at the corners of the cubic unit cell. In cubic crystals the lattice constant has the same value $a$ in all three directions in the crystal, but in other cases, such as in orthorhombic crystals, the lattice constant may have three different values $a$, $b$, and $c$ in the three directions in the crystal.

**Crystal structure.**  The crystal structures of some metals are simply atoms placed at the points of a space lattice; the face-centered cubic structure of copper is an example. Sodium chloride, Fig. 1.9(a), is a more complicated crystal structure. It might at first appear that this structure is simple cubic, but further consideration

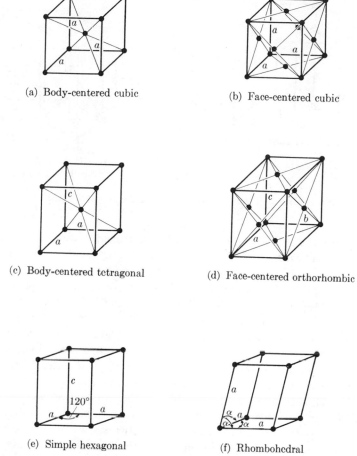

(a) Body-centered cubic

(b) Face-centered cubic

(c) Body-centered tetragonal

(d) Face-centered orthorhombic

(e) Simple hexagonal

(f) Rhombohedral

**Fig. 1.14** Unit cells illustrating six of the more important space lattices in metals. The letters $a$, $b$, and $c$ represent the lattice constants.

reveals that the points corresponding to a simple-cubic space lattice are not identical, since two adjacent points are occupied by a sodium ion and by a chloride ion. Actually, the space lattice in this case is face-centered cubic and each lattice point is associated with the same structural unit, a sodium ion plus a chloride ion. In Fig. 1.9(a) the sodium ions may be considered to mark the points of a space lattice, with each lattice point occupied by a sodium ion and, for example, the chloride ion to its right.

The *hexagonal close-packed* structure, previously described by a hard-sphere model, characterizes magnesium and a number of other pure metals. This structure is also an example of a group of atoms at a lattice point. Figure 1.15(a) shows

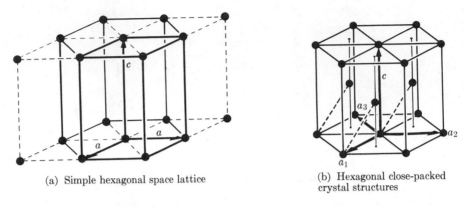

(a)  Simple hexagonal space lattice

(b)  Hexagonal close-packed crystal structures

**Fig. 1.15**  Important features of the hexagonal close-packed crystal structure.

how a cell with hexagonal symmetry can be visualized within a set of four unit cells of the simple hexagonal space lattice, Fig. 1.14. However, magnesium atoms do not occupy merely the points of this *space lattice* but also the intermediate positions that are shown in the sketch of the hexagonal close-packed *crystal structure*, Fig. 1.15(b). As in the NaCl structure, a pair of atoms (connected by a dashed line in the figure) can be associated with a point in the space lattice. Figure 1.11(b) can help explain the preference of atoms for the close-packed structure, in terms of the packing of spheres. The simple hexagonal lattice corresponds to the placing of the upper layer of atoms, A, directly on the lower A layer (layer B being omitted). Visualization of this stacking in terms of marbles shows that it tends to be unstable; marbles in the upper layer would slip down into the valleys in the lower layer. In contrast to this unstable arrangement, the hexagonal close-packed structure (Fig. 1.11(b)) corresponds to the occurrence of the intermediate layer of atoms, B, in such a position that the valleys are occupied, both in the lower and the upper A layers.

For some purposes crystal structures are advantageously described using a *primitive cell*, a unit cell with lattice points at corners only. For example, the lattice in Fig. 1.13(a) is built up of primitive cells. The unit cells of the body-centered cubic and face-centered cubic space lattices, Fig. 1.14, are evidently not primitive cells, since there are centered lattice points as well as corner ones. Problem 13 shows, however, that rhombohedral primitive cells can be used as alternative descriptions of these lattices. A principal advantage in using primitive cells in these two cases is the fact that each cell contains one atom. Figure 1.13(a) can be used to justify this statement. (Show that each of the eight corner atoms of a given cell is shared by a total of eight cells.) Similarly, it can be verified that there are two atoms per cell in the body-centered cubic and hexagonal close-packed structures and four atoms per cell in the face-centered cubic structure. The primitive cell for the hexagonal close-packed structure is the simple hexagonal unit cell, Fig. 1.15(a).

TABLE 1.1

Classification of Space Lattices by Crystal System

| Crystal system | Axial lengths and interaxial angles | Space lattice |
|---|---|---|
| Cubic | Three equal axes at right angles $a = b = c, \alpha = \beta = \gamma = 90°$ | Simple cubic<br>Body-centered cubic<br>Face-centered cubic |
| Tetragonal | Three axes at right angles, two equal $a = b \neq c, \alpha = \beta = \gamma = 90°$ | Simple tetragonal<br>Body-centered tetragonal |
| Orthorhombic | Three unequal axes at right angles $a \neq b \neq c, \alpha = \beta = \gamma = 90°$ | Simple orthorhombic<br>Body-centered orthorhombic<br>Base-centered orthorhombic<br>Face-centered orthorhombic |
| Rhombohedral | Three equal axes, equally inclined $a = b = c, \alpha = \beta = \gamma \neq 90°$ | Simple rhombohedral |
| Hexagonal | Two equal axes at 120°, third axis at right angles $a = b \neq c, \alpha = \beta = 90°, \gamma = 120°$ | Simple hexagonal |
| Monoclinic | Three unequal axes, one pair not at right angles $a \neq b \neq c, \alpha = \gamma = 90° \neq \beta$ | Simple monoclinic<br>Base-centered monoclinic |
| Triclinic | Three unequal axes, unequally inclined and none at right angles $a \neq b \neq c, \alpha \neq \beta \neq \gamma \neq 90°$ | Simple triclinic |

A logical extension of the use of lattice vectors, Fig. 1.13(b), is the classification of the fourteen space lattices by *crystal system*, according to the relative lengths of the vectors (or *axes*) and the interaxial angles, Table 1.1. For example, three of the fourteen space lattices can be described by the simple system of equal axes at right angles (cubic crystal system). This geometrical simplicity is the reason that the face-centered cubic array of points is ordinarily described using cubic axes. However, there are many ways of erecting systems of axes for a given space lattice; the primitive cell for the face-centered cubic space lattice uses the inclined axes of the rhombohedral system, for example. Two other concepts are widely

**TABLE 1.2**  Crystal-Structure and X-ray Data for a Number of Metals

| Metal | Crystal structure | Lattice constants, Å | Kα Characteristic x-ray wavelength, Å | Absorption edges, Å | | μ/ρ Mass absorption coefficient, m²/kg Radiation being absorbed, Å | | | |
|---|---|---|---|---|---|---|---|---|---|
| | | | | K | L | 0.560 | 0.710 | 1.656 | 2.287 |
| Aluminum | FCC | 4.05 | 8.32 | 7.94 | | 0.274 | 0.530 | 5.84 | 14.9 |
| Chromium | BCC | 2.89 | 2.287 | 2.07 | | 1.57 | 3.04 | 31.6 | 8.99 |
| Cobalt | HCP | $c = 4.07$ $a = 2.51$ | 1.786 | 1.60 | | 2.18 | 4.16 | 5.44 | 12.6 |
| Copper | FCC | 3.62 | 1.538 | 1.38 | | 2.64 | 4.97 | 6.50 | 15.4 |
| Gold | FCC | 4.08 | 0.181 | 0.153 | 1.04 | 6.67 | 12.8 | 26.0 | 53.7 |
| Iron | BCC | 2.87 | 1.933 | 1.74 | | 1.99 | 3.83 | 39.7 | 11.5 |
| Lead | FCC | 4.95 | 0.166 | 0.140 | 0.95 | 7.44 | 14.1 | 29.4 | 58.5 |
| Magnesium | HCP | $c = 5.21$ $a = 3.21$ | 9.87 | 9.50 | | 0.227 | 0.438 | 4.79 | 12.0 |
| Molybdenum | BCC | 3.15 | 0.710 | 0.618 | | 7.07 | 2.02 | 19.7 | 43.9 |
| Nickel | FCC | 3.52 | 1.656 | 1.48 | | 2.50 | 4.74 | 6.10 | 14.5 |
| Platinum | FCC | 3.92 | 0.184 | 0.158 | 1.07 | 6.42 | 12.3 | 24.8 | 51.8 |
| Silver | FCC | 4.09 | 0.559 | 0.484 | 3.69 | 1.48 | 2.86 | 27.6 | 58.5 |
| Tin | BCT | $c = 3.18$ $a = 5.83$ | 0.491 | 0.424 | 3.15 | 1.74 | 3.33 | 33.2 | 68.1 |
| Titanium | HCP | $c = 4.68$ $a = 2.95$ | 2.75 | 2.50 | | 1.18 | 2.37 | 24.7 | 60.3 |
| Tungsten | BCC | 3.16 | 0.210 | 0.178 | 1.21 | 5.46 | 10.5 | 20.9 | 45.6 |
| Zinc | HCP | $c = 4.95$ $a = 2.67$ | 1.433 | 1.28 | | 2.82 | 5.48 | 7.21 | 16.9 |

BCC = body-centered cubic    FCC = face-centered cubic    HCP = hexagonal close-packed    BCT = body-centered tetragonal    Å = $10^{-10}$ meter

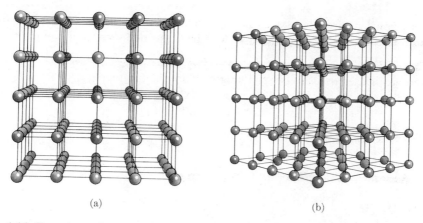

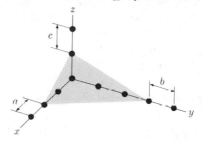

(a)                             (b)

**Fig. 1.16** Two views of a portion of a simple cubic lattice. The {100} planes are the prominent vertical planes in (a), while the (110) planes are the prominent vertical planes in (b). (By permission from *Introduction to Physical Metallurgy* by L. R. VanWert. McGraw-Hill.)

**Fig. 1.17** The intercepts of a plane of atoms on the three crystal axes. Of the infinity of atoms arrayed on the space lattice, only a few along the crystal axes are indicated.

used in classifying more complex crystal structures. The 32 *point groups* used in describing minerals are based on the external symmetry of crystals; the sixfold symmetry of hexagonal quartz crystals is an example. Consideration of the possible ways of placing the point groups on the various space lattices leads to the concept of *space groups*, of which there are 230. Fortunately, it is possible to specify most metal structures merely by a space lattice and one or more lattice constants. The crystal structures of a number of common metals are given in Table 1.2.

**Miller indices.** Special planes and directions within metal crystal structures play an important part in plastic deformation, hardening reactions, and other aspects of metal behavior. A qualitative picture of lattice planes is given by Fig. 1.16. However, the quantitative descriptions needed for many purposes are best supplied by sets of numbers that identify given planes of atoms; for example, the vertical and horizontal planes in Fig. 1.16(a) are {100} type planes, while the vertical planes in Fig. 1.16(b) are (110) planes. These numbers are the *Miller indices* of the planes and can be used directly in x-ray and other analyses. The three steps used in determining the indices of a given plane will be illustrated with the aid of Fig. 1.17.

1. Determine the intercepts of the plane on the three crystal axes. These intercepts are expressed in terms of axial lengths from the origin:

|   | $x$ | $y$ | $z$ | axis |
|---|---|---|---|---|
|   | 2 | 3 | 1 | intercept |

2. Take the reciprocals of these numbers:

$$\frac{1}{2} \qquad \frac{1}{3} \qquad 1$$

3. Reduce the reciprocals to the smallest integers that are in the same ratio:

$$3 \qquad 2 \qquad 6$$

or (326)

Thus the position of the plane of atoms shown in Fig. 1.17 is specified by the Miller indices (326). These indices give only geometrical information about the crystal planes, however, and say nothing about the distribution or kinds of atoms in the plane. On the other hand, a set of Miller indices such as (326) describes not merely a single crystal plane but the entire array of planes parallel to the plane on which the three-step analysis was carried out. For example, a plane having intercepts 4, 6, and 2 would be parallel to the plane shown in Fig. 1.17, and it can be verified that its indices would also be (326). Figure 1.16(a) shows the physical necessity for this identical numbering of parallel planes, since the planes are exactly equivalent.

The plane shown in Fig. 1.17 was convenient for the above analysis because it passed through lattice points on each of the three axes; however, such a plane may not be the first plane out from the origin in the given set of parallel planes. For example, in Fig. 1.18(d) not the first but the second plane passes through lattice points on both the x- and y-axes. When the location of the first plane in a set is required, it can be determined from the fact that its intercepts on the x-, y-, and z-axes are $a/h$, $b/k$, and $c/l$, where $a$, $b$, and $c$ are the axial lengths (lattice constants) and $h$, $k$, and $l$ are the Miller indices of the general plane $(hkl)$. For example, in Fig. 1.18(d) $h = 2$, $k = 1$, and $l = 0$, so that the intercepts of the first plane are $a/2$, $a$, and $\infty$, respectively. The intercept $\infty$ for the z-axis signifies that the plane is parallel to this axis.

The more common planes in cubic crystals are shown in Fig. 1.18, together with their Miller indices. Since the space lattice of atoms is considered to be infinite, and since the origin can be made to coincide with any one of the atoms, no physical difference exists between planes such as (100) and the (100) planes that happen to be on the other side of the arbitrary origin. (Analysis of planes passing *through* the origin is fruitless.) Similarly, (010) planes are equivalent to (100) planes and, in fact, could be made to have the latter indices by merely choosing the coordinates in a different manner. Families of such *equivalent planes* are of sufficient interest to have a special designation, and the symbol

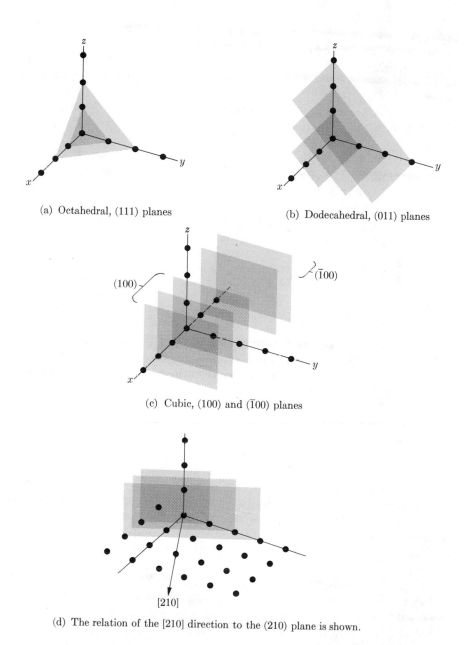

(a) Octahedral, (111) planes

(b) Dodecahedral, (011) planes

(c) Cubic, (100) and ($\bar{1}$00) planes

(d) The relation of the [210] direction to the (210) plane is shown.

**Fig. 1.18** Some of the simpler crystal planes in cubic crystals. Other designations of these planes are (a) octahedral, (b) dodecahedral, and (c) cubic.

{100} stands for the family of planes (100), ($\bar{1}$00), (010), (0$\bar{1}$0), (001), and (00$\bar{1}$), with similar definitions for other families of planes.

A *direction* in a crystal is indicated by bracketed indices, e.g. [210], and in the cubic system the direction is always perpendicular to the plane having the same indices, Fig. 1.18(d). More generally, a line in a given direction, such as [210], can be constructed in the following manner. Draw a line from the origin through the point having the coordinates $x = 2$, $y = 1$, $z = 0$, in terms of axial lengths. This line and all lines parallel to it are then in the given direction. Reciprocals are *not* involved in obtaining the Miller indices of directions. A family of equivalent directions, such as [100], [$\bar{1}$00], [010], [0$\bar{1}$0], [001], and [00$\bar{1}$], is designated as ⟨100⟩.

In the case of hexagonal crystals two equivalent systems of indices are in use. The usual Miller indices are referred to the axes shown in Fig. 1.15(a) and have the customary three numbers. However, Fig. 1.15(b) shows that hexagonal symmetry suggests the use of four axes, the $c$ axis plus three $a$ axes at 120° intervals. The intercepts of a crystal plane on these four axes leads to the four-number Miller-Bravais indices. The two systems of indices are simply related: if $(hkl)$ are the Miller indices, then the Miller-Bravais indices are $(hkil)$, where $i = -(h + k)$. For example, it can easily be verified that the (110) plane in Fig. 1.15(a) would be described as the (11$\bar{2}$0) plane in terms of the axes in Fig. 1.15(b).

**Interstitial spaces.** When structures such as FCC, HCP, and BCC are viewed as hard-sphere models, the void spaces among the spheres are known as *interstices*. These interstitial spaces may serve as locations for atoms in the metal or alloy under certain conditions; for example, carbon in iron may occupy an interstitial position in the iron lattice. The simplest interstice in a close-packed structure, Fig. 1.19(a), exists inside the tetrahedral packing of a triangular atomic unit plus an atom in the valley position. The six straight lines interconnecting the centers of these four atoms form a regular tetrahedron, one type of *coordination polyhedron*. The second sketch shows that the coordination tetrahedron is composed of portions of the four constituent spheres ($A$) plus an interstitial space capable of containing a sphere ($B$) of maximum radius ratio, $r_B/r_A = 0.225$. An octahedral interstice (see Problem 16) is almost twice as large; the maximum radius ratio in this case is 0.414. Because of the manner in which they are defined, the coordination polyhedra pack together to fill space. Figure 1.19(b) shows in exploded views the mode of packing of tetrahedra and octahedra in both the FCC and the HCP structures. Two adjacent interstices are "connected" (in the sense that a small atom might move from one into the other) only if the two corresponding polyhedra share a face. In the FCC structure the only connections are between an octahedral interstice and a tetrahedral interstice. Problem 16 considers the more complex interconnections in the HCP structure.

Coordination polyhedra can also be constructed for the BCC structure, Fig. 1.20, but the interstitial spaces are distorted relative to the regular polyhedra of the close-packed structures. Consequently, the radius ratio of the tetrahedral

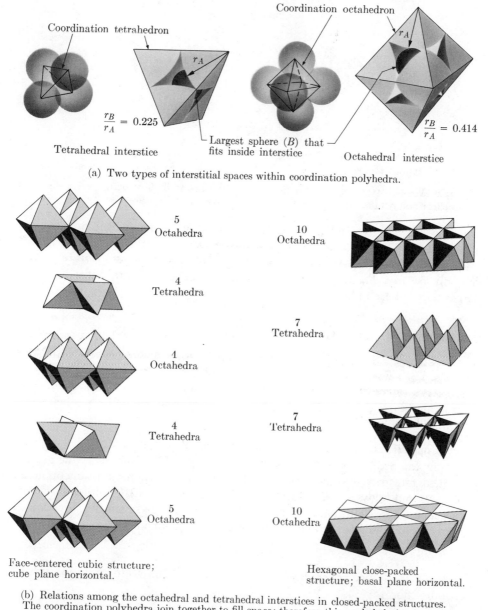

Coordination octahedron

Coordination tetrahedron

$r_A$

$r_A$

$\dfrac{r_B}{r_A} = 0.225$

$\dfrac{r_B}{r_A} = 0.414$

Largest sphere $(B)$ that
fits inside interstice

Tetrahedral interstice

Octahedral interstice

(a) Two types of interstitial spaces within coordination polyhedra.

5
Octahedra

10
Octahedra

4
Tetrahedra

7
Tetrahedra

4
Octahedra

4
Tetrahedra

7
Tetrahedra

5
Octahedra

10
Octahedra

Face-centered cubic structure;
cube plane horizontal.

Hexagonal close-packed
structure; basal plane horizontal.

(b) Relations among the octahedral and tetrahedral interstices in closed-packed structures.
The coordination polyhedra join together to fill space; therefore this exploded view is
employed to reveal the individual polyhedra.

**Fig. 1.19** The use of coordination polyhedra to describe the interstitial spaces in FCC and
HCP structures. (Adapted from L. V. Azároff.)

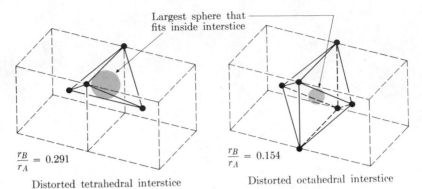

Fig. 1.20 The two types of coordination polyhedron in the BCC structure are a distorted
tetrahedron and octahedron. (Adapted from L. V. Azároff.)

interstice has an increased value, 0.291, while that of the octahedral interstice has
the abnormally low value of 0.154. In some instances the hard-sphere model does
not agree with actual atomic behavior. For example, carbon atoms in BCC iron
actually occupy the octahedral sites, although the tetrahedral sites have the
larger hard-sphere radius.

## QUANTITATIVE OPERATIONS WITH CRYSTAL STRUCTURES

Crystallographic features of metallic structure enter quantitatively into such
subjects as x-ray diffraction, study of imperfections, elastic and plastic deformation,
and mechanisms of strengthening. We now consider two important techniques,
vector algebra and stereographic projection, for performing the mathematical
operations involved.

**Vector operations.** The special terminology employed for vectors in crystallography can be illustrated by some typical examples. For simplicity the present
treatment is restricted to the cubic crystal system, which is commonly encountered
in metals and alloys. Let us begin by proving that the [210] direction is perpendicular to the (210) plane, Fig. 1.18(d). In the cubic system the three lattice
vectors are mutually perpendicular but have the same magnitude,

$$|\mathbf{a}| = |\mathbf{b}| = |\mathbf{c}| = a \qquad (1.15)$$

where $a$ is the lattice constant of the cubic crystal. The vector notation for the
[210] direction is

$$[210] = 2\mathbf{a} + 1\mathbf{b} + 0\mathbf{c} \qquad (1.16)$$

Now, if the scalar (dot) product of two vectors equals zero, $\mathbf{R} \cdot \mathbf{S} = 0$, the two vectors
are perpendicular. Therefore, we choose one* arbitrary vector lying in the (210)

---

*Rigorously, a second vector should also be analyzed.

plane, and show that its scalar product with $(2\mathbf{a} + 1\mathbf{b} + 0\mathbf{c})$ is zero. A vertical line,

$$[001] = 0\mathbf{a} + 0\mathbf{b} + 1\mathbf{c}, \tag{1.17}$$

lies in the (210) plane; the plane through the origin shows the relation most clearly. The desired scalar product is

$$[210] \cdot [001] = (2 \times 0)a^2 + (1 \times 0)a^2 + (0 \times 1)a^2 = 0, \tag{1.18}$$

and therefore [210] and (210) are perpendicular.

A somewhat more difficult problem is the following: What direction is perpendicular to a given plane, say the (326) plane, Fig. 1.17? The solution involves taking the cross product,

$$\mathbf{R} \times \mathbf{S} = \mathbf{T}, \tag{1.19}$$

of two nonparallel vectors $\mathbf{R}$ and $\mathbf{S}$ that lie in the plane in question. By definition of the cross product, the vector $\mathbf{T}$ is then perpendicular to the plane. Two simple choices for $\mathbf{R}$ and $\mathbf{S}$ are vectors parallel to the two edges of the triangular section of the (326) plane sketched in Fig. 1.17;

$$\mathbf{R} = 2\mathbf{a} + 0\mathbf{b} - 1\mathbf{c} = 2a\mathbf{i} + \quad 0\mathbf{j} + 1a\mathbf{k} \tag{1.20}$$

$$\mathbf{S} = 0\mathbf{a} + 3\mathbf{b} - 1\mathbf{c} = \quad 0\mathbf{i} + 3a\mathbf{j} + 1a\mathbf{k}. \tag{1.21}$$

Here the relation of Eq. (1.15) has been used, and the unit vectors $\mathbf{i}$, $\mathbf{j}$, and $\mathbf{k}$ have been introduced to permit the usual determinant-type expansion of the vector product,

$$\mathbf{R} \times \mathbf{S} = \begin{vmatrix} \mathbf{i} & \mathbf{j} & \mathbf{k} \\ 2a & 0 & 1a \\ 0 & 3a & 1a \end{vmatrix} = (3\mathbf{i} + 2\mathbf{j} + 6\mathbf{k})a^2. \tag{1.22}$$

The resulting vector is seen to be in the [326] direction, in agreement with the rule for cubic systems that a plane and a direction with the same indices are perpendicular.

Another common problem is the determination of the angle, $\theta$, between two directions, say [210] and [130]. [The problem for the (210) and (130) *planes* is identical.] The solution involves the scalar product,

$$\mathbf{R} \cdot \mathbf{S} = |\mathbf{R}| \, |\mathbf{S}| \cos \theta. \tag{1.23}$$

The magnitudes $|\mathbf{R}|$ and $|\mathbf{S}|$ are determined by separate scalar products,

$$|\mathbf{R}|^2 = [210] \cdot [210] = (2\mathbf{a} + 1\mathbf{b} + 0\mathbf{c}) \cdot (2\mathbf{a} + 1\mathbf{b} + 0\mathbf{c})$$

$$= 4a^2 + 1a^2 = 5a^2. \tag{1.24}$$

$$|\mathbf{S}|^2 = [130] \cdot [130] = (1\mathbf{a} + 3\mathbf{b} + 0\mathbf{c}) \cdot (1\mathbf{a} + 3\mathbf{b} + 0\mathbf{c})$$

$$= 1a^2 + 9a^2 = 10a^2. \tag{1.25}$$

Similarly,

$$\mathbf{R} \cdot \mathbf{S} = [210] \cdot [130] = (2\mathbf{a} + 1\mathbf{b} + 0\mathbf{c}) \cdot (1\mathbf{a} + 3\mathbf{b} + 0\mathbf{c})$$

$$= 2a^2 + 3a^2 = 5a^2. \qquad (1.26)$$

Substitution of these values in Eq. (1.23) gives the desired expression for the angle $\theta$,

$$\cos \theta = \frac{5}{\sqrt{5}\sqrt{10}}, \qquad \text{or } \theta = 30°. \qquad (1.27)$$

Noncubic crystals can also be treated by vector algebra. For those crystal systems with interaxial angles of 90° (tetragonal and orthorhombic), the methods described here are applicable except that two or three different values must be used for the lattice constants. Problem 17 considers the (201) plane and the [201] direction in tin (tetragonal) and shows that they are not mutually perpendicular in this case. For crystal systems with one or more interaxial angles different from 90°, a scalar product such as Eq. (1.18) results in more than three terms since products of the form $\mathbf{a} \cdot \mathbf{b}$ may be nonzero.

**Stereographic projections.** Essentially, the stereographic projection is a two-dimensional graphical technique for conveniently measuring angles and directions in three-dimensional crystal structures. The crystal to be studied is imagined to be at the center of a reference sphere, Fig. 1.21. Any crystal plane of interest, such as the (001) chosen here as an example, is projected by the following procedure. (1) The normal to the (001) plane is extended until it intersects the reference sphere as the *pole* of the (001) plane. (2) The projection plane is placed tangent to the reference sphere; the point of tangency is at one end of a diameter, the other end of which is the origin of projection. (3) A straight line from the origin causes the (001) pole to project as the (001) plane on the stereographic projection. Typically, the crystal being studied is in an arbitrary orientation, so a given plane merely falls somewhere inside the basic circle of the stereographic projection. Other planes can also be projected using the same procedure; for example, the (010) and (001) planes in Fig. 1.21. Any planes [such as ($\bar{1}$00), (0$\bar{1}$0), and (00$\bar{1}$)] whose poles lie on the "back" half of the reference sphere, however, do not appear on the stereographic projection within the basic circle.

To permit quantitative measurements of points on a stereographic projection, the usual system of latitude and longitude lines is projected stereographically, Fig. 1.22. The resulting reference pattern, known as a *Wulff net*, can then be used to specify the position of any given plane. The (010) plane of Fig. 1.21, for example, when plotted on the Wulff net in Fig. 1.22 is found to be at longitude $\theta = 70°$ and latitude $\phi = 30°$. The most commonly used commercial Wulff nets are about 20 cm in diameter and are divided into 2° intervals.

A basic operation with the Wulff net is the measurement of the angle between two planes, for example, the angle between the (100) and (010) planes in Fig. 1.21. A stereographic projection of the two planes in question is made on transparent

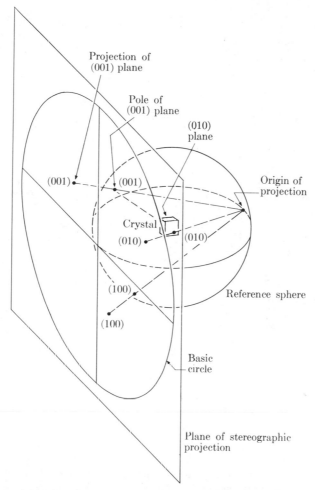

**Fig. 1.21** The method of projection that converts plane normals in a crystal to poles in the reference sphere and finally to points on a stereographic projection.

paper, and this sheet of paper is then placed over the Wulff net and pierced by its central pin, Fig. 1.23(a). The true angle between two points on a stereographic projection can be measured only along a *great circle*; that is, a longitude line or the equator. Consequently, the stereographic projection is rotated about the central pin relative to the Wulff net until the (100) and (010) planes lie on the same longitude line, Fig. 1.23(b). The angle between them can then be read correctly and is found to be 90°, the expected value for {100} type planes in a cubic crystal.

Two kinds of rotations of a stereographic projection are frequently performed. Rotation about the central pin, as in Fig. 1.23, corresponds to a rotation of the crystal about the central axis (containing the origin of projection, Fig. 1.21).

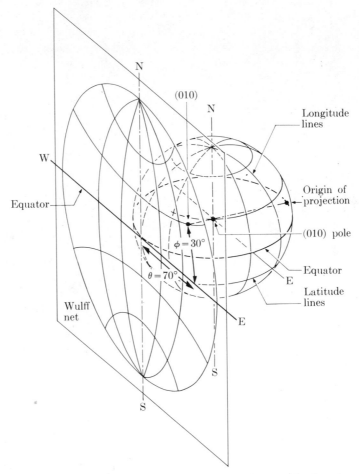

**Fig. 1.22** Projection of a ruled globe to produce the Wulff net. The location of the (010) plane, Fig. 1.21, by its $(\theta, \phi)$ coordinates is shown.

Rotation of the various points in the stereographic projection relative to the underlying Wulff net is accomplished simply by rotation of the tracing paper containing the points. The second kind of rotation is about the north–south axis. In this case, rotation of the crystal by $\theta°$ corresponds to the motion of each point in the stereographic projection by $\theta°$ along the latitude circle on which the point lies. For example, Fig. 1.24(a) shows how the (100) and (010) planes of Fig. 1.23(b) can be rotated 40° about the north–south axis and brought to positions on the basic circle. A suitable curved arrow along a latitude circle indicates the path followed by a given point during the rotation. The final location of the (001) plane at the center of the stereographic projection agrees with the fact that the (001) plane must be at 90° to both the (100) and (010) planes.

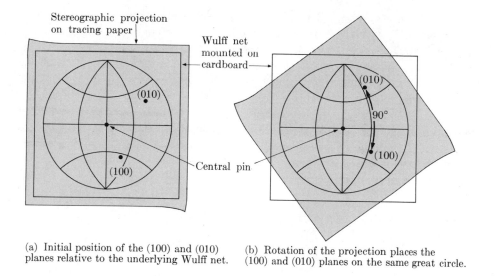

(a) Initial position of the (100) and (010)      (b) Rotation of the projection places the
planes relative to the underlying Wulff net.     (100) and (010) planes on the same great circle.

**Fig. 1.23**  Rotation of the stereographic projection relative to the Wulff net to measure the
angle between the (100) and (010) planes in Fig. 1.21.

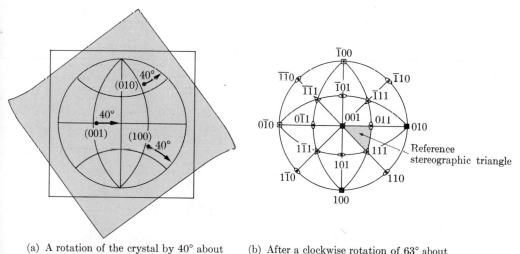

(a) A rotation of the crystal by 40° about        (b) After a clockwise rotation of 63° about
the north-south axis corresponds to these         the central pin, the stereographic projection
displacements of the {100} planes on the          of the {100} planes is in this standard
stereographic projection.                         projection. The {110} and {111} planes
                                                  are shown also.

**Fig. 1.24**  Rotation of points on the stereographic projection about the north-south axis and
the production of a standard projection.

The simple, symmetrical positions of the {100} planes in the stereographic projection of Fig. 1.24(b) characterizes the *standard stereographic projection* of a cubic crystal. To obtain this orientation from that shown in Fig. 1.24(a), we need merely rotate the stereographic projection 63° in the clockwise direction about the central pin. Although for simplicity we have been considering only the {100} planes, other crystal planes can also be shown in a stereographic projection. The {110} and {111} planes are included in Fig. 1.24(b). Because of the high symmetry of a cubic crystal, all significantly different orientations are represented by the points within a single reference stereographic triangle formed by three adjacent (100), (110), and (111) planes, Fig. 1.24(b). Several applications of the stereographic projection are discussed in later chapters.

## PROBLEMS

**1.** Consider the hypothetical experiment in which a microscope is used to measure the position and momentum of an electron moving in the $x$-direction. To minimize the disturbance of the electron, only a single photon of light is used to illuminate it (Fig. 1.25), but if this photon is to enter the microscope objective of half-aperture angle $\theta$, it must be scattered at an angle $\alpha$ that lies in the range $90° - \theta$ to $90° + \theta$.

a) The resolving power of a microscope is $\lambda/(2 \sin \theta)$; that is, distances cannot be measured more accurately than to this limit. Show that $\lambda/(2 \sin \theta)$ is the value of $\Delta x$ in Eq. (1.1) in this instance.

b) The impact of the photon changes the momentum, $p$, of the electron (Compton effect). If the electron is initially at rest, its momentum in the $x$-direction after impact is approximately $(h/\lambda)(1 - \cos \alpha)$, where $\alpha$ is the angle at which the photon is scattered. Evaluate $p$ for the two extreme values of $\alpha$.

c) Since the same image of the electron would be seen in the microscope for any value of $\alpha$ in the range $90° - \theta$ to $90° + \theta$, show that $\Delta p$ (Eq. 1.1) is $(2h \sin \theta)/\lambda$ and that therefore $\Delta x \times \Delta p = h$.

d) Show that even if we were satisfied with only a rough measure of the position of an electron in an atomic system, for example, $\Delta x = 10^{-10}$ m, which is about the radius of an atom, the accompanying uncertainty in velocity would be very large. Use $p = mv$, where $m$ is the rest mass of the electron.

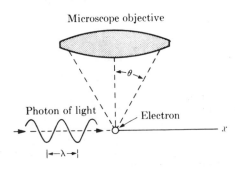

**Fig. 1.25**  Illustration for Problem 1.

**2.** Theory demands that every particle have associated wave properties. Why then does a billiard ball, for example, appear to move in a straight line? [Use Eq. (1.2) and consider that a billiard ball is a "particle" that weighs about 0.2 kg.]

**3.** An *intense* beam of green light ($\lambda = 0.5\,\mu m$) is insufficient to eject a core electron from magnesium metal, although this can be accomplished by a very *weak* beam of x-rays having a wavelength of 1 Å. Show that quantum physics explains this apparent anomaly. [Use Eq. (1.3) and recall that $v = c/\lambda$, where $c$ is the velocity of light.]

**4.** Derive Eq. (1.4) for the electrical potential energy $W$ of an electron of charge $-e$ when it is separated a distance $r$ from a nucleus of charge $+e$. Recall the analogous derivation for the gravitational potential energy of a mass of $m$ kg at a height of $h$ m. In general, $W$ is defined as the work required to take a body from the reference position (where $W = 0$ by convention) to the final position. Therefore, since the gravitational force on a mass $m$ is $mg$, where $g$ is the acceleration due to gravity,

$$W = \int_0^h F\,dx = \int_0^h mg\,dx = mgh.$$

In the analogous integral for electrical potential energy, the conventional reference position, where $W = 0$, is infinite separation of the charges; therefore the integration is from $\infty$ to $r$, the final distance of separation. Also, while the electron is moving from $\infty$ toward $r$, a positive force of $(1/4\pi\varepsilon_0)(e^2/r^2)$ must be exerted on it to maintain the required equilibrium condition at each instant.

**5.** In obtaining a useful expression for the total energy $E$ by adding the potential energy $W$ and the kinetic energy $K$, we must express both $W$ and $K$ in terms of the same quantities. The expression for potential energy, $W = -(1/4\pi\varepsilon_0)(e^2/r)$, is satisfactory, but the usual expression for kinetic energy, $K = \frac{1}{2}mv^2$, should be changed into a form involving $e$ and $r$. (a) The Bohr theory pictures an electron moving in a circular orbit of radius $r$ because the force of electrostatic attraction due to the nucleus just equals the centrifugal force required to keep the electron in its circular orbit. Equate the expressions for these two forces and then re-arrange the resulting equation so that one side of the equation is the quantity $\frac{1}{2}mv^2$. The other side of the equation is then the desired expression for $K$. (b) Add the appropriate expressions for $W$ and $K$ to obtain the expression for $E$ as a function of $r$, Eq. (1.6). (c) Plot $E$ versus $r$ for a range of $r$ values from 0.1 to 10 Å. Convert the $E$ values from joules to electron volts for this plot.

**6.** Apply Eq. (1.3) to determine the frequency of the quantum of radiant energy given off when the electron in an excited hydrogen atom falls from energy state $E_2$ to $E_1$. Recalling the relation $\lambda v = c$, where $c$ is the velocity of light, calculate the corresponding wavelength of the radiation.

**7.** Since momentum, $p = mv$, and kinetic energy, $K = \frac{1}{2}mv^2$, both depend on the two quanti-ties $m$ and $v$, one can obtain an expression for $p$ that involves $K$.

a) Rearrange the equation $K = \frac{1}{2}mv^2$ so that one side of the equation involves only the quantity $mv$ and the other side involves $K$ and $m$, but not $v$. Since $mv$ can be replaced by $p$, show that this is the desired relation between $p$ and $K$.

b) Recalling the relation $E = K + W$, obtain the expression for $p$ that is used in Eq. (1.9).

**8.** The production of x-rays involves the inner, completed groups of electrons in an atom; the $1s$ electrons are especially important. The first step in the process is to eject one of these electrons (from the $1s$ level, for example). This is usually accomplished by bombarding the "target" metal with a beam of high-energy electrons; see Fig. 5.17. The vacant $1s$ level is quickly occupied by one of the other electrons in the atom, most probably by one that falls from the $2p$ level.

a) Use a diagram similar to Fig. 1.2(b) to show qualitatively the change in energy accompanying a $2p \rightarrow 1s$ transition in iron.

b) Calculate the difference in energy between the $2p$ and $1s$ levels in iron from the fact that the wavelength of the x-ray produced by this transition is 1.933 Å.

**9.** a) Explain what a transition element is.

b) Why may a transition element show unusual behavior in alloys?

**10.** The force of electrostatic attraction accounts for most of the binding energy in ionic crystals such as NaCl. This fact can be demonstrated by a rough comparison of the experimentally observed heat of formation, which is about 8 eV per molecule of NaCl, with the electrostatic binding energy of a single chloride ion and a single sodium ion. Use Eq. (1.4) to calculate the latter value for the actual spacing of the ions in the crystal, 2.81 Å. Although only order-of-magnitude agreement can be expected from this crude approximation, quantitative agreement is obtained from calculations based on a more realistic model of the crystal structure of NaCl.

**11.** a) In the planar close-packed array of identical spheres shown in Fig. 1.11(a), consider the two-dimensional plane that passes through the center of the spheres. What fraction of the area of this plane is occupied by the (cross sections of the) spheres? [*Hint*: make the calculation for a simple geometrical unit (such as an equilateral triangle) that is representative of the structure. One can easily take account of the fact that only a certain fraction, say one-sixth, of a given sphere is in the triangle in question.]

b) Using an analogous procedure, calculate the packing fraction of the HCP and FCC arrangements of spheres. The smallest unit of structure in both cases is a tetrahedron (formed by a planar equilateral triangular array plus an atom in the valley position; a total of four atoms). The tetrahedron is constructed through the centers of this four-atom grouping. Alternatively, the analyses can be made for a hexagonal prism (HCP) and for a cube (FCC).

**12.** Show that there are five two-dimensional plane lattices corresponding to the fourteen three-dimensional space lattices. [*Hint*: the least obvious of the five is analogous to the hexagonal unit shown in Fig. 1.11(a).]

**13.** a) Sketch the primirive cell for the face-centered cubic structure within a cubic unit cell. The cell is rhombohedral with $\alpha = 60°$ and lattice constants $1/\sqrt{2}$ those of the cubic cell.

b) The lattice vectors of the rhombohedral primitive cell for the body-centered cubic structure are conveniently drawn from the center atom out to three nonadjacent corner atoms. This makes $\alpha = 109.5°$ and the lattice constants $\sqrt{3}/2$ those of the cubic cell. Your sketch of this primitive cell will extend outside a single body-centered unit cell.

**14.** Sketch planes that have the following Miller indices: (a) (001), (b) $(1\bar{1}0)$, (c) (111).

**15.** Indicate the following *directions* in the corresponding sketches of Problem 14: (a) [210], (b) [111], (c) $[10\bar{1}]$. Draw these directions so that they lie in the corresponding planes.

**16.** a) Show on a sketch of a cube face of the FCC structure the location of an octahedral interstice.

b) Show by suitable sketches the existence of the following interconnections of interstitial spaces in the HCP structure: (1) a pair of tetrahedra; (2) a vertical "string" of octahedra; (3) a tetrahedron with each of three adjacent octahedra.

**17.** Prove that the (201) plane, is *not* perpendicular to the [201] direction in a crystal of tin. The lattice constants of this tetragonal crystal are listed in Table 1.2. [*Hint*: Use Eq. (1.19) to determine the direction that *is* perpendicular to the (201) plane; then use Eq. (1.23) to determine the angle between this direction and the [201] direction.]

**18.** Using the data of Fig. 1–24(a), sketch the path followed by the (100) plane during a rotation of 70° (rather than 40°) about the north–south axis. [*Hint*: when a given (*hkl*) plane "disappears" into the back surface of the reference sphere, the pole of the corresponding negative plane, ($\bar{h}k\bar{l}$), can then be followed on the stereographic projection.]

**19.** Illustrate by suitable qualitative sketches the following series of operations for rotating the (100) plane, Fig. 1.23(a), 30° clockwise about the (010) plane.

a) Rotate (010) to the basic circle, Fig. 1.24(a).

b) Rotate (010) to the north pole.

c) Make the 30° clockwise rotation.

d) Return (010) to its original position by the reverse of (b) and (a). The (100) plane is now in its new position.

**20.** a) For the orientation given in Fig. 1.23(a), locate on a stereographic projection the (011) and (111) planes. Describe the quantitative procedures that might be used to place

b) the (011) plane, and

c) the (111) plane in their correct positions in the standard stereographic projection, Fig. 1.24(b).

# REFERENCES

Barrett, C. S. and T. B. Massalski, *Structure of Metals*, 3rd edn, New York: McGraw-Hill, 1966. Chapters 1 and 2 give additional information on crystallography and stereographic projection, respectively.

Boas, M., *Mathematical Methods in the Physical Sciences*, New York: John Wiley, 1966. Chapter. 5 is a clear treatment of vector calculations.

Cullity, B. D., *Elements of X-ray Diffraction*, Reading, Mass.: Addison-Wesley, 1956. Chapter 2 treats the geometry of crystals and stereographic projection (with numerous examples).

Hume-Rothery, W., *et al.*, *The Structure of Metals and Alloys*, 5th edn, London: Institute of Metals, 1969. Parts I and II discuss in detail the basic electronic structures of metals and alloys and the crystal structures of the elements.

Kelly, A. and G. W. Grove, *Crystallography and Point Defects*, Reading, Mass.: Addison-Wesley, 1970. Chapters 1, 2, and 3 give a more advanced treatment of crystal structures and of the geometry of lattices.

CHAPTER 2

# ELASTIC BEHAVIOR

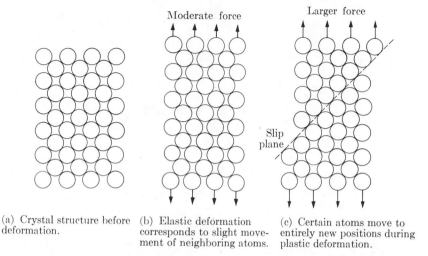

Moderate force

Larger force

Slip plane

(a) Crystal structure before deformation.

(b) Elastic deformation corresponds to slight movement of neighboring atoms.

(c) Certain atoms move to entirely new positions during plastic deformation.

**Fig. 2.1** Schematic illustration of the difference in the nature of elastic deformation and of plastic deformation in a metallic crystal. (After J. C. M. Li.)

Our background in crystal structure gives a useful insight into various aspects of the deformation of metallic materials, including mechanisms, anisotropy, and the relation between ordinary macroscopic behavior and deformation of the separate crystals. In this chapter we study the relatively simple phenomenon of elastic deformation, in which the atoms merely shift a small fraction of an interatomic spacing in response to an applied force, Fig. 2.1(b). Elastic deformation is reversible since the atoms return to their normal positions in the crystal structure when the force is removed. The action of a spring under an applied load is a typical macroscopic example of elastic deformation. If the force applied to a crystal exceeds a certain value, some of the neighboring atoms experience a relative displacement of one or more atomic distances, Fig. 2.1(c). This is the process of *slip* (see Chapter 3) and is one of the principal mechanisms resulting in plastic (or permanent) deformation. The slipped arrangement persists after the deforming force has been removed. Plastic deformation is irreversible and also differs in practice from elastic deformation in permitting large changes in shape rather than only extremely small changes.

## MACROSCOPIC ASPECTS OF ELASTICITY

Elastic behavior of ordinary engineering structures is often a subject of major technical importance. Although elastic deformations are only about one-hundredth as large as those encountered in the plastic range, Fig. 2.2, these small effects interest engineers greatly for two reasons. (1) For most mechanical devices, stresses·must remain within the elastic range to avoid the larger plastic deformations. In this connection, the importance of the yield strength of an alloy (see

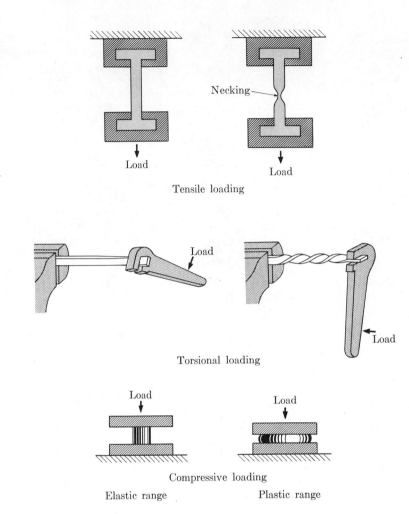

**Fig. 2.2**  The effects of loads that produce stresses in the elastic range and in the plastic range.

Chapter 4) is that it represents the stress below which the deformation is almost entirely elastic.   (2) In many devices even small elastic deformations must be adequately controlled to maintain minimum clearances or to satisfy similar design requirements.  Other aspects of elasticity closely related to metallic structure are stress raisers, residual stresses, and the magnitude of Young's modulus, $E$.  The concepts of stress and strain, which are essential for understanding the elastic behavior of metals, are treated in the following sections.

**Stress.**  Consider a typical metal part in service, such as a turbine wheel.  Forces are applied at the periphery of the wheel, and these forces are then transmitted

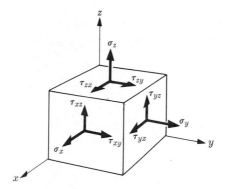

**Fig. 2.3** The stress components that describe a general state of stress at a point.

through the body of the wheel to the shaft of the turbine. Thus the essential phenomenon occurring within the metal is the exertion of force by one portion of the metal on the neighboring portions. This picture of a metal under load suggests the following definition: *a body is in a state of stress when one portion of the body exerts forces on neighboring portions.*

In many engineering problems the state of stress is determined primarily by *surface forces,* such as those acting on the periphery of the turbine wheel, but sometimes other types of forces are important. For example, two types of *body forces* must be considered in many turbine designs. One body force is the centrifugal force that acts on each volume element of the wheel. The other body force is created by differences in temperature within the wheel and the accompanying differences in thermal expansion. These *thermal stresses* are considered in a later section.

In spite of the various factors that can contribute to the state of stress of a body, the most general state of stress can be described in a relatively simple manner. The steps in this description can be given with the aid of Fig. 2.3.

1. A suitably oriented set of Cartesian coordinate axes $(x, y, z)$ is chosen.

2. An infinitesimal volume element is constructed with its edges parallel to the coordinate axes. This volume element is sufficiently small that there is negligible variation in the stress components across it.

3. Three stress components are considered to act on each of the faces of the volume element, one normal stress $(\sigma)$ and two shear stresses $(\tau)$.* Only the positive faces of the volume element can be seen in Fig. 2.3, but the stress components acting on the negative faces must be equal to these, according to (2), although they act in the opposite directions.

---

* A stress, $\sigma$ or $\tau$, is considered to be positive if the corresponding force is in the positive direction and the outward-drawn normal to the plane in question is in the positive direction.

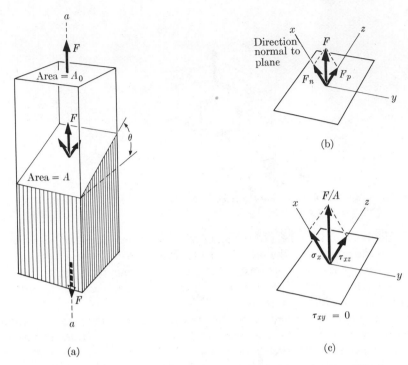

**Fig. 2.4** The action of an axial force $F$ on a bar of square cross section. The hypothetical action of the top portion of the bar on the bottom portion across a plane inclined at the angle $\theta$ is shown in (a). Drawing (b) shows how the force $F$ can be resolved into components $F_n$ and $F_p$, while drawing (c) shows similarly how the stress vector $F/A$ can be resolved into its components $\sigma_x$, $\tau_{xz}$ (and $\tau_{xy}$, which is zero in this instance).

4. It can be shown (see Problem 2) that $\tau_{xy} = \tau_{yx}$, $\tau_{xz} = \tau_{zx}$, and $\tau_{yz} = \tau_{zy}$. Therefore only six stress components are needed to describe the most general state of stress.

The description of a state of stress by the six components $\sigma_x$, $\sigma_y$, $\sigma_z$, $\tau_{yz}$, $\tau_{zx}$, and $\tau_{xy}$ can be used in the most general problems involving static equilibrium. However, it is convenient to consider the special relations that exist when: (1) only surface forces are acting on the body, and (2) the stress is homogeneous (the stress components have the same values at each point in the body).

Figure 2.4 shows the simple case in which an axial force $F$ acts on the ends of a bar of square cross section. Consider the problem of determining the stresses that act across an arbitrary plane, such as the plane inclined at the angle $\theta$. Now, if the force $F$ is to be transmitted along the metal bar, the upper portion of the bar (shown in phantom view in Fig. 2.4(a)) must exert a force equal to $F$ on the lower portion of the bar across the inclined plane. But according to the definition of stress, this force must determine the state of stress, and the method of Fig. 2.3 can be used to determine the stress components. The inclined plane in Fig. 2.4(a) represents one of the planes in Fig. 2.3. If it is taken to be the plane perpendicular to the x-direction, then the appropriate coordinates are shown in Fig. 2.4(b). The axis of the bar, a—a, is *not* one of the coordinates for an inclined plane. The force per unit area, $F/A$, exerted across a given plane is called the stress vector, and the three stress components acting on the plane must combine vectorially to give $F/A$ as their resultant. Figure 2.4(c) shows how this concept is applied to the inclined plane of Fig. 2.4(a). A similar analysis can be made for the other two planes shown in Fig. 2.3 (see Problem 4). It will be seen in Fig. 2.4(c) that one of the shear-stress components, $\tau_{xy}$, is zero, since $F_p$ lies along the z-axis and therefore has no component in the y-direction.

The magnitudes of the stress components acting on a given plane, such as in Fig. 2.4(c), are given by the equations

$$\text{Tensile stress} \qquad \sigma = \frac{F_n}{A}, \qquad\qquad (2.1)$$

$$\text{Shear stress} \qquad \tau = \frac{F_p}{A}, \qquad\qquad (2.2)$$

where $F_n$ is the component of $F$ acting normal to the plane and $F_p$ is the component acting parallel to the plane, Fig. 2.4(b).*

Although a maximum of six stress components are needed to describe the most general state of stress, in most practical problems only one to three stresses are required. This reduction in number may result when some of the stress components are omitted from consideration; for example, the stresses acting on the y- and z-planes were not considered in connection with Fig. 2.4. The number of stress components can also be reduced by a suitable choice of the coordinate axes used in determining the stress components. For example, in a tension test the stress state in the elastic range can be described by a single tensile-stress component, $\sigma$, if the axis of loading is chosen as one of the coordinate axes (see Problem 5). More generally, for any complex condition of loading there is an orientation of coordinate axes (called the *principal axes*) for which the shear-stress components are zero. Only the three tensile-stress components are then needed to describe the most general state of stress. To avoid possible confusion, these *principal stresses* are usually designated $\sigma_1$, $\sigma_2$, and $\sigma_3$, rather than $\sigma_x$, $\sigma_y$, and $\sigma_z$. Most treatments of elastic behavior utilize principal stresses, but in such special problems as the

---

* Equations (2.1) and (2.2) apply when $F_p$ lies along one of the coordinate axes. More generally, the two shear stresses acting on the x-plane are given by

$$\tau_{xz} = \frac{(F_p)_z}{A}, \qquad \tau_{xy} = \frac{(F_p)_y}{A}, \qquad\qquad (2.2a)$$

where $(F_p)_z$ is the component of $F_p$ in the z-direction and $(F_p)_y$ is the component in the y-direction.

elasticity of crystals the use of all six stress components may permit the choice of appropriate coordinate axes simply related to the crystal axes.

**Strain.** Displacements of the points of a metal may be of three kinds:

1. *Relative movement of the points in the body*; this displacement is a measure of the *strain* of the body.

2. Ordinary translation of all the points, which occurs when a block of metal slides along the floor.

3. Rotation of all the points (for example, when a piece of metal turns in a lathe).

In practice, the displacements caused by strain are easily distinguished from those produced by translation and rotation. The last two are not of interest in elasticity calculations.

Just as the *stresses* are obtained by resolving *forces* into convenient components referred to *unit area*, so the *strains* are obtained by resolving *displacements* into convenient components referred to *unit length*. An analysis similar to that shown for stresses in Fig. 2.3 can be used to obtain the following strain components (see Problem 7):

$$\left. \begin{array}{c} \varepsilon_x \\ \varepsilon_y \\ \varepsilon_z \end{array} \right\{ \begin{array}{l} \text{Normal strain components in} \\ \text{the directions of the coordinate} \\ \text{axes } x, y, \text{ and } z, \text{ respectively.} \end{array}$$

$$\left. \begin{array}{c} \gamma_{xy}, \gamma_{xz} \\ \gamma_{yz}, \gamma_{yx} \\ \gamma_{zx}, \gamma_{zy} \end{array} \right\{ \begin{array}{l} \text{Shear-strain components: the first letter specifies the} \\ \text{plane on which shearing occurs;* the second letter} \\ \text{specifies the direction in which shearing occurs.} \end{array}$$

The three normal-strain components $\varepsilon_x, \varepsilon_y,$ and $\varepsilon_z$ correspond to ordinary extensions or contractions in the directions of the respective coordinate axes. A shear-strain component, such as $\gamma_{xy}$, describes a shear about the direction that is *not* used in the description of the component, for example the $z$-direction in the case of a $\gamma_{xy}$ shear.

The physical meaning of these strain components is given in Fig. 2.5 for simple cases in which only a single strain component need be considered. Figure 2.5(a) shows pictorially the definition of the normal strain $\varepsilon_x$:

$$\varepsilon_x = \frac{e_{xx}}{l_x} = \frac{\text{displacement in the } x\text{-direction per unit}}{\text{of the length } l_x \text{ being considered.}} \tag{2.3}$$

Rearrangement of this equation gives

$$e_{xx} = \varepsilon_x l_x. \tag{2.4}$$

---

* Here use is made of the familiar convention of designating a plane by the coordinate axis that is perpendicular to it.

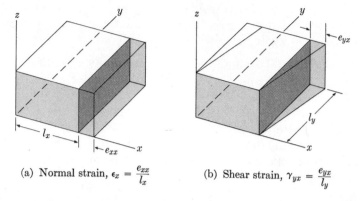

(a) Normal strain, $\epsilon_x = \dfrac{e_{xx}}{l_x}$    (b) Shear strain, $\gamma_{yx} = \dfrac{e_{yx}}{l_y}$

**Fig. 2.5**  Schematic illustration of the physical significance of a normal strain and of a shear strain. The displacements $e_{xx}$ and $e_{yx}$ are shown greatly exaggerated.

That is, when a length $l_x$ experiences a normal strain $\varepsilon_x$, the resulting displacement is in the $x$-direction and has the magnitude $e_{xx}$.  Similarly, Fig. 2.5(b) illustrates the definition of the shear strain $\gamma_{yx}$:

$$\gamma_{yx} = \frac{e_{yx}}{l_y} = \begin{array}{l}\text{displacement in the } x\text{-direction per unit}\\ \text{of the length } l_y \text{ being considered.}\end{array} \tag{2.5}$$

Rearrangement of this equation gives

$$e_{yx} = \gamma_{yx}l_y. \tag{2.6}$$

That is, when a length $l_y$ experiences a shear strain $\gamma_{yx}$, the resulting displacement is in the $x$-direction and has the magnitude $e_{yx}$.

Both the separate straining processes shown in Fig. 2.5 result in a displacement in the $x$-direction.  In fact, the total displacement in the $x$-direction, $e_x$, is the sum of contributions from the normal strain $\varepsilon_x$ and from the two shear strains $\gamma_{yx}$ and $\gamma_{zx}$.  This relation and the corresponding statements for the $y$- and $z$-directions are conveniently written.

$$\begin{aligned}
e_x &= \varepsilon_x x + \gamma_{yx} y + \gamma_{zx} z, \\
e_y &= \gamma_{xy} x + \varepsilon_y y + \gamma_{zy} z, \\
e_z &= \gamma_{xz} x + \gamma_{yz} y + \varepsilon_z z.
\end{aligned} \tag{2.7}$$

These equations describe the displacements at a point $(x, y, z)$ relative to the origin. It will be seen that under these conditions the lengths $l_x$ and $l_y$ in Eqs. (2.4) and (2.6) are given simply by the $x$- and $y$-coordinates of the point.

As in the case of stress, the most general state of strain can be treated using only the three *principal strains* $\varepsilon_1$, $\varepsilon_2$, and $\varepsilon_3$, provided that the strains are defined

in terms of principal axes. The principal axes for stress are usually identical with the principal axes for strain in cases of practical interest. Also, a homogeneous state of stress is usually associated with a homogeneous state of strain.

**Stress-strain relations.** The condition of stress in a body is completely described by the stress components and does not *necessarily* involve a description of the deformation produced by this stress condition. Similarly, the condition of strain in a body is completely described by the strain components; that is, the stress that produced the given strain need not be specified in describing the state of strain. In practice, however, the dependence of strain on stress is of great importance. The simplest type of dependence is given by the usual form of Hooke's law.

$$\varepsilon_x = \frac{1}{E}\sigma_x. \tag{2.8}$$

Here $E$ is Young's modulus, and $\sigma_x$ and $\varepsilon_x$ are the normal stress and strain in the $x$-direction. This equation, however, considers the effect of only a single normal stress. An equation that also considers the effect of stresses in the $y$- and $z$-directions is of more general utility:

$$\varepsilon_x = \frac{1}{E}\sigma_x - \frac{v}{E}\sigma_y - \frac{v}{E}\sigma_z$$

$$= \frac{1}{E}(\sigma_x - v[\sigma_y + \sigma_z]). \tag{2.9}$$

*Poisson's ratio,* $v$, is the ratio of the contractive strain (at right angles to the applied stress) to the strain produced in the direction of a given stress. Thus, Eq. (2.9) allows for the fact that stresses $\sigma_y$ and $\sigma_z$ in the $y$- and $z$-directions contribute to the total strain in the $x$-direction. Even this equation is completely adequate only for *isotropic* substances, that is, for substances that show the same properties in all directions of testing. In anisotropic substances such as metallic single crystals, both $E$ and $v$ vary with direction. Also, even shear stresses can produce normal strains in some anisotropic substances.

Elastic deformations that obey Hooke's law (Hookean deformations) possess the useful property of obeying the *principle of superposition*. This means that the deformation produced by two or more separate forces is the vector sum of the individual deformations caused by each of the forces separately. Similarly, a two- or three-dimensional elastic deformation can be pictured as resulting from a suitable array of forces, each acting in a given coordinate direction. Applications of the superposition principle to complex situations are considered later in this chapter.

A logical extension of these simple stress-strain relations is the *generalized Hooke's law*, which relates all the possible stresses to all the possible strains.

Since six stress components and six strain components* may exist at a point, the most general relations among them are given by the six equations

$$
\begin{aligned}
\varepsilon_x &= S_{11}\sigma_x + S_{12}\sigma_y + S_{13}\sigma_z + S_{14}\tau_{yz} + S_{15}\tau_{zx} + S_{16}\tau_{xy}, \\
\varepsilon_y &= S_{21}\sigma_x + S_{22}\sigma_y + S_{23}\sigma_z + S_{24}\tau_{yz} + S_{25}\tau_{zx} + S_{26}\tau_{xy}, \\
\varepsilon_z &= S_{31}\sigma_x + S_{32}\sigma_y + S_{33}\sigma_z + S_{34}\tau_{yz} + S_{35}\tau_{zx} + S_{36}\tau_{xy}, \\
\gamma_{yz} &= S_{41}\sigma_x + S_{42}\sigma_y + S_{43}\sigma_z + S_{44}\tau_{yz} + S_{45}\tau_{zx} + S_{46}\tau_{xy}, \\
\gamma_{zx} &= S_{51}\sigma_x + S_{52}\sigma_y + S_{53}\sigma_z + S_{54}\tau_{yz} + S_{55}\tau_{zx} + S_{56}\tau_{xy}, \\
\gamma_{xy} &= S_{61}\sigma_x + S_{62}\sigma_y + S_{63}\sigma_z + S_{64}\tau_{yz} + S_{65}\tau_{zx} + S_{66}\tau_{xy}.
\end{aligned}
\tag{2.10}
$$

$S_{11}$, $S_{12}$, etc. are the *elastic compliances* and are constants for a given material. These equations allow for the fact that a given stress component, say $\sigma_x$, *may* contribute to any one of the six strain components. In most practical cases $\sigma_x$ does not contribute to the shear strains $\gamma_{yz}, \gamma_{zx},$ and $\gamma_{xy}$; that is, $S_{41} = S_{51} = S_{61} = 0$. However $\sigma_x$ always contributes to the three normal strains. $S_{11}$ determines the component of the normal strain in the $x$-direction produced by $\sigma_x$, the normal stress in the $x$-direction. Comparison of this term with Eq. (2.8) shows that $S_{11}$ is the reciprocal of Young's modulus $E$ *for this direction of testing*. $S_{21}$ determines the component of the normal strain in the $y$-direction produced by $\sigma_x$. Comparison with Eq. (2.9) shows that $S_{21}$ is equal to $-v/E$ *for these directions of testing*. $S_{31}$ determines the component of the normal strain in the $z$-direction produced by $\sigma_x$. $S_{31}$ is also equal to $-v/E$ for the directions of testing involved. Similar reasoning can be used in connection with $\sigma_y$ and $\sigma_z$, the normal-stress components in the $y$- and $z$-directions, to see the significance of many of the remaining $S$'s.

The compliances $S_{44}$, $S_{55}$, and $S_{66}$ can be understood by comparison with the form of Hooke's law that relates a shear stress $\tau$ and a shear strain $\gamma$,

$$
\gamma = \frac{1}{\mu}\,\tau,
\tag{2.11}
$$

where $\mu$ is the shear modulus. Now, if only the shear stress $\tau_{yz}$ is applied to a body, Eq. (2.10) can be used to determine the shear strain $\gamma_{yz}$ produced by this stress:

$$
\gamma_{yz} = S_{44}\tau_{yz}.
\tag{2.12}
$$

---

* The six shear-strain components $\gamma_{xy}$, $\gamma_{yz}$, etc. used in this chapter are those commonly used in engineering practice. It is not necessary that $\gamma_{xy}$ be equal to $\gamma_{yx}$, even though relations of the type $\tau_{xy} = \tau_{yx}$ are true of the stress components. In fact, usually when an engineering problem involves $\gamma_{xy}$, the conditions are such that $\gamma_{yx} = 0$, and it is for this condition that Eqs. (2.10) are valid. However, for cases in which both shear-strain components defined by a given pair of letters (for example, $\gamma_{xy}$ and $\gamma_{yx}$) have finite values, the only change required in Eqs. (2.10) is the use of the sum $\gamma_{xy} + \gamma_{yx}$ as the left-hand member of the equation in place of the corresponding single term $\gamma_{xy}$.

**TABLE 2.1**
The Elastic Compliances of Several Metal Crystals (The units are $10^{-11}$ m²/N)

| Metal | Crystal structure | $S_{11}$ | $S_{22}$ | $S_{33}$ | $S_{12}$ | $S_{13}$ | $S_{23}$ | $S_{44}$ | $S_{55}$ | $S_{66}$ | $S_{14}, S_{15},$ etc. |
|---|---|---|---|---|---|---|---|---|---|---|---|
| Aluminum | Face-centered cubic | 1.59 | $S_{11}$ | $S_{11}$ | −0.58 | $S_{12}$ | $S_{12}$ | 3.52 | $S_{44}$ | $S_{44}$ | 0 |
| Copper | Face-centered cubic | 1.49 | $S_{11}$ | $S_{11}$ | −0.63 | $S_{12}$ | $S_{12}$ | 1.33 | $S_{44}$ | $S_{44}$ | 0 |
| Iron | Body-centered cubic | 0.80 | $S_{11}$ | $S_{11}$ | −0.28 | $S_{12}$ | $S_{12}$ | 0.86 | $S_{44}$ | $S_{44}$ | 0 |
| Tungsten | Body-centered cubic | 0.26 | $S_{11}$ | $S_{11}$ | −0.073 | $S_{12}$ | $S_{12}$ | 0.66 | $S_{44}$ | $S_{44}$ | 0 |
| Magnesium | Hexagonal | 2.23 | $S_{11}$ | 1.98 | −0.77 | −0.45 | $S_{13}$ | 5.95 | $S_{44}$ | $2(S_{11} - S_{12})$ | 0 |
| Zinc | Hexagonal | 0.84 | $S_{11}$ | 2.87 | +0.11 | −0.78 | $S_{13}$ | 2.64 | $S_{44}$ | $2(S_{11} - S_{12})$ | 0 |
| Tin | Body-centered tetragonal | 1.85 | $S_{11}$ | 1.18 | −0.99 | −0.25 | $S_{13}$ | 5.70 | $S_{44}$ | 13.5 | 0 |
| Iron | Polycrystalline (and therefore isotropic) | 0.50 | $S_{11}$ | $S_{11}$ | −0.15 | $S_{12}$ | $S_{12}$ | $2(S_{11} - S_{12})$ | $S_{44}$ | $S_{44}$ | 0 |

To convert from m²/N to in²/lb multiply by 6.90 × 10³.

Comparison of Eqs. (2.11) and (2.12) shows that $S_{44}$ is the reciprocal of the shear modulus $\mu$ for this direction of testing. Similar reasoning applies to $S_{55}$ and $S_{66}$.

The remaining compliances are zero in most metallurgical applications. This is true, for example, for all the metal crystals in Table 2.1. The large number of zero compliance values for a typical metal, tin, will be shown by Eq. (2.13) in connection with an illustration of the use of the equations governing elastic behavior.

Equations (2.10) are indispensable in treating the elastic behavior of such anisotropic substances as single crystals of metals. The directions $x$, $y$, and $z$ are then the directions of the crystal axes* and the compliances are experimentally determined for use in these directions. In the most general case all the $S$'s exist. However, the number of independent and nonzero terms in Eqs. (2.10) rapidly decreases in going from triclinic crystals to the more common crystal structures that have higher symmetry. Table 2.1 shows the terms that must be considered in dealing with typical metal crystals. (A term such as $S_{21}$ is not shown in the table because it has the same value as $S_{12}$; and in general $S_{ij} = S_{ji}$.) To illustrate the use of the data of Table 2.1, the following set of equations for tin is given in full for $\sigma$ in units of $10^{11}$ N/m².

$$
\begin{aligned}
\varepsilon_x &= \phantom{-}1.85\sigma_x - 0.99\sigma_y - 0.25\sigma_z + \phantom{5.70\tau_{yz}}0 + \phantom{5.70\tau_{zx}}0 + \phantom{13.5\tau_{xy}}0, \\
\varepsilon_y &= -0.99\sigma_x + 1.85\sigma_y - 0.25\sigma_z + \phantom{5.70\tau_{yz}}0 + \phantom{5.70\tau_{zx}}0 + \phantom{13.5\tau_{xy}}0, \\
\varepsilon_z &= -0.25\sigma_x - 0.25\sigma_y + 1.18\sigma_z + \phantom{5.70\tau_{yz}}0 + \phantom{5.70\tau_{zx}}0 + \phantom{13.5\tau_{xy}}0, \\
\gamma_{yz} &= \phantom{-}0 + \phantom{0.99}0 + \phantom{0.25}0 + 5.70\tau_{yz} + \phantom{5.70\tau_{zx}}0 + \phantom{13.5\tau_{xy}}0, \\
\gamma_{zx} &= \phantom{-}0 + \phantom{0.99}0 + \phantom{0.25}0 + \phantom{5.70\tau_{yz}}0 + 5.70\tau_{zx} + \phantom{13.5\tau_{xy}}0, \\
\gamma_{xy} &= \phantom{-}0 + \phantom{0.99}0 + \phantom{0.25}0 + \phantom{5.70\tau_{yz}}0 + \phantom{5.70\tau_{zx}}0 + 13.5\tau_{xy}.
\end{aligned}
\tag{2.13}
$$

Shear stresses do not produce normal strains in this case. It can be seen that a tin crystal has greater "stiffness" in the $z$-direction than in the $x$- or $y$-direction. For example, if the only stress acting on the crystal is a stress of $0.001 \times 10^{11}$ N/m² in the $z$-direction, the normal strain in this direction is

$$
\varepsilon_z = -0 - 0 + 1.18 \times 0.001 + 0 + 0 + 0 = 0.00118. \tag{2.14}
$$

On the other hand, if a single stress of the same magnitude acts in the $x$-direction, the normal strain in the $x$-direction is

$$
\varepsilon_x = 1.85 \times 0.001 - 0 - 0 + 0 + 0 + 0 = 0.00185. \tag{2.15}
$$

Thus the effect produced by stress of a given magnitude depends on the direction of testing.

---

* If the crystal axes are not mutually perpendicular, then the Cartesian $x$-, $y$-, $z$-axes are oriented in a standard manner that reflects a maximum degree of crystal symmetry.

The values of elastic compliances listed in Table 2.1 are determined for directions that coincide with the crystal axes of the metal crystal. For example, the $S_{11}$ given for iron refers only to the normal strain produced by a normal stress applied along one of the cube axes, [100]. If a normal stress were applied in a different direction (for example a [111] direction in an iron crystal), a different compliance value, $S'_{11}$, would relate this stress to the resulting strain. An $S'_{11}$ value can be calculated for any direction of application of a normal stress, and the results for iron show that the compliance in a [111] direction has the smallest value, $0.35 \times 10^{-11}\,\mathrm{m^2/N}$, while the compliance in a [100] direction has the largest value, $0.80 \times 10^{-11}\,\mathrm{m^2/N}$. Thus the stiffness of an iron crystal changes by more than a factor of two as the direction of testing is varied.

**Continuum mechanics.**  Many important problems in elasticity involve situations more complex than the homogeneous strains considered in the preceding sections. In particular, the stresses and strains may vary with position and with time, and differences in temperature may exert an influence. The relations between stress and strain under these more general conditions are known as *constitutive relations*; Hooke's law can be considered an especially simple example. A few basic ideas from continuum mechanics are summarized here as the basis for deriving some essential results for metallic structures. Detailed derivations and additional details can be found in the reference by Timoshenko and Goodier listed at the end of the chapter.

*Equilibrium of stresses.*  If the stresses within a body vary with position, the equilibrium of forces on a given element of volume requires that the following relation hold for the *x*-direction;

$$\frac{\partial \sigma_x}{\partial x} + \frac{\partial \tau_{yx}}{\partial y} + \frac{\partial \tau_{zx}}{\partial z} + f_x = 0. \tag{2.16}$$

Two corresponding equations hold for the *y*- and *z*-directions. The force $f_x$ is a body force within the element of volume. The most common example is a reaction to acceleration, $f_x = -ma_x$, and causes Eq. (2.16) to take the form,

$$\rho \frac{\partial^2 u}{\partial t^2} = \frac{\partial \sigma_x}{\partial x} + \frac{\partial \tau_{yx}}{\partial y} + \frac{\partial \tau_{zx}}{\partial z}. \tag{2.17}$$

The density of the material is $\rho$ and $u$ is the displacement* of a given point in the body in the *x*-direction ($v$ and $w$ represent the displacements in the *y*- and *z*-directions, respectively). At static equilibrium (in the absence of electrical or similar body forces),

$$\frac{\partial \sigma_x}{\partial x} + \frac{\partial \tau_{yx}}{\partial y} + \frac{\partial \tau_{zx}}{\partial z} = 0. \tag{2.18}$$

---

* The corresponding displacement $e_{xx}$ used previously (see Fig. 2.5) is less general, since the effects of rigid-body translations and rotations are excluded.

*Compatibility of strains.* If the strains vary with position in a body, the variations must satisfy the following *compatibility conditions* since the body is a continuous medium:

$$\frac{\partial^2 \gamma_{xy}}{\partial x \partial y} = \frac{\partial^2 \varepsilon_x}{\partial y^2} + \frac{\partial^2 \varepsilon_y}{\partial x^2}, \tag{2.19}$$

$$2\left(\frac{\partial^2 \varepsilon_x}{\partial y \partial z}\right) = \frac{\partial}{\partial x}\left(-\frac{\partial \gamma_{yz}}{\partial x} + \frac{\partial \gamma_{zx}}{\partial y} + \frac{\partial \gamma_{xy}}{\partial z}\right). \tag{2.20}$$

Two additional equations analogous to Eq. (2.19) can be written for $\gamma_{yz}$ and $\gamma_{zx}$ and also two equations analogous to Eq. (2.20) for $\varepsilon_y$ and $\varepsilon_z$. Therefore, the compatibility of strains leads to a total of six relations among the strains.

*Constitutive equations with Lamé constants.* For the solution of many problems in continuum mechanics, the generalized relation between stress and strain is needed in a form significantly different from Eq. (2.10). First, the strains (rather than the stresses) are the known quantities or independent variables; therefore, a typical equation in the analog of Eq. (2.10) is,

$$\sigma_x = C_{11}\varepsilon_x + C_{12}\varepsilon_y + C_{13}\varepsilon_z + C_{14}\gamma_{yz} + C_{15}\gamma_{zx} + C_{16}\gamma_{xy}. \tag{2.21}$$

Each constant $C_{ij}$ is known as an *elastic stiffness* and is related to the elastic compliances $S_{ij}$ for the metal in question. Actually, the full array of $C_{ij}$ values is seldom employed in typical calculations; instead, the material is assumed to be isotropic and therefore to be completely characterized by only two elastic constants, $\lambda$ and $\mu$, known as the Lamé constants. The complete set of constitutive equations is then

$$\begin{aligned}
\sigma_x &= (\lambda + 2\mu)\varepsilon_x + & \lambda\varepsilon_y + & \lambda\varepsilon_z \\
\sigma_y &= \lambda\varepsilon_x + (\lambda + 2\mu)\varepsilon_y + & \lambda\varepsilon_z \\
\sigma_z &= \lambda\varepsilon_x + & \lambda\varepsilon_y + (\lambda + 2\mu)\varepsilon_z \\
\tau_{yz} &= \mu\gamma_{yz} \\
\tau_{zx} &= \mu\gamma_{zx} \\
\tau_{xy} &= \mu\gamma_{xy}.
\end{aligned} \tag{2.22}$$

The shear modulus is $\mu$, defined previously in Eq. (2.11), and is related as follows to Young's modulus, $E$, and Poisson's ratio, $v$

$$\mu = \frac{E}{2(1 + v)}. \tag{2.23}$$

The corresponding relation for $\lambda$ is

$$\lambda = \frac{vE}{(1 + v)(1 - 2v)}. \tag{2.24}$$

An example of the use of the constitutive equations, Eq. (2.22), is the following treatment of stresses associated with dislocations.

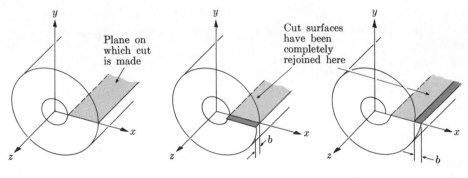

(a) Reference state is a stress-free cylindrical volume.

(b) Rejoining of the cut surfaces in this position produces a screw dislocation.

(c) Rejoining the cut surfaces in this position produces an edge dislocation.

**Fig. 2.6** Schematic illustration of the nature of Volterra dislocations in an isotropic continuum.

**Elastic analysis of dislocations.** Of primary significance among the various crystalline defects (to be studied in Chapter 3) is the *dislocation*. The configuration of atoms that characterizes a dislocation is a principal feature for many purposes; this aspect will be considered extensively throughout the balance of the book. Alternatively, dislocations can be viewed macroscopically as "disturbances" in an otherwise continuous, isotropic medium. The latter view is employed here to make an analysis of elastic strains and stresses applicable to dislocations in crystalline metals.

In 1907 Volterra gave a formal method for determining the distribution of stresses around the type of imperfection that was recognized twenty years later as existing in crystals. His method employs a stress-free cylindrical volume, Fig. 2.6(a). An axial hole avoids the occurrence of mathematical complications in the course of the analysis. The cylinder is pictured as being cut in the manner shown. After the upper half plane is displaced with respect to the lower half plane in either of the two ways shown in Fig. 2.6(b) and (c), the isotropic medium is pictured as being rejoined perfectly along the cut surfaces. We can now determine the resulting distribution of elastic stresses, beginning with the simpler case of the screw dislocation.

*Screw dislocation.* We will employ the procedure of continuum mechanics, starting with the known values of the displacements, $u$, $v$, and $w$ in the $x$-, $y$-, and $z$-directions, respectively. Since the displacement at the cut surface for a screw dislocation, Fig. 2.6(b), is only in the $z$-direction,

$$u = 0 \quad \text{and} \quad v = 0, \tag{2.25}$$

for all points in the body. The displacement in the $z$-direction is readily seen to have the value $w = b$ for those points that lie in the upper plane of the cut. For other points the answer is not so obvious, but the correct expression turns out to be

$$w = \frac{b}{2\pi} \tan^{-1}\left(\frac{y}{x}\right).$$

(2.26)

To obtain the expressions for the strains from the displacements given by Eqs. (2.25) and (2.26), one uses differential expressions of the following type:

$$\varepsilon_x = \frac{\partial u}{\partial x} \qquad \gamma_{yx} = \frac{\partial u}{\partial y} + \frac{\partial v}{\partial x}.$$

(2.27)

These derivatives are equivalent to ratios of Eqs. (2.3) and (2.5), respectively, for homogeneous strains. The strains associated with a screw dislocation are readily evaluated and have the values

$$\varepsilon_x = \varepsilon_y = \varepsilon_z = \gamma_{yx} = 0,$$

$$\gamma_{xz} = -\frac{b}{2\pi}\left(\frac{y}{x^2 + y^2}\right),$$

$$\gamma_{yz} = +\frac{b}{2\pi}\left(\frac{x}{x^2 + y^2}\right).$$

(2.28)

The fact that these strains satisfy the compatibility equations (Problem 10) helps substantiate the correctness of the displacements given in Eqs. (2.25) and (2.26). The stresses corresponding to the strains listed in Eq. (2.28) are given directly by the constitutive equations, Eq. (2.22).

$$\sigma_x = \sigma_y = \sigma_z = \tau_{yx} = 0,$$

$$\tau_{xz} = -\frac{\mu b}{2\pi}\left(\frac{y}{x^2 + y^2}\right),$$

$$\tau_{yz} = +\frac{\mu b}{2\pi}\left(\frac{x}{x^2 + y^2}\right).$$

(2.29)

Problem 11 verifies that this set of stresses satisfies the equilibrium of stresses. Typical constant-stress contours for $\tau_{xz}$ and $\tau_{yz}$ are shown in Fig. 2.7(a) and (b). The stresses of Eq. (2.29) are easily converted from Cartesian to cylindrical coordinates through the standard relations,

$$r^2 = x^2 + y^2,$$

$$\tan \theta = y/x,$$

$$z = z.$$

(2.30)

The elastic displacements parallel to $r$, $\theta$, and $z$ are then denoted $u_r$, and $u_\theta$, and $w$, respectively. In cylindrical coordinates Eq. (2.29) has the especially simple form,

$$\sigma_r = \sigma_\theta = \sigma_z = \tau_{r\theta} = \tau_{zr} = 0,$$

$$\tau_{\theta z} = \frac{\mu b}{2\pi r}.$$

(2.31)

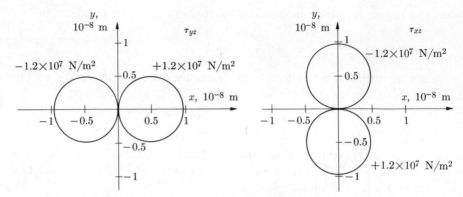

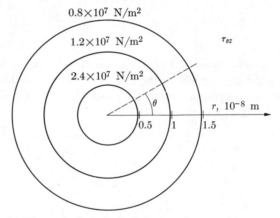

(a) The two shear-stress components referred to cartesian
coordinates. [To convert from N/m² to dyne/cm² multiply
by 10. To convert from N/m² to lb/in² multiply by
$1.450 \times 10^{-4}$.]

(b) The equivalent single shear stress in
cylindrical coordinates, plus two additional
contours.

**Fig. 2.7** Representative constant-stress contours around a screw dislocation directed out of
the plane of the page. The values are for aluminum: $\mathbf{b} = \frac{1}{2}[110]$ and $b = 2.86 \, \text{Å}$. [Data of
H. M. Yoo and B. T. M. Loh.]

The variation of shear stress in the vicinity of a screw dislocation, Eq. (2.31),
can be visualized in the convenient manner shown in Fig. 2.7(b). The single nonzero shear stress, $\tau_{\theta z}$, acts in the z-direction on a plane perpendicular to the $\theta$
coordinate (i.e., on a plane analogous to that on which the cut was made in Fig.
2.6). The magnitude of $\tau_{\theta z}$ decreases as $1/r$ with increasing distance from the center
of the screw dislocation. Since Eq. (2.31) requires that the stress approach infinity
as r approaches zero, the *core* region near $r = 0$ is omitted from this calculation and
is treated instead by means other than elasticity theory (see Chapter 3).

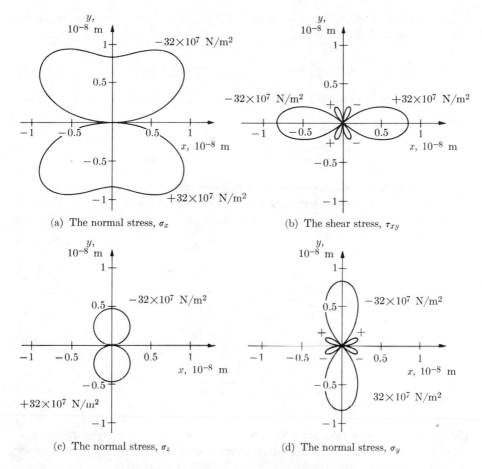

(a) The normal stress, $\sigma_x$

(b) The shear stress, $\tau_{xy}$

(c) The normal stress, $\sigma_z$

(d) The normal stress, $\sigma_y$

**Fig. 2.8** Representative constant-stress contours around an edge dislocation directed into the plane of the page. For the normal stress components, a tensile stress is $(+)$ and a compressive stress is $(-)$. The values are for tungsten: $\mathbf{b} = \frac{1}{2}[111]$ and $b = 2.74\,\text{Å}$. (Data of H. M. Yoo and B. T. M. Loh.)

*Edge dislocation.* Adequate analysis of the strains and stresses near an edge dislocation requires the specialized methods of elasticity theory, but the final quantitative results can be understood from the following simplified development. Comparison of the edge and screw dislocations in Fig. 2.6 suggests that, by analogy with Eq. (2.26), the displacement, $u$, in the $x$-direction might be,

$$u = \frac{b}{2\pi} \tan^{-1}\left(\frac{y}{x}\right).$$
(2.32)

This expression cannot be adequate, however, since the corresponding strains do not satisfy the compatibility condition of Eq. (2.20). The correct expression for $u$ involves an additional term,

$$u = \frac{b}{2\pi}\left[\tan^{-1}\frac{y}{x} + \frac{xy}{2(1-v)(x^2+y^2)}\right]. \tag{2.33}$$

Also, displacements in the $y$-direction, $v$, must occur (although they are not evident in Fig. 2.6(c)):

$$v = -\frac{b}{2\pi}\left[\frac{1-2v}{4(1-v)}\ln(x^2+y^2) + \frac{x^2-y^2}{4(1-v)(x^2+y^2)}\right]. \tag{2.34}$$

No displacements occur in the $z$-direction; $w = 0$.

The procedure used above for the screw dislocation can be employed to determine the strains around an edge dislocation, and finally to obtain the following expressions for the stresses:

$$\sigma_x = -\frac{\mu b}{2\pi(1-v)}\frac{y(3x^2+y^2)}{(x^2+y^2)^2},$$

$$\sigma_y = \frac{\mu b}{2\pi(1-v)}\frac{y(x^2-y^2)}{(x^2+y^2)^2},$$

$$\sigma_z = v(\sigma_x + \sigma_y) = -\frac{\mu v b y}{\pi(1-v)(x^2+y^2)}, \tag{2.35}$$

$$\tau_{xy} = \frac{\mu b}{2\pi(1-v)}\frac{x(x^2-y^2)}{(x^2+y^2)^2},$$

$$\tau_{yz} = \tau_{zx} = 0.$$

The occurrence of a Poisson contraction, specified by $v$, requires a stress, $\sigma_z$, in the $z$-direction to maintain zero displacement in this direction. Typical contours of the four stresses given in Eq. (2.35) are shown in Fig. 2.8.


## ELASTIC PROPERTIES OF ENGINEERING ALLOYS

We now consider several factors that influence the behavior of metallic materials in typical conditions of service. As was mentioned previously, ordinary alloys can usually be treated as isotropic (rather than anisotropic, like single crystals). The alloys are made up of millions of very small crystals, and if the crystals are randomly oriented, the average elastic properties are the same in all directions. The value of $S_{11}$ for an ordinary (polycrystalline) bar of iron given in Table 2.1 is seen to lie between the maximum and minimum values for an iron single crystal. As a result of casting, rolling, or heat-treating processes, however, the grains in a polycrystalline bar may assume very nearly identical orientations, in which case the bar is said to have a *preferred orientation* of its crystals. Figure 2.9 shows that a suitable preferred orientation can greatly increase the stiffness of an iron bar.

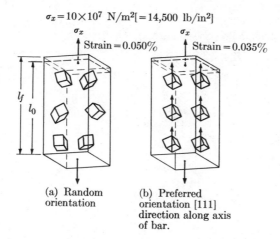

$\sigma_z = 10 \times 10^7 \ \text{N/m}^2 [= 14,500 \ \text{lb/in}^2]$

(a) Random orientation

(b) Preferred orientation [111] direction along axis of bar.

**Fig. 2.9** The effect of a possible preferred orientation of the crystals in an iron bar in changing the elastic behavior of a polycrystalline alloy. The orientations of the crystal axes of the separate grains in each bar are represented by the small cubes.

In this case the orientation of the small cubes in Fig. 2.9(b) indicates that all the grains in the bar have a [111] direction along the direction of loading. Therefore, the extension in this direction is

$$\varepsilon_x = S'_{11}\sigma_x = (0.35 \times 10^{-11} \ \text{m}^2/\text{N})\sigma_x$$
$$= (0.35 \times 10^{-11} \ \text{m}^2/\text{N}) \times (10 \times 10^7 \ \text{N/m}^2) \qquad (2.36)$$
$$= 0.00035 = 0.035\%.$$

On the other hand, using $20 \times 10^{10} \ \text{N/m}^2$ ($29 \times 10^6 \ \text{lb/in}^2$) as the value of Young's modulus $E$ for ordinary polycrystalline iron gives

$$\varepsilon_\lambda = \frac{\sigma_x}{E} = \frac{10 \times 10^7 \ \text{N/m}^2}{20 \times 10^{10} \ \text{N/m}^2} = 0.00050 = 0.050\% \qquad (2.37)$$

as the extension produced by the same stress in the absence of the preferred orientation, Fig. 2.9(a).

The most important elastic property of commercial alloys is Young's modulus $E$. The range of values of Young's modulus in commercial alloys is shown in Table 2.2 and is seen to be large. Since the deflection of a stressed member is usually inversely proportional to the $E$ value of the metal used, the deflection for a given load can be decreased by a factor of 7.5 by changing from magnesium to molybdenum.

Except for preferred orientation or substantial changes in composition, no method is known for significantly increasing the Young's modulus of a given material. Thus if a steel rod of a given size deflects too much under load, the amount

**TABLE 2.2**

Elastic Properties of Some Polycrystalline Metals and Alloys

| Material | Young's modulus $E$ at 20°C, N/m$^2$ | Temperature coefficient $dE/dT$ in the range 0–100°C, N/m$^2$ per °C |
|---|---|---|
| Aluminum | $7 \times 10^{10}$ | $-3.0 \times 10^7$ |
| Beryllium | $26 \times 10^{10}$ | $-4.0 \times 10^7$ |
| Cobalt | $21 \times 10^{10}$ | $-5.1 \times 10^7$ |
| Copper | $11 \times 10^{10}$ | $-3.8 \times 10^7$ |
| Iron | $20 \times 10^{10}$ | $-5.5 \times 10^7$ |
| Lead | $1.8 \times 10^{10}$ | $-1.9 \times 10^7$ |
| Magnesium | $4.3 \times 10^{10}$ | $-18 \times 10^7$ |
| Molybdenum | $32 \times 10^{10}$ | $-4.3 \times 10^7$ |
| Nickel | $21 \times 10^{10}$ | $-6.9 \times 10^7$ |
| Osmium | $56 \times 10^{10}$ | ——— |
| Titanium | $10 \times 10^{10}$ | $-6.9 \times 10^7$ |
| Tungsten | $41 \times 10^{10}$ | $-4.1 \times 10^7$ |
| Sintered carbide (94% WC, 6% Co) | $69 \times 10^{10}$ | $-3.2 \times 10^7$ |
| Elinvar (36% Ni, 12% Cr + W, balance Fe) | $17 \times 10^{10}$ (yield strength, $45 \times 10^7$ N/m$^2$) | 0 |
| Invar (36% Ni, balance Fe) | $14 \times 10^{10}$ (yield strength, $35 \times 10^7$ N/m$^2$) | $+6.9 \times 10^7$ |
| Ni-Span C (42% Ni, 5.5% Cr, 2.5% Ti, balance Fe) | $19 \times 10^{10}$ (yield strength, $124 \times 10^7$ N/m$^2$) | 0 |

To convert from N/m$^2$ to lb/in$^2$ multiply by $1.450 \times 10^{-4}$.

of this deflection cannot be decreased by heat treatment or ordinary alloying. On the other hand, the value of $E$ may be somewhat *decreased* by strengthening procedures discussed in later chapters (precipitation hardening, eutectoid decomposition, cold-working) or other changes that produce internal stresses in the alloy. Increasing the operating temperature also tends to decrease Young's modulus, as shown in Fig. 2.10. This temperature effect may be significant in devices such as Bourdon tubes and precision springs. In these instances special alloys with low temperature coefficients, $dE/dT$, are used, for example *Elinvar* and *Ni-Span-C*, Table 2.2. The magnitude of the change of Young's modulus with temperature for iron is shown by the following calculation. Between 0 and

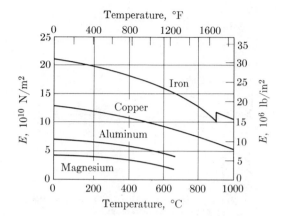

**Fig. 2.10** The effect of temperature on the value of Young's modulus for several pure metals. (W. Köster.)

100°C Young's modulus decreases by

$$100 \times \frac{dE}{dT} = 100(5.5 \times 10^7)$$

$$= 0.55 \times 10^{10} \text{ N/m}^2 \, [0.78 \times 10^6 \text{ lb/in}^2]$$

(2.38)

whereas the values for Elinvar and Ni-Span-C are essentially constant over this temperature range.

Large changes in chemical composition produce variations in Young's modulus. If pure metal $A$ is alloyed with metal $B$, Young's modulus changes in a roughly linear manner from the value, $E_A$, for pure metal $A$ toward a second value, $E_P$. $E_P$ is the Young's modulus of the crystalline material (phase) resulting from a "terminal" amount of alloying. For example, if lead is alloyed with magnesium (the phase diagram describing this alloy system is Fig. 7.26) a new crystalline substance, the $\beta$ phase, occurs when 19% magnesium has been alloyed with the lead. In this case, Young's modulus increases roughly linearly between $E_{Pb}$ and $E_\beta$ as the magnesium content of the alloy increases from 0% to 19%. Above 19% magnesium a different linear relation holds since another crystalline material, metallic magnesium, begins to appear in the alloy. In other alloy systems, such as copper-nickel (Fig. 7.5) or aluminum-silicon (Fig. 7.13), the terminal amount of alloying occurs at 100% of the alloying element. Young's modulus then varies roughly linearly with composition between $E_A$ at pure metal $A$ to $E_B$ at pure metal $B$.

## ELASTIC STRESSES IN DESIGN

The chief design principle concerning elasticity is the following: *the working stress must lie within the elastic range.* The preceding sections have considered the possible values of Young's modulus in this range, and Chapter 4 defines a practical maximum

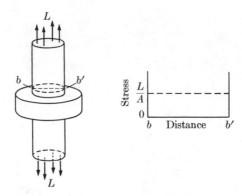

(a) Nominal stress at section b-b' caused by the
load L. A is the area of section b-b'.

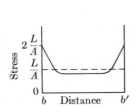

(b) Actual stress at section b-b'
produced by stress raisers.

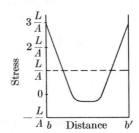

(c) Possible stress at b-b' caused by the
stress raiser plus a hardening heat
treatment.

**Fig. 2.11** Schematic illustration of the difference between nominal stress and actual stress in
a steel part subjected to a load L.

stress limit to the elastic range, the yield strength. All this information is of little
value, however, if the actual *working stress* is not known. The stress calculated in
the usual manner (total load divided by the total cross-sectional area) is the
*nominal stress.* Its magnitude for the section b-b' is shown in Fig. 2.11(a), but for a
number of reasons the actual working stress may differ from this nominal stress.

1. The shape of the part may cause the stress to be higher in certain portions
of the cross section. Such a shape effect is called a *stress raiser*. Figure 2.11(b)
shows that the change in cross section near b-b' has a stress-raising effect which
causes the actual stress to differ significantly from the nominal stress.

2. Even in the absence of stress raisers the nominal stress may differ from the
actual stress as a result of the presence of *residual stresses*. These are stresses
produced by differences in plastic deformation of the metal, as described below.
They can arise during casting, welding, heat-treating, or cold-working operations.

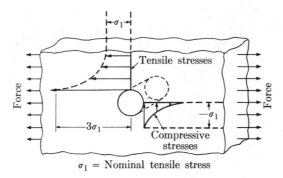

$\sigma_1$ = Nominal tensile stress

**Fig. 2.12** The stress-raising effect of a small hole through a flat plate subjected to a tensile force.

Figure 2.11(c) shows a possible stress pattern resulting from the combined effects of the stress raiser and a hardening heat treatment.

**Stress raisers.** Both calculations and actual measurements demonstrate that the stress in the region of a discontinuity may be different from the nominal stress. An example of this stress-raising effect is shown in Fig. 2.12, where a high tensile stress exists at the top of the hole and a compressive stress is produced at the side. (The magnitudes of only certain stress components at only a few positions in the plate are indicated in this figure.) The high tensile stresses at the top (and bottom) of the hole can be roughly visualized as being caused by the inability of the stress to pass across the hole. The fibers adjoining the hole must carry this portion of the stress in addition to their normal share. Such increases in stress can be described in terms of a *stress concentration factor K*, defined as follows:

$$K = \frac{\text{maximum actual stress}}{\text{nominal stress}}. \qquad (2.39)$$

In Fig. 2.12 the stress concentration factor is $K = 3$. The $K$ values for a number of important stress raisers are given in Fig. 2.13. Local concentration of stress is especially important in the *fracture* of metals, and the significant stress $\sigma_{max}$ is then $\sigma_{max} = K\sigma$; this topic is treated in Chapter 4.

Because many alloys undergo plastic deformation when acted on by sufficiently high stresses, the stress values corresponding to these stress concentration factors are often not fully developed. Ductile materials can often deform locally in the highly stressed regions and thus partially relieve the stresses. In general, stress raisers are not dangerous in *ductile* materials subjected to *static* loads. When the stresses rapidly alternate from tension to compression, however, even ductile materials are greatly affected by stress raisers. Most brittle alloys are sensitive to stress concentration effects under all conditions of loading. An interesting excep-

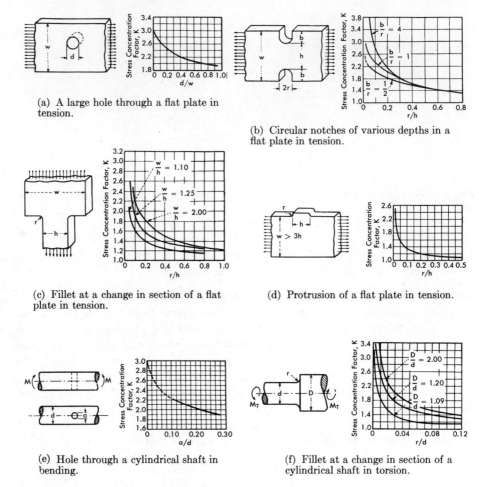

(a) A large hole through a flat plate in tension.

(b) Circular notches of various depths in a flat plate in tension.

(c) Fillet at a change in section of a flat plate in tension.

(d) Protrusion of a flat plate in tension.

(e) Hole through a cylindrical shaft in bending.

(f) Fillet at a change in section of a cylindrical shaft in torsion.

**Fig. 2.13** The stress-raising effect of several machine element shapes. (Courtesy G. H. Neugebauer and *Product Engineering*.)

tion is gray cast iron, in which the graphite flakes that make it brittle also act like internal stress raisers that are much more severe than small notches cut into the surface.

**Residual stresses.** Almost every heat-treating shop has its story about a freshly hardened piece of steel that flew apart while merely resting on a bench. A similar phenomenon occasionally occurs on a larger scale, as in the case of a welded steel bridge that broke in two with a loud noise and collapsed at a time when there was no load on it. These examples demonstrate clearly that metal parts often experience stresses in addition to those produced by the external load. Since all stresses influence the behavior of the metal in service, it is important to understand their natures and the circumstances that cause them. For the purpose of this discussion we distinguish between *body stress*, which usually varies on a macroscopic scale, and *textural stress* that arises because of microscopic inhomogeneities in a metal.

The stress produced by an external load is one kind of body stress, but similar body stresses in the elastic range can also arise under the following circumstances:

1. Uniform thermal expansion in a bar that is fixed at the two ends (see Problem 14) or nonuniform expansion even in an unrestrained part. The constitutive relation for isotropic materials [corresponding to Eq. (2.9)] that includes the effect of thermal expansion is,

$$\varepsilon_x = \frac{1}{E}\sigma_x - \frac{v}{E}\sigma_y - \frac{v}{E}\sigma_z + \alpha\Delta T. \tag{2.40}$$

$\Delta T$ is the change in temperature relative to the reference state and $\alpha$ is the coefficient of linear expansion.

2. Imposition of restraints at the surface of a metal as a result of a chemical treatment, such as nitriding or electroplating.

Stresses of this kind are called *contingent* body stresses, because their existence is dependent (contingent) on the continuing presence of a source of stress, such as external load, nitrided layer, etc. These stresses disappear when the source is removed.

*Residual stress* is the second type of body stress. As its name suggests, it remains after its source has been removed. Several examples of the generation of residual stress will now be discussed: in every instance the cause of the retention of the stress is the occurrence of inhomogeneous plastic deformation.

Plastic deformation at relatively low temperatures (called cold-working) always produces residual stresses. However, the resulting stress pattern varies greatly, depending on the type of working that is employed. If the metal is only lightly worked, so that plastic deformation is limited to regions near the surface, the surface layers are left with compressive stresses, see Problem 15. Figure 2.14(a) gives an example of this pattern of residual stresses, which is produced by processes such as shot-peening, light cold-drawing, and surface rolling. Usually if the working is severe, so that deformation occurs throughout the cross section of the metal, the pattern of residual stresses is similar to that shown in Fig. 2.14(b) for a cold-drawn rod, and there is tensile stress in the surface layers. Tensile residual stresses in the surface layers can seriously reduce the resistance of a metal to cracking; for example, in endurance testing. Therefore one of the processes of light plastic deformation is often used to convert tensile residual stress in the surface layer to compressive stress. Figure 2.14(c) shows the action of surface rolling in improving the pattern of residual stresses originally present in a cold-drawn steel rod.

Another major source of residual stresses is nonuniform change in volume because of thermal effects. In many cases, such as fusion welding, several thermal factors may act simultaneously to produce the final, complex pattern of residual

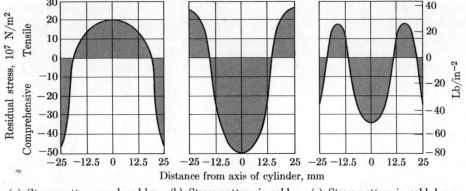

(a) Stress pattern produced by surface rolling; surface layers are in compression.

(b) Stress pattern in cold-drawn steel; surface layers are in tension.

(c) Stress pattern in cold-drawn steel given a final surface rolling; surface layers are in compression.

**Fig. 2.14** Representative patterns of longitudinal residual stress plotted across the diameter of a 50 mm steel rod.

stress. These factors include (1) thermal expansion on heating or cooling, (2) a change in volume, and (3) a change from the liquid state to the solid state. Each of these three sources of residual stress will be discussed separately.

Consider a hot, solid metal rod that is rapidly cooled, for example, by being plunged into water. The surface cools more rapidly than the center and therefore shrinks faster. This nonuniform shrinkage tends to cause plastic compression of the hotter, softer interior of the rod to preserve "matching" of adjoining portions of the rod. After the surface of the rod becomes cool, the interior continues to contract as its temperature continues to fall. When both the interior and the surface of the rod are able to support large stresses, this contraction of the interior produces compressive stresses at the surface of the rod. Thus the final pattern of residual stress is similar to that shown in Fig. 2.14(a). The magnitude of the longitudinal stress at the surface produced by severe quenching is about $-20 \times 10^7 \, \text{N/m}^2$ $[-30,000 \, \text{lb/in}^2]$ in iron and about half this value in nonferrous metals, such as aluminum. As would be expected, use of a milder quench reduces these residual stresses. This beneficial effect is especially pronounced in aluminum, where the use of oil quenching almost eliminates residual stresses.

The most important example of a phase change involving expansion is the formation of martensite from austenite in the hardening of steel (discussed in Chapter 10). An argument similar to that used above for contraction during cooling can be used to show that the expansion accompanying martensite formation should result in tensile residual stresses at the surface and compressive stresses in the interior. This pattern, similar to Fig. 2.14(b), can be produced experimentally if special care is taken to minimize effects due to thermal contraction. The maximum tensile stress at the surface is then about $55 \times 10^7 \, \text{N/m}^2$ $[80,000 \, \text{lb/in}^2]$.

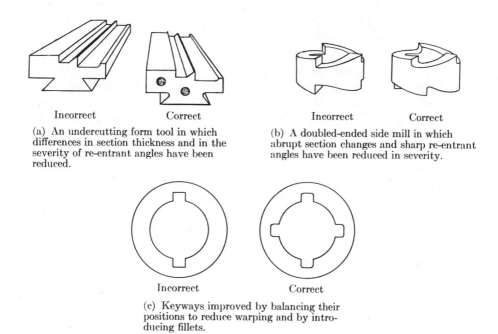

Incorrect                    Correct

(a) An undercutting form tool in which differences in section thickness and in the severity of re-entrant angles have been reduced.

Incorrect                    Correct

(b) A doubled-ended side mill in which abrupt section changes and sharp re-entrant angles have been reduced in severity.

Incorrect                    Correct

(c) Keyways improved by balancing their positions to reduce warping and by introducing fillets.

**Fig. 2.15** Examples of correct and incorrect design of steel parts that are to be quenched during heat treatment. (Courtesy *Tool Steel Simplified*, published by Carpenter Steel Company, Reading, Pennsylvania.)

In actual hardening of steel the final pattern of residual stress is determined by a combination of thermal contraction and martensitic expansion. Consequently the patterns vary greatly, depending on the steel and the details of the heat treatment used, and the surface layers can be either in tension or under compression.

Although surface stresses are developed even in such simple shapes as a cylindrical bar, they are intensified by the presence of discontinuities in the steel part being quenched. Typical causes of distortion, cracking, or breakage of steel parts during heat treatment are sharp corners, re-entrant angles, and abrupt changes in section. Figure 2.15 shows the improvement in the design of typical steel parts by the reduction of these discontinuities. When the part to be heat treated is unavoidably complex in shape, the stresses can be greatly reduced by the use of a milder quenching medium (oil instead of water, for example).

The residual stresses in hardened steel parts are largely removed if the steel is given a high-temperature tempering treatment, for example at 600°C [1100°F]. More generally, residual stresses existing in any alloy for any of the reasons discussed above can be reduced to a safe level by a *stress-relief* heat treatment in which the alloy is heated to a temperature at which local plastic flow occurs under the action of the residual stresses. Typical temperatures for stress relief in one hour are: brass, 250°C [500°F]; nickel alloys, 500°C [900°F]; stainless steel, 850°C [1600°F].

In designing parts to be cast, an additional stress problem must be considered; that is, the development of stresses caused by shrinkage during solidification of the molten alloy. These stresses may become great enough to cause the formation of cracks while the casting is solidifying, a phenomenon called *hot tearing*. Although section discontinuities are also important in this case, an additional factor is the solidification behavior of the alloy. In most cases a solidifying alloy becomes a "connected solid" through interlocking of the previously independent dendrites only after solidification is about 80% complete. Cracking can occur, therefore, only after this connection has been accomplished. Tests have shown that the tendency toward hot tearing is related to the length of the temperature interval over which the remaining solidification occurs. If approximately the final 20% solidification occurs at constant temperature [because of the presence of eutectic liquid; see Chapter 7] then hot tearing is greatly reduced.

In contrast to the body stresses considered above, which tend to vary in a characteristic fashion with macroscopic position in a metal bar (Fig. 2.14), *textural stresses* are localized about certain microstructural features of the metal. The stress field that exists around a dislocation is a good example of this type of stress. Textural stresses also arise across the boundary between adjacent grains during plastic deformation, because of differences in deformation behavior. Similar grain-boundary stresses can be produced by anisotropic expansion of grains as a result of magnetic effects (magnetostriction) or anisotropic thermal expansion. When a second phase forms within an existing solid phase, large textural stresses are created in both phases. Although many details of textural stresses remain to be studied, these stresses can be comparable in magnitude to body stresses and can profoundly influence the behavior of metals.

## PROBLEMS

**1.** Consider a cantilever spring of length $l$ and area $A$. Displacement of the free end by the amount $d$ produces an elastic restoring force $F$ given by the spring equation,

$$F = Kd, \tag{2.41}$$

where $K$ is a constant characteristic of the spring in question. Using the definitions, $\tau = F/A$ and $\gamma = d/l$, for shear stress and strain, show that Eq. (2.41) can be put in the form of Hooke's law, $\tau = \mu\gamma$ [see Eq. (2.11)], provided that the shear modulus, $\mu$, is given by $Kl/A$.

**2.** Make a sketch similar to Fig. 2.3, but show on it the stress components acting on the minus x- and y-faces as well as those shown in the figure acting on the corresponding positive faces. A stress component on a negative face is numerically equal to the corresponding component on the positive face but is pointed in the opposite direction. Since the element of volume being considered is at equilibrium, the algebraic sum of the moments of the forces about an axis, such as the z-axis, must be zero.

a) Obtain an expression for this sum and show that it will be zero if $\tau_{xy} = \tau_{yx}$.

b) What conclusion can be drawn about $\tau_{yz}$, $\tau_{zy}$, $\tau_{zx}$, and $\tau_{xz}$?

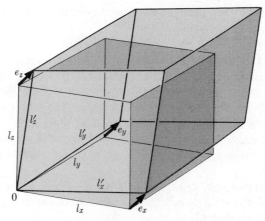

**Fig. 2.16** Illustration for Problem 7.

**3.** Consider a square bar 0.25 m long and 100 mm² in cross-sectional area hanging vertically from a support with a 50 kg load attached to its lower end. What are the magnitudes of the shear stress and the tensile stress on the following planes, which are considered to be at about the center of the bar along its length?

a) A plane perpendicular to the axis of the bar, i.e., a plane whose normal makes an angle of 0° with the axis of the bar. A plane whose normal makes the following angle with the axis of the bar;

b) 30°,          c) 40°,          d) 45°,          e) 0.

f) Plot the magnitudes of shear stress and tensile stress as a function of the angle between the plane normal and the axis of the bar.

g) What is the meaning of the statement, "In a tension specimen the maximum shear stress exists on a plane at 45° to the axis of loading"?

**4.** Obtain expressions involving $F$, $A_0$, and $\theta$ for all the stress components in the bar shown in Fig. 2.4(a) referred to the system of axes shown in Fig. 2.4(c). The $y$-face of the volume element (see Fig. 2.3) is conveniently taken as lying in one of the lateral surfaces of the bar. For the purpose of this analysis the $z$-face is treated analogously to the $x$-face (Fig. 2.4(a)), except that the angle involved is $90° - \theta$.

**5.** Obtain expressions for the stress components in the bar shown in Fig. 2.4(a) referred to the following system of axes: the $x$-axis is the axis of the bar and the $y$- and $z$-axes are perpendicular to the lateral faces of the bar.

**6.** Explain why you would use the axes of Problem 4 or of Problem 5 in treating each of the following phenomena.

a) A long rod is loaded in tension and the resulting elongation is to be measured.

b) When a long rod is loaded in tension, it is observed that at a critical load faint markings appear on the surface. Their traces make an angle of 45° with the axis of the bar.

**7.** Consider a rectangular parallelepiped described by the edge lengths $l_x$, $l_y$, and $l_z$, (Fig. 2.16) that experiences elastic (small) deformation into the parallelepiped described by the edge lengths $l'_x$, $l'_y$, and $l'_z$. This arbitrary deformation is completely described by the three displacement vectors $e_x$, $e_y$, and $e_z$, but it is more convenient to work with the three components of each of these vectors.

a) Show how a displacement vector $e_x$, for example, can be resolved into components $e_{xx}$, $e_{xy}$, and $e_{xz}$ along the three axes $l_x$, $l_y$, and $l_z$.

b) Explain how the *displacement* components, such as $e_{xx}$, $e_{xy}$, and $e_{xz}$, which have the disadvantage of depending on the value chosen for $l_x$, can be converted into the *strain* components $\varepsilon_x$, $\gamma_{xy}$, and $\gamma_{xz}$, which are conveniently referred to a unit length.

**8.** If a body, such as a steel sphere one meter in diameter, is placed in a liquid that is under pressure, the body is said to be under hydrostatic pressure. The components of the strain tensor then have the values $\varepsilon_x = \varepsilon_y = \varepsilon_z = -k$, $\gamma_{yz} = \gamma_{zx} = \gamma_{xy} = 0$, when, as in this case, the body is isotropic.

a) If $E = 20 \times 10^{10}$ N/m² and $v = 0.3$, use Eq. (2.9) to show that $k = 10^{-3} = 0.001$ when the pressure on the liquid is $50 \times 10^7$ N/m². [Note that the set of coordinate axes can be chosen at random because of symmetry.]

b) Explain why Eq. (2.8) cannot be used to determine $k$ in (a).

c) If the center of the steel sphere is chosen as the origin of coordinates, what is the *displacement* with respect to this origin of the point having the coordinates $x = 0.433$ m, $y = 0$, $z = 0$?

d) As in (c), what is the displacement of the point having the coordinates $x = 0.250$ m, $y = 0.250$ m, $z = 0.250$ m?

e) Show that the magnitudes of the displacements in (c) and (d) are equal, and that both are directed toward the center. That is, (c) and (d) represent the solution of the same problem but with two different sets of axes.

**9.** Use Eqs. (2.9) and (2.10) and the data given in Table 2.1 to determine the value of Poisson's ratio for polycrystalline iron.

**10.** By suitably differentiating the strains given in Eq. (2.28) show that they satisfy the six compatibility equations [Eqs. (2.19) and (2.20) and their analogs].

**11.** By suitably differentiating the stresses given in Eq. (2.29) show that they satisfy the three equations for static equilibrium [Eq. (2.18) and its analogs].

**12.** Consider the following problem in a metal ladder designed for lightness of weight. Assume that aluminum was used as the material for building such a lightweight ladder and that the ladder was found to deflect excessively under a man's weight, although no permanent bending took place. Consider some of the following possible remedies for this condition:

a) use of a higher-strength aluminum alloy,

b) use of steel in place of aluminum,

c) use of magnesium in place of aluminum,

d) redesign of the ladder.

**13.** A flywheel shaft is designed so that a bearing section 25 mm in diameter is adjacent to a shaft section 50 mm in diameter. The radius of the fillet connecting these two sections is 2 mm.

a) If the nominal shear stress in the surface fibers of the bearing section is $6 \times 10^7$ N/m², what is the actual maximum stress?

b) Compare the reduction of the stress concentration factors produced by
    1) increasing the fillet radius to 4 mm,
    2) decreasing the shaft diameter from 50 to 30 mm.

**14.** Consider a thin steel rod that is rigidly fixed at both ends. Calculate the stress that is produced in the rod by lowering its temperature from 100 to 0°C. It can be assumed that the rod is not able to contract and that the tendency toward contraction (given quantitatively by $\alpha$ in Table 12.4) is just balanced by a stress tending to cause extension.

**15.** It will be shown in Chapter 4 that plastic deformation begins when a certain value of stress is obtained. Looking at this another way, we could say that plastic deformation tends to prevent stress from rising above this value. Now, assume for simplicity that the elongations produced by surface rolling, Fig. 2.14(a), vary linearly from zero at the center of the bar to a maximum in the surface layers. These elongations produce only elastic stresses in the central portion of the bar, but they ordinarily cause plastic deformation of the surface layers. Therefore the stress can be pictured as rising linearly from the center of the bar to a plateau in the surface layers where plastic deformation has occurred.

a) First consider the simple case of extremely light surface rolling in which even the surface layers of the bar are subjected only to stresses within the elastic region. Sketch the distribution of stress under load and explain what happens to this distribution when the load is removed.

b) Now apply the same analysis to an example of surface rolling in which the outer 10% of the bar undergoes plastic deformation. Sketch the distribution of stress under load. When the load is removed, the stresses tend to return the bar to its original shape. Show that the undeformed central part of the bar is restrained by the plastically elongated surface layers. Sketch the final distribution of residual stresses that represents a balance between the unrelieved tensile stresses in the central part of the bar and the balancing compressive stresses produced in the surface layers.

## REFERENCES

Cottrell, A. H., *Mechanical Properties of Matter*, New York: John Wiley, 1964. Chapter 4 gives a broad, elementary picture of elasticity of solids.

McClintock, F. A. and A. S. Argon, Eds., *Mechanical Behavior of Materials*, Reading, Mass.: Addison-Wesley, 1966. Chapters 1 through 4 give a thorough treatment of the subject matter of the present chapter, including the structural aspects of metals and other materials.

Nye, J. F., *Physical Properties of Crystals*, London: Oxford University Press, 1960. Describes mathematical procedures for treating anisotropic properties of crystals, such as elasticity.

Timoshenko, S. P. and J. N. Goodier, *Theory of Elasticity*, 3rd edn, New York: McGraw-Hill, 1970. Thorough treatment of stress, strain, equilibrium of stress, and compatibility of strains.

# CHAPTER 3

# IMPERFECTIONS IN CRYSTALS

The perfect, long-range crystallinity of metals is, in fact, disrupted by imperfections of various kinds. The major imperfections (dislocations and point defects) affect an extremely small fraction of the atoms, and yet they are crucial in determining such important bulk properties as strength, hardness, and ductility—known as *structure-sensitive properties*. Other properties, like density and Young's modulus, are called structure-insensitive because they are determined by the essential crystalline structure and are almost unaffected by the small fraction of imperfections. In this chapter, we will consider the atomic nature of the various imperfections and discuss the mechanisms by which they influence such processes as plastic deformation. Examples of improvement of the properties of metals through control of imperfections will be described in subsequent chapters.

## TYPES OF IMPERFECTIONS

Metals and alloys for technical applications contain imperfections of many varieties. A brief survey of the various types will give a useful background for understanding both the predominant effects of each kind of imperfection and the interactions among the various types of imperfection.

**Thermal vibrations.** Even an otherwise completely regular atomic lattice of atoms has its atoms displaced from their ideal locations by thermal vibrations. The following data show the magnitude of this effect at room temperature.

| Metal | Frequency of vibration, $\nu$, 1/s | Average amplitude of vibration, Å |
|---|---|---|
| Sodium | $4 \times 10^{12}$ | 0.5 |
| Aluminum | $7 \times 10^{12}$ | 0.2 |
| Copper | $6 \times 10^{12}$ | 0.15 |
| Molybdenum | $7 \times 10^{12}$ | 0.1 |

The frequency of vibration is almost independent of temperature, but the amplitude increases with increasing temperature. For copper, the amplitude near room temperature is about one-half the value near the melting point and about twice the value near 0 K. Since the atoms interact with one another, they tend to vibrate in synchronism; that is, groups of atoms tend to move in the same direction, somewhat as waves on the ocean. However, the wave motion of these elastic displacements in a crystal is constrained by the periodic spacing of the atoms to have wavelengths that are simple multiples of the atomic spacing. This concept is analogous to that used in the electron theory in Chapter 1 and leads to a similar quantizing effect. Here the term *phonon* is used to describe one of the quantized elastic waves. Advanced treatments of effects due to thermal vibrations of crystal lattices are conveniently made in terms of collisions of phonons with impediments such as other phonons, photons, or electrons.

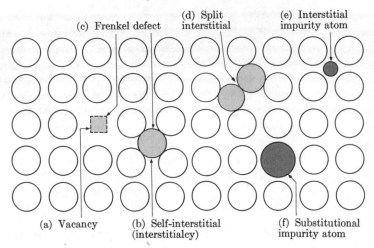

**Fig. 3.1** Schematic illustrations of several point defects in metal crystals.

**Point defects.** In contrast to the slight displacements associated with thermal vibrations, other types of imperfections represent actual structural defects in the crystal lattice of the metal. These structural imperfections are conveniently classified as *point, line,* and *planar* defects in view of their basic geometric configurations. We begin with a discussion of point defects, which are the simplest of the three.

The *vacancy* (or Schottky defect) is simply a vacant atomic position in a crystal structure, Fig. 3.1(a). A thermodynamic argument (see Chapter 6) shows that vacancies are a normal constituent of metals and alloys. Since the atoms surrounding a vacancy experience a slight displacement into the empty lattice site, a vacancy is a center of approximately spherical distortion in the lattice. The inverse of the vacancy is the *self-interstitial* (or interstitialcy), Fig. 3.1(b), in which a metal atom squeezes into the space between normal lattice positions. Because such interstitial spaces are quite small (as shown previously in Fig. 1.19(a)), a self-interstitial produces much greater distortion of the surrounding lattice than does a vacancy. Correspondingly, the energy associated with a self-interstitial is much greater, and therefore the number of self-interstitials in an ordinary metal is negligibly small compared to the number of vacancies. Under unusual circumstances, however, such as exposure of a metal to high-energy neutrons in a nuclear reactor, self-interstitials can be produced in significant numbers. If an atom is removed from a normal lattice position (thus producing a vacancy) and forced into an interstitial position, the resulting pair of point defects is termed a *Frenkel defect*, Fig. 3.1(c). A self-interstitial, Fig. 3.1(b), can also exist in an alternative form, the split interstitial, Fig. 3.1(d).

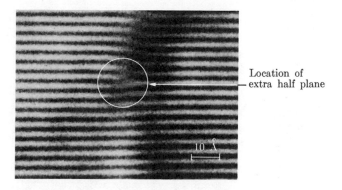

Location of
extra half plane

**Fig. 3.2** Electron-microscopic image of a dislocation in the crystal lattice of germanium. (Courtesy V. A. Phillips and J. A. Hugo, General Electric Research and Development Center. By permission from *Acta Metallurgica*.)

Point defects can also involve impurity atoms. The vacancy remains essentially unchanged except that its energy varies slightly with change in the average chemical composition of its surroundings. Interstitial impurity atoms, Fig. 3.1(e), especially if their ions are small (boron, carbon, hydrogen, nitrogen, or oxygen), can be present in large concentration in a metallic matrix and can greatly influence its properties. A well-known example is interstitial carbon in iron, which can attain concentrations as high as 10 atomic percent. Impurity atoms of normal size generally substitute for the matrix atom on the crystal lattice, Fig. 3.1(f); they are then called substitutional impurity atoms. A fuller discussion of the behavior of impurity (or "alloying") elements in metals and alloys will be given in Chapter 6.

**Dislocations.** The macroscopic Volterra dislocation, whose elastic stresses were discussed in Figs. 2.6–2.8, has its analog at the atomic level in actual crystal structures. Experimental evidence of various kinds demonstrates the existence of dislocations in crystals, but especially direct proof is given by electron micrographs of unusually high magnification, Fig. 3.2. Such micrographs reveal directly the "extra plane" of atoms that characterizes a dislocation. In the metallurgical literature the term dislocation always refers to the crystalline imperfection; the principal use of the macroscopic Volterra dislocation is to furnish a suitable treatment of the corresponding strains and stresses.

An edge dislocation has the essential configuration sketched in Fig. 3.3(b). An extra half plane of atoms is pictured as extending above the symbol, ⊥, which marks the location of the dislocation line at the center (or core) of the dislocation. In Fig. 3.3(b) the dislocation line extends out of the crystal, perpendicular to the plane of the drawing. By analogy with the Volterra dislocation, Fig. 2.6(c), the influence of a dislocation extends far beyond the dislocation line as fields of strain and stress.

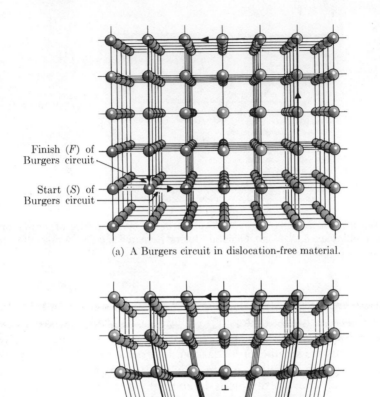

Finish (F) of Burgers circuit

Start (S) of Burgers circuit

(a) A Burgers circuit in dislocation-free material.

F

Burgers vector, b

S

x

a

(b) The same Burgers circuit passing through dislocation-free material, but encircling a dislocation of Burgers vector, b.

**Fig. 3.3** Description of a dislocation in terms of its Burgers vector.

*Burgers vector.* The quantitative description of a dislocation involves a *Burgers circuit.* This circuit starts at some reference atom, *S*, and traces some simple atom-by-atom path in the crystal structure, Fig. 3.3(a), such that the circuit ends at a finish point, *F*, that is identical with the start, *S*, *if the encircled region of crystal structure contains no\* dislocations.* When a Burgers circuit encloses one

---

\* If the encircled region contains two or more dislocations whose individual Burgers vectors sum vectorially to zero, the Burgers circuit will also close. Problem 1 considers a simple example.

(or more) dislocations, it determines the magnitude of the dislocations in terms of the (net) Burgers vector, **b**. The sense (+ or −) of **b** depends on the convention employed; for example, the following (often designated *SF/RH*):

1. Choose a positive direction, defined by the unit vector, **t**, along the dislocation line in question; the direction of **t** is out of the page in Fig. 3.3(b).

2. Use the right-hand-screw convention (*RH*) to choose the proper direction of the Burgers circuit; it is counterclockwise in Fig. 3.3(b).

3. The same Burgers circuit that closes in Fig. 3.3(a) *does not close* in encircling the dislocation in Fig. 3.3(b). The Burgers vector, **b**, is drawn from point *S* to point *F*. Since the vector **b** in Fig. 3.3(b) has a positive (+) sense [points in the positive x-direction], the dislocation ⊥ is a positive edge dislocation in the convention being employed here. If the extra plane of atoms were located in the bottom half of the sketch of Fig. 3.3(b), the corresponding symbol would be ⊤. Application of the same Burgers circuit would then give a Burgers vector pointing in the opposite direction and the dislocation would be a negative edge dislocation.

The lattice vectors ($\hat{\mathbf{a}}, \hat{\mathbf{b}}, \hat{\mathbf{c}}$) of the crystal are used to specify the magnitude and direction of a given Burgers vector. For example, in Fig. 3.3(b)

$$\mathbf{b} = 1\hat{\mathbf{a}} + 0\hat{\mathbf{b}} + 0\hat{\mathbf{c}}. \tag{3.1}$$

This vector might also be represented merely by its components,

$$\mathbf{b} \equiv (1, 0, 0). \tag{3.2}$$

In the cubic crystal system a further simplification is achieved by combining the Miller indices of the direction in question, [100], and the lattice parameter, $a$:

$$\mathbf{b} = a[100]. \tag{3.3}$$

This conventional representation is seen to follow directly from Eq. (3.1). A somewhat more complex example of this type of representation is

$$\mathbf{b'} = -\tfrac{1}{2}\hat{\mathbf{a}} + \tfrac{1}{2}\hat{\mathbf{b}} + \tfrac{1}{2}\hat{\mathbf{c}}, \tag{3.4}$$

which can be written as

$$\mathbf{b'} = \frac{a}{2}[\bar{1}11]. \tag{3.5}$$

Because Miller indices are always integers, ($\tfrac{1}{2}$) has been factored from each of the three components and combined with the lattice parameter. The addition of two Burgers vectors, say $\mathbf{b''} = \mathbf{b} + \mathbf{b'}$, can be performed using the ordinary vector representations, Eqs. (3.1) and (3.4), as described in Problem 2. The equivalent operation can be performed with the conventional representations, however,

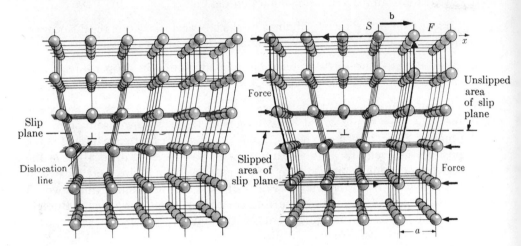

(a) An edge dislocation in a crystal structure.

(b) The dislocation has moved one lattice spacing under the action of a shear force.

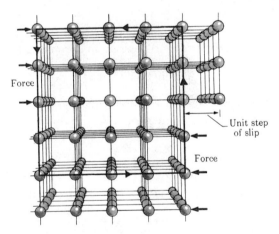

(c) The dislocation has reached the edge of the crystal and produced unit slip.

**Fig. 3.4**  The motion of an edge dislocation and the production of a unit step of slip at the surface of the crystal.  The Burgers circuits in this figure and in Fig. 3.3 are equivalent, even though different starting points $S$ are used.

simply by addition of the indices (multiplied by the prefactor, if any),

$$\mathbf{b}'' = \mathbf{b} + \mathbf{b}'$$

$$= a[100] + \frac{a}{2}[\bar{1}11]$$

$$= a[(1 - \tfrac{1}{2}), (0 + \tfrac{1}{2}), (0 + \tfrac{1}{2})]$$

$$= a[(\tfrac{1}{2}), (\tfrac{1}{2}), (\tfrac{1}{2})]$$

$$= \frac{a}{2}[111]. \tag{3.6}$$

Among the important operations involving the Burgers vector is the determination of the "character" of a given dislocation line: edge, screw, or mixed. If the Burgers vector, $\mathbf{b}$, is *perpendicular* to the direction, $\mathbf{t}$, of the dislocation line (for example, in Fig. 3.3(b)), the dislocation is of edge character. If the Burgers vector is *parallel* to the dislocation line (as in Fig. 3.5(b)) the dislocation is of screw character. Intermediate orientations of $\mathbf{b}$ and $\mathbf{t}$ correspond to dislocations of *mixed* character, as discussed below.

*Deformation by movement of dislocations.* A suitable shear force, Fig. 3.4(b), causes a dislocation to move through a crystal until a unit step of *slip* is produced at the surface, Fig. 3.4(c). The final result is a permanent change in shape of the crystal; that is, plastic deformation. If many dislocations move through a crystal in various directions, large general changes in shape are produced in the manner described in the next chapter. Consideration of Fig. 3.4(b) shows that a dislocation line can be pictured as the boundary between an area of crystal plane on which slip has already occurred and an area that is still unslipped. The Burgers vector, Fig. 3.3(b), is the amount of slip that occurs in each step of the process of slip; the essential phenomenon is the movement of the dislocation line (described by the unit vector $\mathbf{t}$) over the distance $\mathbf{b}$. Consequently, the *slip plane* corresponding to this motion of the dislocation line is the plane containing both $\mathbf{b}$ and $\mathbf{t}$. This plane can be described by the direction perpendicular (normal) to it:

$$\text{Normal to slip plane} = \mathbf{b} \times \mathbf{t}. \tag{3.7}$$

This is the vector cross product employed previously in Chapter 1.

A curved dislocation line may have regions of edge, screw, or mixed character. In Fig. 3.5(a) such a dislocation line is the boundary of the slipped area of a slip plane. One imagines that a certain shear force, $F$, caused this limited amount of slip in one portion of the block of crystal. The unit process of slip, described by the Burgers vector, $\mathbf{b}$, has been the same everywhere in the slipped region. Consequently, the dislocation line has the same Burgers vector at every point along its length. Let us now consider why the atomic displacements adjacent to the dislocation line have various aspects depending on the position along the curved

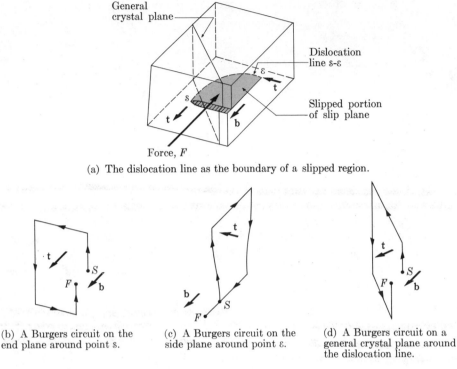

General
crystal plane

Dislocation
line s-ε

ε

t

s

t

b

Slipped portion
of slip plane

Force, F

(a) The dislocation line as the boundary of a slipped region.

t

S

F    b

t

b

S

F

t

S

F    b

(b) A Burgers circuit on the
end plane around point s.

(c) A Burgers circuit on the
side plane around point ε.

(d) A Burgers circuit on a
general crystal plane around
the dislocation line.

**Fig. 3.5**  Various aspects of a curved section of dislocation line, $S$–$\mathscr{E}$.

line. A Burgers circuit on the end face of the crystal around point $S$ of the dislocation line, Fig. 3.5(b), reveals a screw-type dislocation (compare Fig. 2.6(b)). The Burgers vector is, necessarily, in the direction of slip. At point $S$ the direction, **t**, of the dislocation line is parallel to **b**, the requirement for a dislocation of pure screw character.  Unlike an edge dislocation, which has an extra half plane, a screw dislocation consists of a spiral ramp of crystal planes encircling the line of the dislocation.  In the ball model of Fig. 3.6 the screw-dislocation line is vertical, and the ramp circles it in an essentially horizontal plane.

A Burgers circuit around point $\mathscr{E}$ on the side face, Fig. 3.5(c), is the same as that considered in detail in Fig. 3.3.  Although the geometrical shape of the circuit is different in character from that for the screw orientation of this dislocation line, the Burgers vector is identical in the two cases.  The ball model of Fig. 3.7 shows how the atomic configuration characteristic of the edge orientation (an extra plane of atoms) gradually changes into a helical ramp of crystal planes as the dislocation line makes the transition from edge to screw character.  A Burgers circuit around the dislocation line of any character, Fig. 3.5(d), can be made on any nonparallel crystal plane, such as the general plane indicated in Fig. 3.5(a).

Direction of
t   screw-dislocation line

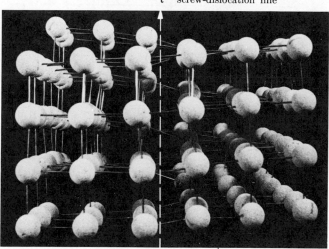

**Fig. 3.6** Atomic model of a simple cubic crystal structure containing a screw dislocation. The spiral ramp of crystal planes encircles the (vertical) dislocation line. (Photograph courtesy of R. de Wit, Naval Research Laboratory.)

The geometrical shape of the circuit is then generally different from both the usual screw and the usual edge circuits, but the Burgers vector is again identical.

This brief introduction to dislocations gives a basis for understanding several of their important geometrical properties.

1. *A dislocation cannot terminate within a crystal.* Since a dislocation line is the boundary of a slipped region in a crystal, Fig. 3.5(a), the line generally runs from one point on the surface of the crystal to another such point. Another possibility, Fig. 3.19, is the formation of a closed loop of dislocation line within the crystal.

2. *The Burgers vector, **b**, is constant along a given dislocation line.* This property results from the uniformity of the process of slip within the region bounded by a given dislocation line.

3. *When two dislocations, **b**$_1$ and **b**$_2$, combine, the result is a dislocation with Burgers vector* (**b**$_1$ + **b**$_2$). A simple example is the combination of one process of slip with another to produce either 2**b**, or zero (see Problem 3). The same choice of positive direction for **t** must be made when parallel dislocations are being combined: if **t** is changed from ( + ) to ( − ), **b** changes sign also.

4. *A dislocation line in direction **t** with Burgers vector **b** defines a slip plane normal to **b** × **t**.* Specific crystallographic slip planes, such as the {111} planes in

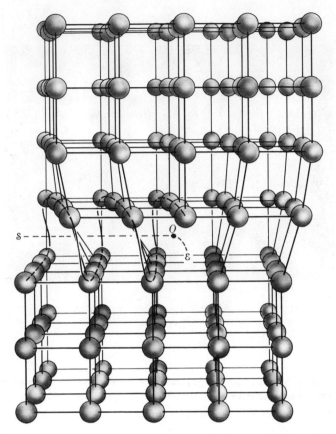

**Fig. 3.7** Atomic model of a simple cubic lattice containing a curved dislocation. Near point $O$ within the lattice, the dislocation makes a gradual transition from edge character ($\mathcal{E}$) to screw character ($\mathcal{S}$).    The helical ramp is indicated.

FCC metals, are examples of this property of most dislocations.  For a pure screw dislocation, however, $\mathbf{b} \times \mathbf{t} = 0$ and the dislocation can move on any crystal plane.

These geometrical properties of dislocations will be used often in later sections.

## BEHAVIOR OF DISLOCATIONS

Dislocations influence the engineering properties of metals through such aspects of their behavior as the creation of new dislocations, interaction among dislocations, and interaction between a dislocation and another feature of the metallic structure (for example, a second-phase particle).   As a first step toward understanding the general behavior of dislocations, we begin with a consideration of the energies and forces involved.

**Energy of a dislocation.** Since a dislocation strains the crystal lattice in its vicinity, it increases the energy of a metal. The total increase in energy can be divided into two parts. (1) The core of the dislocation (the region within about a radius $r_0 = 2|\mathbf{b}| = 2b$ of the dislocation line) contributes about 10% of the energy. An exact determination is difficult because the atomic displacements at the core are too large to justify use of elasticity theory. (2) The elastic strains in the balance of the crystal (principally out to a radius of about $R = 2000b$) account for the remaining 90% of the self-energy of a dislocation in a typical polycrystalline metal. Consequently, macroscopic elasticity theory is a suitable tool for studying the energetics of dislocations.

To obtain an estimate of the self-energy of a dislocation, we can apply the basic equation for work involving force, $F$, and distance, $s$:

$$\text{Work} = \int_0^{s_1} F\, ds = V \int_0^{s_1} \frac{F}{A} \frac{ds}{l} = V \int_0^{\varepsilon_1} \sigma\, d\varepsilon. \tag{3.8}$$

The volume, $V$, of the crystal is its area, $A$, times its length, $l$. Since each of the stresses, $\sigma$, does work in producing its corresponding strain, the total strain energy, $E$, per unit volume of crystal is the sum of the effect of all six stresses. We imagine that the initially strain-free crystal is gradually brought to the state of strain associated with the dislocation by a simultaneous increase in all the stresses. Since the strains depend linearly on the stresses, the result of the integration of Eq. (3.8) is,

$$E = \tfrac{1}{2}(\sigma_x \varepsilon_x + \sigma_y \varepsilon_y + \sigma_z \varepsilon_z + \tau_{xy} \gamma_{xy} + \tau_{yz} \gamma_{yz} + \tau_{zx} \gamma_{zx}). \tag{3.9}$$

The factor $\tfrac{1}{2}$ enters during integration since the average stress is only half the final stress. This value of $E$ applies only to those portions of the crystal where the magnitudes of the stresses are $\sigma_x, \sigma_y$, etc.

For a screw dislocation the strain energy, or self-energy, can be obtained easily since the only nonzero strains and stresses are $\gamma_{xz} = \tau_{xz}/\mu$ and $\gamma_{yz} = \tau_{yz}/\mu$, Eqs. (2.28) and (2.29). Substitution of these values of stress and strain in Eq. (3.9) gives for unit length of screw dislocation,

$$E = \frac{1}{2\mu}(\tau_{xz}^2 + \tau_{yz}^2) \tag{3.10}$$

or, with the values of shear stress given by Eq. (2.29),

$$E = \frac{1}{2\mu}\left(\frac{\mu b}{2\pi}\right)^2 \left(\frac{1}{x^2 + y^2}\right) \tag{3.11}$$

and in cylindrical coordinates,

$$E = \frac{1}{2\mu}\left(\frac{\mu b}{2\pi}\right)^2\left(\frac{1}{r^2}\right). \tag{3.12}$$

Since $E$ varies with distance, $r$, from the dislocation line, we must perform the following integration to obtain the self-energy, $E_S$, per unit length of screw dislocation in the cylindrical region of radius $R$,

$$E_S = \int_V E\, dV = \frac{1}{2\mu}\left(\frac{\mu b}{2\pi}\right)^2 \int_{r_0}^R \frac{2\pi r\, dr}{r^2} = \frac{\mu b^2}{4\pi}\ln\frac{R}{r_0}. \tag{3.13}$$

Using the estimates $r_0 = 2b$ and $R = 2000b$ given above, we obtain the value,

$$E_S \simeq \mu b^2/2. \tag{3.14}$$

The appropriate value for $R$ is the average distance at which the stress field of the dislocation has decreased to the value produced by neighboring dislocations. Since this distance is equal to half the spacing of adjacent dislocations, the value of $R$ and therefore of $E_S$ increases as the density of dislocations decreases. Because of the logarithmic function in Eq. (3.13), however, $E_S$ increases only by 50 % for a decrease of dislocation density from the normal value, $10^{12}/m^2$, to the value $10^{10}/m^2$ characteristic of "relatively perfect" metal crystals. For highly imperfect crystals ($10^{16}/m^2$), $E_S$ decreases to about $\mu b^2/4$.

A typical numerical value for $E_S$ is obtained from Eq. (3.14) using $\mu = 0.7 \times 10^{11}\, N/m^2$ for the shear modulus (the value for iron) and $b = 3 \times 10^{-10}\, m$:

$$E_S \simeq 3 \times 10^{-9}\, J/m\ [\text{joule/meter}]. \tag{3.15}$$

Assuming that a line of atoms one meter in length in a typical crystal contains $3 \times 10^9$ atoms, we obtain

$$E_S \simeq 10^{-18}\, J/\text{atom} \simeq 6\, eV/\text{atom}. \tag{3.16}$$

Experimentally determined energies fall in the range of 3 to 10 eV for various metals. Such energies include dislocations in both screw and edge orientations. The self-energy, $E_E$, of unit length of edge dislocation can be calculated by the method used for the screw dislocation: the result is

$$E_E = E_S/(1 - v). \tag{3.17}$$

Since Poisson's ratio, $v$, is about $\frac{1}{3}$ for typical metals, $E_E \simeq \frac{3}{2}E_S$.

**Slip directions and planes.**  Since the self-energy of a dislocation line varies as the square of the Burgers vector, dislocations with the smallest possible Burgers vector tend to occur. For example, the process of slip occurs in the direction of closest atomic spacing in almost all metals and alloys. Examination of Fig. 1.8 shows that the atoms in FCC metals are most closely packed (actually, in contact in a hard-sphere model) along $\langle 110 \rangle$ directions. Consequently, slip in these directions

**TABLE 3.1**

Slip Systems for Representative Metals

| Metal | Crystal structure | Slip plane | Slip direction |
|---|---|---|---|
| Aluminum | FCC | $\{111\}$ | $\langle 10\bar{1}\rangle$ |
| Copper | FCC | $\{111\}$ | $\langle 10\bar{1}\rangle$ |
| Gold | FCC | $\{111\}$ | $\langle 10\bar{1}\rangle$ |
| Tungsten | BCC | $\{112\}$ | $\langle 11\bar{1}\rangle$ |
| Molybdenum | BCC | $\{112\}$ | $\langle 11\bar{1}\rangle$ |
| | | $\{1\bar{1}0\}$ | $\langle 11\bar{1}\rangle$ |
| $\alpha$-Iron | BCC | $\{112\}$ | $\langle 11\bar{1}\rangle$ |
| | | $\{1\bar{1}0\}$ | $\langle 11\bar{1}\rangle$ |
| | | $\{123\}$ | $\langle 11\bar{1}\rangle$ |
| Cadmium | HCP | $\{0001\}$ | $\langle 11\bar{2}0\rangle$ |
| Zinc | HCP | $\{0001\}$ | $\langle 11\bar{2}0\rangle$ |
| | | $\{11\bar{2}2\}$ | $\langle 11\bar{2}3\rangle$ |
| Magnesium | HCP | $\{0001\}$ | $\langle 11\bar{2}0\rangle$ |
| | | $\{11\bar{2}2\}$ | $\langle 10\bar{1}0\rangle$ |
| | | $\{10\bar{1}1\}$ | $\langle 11\bar{2}0\rangle$ |
| Titanium | HCP | $\{10\bar{1}0\}$ | $\langle 11\bar{2}0\rangle$ |
| | | $\{0001\}$ | $\langle 11\bar{2}0\rangle$ |
| | | $\{10\bar{1}1\}$ | $\langle 11\bar{2}0\rangle$ |

involves the smallest possible Burgers vector.* Directions of closest packing are also the slip directions in HCP metals, $\langle 11\bar{2}0\rangle$, and in BCC metals, $\langle 111\rangle$.

Although the plane on which slip occurs is a prominent feature of the slip process, the slip plane is more variable than is the slip direction. A primary requirement of any slip plane is that it be parallel to the corresponding slip direction. This requirement is not very restrictive, however, since a large array of planes (called planes of a zone) are parallel to any given direction (called the *zone axis*). Problem 6 shows that planes of the $\{100\}$, $\{110\}$, and $\{111\}$ types are among the planes in the [110] zone. In FCC metals the most closely packed planes are the $\{111\}$, and these serve most commonly as the slip plane. Actual manipulation of hard-sphere models (see Problem 7) reveals that slip on a (111) plane involves less elastic distortion and also permits a lowering of the self-energy through decomposition of the total dislocation into two partials. The combination of a slip plane and a slip direction is known as a *slip system* and is designated $\{111\} \langle 10\bar{1}\rangle$, for example, for FCC metals. (These symbols can be read: one-one-one type planes in the one-zero one-minus type directions.) The data of Table 3.1 indicate that BCC and HCP metals also tend to slip on close-packed planes, (110) and (0001), but that numerous exceptions are observed. Furthermore, a given metal may

---

* The *total* Burgers vector can split into two *partials*, however. This aspect is considered on p. 100.

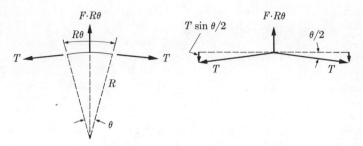

(a) External forces required to maintain a small segment of dislocation line in equilibrium.

(b) The corresponding balance-of-force diagram, including the vertical components.

**Fig. 3.8** Sketch for analyzing the force, $F$, per unit length required to maintain a radius of curvature, $R$, in a dislocation line.

change from the close-packed plane to some alternative slip plane when subjected to sufficiently severe restraints.

**Line tension of a dislocation.** The energy per unit length ($E$, joule/m) of a dislocation line corresponds to a force ($T'$, newton) tending to shorten the length of the line. This correspondence can be seen from a dimensional analysis,

$$\text{joule/meter} = (\text{newton} \cdot \text{meter})/\text{meter} = \text{newton}.$$

The force, $T'$, is directed along the dislocation line and is called the *line tension* of a dislocation. A force $T = -T'$ must be applied to a dislocation line to overcome the line tension and to cause the dislocation to maintain or increase its length. An expression for the line tension can be obtained by the type of argument used with Eqs. (3.13) and (3.14). The increase in self-energy $\Delta E_S$, associated with an increase in length $\Delta l$ of a dislocation, is released when shortening occurs. Therefore, the restoring force $T'$ is given as

$$T' = -\frac{\Delta E_S}{\Delta l} = -\frac{\mu b^2}{4\pi} \ln\left(\frac{R}{r_0}\right). \tag{3.18}$$

With the usual estimates for $R$ and $r_0$, the approximate value of $T'$ for both screw and edge dislocations is

$$T' \simeq -\mu b^2. \tag{3.19}$$

By analogy with a rubber band, if a dislocation is in a bowed configuration, Fig. 3.8, the line tension produces a force, $F'$, per unit length tending to straighten the line (and thereby decrease its length). An equal opposing force, $F = -F'$, must act to maintain a given curvature in the dislocation line. An expression for $F$ can be obtained by analyzing the forces acting on an increment, $R\theta$, of bowed dislocation line. The balance of the vertical components (see Problem 8) leads to the result,

$$F = T/R \simeq \mu b^2/R. \tag{3.20}$$

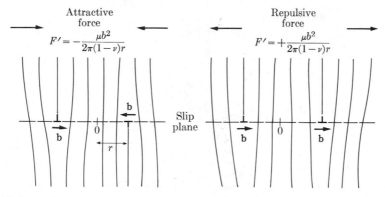

(a) Attraction of a pair of dislocations of opposite sign.

(b) Repulsion of a pair of dislocations of the same sign.

**Fig. 3.9** Schematic illustration of (a) attraction, and (b) repulsion of a pair of edge dislocations on the same slip plane.

The bowing of dislocations around hard particles in a metal is an important aspect of the strengthening of metals (Chapter 10).

**Forces between dislocations.** We now consider the effect of one dislocation on a neighboring dislocation. Because the general treatment involving forces is rather abstract, we begin with two special cases that can be visualized in terms of energies and which give a useful physical picture. Two dislocations on the same slip plane, Fig. 3.9(a), can combine and annihilate each other provided their Burgers vectors are equal but opposite. Annihilation would cause a decrease in energy [see Eqs. (3.13) and (3.17)]

$$\Delta E = \frac{\mu b^2}{2\pi(1 - v)} \ln \frac{R}{r_0}.$$ (3.21)

Therefore, the energy of the system can be pictured as increasing by the amount given by Eq. (3.21) as $r$ increases from $r_0$ to the normal half-distance, $R$, between dislocations. Since $E$ is a potential energy, we can obtain an expression for the force of attraction,

$$F' = -\frac{d(\Delta E)}{dr} = -\frac{d}{dr}\left[\frac{\mu b^2}{2\pi(1 - v)} \ln \frac{r}{r_0}\right] = -\frac{\mu b^2}{2\pi(1 - v)r}.$$ (3.22)

A similar analysis applies for two edge dislocations of the same sign, Fig. 3.9(b), except that the energy *increases* if the two dislocations combine. The Burgers vector is then 2b, and the factor $(2b)^2$ in the energy term causes the energy of the resulting dislocation to be larger by $\Delta E$, Eq. (3.21), than the sum of the energies of the two separate dislocations. Consequently, the calculation of Eq. (3.22) gives the opposite sign, and the force of interaction is a repulsion in this case.

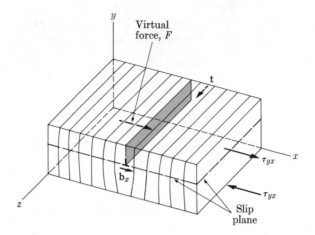

**Fig. 3.10** Schematic illustration of the virtual force, $F$, on an edge dislocation corresponding to the action of a shear stress on the slip plane of the dislocation.

*Peach–Koehler Equation.* In general the force between two dislocations depends not only on their Burgers vectors but also on the direction, **t**, of each dislocation line and on the orientation of the two lines in three dimensions. This general case can often be analyzed in terms of stresses. Consider first the effect of an applied shear stress, $\tau_{yx}$, on an edge dislocation **b** having its line direction, **t**, in the positive $z$-direction, Fig. 3.10. Comparison with Fig. 3.4 shows that the shear stress causes the dislocation to move in the positive $x$-direction. The virtual force, $F_x$, acting on unit length of the dislocation line to cause this movement can be readily derived by the following reasoning. If a constant shear stress, $\tau_{yx}$, causes an incremental strain, $\Delta\gamma_{yx}$, in a unit cube of material, an argument similar to that employed with Eq. (3.8) shows that the work done is,

$$\text{Work} = \tau_{yx}\Delta\gamma_{yx}. \tag{3.23}$$

Let the incremental strain be that corresponding to the passage of a dislocation with Burgers vector, $b_x$, across the unit cube,

$$\Delta\gamma_{zx} = \frac{\text{displacement}}{\text{unit length}} = \frac{b_x}{1} = b_x. \tag{3.24}$$

The amount of work given by Eq. (3.23) can then be pictured as resulting from the action of a virtual force, $F_x$, in moving the dislocation line across the unit cube in the $x$-direction,

$$\text{Work} = \tau_{yx}b_x = F_x \, (1 \text{ meter}). \tag{3.25}$$

Therefore,

$$F_x = \tau_{yx}b_x, \tag{3.26}$$

where $F_x$ is a force per unit length of dislocation line.

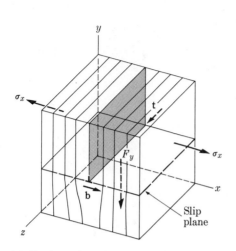

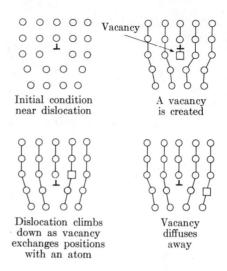

Initial condition
near dislocation

A vacancy
is created

Dislocation climbs
down as vacancy
exchanges positions
with an atom

Vacancy
diffuses
away

(a) Configuration of an edge dislocation for which the "climb" force, $F_y$, occurs.

(b) The motion of vacancies permits an edge dislocation to climb out of its slip plane.

**Fig. 3.11** The action of a normal stress, $\sigma_x$, in producing a virtual force, $F_y$, that causes an edge dislocation to climb.

Applied stresses interact with dislocations in two other essential ways. First, a shear stress, $\tau_{zx}$, exerts a virtual force,

$$F_y = \tau_{zx}b_x, \tag{3.27}$$

on a *screw* dislocation. The derivation is similar to that of Eq. (3.26), see Problem 9. Second, a normal stress, $\sigma_x$, exerts a virtual force,

$$F_y = -\sigma_x b_x \tag{3.28}$$

on a suitably oriented edge dislocation, Fig. 3.11(a). The derivation of this expression is analogous to that given above provided that the dislocation line moves in a novel manner. In this case, *the edge dislocation moves perpendicular to its slip plane*, a process known as *climb*. The mechanism of climb involves the motion (diffusion) of vacancies away from the dislocation line (causing negative climb), Fig. 3.11(b), or toward the dislocation line (causing positive climb).

In the general case in which both the line direction, **t**, and the Burgers vector, **b**, have arbitrary orientations, there are 18 separate force terms (such as $F_x = \tau_{yz}b_z t_z$). For **t** in a given direction (say the z-direction), *two* force terms are given by each of Eqs. (3.26), (3.27), and (3.28) for suitable choices of stress and **b**—a total of six terms (Problem 10). Since the unit vector, **t**, in general has components $t_x$, $t_y$, and $t_z$ along the three coordinate axes, a total of $3 \times 6 = 18$ terms results. All of these 18 terms are conveniently represented in vector notation in the form known as the Peach–Koehler equation,

$$F = G \times t = \begin{vmatrix} i & j & k \\ G_x & G_y & G_z \\ t_x & t_y & t_z \end{vmatrix}$$

$$= (G_y t_z - G_z t_y)i + (G_z t_x - G_x t_z)j + (G_x t_y - G_y t_x)k. \tag{3.29}$$

Here $i, j$, and $k$ are the usual unit vectors along the $x$-, $y$-, and $z$-directions, and the symbols $G_x$, etc., merely represent the sum of terms

$$G_x = \sigma_x b_x + \tau_{xy} b_y + \tau_{xz} b_z$$
$$G_y = \tau_{yx} b_x + \sigma_y b_y + \tau_{yz} b_z$$
$$G_z = \tau_{zx} b_x + \tau_{zy} b_y + \sigma_z b_z. \tag{3.30}$$

The vector cross product of Eq. (3.29) also emphasizes that the total virtual force, $F$, acts in a direction perpendicular to the direction, $t$, of the dislocation line.

*Two Parallel Dislocations.* As an example of the use of Eq. (3.29), let us calculate the force between two parallel screw dislocations in the $z$-direction. We picture the stress field of the first dislocation, Eq. (2.29), as acting on the Burgers vector, $b_2 = b_2 k$ of the second dislocation with $t$ in the positive $z$-direction ($t_x = 0$, $t_y = 0, t_z = 1$). The only nonzero terms in Eq. (3.29) are,

$$F = \tau_{yz} b_2 i - \tau_{xz} b_2 j \tag{3.31}$$

and therefore,

$$F = \frac{\mu b_1 b_2}{2\pi(x^2 + y^2)} (xi + yj). \tag{3.32}$$

If the relative location of the two dislocations is described not by $x$ and $y$ but by cylindrical coordinates $(r, \theta)$, Eq. (3.32) becomes,

$$F = \frac{\mu b_1 b_2}{2\pi r} r, \tag{3.33}$$

where $r$ is the unit vector in the $r$ direction. A positive value of $F$ represents a repulsive force (between dislocations of like sign). If the Burgers vectors are of opposite sign, $F$ is negative and the two dislocations experience an attraction.

Calculation of the force between two parallel edge dislocations proceeds in the same manner (Problem 11), and leads to the expression,

$$F = \frac{\mu b_1 b_2}{2\pi(1 - v)(x^2 + y^2)^2} [x(x^2 - y^2)i + y(3x^2 + y^2)j]. \tag{3.34}$$

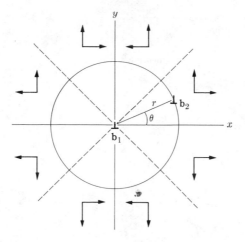

**Fig. 3.12** Schematic illustration of the changes in sign of the force components, $F_x$ and $F_y$, of two parallel edge dislocations of the same sign as a function of their angular relationship, $\theta$.

An alternative scalar form of this equation uses the $(r, \theta)$ coordinates,

$$F_x = \frac{\mu b_1 b_2}{2\pi(1 - v)r} \cos \theta(\cos^2 \theta - \sin^2 \theta), \qquad (3.35)$$

$$F_y = \frac{\mu b_1 b_2}{2\pi(1 - v)r} \sin \theta(1 + 2 \cos^2 \theta). \qquad (3.36)$$

The signs (as well as the magnitudes) of $F_x$ and $F_y$ change with variation in their angular orientation in the manner shown in Fig. 3.12. At angles near 90° (or 270°) the $F_x$ component tends to align two edge dislocations vertically. This action is responsible for the occurrence of aggregates of parallel dislocations in such configurations as the low-angle boundaries discussed below.

## REACTIONS BETWEEN DISLOCATIONS

In view of the energies of dislocations and the associated virtual forces, two (or more) dislocations tend to interact and perhaps produce a new dislocation by a reaction such as

$$\mathbf{b}_1 + \mathbf{b}_2 = \mathbf{b}_3. \qquad (3.37)$$

In the absence of other influences (external forces, thermal energy, etc.), a requirement for the occurrence of this reaction is a net decrease in the self-energies; that is,

$$\mu b_1^2 + \mu b_2^2 > \mu b_3^2. \qquad (3.38)$$

Since the shear modulus is a common factor, only the squares of the Burgers

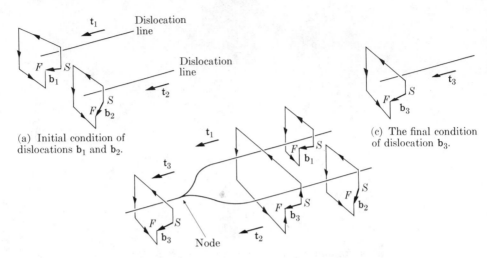

(a) Initial condition of dislocations $b_1$ and $b_2$.

(c) The final condition of dislocation $b_3$.

(b) The dislocation reaction in progress. A node exists.

**Fig. 3.13** Schematic illustration of three stages of the dislocation reaction $b_1 + b_2 = b_3$ as it occurs in the course of time. Only portions of the dislocation lines are shown.

vectors need be considered. Kinetic factors determine whether an energetically possible reaction actually occurs: These factors are considered in later chapters. In Problem 3 we have already considered the simple case of the annihilation of two parallel edge dislocations of opposite sign on the same slip plane.

**Dislocation nodes.** The essential portion of a dislocation reaction occurs at the junction or *node* of the dislocation lines involved. In the example of Fig. 3.13, dislocation lines 1 and 2 must first approach each other, Fig. 3.13(a). When they meet, Fig. 3.13(b), a node forms and is the site at which the reaction of Eq. (3.37) occurs. As the reaction proceeds, portions of lines 1 and 2 are converted into dislocation line 3. Since the Burgers vector must be conserved within a crystal (see p. 83), a Burgers circuit encircling both lines 1 and 2, Fig. 3.13(b), must yield $b_3 = b_1 + b_2$. The same value is also given by a Burgers circuit beyond the node; that is, around dislocation line 3. Frank's rule for the dislocations at a node states:

$$(b_1 + b_2 + \cdots)_{\text{entering}} = (b_3 + b_4 + \cdots)_{\text{leaving}} . \tag{3.39}$$

The direction of each dislocation line determines whether that dislocation is entering or leaving the node in question. If all directions are arbitrarily chosen as entering (or all leaving) the node, Eq. (3.39), takes the form,

$$\sum_{i=1}^{n} b_i = 0. \tag{3.40}$$

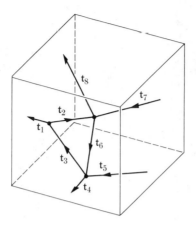

**Fig. 3.14**  Schematic illustration of a network of dislocation lines within a portion of a crystal. Three nodes are shown.

Nodes have varying degrees of stability.  The node in Fig. 3.13(b) is unstable and disappears when the reaction has been completed.  Stable nodes are produced under many conditions; for example, in the reaction of nonparallel dislocations or of dislocations on different slip planes.  Complex networks, Fig. 3.14, containing many relatively stable nodes occur commonly in actual metals.  The nodes may be stable because of an equilibrium of forces or because of kinetic barriers to further reaction.  The application of Frank's rule to the three nodes in Fig. 3.14 gives,

$$\mathbf{b}_3 = \mathbf{b}_1 + \mathbf{b}_2$$

$$\mathbf{b}_2 + \mathbf{b}_7 = \mathbf{b}_6 + \mathbf{b}_8 \tag{3.41}$$

$$\mathbf{b}_5 + \mathbf{b}_6 = \mathbf{b}_3 + \mathbf{b}_4.$$

Here $\mathbf{b}_1$ is the Burgers vector of line $\mathbf{t}_1$, etc.  Substantially more force is required to disturb a stable network of dislocations than to move an isolated dislocation line.  Consequently, arrays of dislocations play an important role in the strengthening of metals.

**Kinks, jogs, and intersections of dislocations.**  Two different types of short segment, kinks and jogs, can connect one straight portion of dislocation line with a second portion offset by a small amount, Fig. 3.15.  If the offset portion lies in the original slip plane, then the connecting segment is called a *kink*.  If the straight portions of the dislocation line are in edge orientation, Fig. 3.15(a), then the kink is of screw character.  In view of the definitions given earlier, an offset portion of an edge dislocation is connected to the main dislocation line by a $+/-$ pair of kinks. Problem 12 considers the analogous situation in which the main dislocation line is

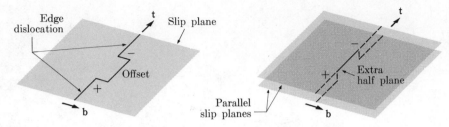

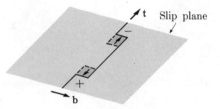

(a) An edge dislocation with a $+/-$ pair of kinks of screw character.

(b) An edge dislocation with a $+/-$ pair of jogs, also of edge character.

(c) Spreading of a $+/-$ pair of kinks corresponds to an advance of the dislocation line. The shaded areas indicate the change from the position in (a).

**Fig. 3.15**  Schematic illustrations of kinks and jogs in dislocation lines.

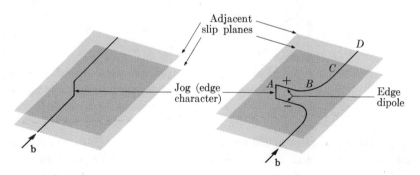

(a) Initial condition of a screw dislocation containing a jog.

(b) As the main part of the dislocation moves, the immobile jog causes an edge dipole to form.

**Fig. 3.16**  The action of a jog in pinning a screw dislocation.

screw type, the kinks then having an edge character. If the offset portion lies, not in the original slip plane, but in a parallel slip plane, Fig. 3.15(b), then the connecting segment is called a *jog*. A jog in a line of edge character has a direction, $t$, that is perpendicular to $b$; therefore, the jog is also edge character. Jogs in screw dislocations, however, are of opposite (edge) character (Problem 13).

Kinks affect the mobility of the original dislocation line quite differently from jogs. When a dislocation moves (glides) on its slip plane, usually the dislocation does not remain perfectly straight but instead develops some degree of curvature. Since curvature can be described by a suitable array of kinks (each perhaps only of length $b$; that is, about one atomic diameter), kinks are a common feature in the motion of dislocations. They are mobile on the original slip plane. Also, after a pair of kinks has been produced, they can spread laterally, Fig. 3.15(c), or, of course, can approach each other and be annihilated. Whereas kinks are a normal accompaniment to the motion of dislocations, jogs usually impede mobility since they do not ordinarily lie in a slip plane. The jogs form initially as a result of: (1) intersection of two dislocations, or (2) climb of a portion of a straight dislocation line to produce an offset portion, Fig. 3.15(b). * The latter process involves the motion of vacancies, Fig. 3.11(b), and results in an upward extension of the extra half plane of atoms in the offset region.

To see how a jog can impede the motion of a dislocation, consider a jog (edge character) joining two segments of screw-dislocation line, Fig. 3.16(a). As the screw dislocations move on their slip planes, each is *pinned* at the location of the jog because the edge dislocation is not on a slip plane and therefore cannot glide. The result of this pinning action is the formation of an edge dipole, Fig. 3.16(b), consisting of a pair of positive and negative edge dislocations. The screw dislocation can continue to glide, but only at the expense of creating additional length of edge dipole. Alternatively, the original jog can move by climb, but this is an impractically slow process during deformation at low temperatures.

An important mechanism for forming kinks or jogs is the intersection of a moving dislocation line with another dislocation line in the crystal. Consider first the simple case in which both dislocation lines are edge character and are mutually perpendicular. If the two Burgers vectors are perpendicular, Fig. 3.17(a), then the downward motion of dislocation 1 causes an edge jog equal to $b_1$ to appear in dislocation line 2 after intersection, Fig. 3.17(b). (Dislocation line 1 is unaffected by the intersection because $b_2$ is parallel to the direction of line 1.) The jog is *glissile* (i.e., able to glide) because it lies in a slip plane, though it can move only parallel to its Burgers vector. Intersection produces quite a different result if the two Burgers vectors are parallel, Fig. 3.17(c). A kink of screw character appears in each of the dislocation lines after intersection, Fig. 3.17(d). In general, intersections can involve dislocation lines of any character, and the two lines need not be on perpendicular planes (see Problem 14).

An important type of intersection is one that produces a *sessile jog* (i.e., a segment of dislocation unable to glide). An example, Fig. 3.18, is the intersection of

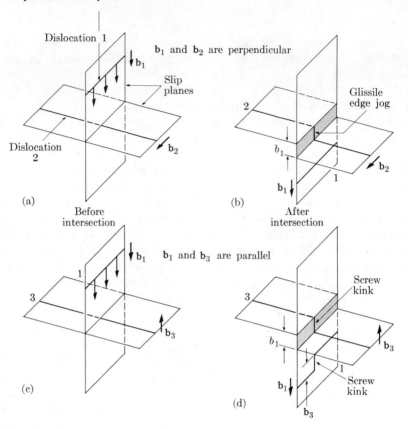

**Fig. 3.17** Schematic illustration of the formation of jogs and kinks by the intersection of two mutually perpendicular edge-dislocation lines.

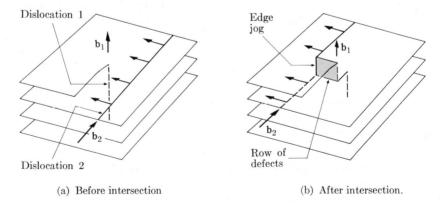

**Fig. 3.18** Schematic illustration of the intersection of two screw dislocations to produce an edge jog in the moving dislocation.

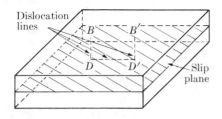

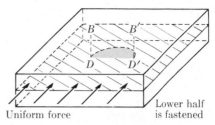

(a) Initial condition, with the dislocation line $O$—$O'$ leaving the slip plane at $B$ and at $B'$.

(b) A uniform force tending to shear the top half of the crystal causes $O$—$O'$ to begin sweeping over the slip plane by bowing outward.

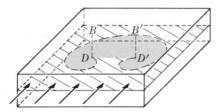

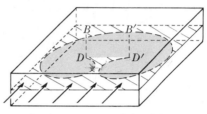

(c) As the dislocation line continues to expand it spirals about $O$ and $O'$.

(d) Further expansion of the dislocation line brings the two spiral sections together.

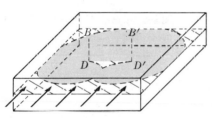

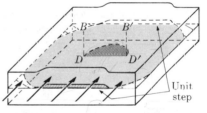

(e) As the two sections join they produce an outer loop that continues to expand and an inner line similar to $O$—$O'$ in (a).

(f) The outer loop produces a unit step as it leaves the surface of the crystal; $O$—$O'$ begins the cycle again.

**Fig. 3.19** The Frank–Read mechanism for the production of complete dislocation loops.

a moving screw dislocation with a stationary screw dislocation. The resulting edge jog is able to glide along dislocation 2 but cannot glide in the direction of motion of this dislocation. Dislocation 2 can continue to move after intersection only by forming an edge dipole, Fig. 3.16, or by producing a row of vacancies or interstitialcies. In either case, motion of the dislocation is severely impeded, corresponding to significant strengthening of the crystal.

**Dislocation sources.** A simple calculation (see Problem 15) shows that the total number of dislocations present in a typical metal could account for only a small permanent deformation of the metal. Since dislocations, in fact, account for extremely large deformations, dislocations must be continuously generated in a metal during the course of mechanical working (and also of other processing, such as heat treatments). Several mechanisms for the generation of dislocations prob-

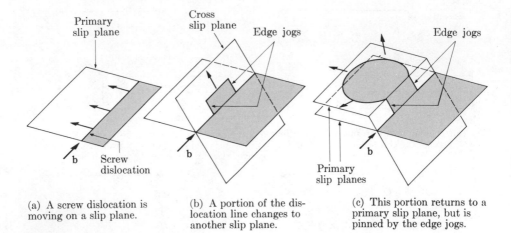

(a) A screw dislocation is moving on a slip plane.

(b) A portion of the dislocation line changes to another slip plane.

(c) This portion returns to a primary slip plane, but is pinned by the edge jogs.

**Fig. 3.20**  Schematic illustration of the formation of a Frank–Read source by double cross-slip of a screw dislocation.

ably operate simultaneously.  An example is the Frank–Read mechanism, Fig. 3.19, which produces dislocation loops.  The dislocation line $B$—$O$—$O'$—$B'$, Fig. 3.19(a), is pictured as having a segment $O$—$O'$ on the slip plane in question, $O$ and $O'$ being points at which the segment is pinned.  A shear stress causes the segment to bow and eventually to form a complete loop, as described in the caption to the figure.  As each loop is formed, it is free to move on the slip plane and to pass out through the surface of the crystal, thus producing a step of plastic deformation.

The term *Frank–Read source* refers to a segment of mobile dislocation line pinned at its ends, such as in Fig. 3.19(a).  The production of Frank–Read sources is evidently essential, and one important mechanism is shown in Fig. 3.20.  Cross slip of a screw dislocation produces sessile edge jogs, Fig. 3.20(b), which then serve as anchor points during the formation of Frank–Read loops on the primary slip plane, Fig. 3.20(c).  Many other mechanisms for the "multiplication" of dislocations also exist.  Furthermore, dislocations can be generated at the surface of a crystal at such points of stress concentration as notches, scratches, and cracks.

**Partial dislocations.**  For an adequate study of deformation and strengthening of metals, we must distinguish between total (or perfect) dislocations and partial (or imperfect) dislocations.  As we have seen, the passage of a total dislocation through a crystal leaves the perfection of the crystal unimpaired; that is, each atom is shifted from one normal position in the lattice to an adjacent normal position by the passage of a total dislocation.  Passage of a partial dislocation, on the other hand, shifts an atom only a fractional amount and therefore leaves behind a planar region of imperfect crystal.

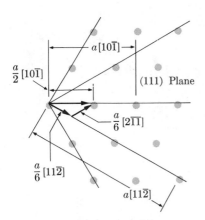

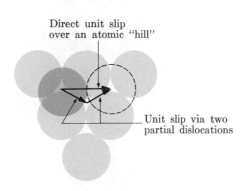

Direct unit slip
over an atomic "hill"

Unit slip via two
partial dislocations

(a) The total and partial Burgers vectors involved in unit slip in the slip direction.

(b) Unit slip of an atom in the slip direction by two alternative routes.

**Fig. 3.21**   Description of the mechanism of the slip process in FCC metals.

The principal reason for the splitting of a total dislocation, $\mathbf{b_T}$, into two partials, $\mathbf{b_1}$ and $\mathbf{b_2}$, is the accompanying decrease in energy, Eq. (3.38),

$$b_T^2 > b_1^2 + b_2^2. \tag{3.42}$$

As a typical example we consider the partial dislocations that take part in the process of slip in FCC metals. Recall from Table 3.1 that slip occurs on (111) planes and in $[10\bar{1}]$ directions. The total Burgers vector for the slip process is $(a/2)[10\bar{1}]$, Fig. 3.21(a), and leaves the crystal structure with unimpaired stacking. In particular, the stacking sequence of successive (111) planes remains $ABCABC$. One disadvantage of this total Burgers vector, however, is that its energy is higher than the sum of the two partial Burgers vectors, $(a/6)[11\bar{2}] + (a/6)[2\bar{1}\bar{1}]$, that are equivalent to $(a/2)[10\bar{1}]$. These two partials are examples of *Shockley partials*, characterized by being glissile. Substitution of the appropriate Burgers vectors in Eq. (3.42) gives,

$$\frac{a^2}{4}(1^2 + 0^2 + 1^2) > \frac{a^2}{36}(1^2 + 1^2 + 2^2) + \frac{a^2}{36}(2^2 + 1^2 + 1^2)$$

$$\frac{a^2}{2} > \frac{a^2}{3}. \tag{3.43}$$

Therefore, the total Burgers vector tends to dissociate into two partials,

$$\mathbf{b_T} = \mathbf{b_1} + \mathbf{b_2} \tag{3.44}$$

$$\frac{a}{2}[10\bar{1}] = \frac{a}{6}[11\bar{2}] + \frac{a}{6}[2\bar{1}\bar{1}]. \tag{3.45}$$

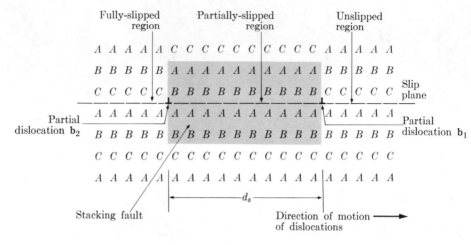

**Fig. 3.22** Schematic illustration of the stacking fault existing on the portion of slip plane traversed only by the first (leading) partial dislocation. The figure represents a side view of a stack of (111) planes in an FCC metal.

The slip mechanism involving partial dislocations has the additional energetic advantage that the path followed by the moving atom, Fig. 3.21(b), follows "valleys" in the (111) planes rather than proceeding over an atomic "hill."

*Stacking Faults.* The planar region of imperfection produced by the passage of a single partial dislocation is called a stacking fault, Fig. 3.22. In the region of the slip plane traversed by partial dislocation $\mathbf{b}_1$ (but not yet by $\mathbf{b}_2$) the stacking sequence of close-packed planes changes from $ABCABC$ to $ABAB$, the sequence characteristic of HCP metals. An additional energy, $\gamma_s$ J/m$^2$, is needed to create a region of HCP structure in an FCC metal. This stacking-fault energy tends to decrease the spacing, $d_s$, between the leading ($\mathbf{b}_1$) and trailing ($\mathbf{b}_2$) partial dislocations. Table 3.2 gives the stacking-fault energies of typical metals and the corresponding equilibrium spacing (see Problem 16).

In HCP metals dissociation of total dislocations into partials also occurs. The smallest Burgers vector for a total dislocation on the (0001) plane is

$$\mathbf{b}_T = \frac{a}{3}[11\bar{2}0]. \tag{3.46}$$

The appropriate dissociation reaction is,

$$\mathbf{b}_T = \mathbf{b}_1 + \mathbf{b}_2, \tag{3.47}$$

$$\frac{a}{3}[11\bar{2}0] = \frac{a}{3}[10\bar{1}0] + \frac{a}{3}[01\bar{1}0]. \tag{3.48}$$

A construction analogous to Fig. 3.22 (see Problem 17) shows that the passage

**TABLE 3.2**

Stacking-Fault Data for Representative Metals

| Metal | Stacking-fault energy, $\gamma_s$, J/m$^2$ | Equilibrium spacing of partial dislocations, $d_s$, Å |
|---|---|---|
| Silver | 0.02 | 18.7 |
| Gold | 0.06 | 5.7 |
| Copper | 0.07 | 6.8 |
| Aluminum | 0.20 | 1.5 |
| Nickel | 0.40 | 2.0 |

To convert from J/m$^2$ to erg/cm$^2$ multiply by $10^3$.

of the leading partial dislocation ($\mathbf{b}_1$) along a slip plane produces a region of stacking fault with stacking sequence $ABAB\,CACA$; the dashed line indicates the location of the slip plane. The stacking-fault energy can be considered to be associated with the region of FCC stacking, $ABCA$. Passage of the trailing partial dislocation ($\mathbf{b}_2$) restores the normal stacking in the region of the lattice in question and completes one full unit of slip. In the BCC and diamond cubic structures, partial dislocations also occur but the slip process presents some complications. The stacking-fault energies in these structures are so high that the distance separating the two partial dislocations is only $d_S \simeq b$. Under these conditions, dissociation is of doubtful significance.

*Reactions between Partial Dislocations.* The individual partial dislocations can undergo reactions analogous to those described earlier for total dislocations. A principal complication, however, may be the presence of the "partner" partials, still connected by a region or "ribbon" of stacking fault. Also, the result of the reaction of two partial dislocations is often a sessile dislocation; that is, a dislocation that is unable to move on its slip plane. Sessile dislocations not only immobilize the dislocations taking part in the reaction but may also block other dislocations from moving on adjacent slip planes.

A typical reaction in FCC metals involves two pairs of dissociated dislocations moving on intersecting slip planes, Fig. 3.23(a). When the leading partial dislocations meet, the resulting reaction may or may not be energetically possible depending on the exact Burgers vectors involved (see Problem 18). For the Burgers vectors shown in Fig. 3.23(a) the reaction is

$$\mathbf{b}_1 + \mathbf{b}_3 = \mathbf{b}_{1,3}, \tag{3.49}$$

$$\frac{a}{6}[2\bar{1}1] + \frac{a}{6}[\bar{1}2\bar{1}] = \frac{a}{6}[110]. \tag{3.50}$$

This reaction occurs with a decrease in energy, but the resultant dislocation is

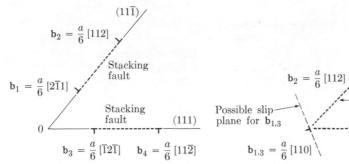

(a) Two pairs of partial dislocations on intersecting {111} planes advance toward point 0, where the leading partials will react.

(b) The result of the reaction is an immobile dislocation, known as a *stair-rod* dislocation in this instance.

**Fig. 3.23**  The reaction of partial dislocations in an FCC metal to produce a sessile dislocation. The figures represent edge-on views of two {111} planes.

sessile. The theoretically possible slip plane [given by $\mathbf{b}_{1,3} \times \mathbf{t}$, Eq. (3.7)], does not coincide with the operative slip planes, {111}, in FCC metals. Therefore, dislocation $\mathbf{b}_{1,3}$ is sessile (immobile), and also anchors the two remaining partial dislocations tied to it by ribbons of stacking fault. The entire array forms a roofed surface in three dimensions and is known as a Cottrell–Lomer lock.

## GRAIN BOUNDARIES

Because many types and gradations of grain boundaries exist, we will include under this term: (a) any two-dimensional surface within a (single-phase) metal, which (b) separates two regions differing in orientation of their crystal axes. The *degree* of difference in orientation distinguishes among such types of boundaries as:

1.  *Sub-boundaries*, having such a small difference that the boundaries are considered to exist within a given grain;

2.  *Small-angle boundaries*, in which the two adjacent grains differ in orientation by less than about 10°;

3.  *High-angle boundaries* (ordinary grain boundaries), with large angular differences between grains.

The *type* of difference in orientation distinguishes among such classifications of boundaries as:

1.  *Tilt boundaries*, in which the misorientation can be pictured as a relative tilting of the two lattices;

2.  *Twist boundaries*, in which the misorientation can be pictured as a relative rotation of the two lattices;

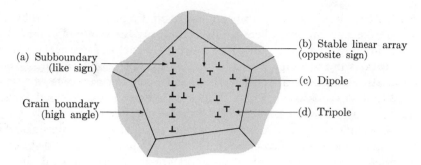

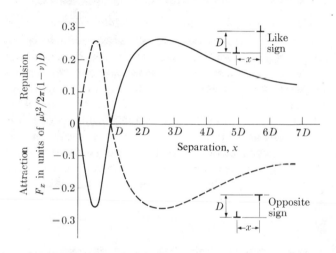

**Fig. 3.24** Schematic illustrations of typical arrays of parallel edge dislocations that can exist within the body of a grain.

**Fig. 3.25** Plot of the force component, $F_x$, between parallel edge dislocations as a function of their relative separation in the $x$-direction (after A. H. Cottrell).

3. *Coincidence boundaries,* in which atomic positions common to both of the neighboring grains constitute some fraction of the total boundary area.

Grain boundaries are studied from many points of view—structural, energetic, microscopic, etc.—but dislocations offer a basis for a broad understanding of the nature and properties of grain boundaries.

**Arrays of dislocations.** When the number and types of dislocations present within a single grain of metal are relatively small, a simple analysis may explain the configurations assumed by the dislocations. For example, Eq. (3.35) for the $x$ component, $F_x$, of the force between parallel edge dislocations can be used to explain why such dislocations tend to align vertically to form a sub-boundary,

Fig. 3.24(a). If neighboring dislocations are on slip planes separated by a distance $y = D$, the $F_x$ component between dislocations of the same sign plots as the solid curve in Fig. 3.25 as a function of $x$. Since an attractive force acts on neighboring dislocations in the range $x = 0$ to $x = D$, a stable configuration of an array of dislocations is a linear alignment (planar wall in three dimensions) that constitutes a sub-boundary within the grain. A slight misorientation, described quantitatively by Eq. (3.51), tends to exist across even a sub-boundary.

If neighboring dislocations in a parallel array have opposite signs, the sign of $F_x$ changes and the dashed curve in Fig. 3.25 applies. Adjacent dislocations then experience an attractive force at distances greater than $x = D$, and in principle a linear array at 45°, Fig. 3.24(b), would be stable. In practice, aggregates of dislocations with opposing signs usually form random arrays called dipoles, tripoles, and multipoles, depending on the number of dislocations in the array.

**Tilt and twist boundaries.**  The sub-boundary of Fig. 3.24(a) is actually an example of a symmetric tilt boundary. If two grains, initially in the same orientation, are pictured as tilted symmetrically about an axis parallel to the direction of the edge dislocation, Fig. 3.26(a), the two lattices can be maintained continuous by the periodic insertion of an edge dislocation. The resulting vertical array of parallel dislocations of like sign is identical with that of Fig. 3.24(a). This type of boundary can be seen under the microscope, Fig. 3.26(b), as a uniformly spaced line of etch pits. The presence of each dislocation at the surface of a specimen is revealed because of the faster rate of chemical attack by the etching reagent in the strained area of the crystal near the core of the dislocation. As can be seen from Fig. 3.26(a), the distance, $D$, separating adjacent dislocations in the boundary is

$$D = \frac{b}{2 \sin \theta/2} \simeq b/\theta. \tag{3.51}$$

As the angle, $\theta$, of a symmetric tilt boundary increases, the energy, $\gamma$, of the boundary must change because: (1) the number of dislocations increases as $\theta/b$; and (2) the energy of each dislocation [see Eq. (3.17)],

$$E_E = \frac{\mu b^2}{4\pi(1 - v)} \ln \frac{D}{2r_0} \tag{3.52}$$

decreases, since $D \simeq b/\theta$. If $r_0 = 2b$, and the energy of unit length of dislocation core is $E_c$, the energy of unit area of boundary becomes,

$$E_b \simeq \frac{\theta}{b}\left[ E_c + \frac{\mu b^2}{4\pi(1 - v)} \ln \frac{1}{4\theta} \right] \tag{3.53}$$

$$\simeq \theta(A - B \ln \theta). \tag{3.54}$$

An equation of this type fits the experimental data of Fig. 3.27 below about 8°, but the predicted maximum is not observed. At higher angles the dislocations

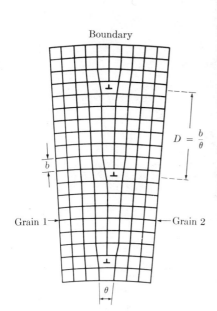

Boundary

$D = \dfrac{b}{\theta}$

Grain 1

Grain 2

$b$

$\theta$

(a) A tilt boundary as a succession of added crystal planes. The corresponding edge dislocations are indicated. (After Burgers.)

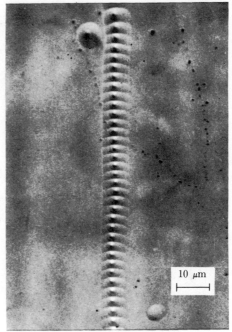

10 $\mu$m

(b) Photomicrograph of a tilt boundary in germanium showing the regularly spaced dislocations.

**Fig. 3.26** The nature of tilt boundaries in metals. (Photomicrograph courtesy F. L. Vogel, Jr., Bell Telephone Laboratories.)

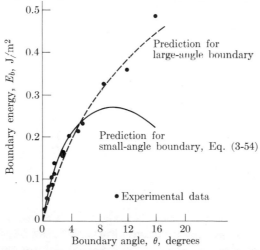

**Fig. 3.27** Comparison of experimental data on the energy of a tilt boundary in copper with the theoretical curve for small-angle boundaries. (Data of N. A. Gjostein and F. N. Rhines.) [To convert from J/m$^2$ to erg/cm$^2$ multiply by 10$^3$.]

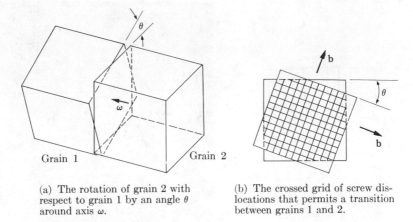

(a) The rotation of grain 2 with respect to grain 1 by an angle $\theta$ around axis $\omega$.

(b) The crossed grid of screw dislocations that permits a transition between grains 1 and 2.

**Fig. 3.28** Schematic illustration of a pure twist boundary. (After J. P. Hirth and J. Lothe.)

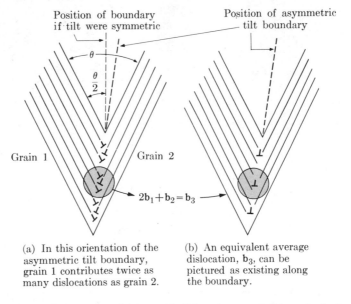

(a) In this orientation of the asymmetric tilt boundary, grain 1 contributes twice as many dislocations as grain 2.

(b) An equivalent average dislocation, $b_3$, can be pictured as existing along the boundary.

**Fig. 3.29** Sketches of the array of (a) actual dislocations, and (b) equivalent average dislocations in an illustrative asymmetric tilt boundary.

presumably have the more complex arrangements characteristic of large-angle grain boundaries. A pure twist boundary, Fig. 3.28(a), can be pictured as resulting from the rotation of one grain relative to its neighbor about an axis perpendicular to the plane of the boundary. The corresponding displacements are equivalent to two separate shears at right angles. Since each shear can be produced by a set of

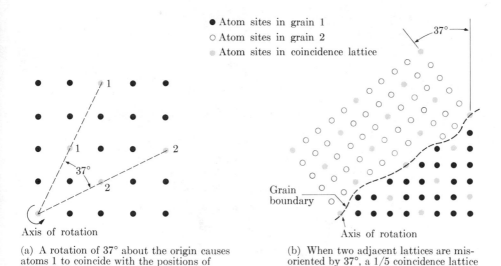

- ● Atom sites in grain 1
- ○ Atom sites in grain 2
- ◉ Atom sites in coincidence lattice

Axis of rotation

Axis of rotation

(a) A rotation of 37° about the origin causes atoms 1 to coincide with the positions of atoms 2.

(b) When two adjacent lattices are misoriented by 37°, a 1/5 coincidence lattice exists.

**Fig. 3.30**  Illustration of a coincidence lattice in a two-dimensional square lattice.

parallel screw dislocations (see Problem 19), two such sets forming a crossed grid, Fig. 3.28(b), constitute the twist boundary joining grains 1 and 2.

Typical tilt and twist boundaries are asymmetric; that is, the total angle of misorientation is not divided equally between grains 1 and 2. In Fig. 3.29 the boundary angle has been chosen so that twice as many edge dislocations, $b_1$, are contributed to the boundary by grain 1 as by grain 2. The average Burgers vector, $b_3$, of a given section of boundary is therefore,

$$2b_1 + b_2 = b_3. \tag{3.55}$$

This reaction need not actually occur, but $b_3$ represents the average Burgers vector perpendicular to the asymmetric boundary. A similar analysis can be made of an asymmetric twist boundary except that at least three sets of screw dislocations constitute the actual array in this case.

**Coincidence lattice.**  A grain boundary is a region of relatively high energy because the atoms in the boundary do not fit perfectly into the repetitive crystal lattice of either adjacent grain. For certain relative orientations of two grains, however, a certain fraction, $1/n$, of the atomic sites in both grains have corresponding locations and these special sites are said to form a *coincidence lattice*. In a square lattice, Fig. 3.30(a), a rotation of 37° about a lattice point causes certain atoms, 2, to coincide with the position of atoms, 1, in the unrotated lattice. Therefore, if grain 2 (rotated 37° relative to grain 1) is adjacent to grain 1, Fig. 3.30(b), an array of coincidence sites extends uniformly from grain 1 into grain 2, unaffected by the presence of the grain boundary. One atom out of five is a coincidence site in this

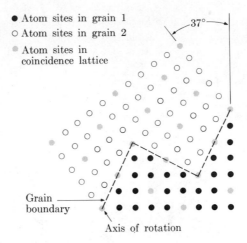

**Fig. 3.31**  The grain boundary in Fig. 3.30(b) can pass through many more coincidence sites by changing to this zig-zag path.

case, so the term $\frac{1}{5}$ coincidence lattice is applied to the array.  In three-dimensional crystal structures, coincidence lattices in three dimensions exist after specific rotations about a given crystal axis.  Some examples with relatively high densities of coincident sites are listed in Table 3.3.

**TABLE 3.3**

Important Coincidence Lattices in BCC and FCC Metals

| Crystal structure | Axis of rotation | Least angle of rotation | Density of coincident sites |
|---|---|---|---|
| Body-centered cubic | [100] | 36.9 | $\frac{1}{5}$ |
| | [110] | 70.5 | $\frac{1}{3}$ |
| | [110] | 38.9 | $\frac{1}{9}$ |
| | [111] | 60.0 | $\frac{1}{3}$ |
| | [111] | 38.2 | $\frac{1}{7}$ |
| Face-centered cubic | [100] | 36.9 | $\frac{1}{5}$ |
| | [110] | 38.9 | $\frac{1}{9}$ |
| | [111] | 22.0 | $\frac{1}{7}$ |
| | [111] | 38.2 | $\frac{1}{7}$ |

Two modifications in a grain boundary can permit it to profit from the low energy of coincidence sites.  First, rather than following a relatively straight path between the two crystal structures, Fig. 3.30(b), the grain boundary can follow a

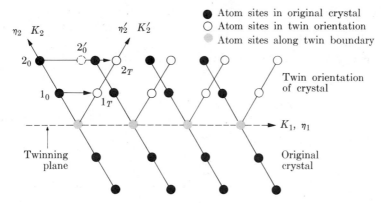

**Fig. 3.32** Schematic illustration of a reflection twin and of the twinning transformation that produces the twin orientation in a portion of the original crystal.

zig-zag path along directions containing a high density of coincidence sites, Fig. 3.31. Second, the coincidence-lattice model can be augmented by the presence of a sub-boundary, composed of an array of dislocations. The sub-boundary permits the existence of coincidence sites in the grain boundary even though the two grains deviate as much as 6° from the ideal orientation difference listed in Table 3.3.

**Twin boundaries.** A special relative orientation of two neighboring grains is the *twin* relationship. A coincidence lattice (with a density of coincidence points as high as $\frac{1}{2}$) may exist in the pair of twin crystals. In a reflection twin, Fig. 3.32, the *lattice* points in the twin crystal are mirror images across the twinning plane of the lattice points in the original crystal. Not every *atomic* position in the structure need be mirrored, however. Another type of twin, the rotation twin, converts the original orientation into the twin orientation by rotation about an axis lying either in the twinning plane or perpendicular to this plane. Twins in BCC, FCC, and HCP metals satisfy the criteria for both reflection and rotation twins.

The mechanism that produces a twin orientation requires that a given atom move only a fraction of a lattice spacing, even though large overall deformations of a metal can be accomplished by twinning. In the first row of atoms above the twinning plane in Fig. 3.32, a typical atom (labeled $1_0$) moves a short distance into the twin orientation, $1_T$. The atoms in row 2 (and higher rows) are also carried along by the motion of the atoms in the first row and are moved from position $2_0$ to $2_0'$. Repetition of the twinning transformation in the second row now carries the atom shown to the twin position, $2_T$. A twinning transformation is completely described by four *twinning elements*, $K_1$, $\eta_1$, $K_2$, $\eta_2$, each specified by its Miller indices. $K_1$ is the twinning plane, and $\eta_1$ is the crystallographic direction in which twinning (shearing) occurs. The lattice points on the twinning plane are evidently unaffected (undistorted) by the occurrence of twinning. Examination of Fig. 3.32 shows that the plane $K_2'$ in the twin crystal is the same as the plane $K_2$ in the original crystal. $K_2$ is known as the second undistorted plane, and $\eta_2$ is the direction in which shearing occurs on this plane. The amount of shear strain, $S$, that occurs during twinning is (Problem 20),

$$S = 2 \cot 2\varphi, \tag{3.56}$$

where $2\varphi$ is the angle between $\eta_1$ and $\eta_2'$. Table 3.4 lists the twinning elements and shear strain for important twinning processes in typical metals.

**TABLE 3.4**

Twinning Parameters for Important Twinning Processes in Representative Metals

| Metal | Structure | Twinning elements | | | | Shear strain, S |
|---|---|---|---|---|---|---|
| | | $K_1$ | $\eta_1$ | $K_2$ | $\eta_2$ | |
| Copper | FCC | $\{111\}$ | $\langle11\bar{2}\rangle$ | $\{11\bar{1}\}$ | $\langle112\rangle$ | 0.707 |
| $\alpha$-Iron | BCC | $\{112\}$ | $\langle11\bar{1}\rangle$ | $\{11\bar{2}\}$ | $\langle111\rangle$ | 0.707 |
| Cadmium* | HCP (c/a = 1.886) | $\{10\bar{1}2\}$ | $\langle\bar{1}011\rangle$ | $\{\bar{1}012\}$ | $\langle10\bar{1}1\rangle$ | 0.175 |
| Zinc | HCP (c/a = 1.856) | $\{10\bar{1}2\}$ | $\langle\bar{1}011\rangle$ | $\{\bar{1}012\}$ | $\langle10\bar{1}1\rangle$ | 0.143 |
| Magnesium | HCP (c/a = 1.623) | $\{10\bar{1}2\}$ | $\langle\bar{1}011\rangle$ | $\{\bar{1}012\}$ | $\langle10\bar{1}1\rangle$ | −0.131 |
| $\alpha$-Titanium | HCP (c/a = 1.587) | $\{10\bar{1}2\}$ | $\langle\bar{1}011\rangle$ | $\{\bar{1}012\}$ | $\langle10\bar{1}1\rangle$ | −0.175 |
| Bismuth | Rhombohedral | $\{110\}$ | $\langle00\bar{1}\rangle$ | $\{001\}$ | $\langle110\rangle$ | 0.118 |
| $\beta$-Tin* | Tetragonal | $\{301\}$ | $\langle\bar{1}03\rangle$ | $\{\bar{1}01\}$ | $\langle101\rangle$ | 0.119 |

* Other types of twinning processes are possible in all of the structures listed and actually occur commonly in HCP and tetragonal crystals.

Two types of boundaries can exist between a pair of grains in twin orientation, incoherent twin boundaries and coherent twin boundaries. If the boundary is *not* the twinning plane, the lattice sites in the boundary are not *all* common to both grains (although usually one-half of them are coincidence sites). Such a boundary is called *incoherent* and has a relatively high energy, about half that of an ordinary grain boundary. Most twin boundaries are of the second type and appear as straight lines in a microstructure. These boundaries coincide with the twinning plane and are coherent, since every lattice site in the boundary is common to both grains. Nearest-neighbor lattice sites are in the same position as in a normal crystal, and only second-nearest neighbors are in abnormal positions. In an FCC metal, for example, the atomic positions across a twin boundary are the same as across a stacking fault. Therefore, the energy of a (coherent) twin boundary is low—only about one-tenth that of an ordinary grain boundary.

Twins form either by a process of growth or as a result of deformation. Those that form by growth usually do so during annealing (a high-temperature heat treat-

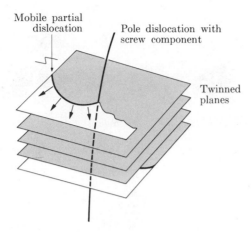

**Fig. 3.33** Schematic illustration of the pole mechanism for the generation of a deformation twin.

ment) and are generally called *annealing twins*. *Deformation twins* develop during plastic deformation and play an especially important role in HCP metals by serving as an additional deformation mechanism to augment the limited slip processes (see Chapter 4). The pole mechanism, Fig. 3.33, by which deformation twins are believed to form, pictures a single partial dislocation moving successively from plane to plane via a spiral ramp. The pole dislocation must have a screw component of such a nature that one of its partials can lie entirely in a possible twinning plane. A partial of this type is the "mobile partial dislocation" shown in Fig. 3.33. The mobile partial dislocation can sweep around the pole dislocation and trace out the helical ramp shown. The volume of material traversed in this manner is converted into a twinned orientation relative to the perfect lattice.

**General grain boundaries.** When two grains (growing in a solidifying metal, for example) come into contact, the resulting boundary is quite arbitrary since any planar portion of a boundary is determined by five variables or *degrees of freedom*. These consist of three variables describing the orientation of the second grain relative to the first, plus two variables describing the orientation of the planar boundary relative to the first grain. Evidently only a small fraction of such random boundaries satisfy the restrictive conditions governing the "special boundaries" discussed above. The majority of boundaries are therefore ordinary (high-angle) grain boundaries.

When an array of grains is held at a sufficiently high temperature that a limited amount of adjustment of the grain boundaries can occur, this adjustment takes two principal forms. First, the atoms at the boundary assume positions that tend to minimize the local elastic strains. A series of ledges and facets, Fig. 3.34, can

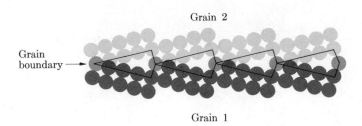

Grain 2

Grain
boundary →

Grain 1

**Fig. 3.34** A ledge-and-facet structure that can reduce the strain energy in a grain boundary. (After H. Gleiter.)

accomplish this purpose, and this type of surface has been observed microscopically. Second, the array of grains changes its geometry to reduce the total amount of grain boundary area. This and other aspects of grain boundaries are considered in detail in later chapters.

## PROBLEMS

**1.** Make a two-dimensional sketch analogous to the front plane of Fig. 3.3(b) but containing both a positive edge dislocation, ⊥, and a negative edge dislocation, ⊤. For convenience, separate ⊥ and ⊤ by two atom distances both vertically and horizontally.

a) Make a Burgers circuit around both ⊥ and ⊤ and show that the resulting Burgers vector is zero.

b) Make individual Burgers circuits around each of the two dislocations and show that the resulting Burgers vectors are equal but opposite.

**2.** To obtain the vector sum, $\mathbf{R} + \mathbf{S} = \mathbf{T}$, one merely adds algebraically the respective components of $\mathbf{R}$ and $\mathbf{S}$ to obtain the corresponding component of $\mathbf{T}$. Add vectors $\mathbf{b}$ and $\mathbf{b}'$, Eqs. (3.1) and (3.4), and show that the resultant vector, $\mathbf{b}''$, can be represented by Eq. (3.6).

**3.** Consider the two areas of slip, $A$ and $A'$, on the same horizontal plane, Fig. 3.35. As slip continues, the two areas first begin to combine as points $\mathscr{E}$ and $\mathscr{E}'$ join.

a) Show that the two segments of (edge) dislocation near $\mathscr{E}$ and $\mathscr{E}'$ combine according to the reaction, $\mathbf{b} + (-\mathbf{b}) = 0$.

[*Note*: if the positive direction for $\mathbf{t}$ is chosen as being *into* the crystal for the two dislocation lines originating at both $\mathscr{E}$ and $\mathscr{E}'$, then a Burgers circuit gives $(\mathbf{b})_{\mathscr{E}'} = -(\mathbf{b})_{\mathscr{E}}$.]

b) When the two segments of screw dislocation (emanating from $\mathrm{S}$ and $\mathrm{S}'$, respectively) are eventually in the process of combining, explain why the two segments are considered to have equal (not opposite) Burgers vectors.

[*Note*: recall the requirement on the sense of $\mathbf{t}$ for two dislocations being compared.]

c) Make a sketch similar to Fig. 3.35 but showing two slip processes whose combination corresponds to the dislocation interaction, $\mathbf{b} + \mathbf{b} = 2\mathbf{b}$.

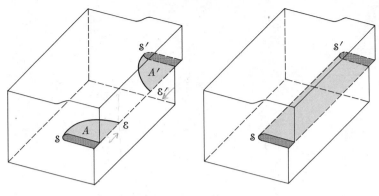

(a) Early stage of slip when two separate areas exist.

(b) At a later stage the two areas have combined.

**Fig. 3.35**  Illustration analogous to Fig. 3.5(a) for use in Problem 3.

**4.** In the text, Eq. (3.13) was evaluated for a dislocation density of $10^{12}/m^2$, and the value given in Eq. (3.14) was obtained.

a) Using a suitable sketch, determine the value of $R$ for a dislocation density of $10^{10}/m^2$.

b) Calculate the corresponding value of $E_S$.

**5.** Using a procedure similar to that employed in obtaining Eq. (3.13), derive the corresponding expression for $E_E$, Eq. (3.17).

**6.** a) Read from the standard stereographic projection, Fig. 1.24(b), the planes that belong to the [110] zone in a cubic crystal.

[*Note*: the *poles* of the planes are 90° from the zone axis.]

b) Many other planes also belong to the [110] zone. Which of the following planes belong to it? $(1\bar{1}2)$, $(0\bar{1}2)$, $(\bar{1}13)$, $(1\bar{3}2)$, $(\bar{2}21)$.

**7.** Using hard-sphere models of pairs of (100), (110), and (111) planes in the FCC crystal structure, observe the slip of one plane over another plane of the same type in the [110] direction.

a) List the planes in order of increasing amount of lateral separation of the pair of planes required for slip to be accomplished.

b) For which type of plane does slip tend to occur by a zig-zag motion, rather than directly in the [110] direction? This is the basis for the decomposition into partial dislocations illustrated in Fig. 3.21.

**8.** Derive Eq. (3.20) making use of Fig. 3.8.

a) Explain why the angle in Fig. 3.8(b) is $\theta/2$.

b) Obtain the equation for the balance of the vertical components of force.

c) Using the appropriate approximation for $\sin(\theta/2)$, obtain Eq. (3.20).

**9.** Using the geometry sketched in Fig. 3.36, derive Eq. (3.27) by the method employed in the text to obtain Eq. (3.26).

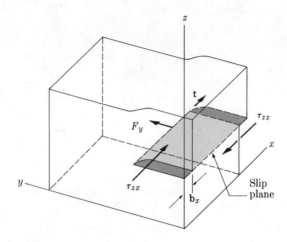

**Fig. 3.36** Sketch analogous to Fig. 3.10 for use with Problem 9.

**10.** With the aid of suitable sketches, show the other three force terms analogous to those of Eqs. (3.26), (3.27), and (3.28). Do not change the line direction, **t**.

**11.** Employing the procedure used in the text to obtain Eq. (3.32), derive the corresponding expression for parallel edge dislocations, Eq. (3.34). Using the relations $(x^2 + y^2)^{1/2} = r$, $x/r = \cos\theta$, and $y/r = \sin\theta$, obtain Eqs. (3.35) and (3.36) from Eq. (3.34).

**12.** Using Fig. 3.15(a) as a model, sketch a screw-dislocation line containing a parallel offset. Show that the kinks are of edge character in this case. Indicate which kink is $(+)$ and which is $(-)$ for the directions of **b** and **t** that you have chosen.

**13.** Using Fig. 3.15(b) as a model, show that a screw-dislocation line has jogs of edge character.

**14.** Using "before-and-after" sketches similar to those of Fig. 3.17, determine the result of the following intersections, assumed to occur on mutually perpendicular slip planes.

a) Two screw dislocations with parallel Burgers vectors.

b) Two screw dislocations with anti-parallel Burgers vectors.

c) One edge and one screw dislocation with parallel Burgers vectors.

d) One edge and one screw dislocation with perpendicular Burgers vectors.

**15.** Consider a cylindrical crystal 1 mm$^2$ in area and $10^2$ mm long subjected to a tensile stress.

a) If a single dislocation line on a plane 45° to the cylindrical axis moves entirely across the specimen and passes out the opposite surface, what elongation of the specimen has been produced? Assume $\mathbf{b} = 2 \times 10^{-10}$ m.

b) A typical dislocation density in a metal is $10^{14}$ lines/m$^2$. Calculate the total deformation that would be produced if all of these lines moved and passed out of the crystal under the influence of the stress. Assume that no new dislocations are created.

c) What is the corresponding normal strain?

**16.** Show that the equilibrium spacing, $d_s$, between two partial dislocations is given approximately by

$$d_s = \frac{\mu b^2}{24\pi\gamma} \tag{3.57}$$

where $\mu$ is the shear modulus, $b$ is the magnitude of the Burgers vector, and $\gamma$ is the stacking-fault energy. [*Hint*: Assume that $d_s$ is determined by a balance between the mutual elastic repulsion of the partials and the energy per unit length of the stacking fault that forms between the separated partials.]

**17.** Construct a figure analogous to Fig. 3.22 except for slip in HCP metals. Show that the stacking sequence in the stacking fault is: $ABAB|ABCA|CACA$.

**18.** Consider the interaction of various partial dislocations on intersecting {111} planes (see Fig. 3.23). Let one dislocation be on a (111) plane and the other on (11$\bar{1}$). Distinguish between Burgers vectors that differ in sign.

a) How many reactions are possible?

b) Half of these reactions result in a decrease in energy according to Frank's rule; write the equations for these reactions. [See Eq. (3.50).]

**19.** Show that Eq. (3.51) represents also the spacing between parallel screw dislocations that produce a twist boundary with rotation $\theta$ about the normal to the boundary. [*Hint*: Consider each set of dislocations separately.]

**20.** Using the definition of shear strain given by Eq. (2.5), show that the strain corresponding to the motion of an atom from position $1_0$ to $1_T$, Fig. 3.32, is given by Eq. (3.56).

**21.** a) On a standard (001) stereographic projection of the "original" crystal of copper, show the location of $K_1$, $K_2$ and $\eta_1$, $\eta_2$. Also show $K_2'$ and $\eta_2'$ in the twin crystal. Use the data of Table 3.4.

b) Repeat separately for $\alpha$-iron.

**22.** a) Why must the sweeping dislocation in the pole mechanism be a partial dislocation?

b) Suggest specific dislocations that could produce twinning by this mechanism in an FCC metal.

## REFERENCES

Cottrell, A. H., *Dislocations and Plastic Flow in Crystals*, London: Oxford University Press, 1956. Chapters 1, 2, and 3 treat comprehensively the basic properties of dislocations.

Gilman, J. J., *Micromechanics of Flow in Solids*, New York: McGraw-Hill, 1969. Chapter 4, "Dislocation Geometry," presents the subject from a different point of view and at a more advanced level.

Hirth, J. P. and J. Lothe, *Theory of Dislocations*, New York: McGraw-Hill, 1968. This book is an advanced treatment of virtually every aspect of dislocations.

Honeycombe, R. W. K., *The Plastic Deformation of Metals*, London: Edward Arnold, 1968. Chapter 8, "Other Deformation Processes in Crystals," discusses the mechanism of twinning in various crystal structures.

Kelly, A. and G. W. Groves, *Crystallography and Crystal Defects*, Reading, Mass.: Addison-Wesley, 1970. Chapters 7 through 10 give an advanced treatment of imperfections in crystals.

Nabarro, F. R. N., *Theory of Crystal Dislocations*, London: Oxford University Press, 1967. This book gives a comprehensive coverage of the entire subject of dislocations.

Weertman, J. and J. R. Weertman, *Elementary Dislocation Theory*, New York: Macmillan, 1964. Presents a more detailed discussion at an elementary level of many topics covered in this chapter.

CHAPTER 4

# PLASTIC DEFORMATION AND FRACTURE

## INTRODUCTION

A properly designed machine member operates in the *elastic* range of stresses, but the *plastic* behavior of metals is also important. The explanation is simple. The more efficiently a given alloy is used in design, the more closely its operating stress approaches the limit of elastic behavior, and therefore the more necessary is a knowledge of its behavior in the "beyond-elastic" stress range. Although it is not possible to ensure that a machine member will not be subjected to extraordinarily high stresses, even above the safety factor invariably employed, it *is* possible to make reasonably certain that the result of such misuse is only an inconvenience, not a catastrophe. For example, an automobile rounding a curve at high speed may strike its wheel against the curbing. If the resulting stress is above the elastic range, the wheel may be bent if plastic deformation occurs. The inconvenience of this result is far preferable to the alternative of *fracture*, which would result in the breaking of the wheel and the possibility of a serious accident. Metallic plasticity is also involved in the forming operations that are carried out on all wrought metals. Hot rolling or forging of steel, even at the extremely high temperatures used in practice, must be done at a rate determined by the deformation characteristics of the individual alloy. Cold forming operations, so important in the mass production of many metal parts, are even more restricted by plastic properties. Although the quantitative application of the theory of plasticity to industrial forming operations is complex, a basis for understanding these applications is furnished by the subject matter of this chapter.

The behavior of an alloy subjected to stresses above the elastic range may be divided as follows:

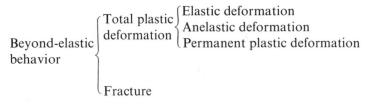

Actually, it is difficult to separate sharply the phenomena of plasticity and of fracture, since few commercially important alloys fracture without first undergoing some plastic deformation. Typical behavior is shown in Fig. 4.1, a complete stress-strain curve. Such a curve is obtained by a tension test, in which the test bar is elongated by the relative movement of two crossheads, and the load necessary to cause this elongation is indicated on a dial gage. While the test bar is being pulled until it fractures, data on changes in length and diameter are taken with the load data. Stress and strain values calculated from these data are then plotted to give the stress-strain curve. The strain in the elastic range is very small, and the stress-strain curve appears in this figure to rise on the stress axis itself until the range for plastic deformation is reached. The strain then increases rapidly with increasing stress until fracture occurs.

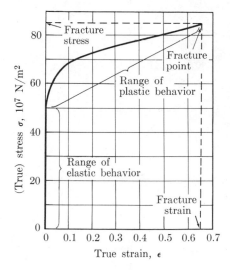

Fig. 4.1 Typical stress-strain curve for a metal subjected to a tension test.

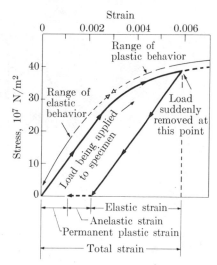

Fig. 4.2 Initial portion of a typical stress-strain curve, showing the effect of suddenly removing the load.

All the strain·referred to in Fig. 4.1 is essentially permanent plastic strain. For some purposes, however, it is necessary to distinguish the very small components of elastic strain and of anelastic strain present in the usual experimental measurement of the total strain. This distinction can be made with the aid of Fig. 4.2, which shows the initial portion of a stress-strain curve. When the load is removed from a plastically deformed specimen, the total strain is immediately reduced by the amount $\sigma/E$, the *elastic strain*, where $E$ is Young's modulus and $\sigma$ is the stress corresponding to the load that acted on the specimen. Even the remaining strain is not all permanent plastic strain, since a portion of it, the *anelastic strain*, disappears *gradually* after the load has been removed. When the total deformation is small, as in Fig. 4.2, the elastic and anelastic portions are significant, but they can be neglected in studying large plastic deformation of the kind shown in Fig. 4.1, and with which we will be mainly concerned in this chapter.

To understand the varied aspects of plastic deformation, we will need to employ two distinctly different approaches, dislocation theory and continuum mechanics. Dislocation theory explains the mechanisms by which deformation occurs and describes the accompanying structural changes. Continuum mechanics uses empirical constants to describe such technical forming processes as rolling or extrusion. Since the two approaches are so radically different, no confusion will be caused by using first one and then the other, depending on the subject being treated.

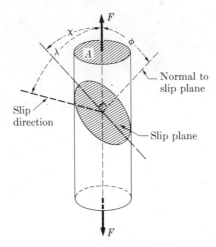

**Fig. 4.3**  Description of the angles $\lambda$, $\chi$, and $\phi$ used in determining the critical stress for slip.

## DEFORMATION OF SINGLE CRYSTALS

Although with few exceptions technical alloys are polycrystalline, a study of single crystals gives a useful insight into the nature of plastic deformation. The process of slip is of major importance in most metals and therefore is the principal topic here. Other deformation mechanisms are discussed at the end of this section.

**Critical resolved shear stress.**  At normal temperatures a single crystal deforms most easily by the gliding (slip) of dislocations along a set of parallel slip planes. Gliding tends to occur, however, only on specific slip systems, Table 3.1, and only when a certain minimum stress, the *critical resolved shear stress*, $\tau_0$, has been attained. The concept of a critical resolved shear stress can be understood with the aid of Fig. 4.3. If a cylindrical crystal of cross-sectional area $A$ is acted on by an axial force $F$, only the component of this force in the slip direction is effective in moving the dislocation. This force component is $F \cos \lambda$, and when it is divided by the area of the slip plane, $A/\cos \varphi$, the corresponding shear stress $\tau$ is obtained (see Eq. 2.2). Thus,

$$\tau = \frac{F}{A} \cos \lambda \cos \varphi = \sigma \cos \lambda \sin \chi, \qquad (4.1)$$

where $\sigma = F/A$ is the tensile stress and $\chi = 90° - \varphi$. Detectable deformation occurs in a given metal when $\tau$ reaches a definite, critical value $\tau_0$. Representative values of this *critical resolved shear stress* are given in Table 4.1. The magnitude of $\tau_0$ for a given base metal is affected by solid-solution alloying, as shown by the data for the silver-gold system, and also by temperature. An increase in temperature causes a moderate decrease in $\tau_0$ for a given slip system. A more important

**TABLE 4.1**

Typical Values of Critical Resolved Shear Stress, $\tau_0$, for Slip in Metal Crystals*

| Metal | Temp., °C | $\tau_0$, $10^5 \text{N/m}^2$† | Remarks |
|---|---|---|---|
| | | Face-centered cubic metals | |
| Cu | 20 | 0.10 | |
| Ag | 20 | 0.06 | Silver and gold form a complete |
| 80 Ag-20 Au | 20 | 0.35 | series of solid solutions. |
| 50 Ag-50 Au | 20 | 0.51 | $\tau_0$ has a maximum value at |
| 20 Ag-80 Au | 20 | 0.35 | about the 50 atomic percent |
| Au | 20 | 0.09 | composition. |
| | | Close-packed hexagonal metals | |
| Zn | 20 | 0.03 | These values are for slip on |
| Mg | 20 | 0.08 | the (0001) plane in the [$11\bar{2}0$] |
| Mg | 330 | 0.07 | direction. |
| Mg | 330 | 0.400 | This value is for slip on ($10\bar{1}1$) type planes in the [$11\bar{2}0$] direction. |

* Data from C. S. Barrett, *Structure of Metals*, McGraw-Hill, 1952.
† To convert from $\text{N/m}^2$ to $\text{lb/in}^2$, multiply by $1.450 \times 10^{-4}$.

effect of increased temperature, however, is the activation of an additional slip system in some metals. For example, above about 225°C magnesium crystals can slip on ($10\bar{1}1$) type planes as well as on the usual (0001) plane. This phenomenon explains why magnesium alloys can be more severely deformed at slightly elevated temperatures (warm-working) than at room temperature (cold-working).

**Multiple slip.** Slip begins first on the (primary) slip system for which $\tau_0$ is first attained. This slip system, however, is only one of several equivalent systems, Table 3.1. For example, 12 systems of the type {111} $\langle 10\bar{1}\rangle$ are operative in FCC metals. After primary slip has occurred to a certain extent, the geometrical factors in Eq. (4.1) may change sufficiently to cause $\tau_0$ to be reached on a second (conjugate) slip system. Figure 4.4 explains why the orientation of a crystal changes relative to the axis of loading during the course of primary slip. The stereographic projection permits a convenient quantitative treatment of this effect. In Fig. 4.5, point 1 represents the position of the axis of loading relative to the orientation of an FCC crystal. Consideration of Eq. (4.1) shows that when point 1 is within the indicated standard stereographic triangle, slip occurs on the (111) plane in the [$\bar{1}01$] direction. Since this is primary slip, the terms *primary plane* and *primary direction* are used to describe the two elements of the primary slip system, (111) [$\bar{1}01$]. As deformation proceeds, the axis of loading tends to move (along the

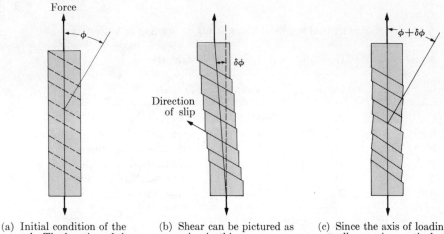

(a) Initial condition of the crystal. The location of the active primary slip plane is shown.

(b) Shear can be pictured as occurring in this manner on each of the active planes.

(c) Since the axis of loading actually remains vertical, the angle $\varphi$ changes significantly.

**Fig. 4.4** Schematic illustration of the change in the geometry of testing that occurs during the course of plastic deformation.

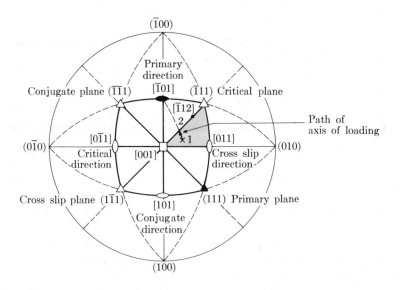

**Fig. 4.5** Stereographic representation of the path of the axis of loading relative to the orientation of an FCC crystal undergoing primary slip. Point 1 is the initial orientation of the axis of loading.

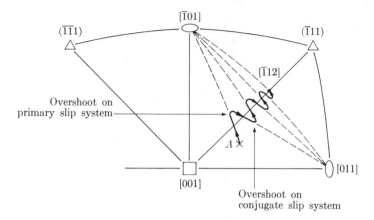

**Fig. 4.6** Stereographic representation of alternate periods of primary and conjugate slip that usually occur during the stage of double slip in an FCC crystal.

broken line, a great circle) toward the direction of slip. When the axis reaches point 2, however, the resolved shear stress $\tau$, Eq. (4.1), on the conjugate slip system, $(\bar{1}\bar{1}1)[101]$, becomes equal to that on the primary slip system.

Theoretically, the primary and conjugate slip systems should both be active, beginning at point 2, and the axis of loading should move from point 2 toward the $[\bar{1}12]$ direction. Actually, slip usually continues on the primary slip system until a certain amount of "overshoot" has occurred, Fig. 4.6. Slip then shifts to the conjugate slip system and continues until overshoot again develops. In this manner the axis of loading reaches the $[\bar{1}12]$ direction, at which point no further change in orientation should occur (see Problem 5).

The geometric factors, $\lambda$ and $\chi$, in Eq. (4.1) strongly influence the magnitude of applied stress, $F/A$, necessary to achieve a given value of the resolved shear stress, $\tau$. In cubic crystals this influence is moderated by the availability of alternative slip systems, for which the geometric factors have different values. In crystals of lower symmetry, however, a change in slip system may not be possible, and then geometric effects can be important. At the most favorable value of the geometric factors, $\sin \chi \cos \lambda = 0.5$, a stress of magnitude $\sigma = 2\tau_0$ causes glide to begin, Fig. 4.7. For less favorable values of the geometric factors, the tensile stress needed to initiate glide may become much greater than $2\tau_0$.

**Resolved shear strain.** To interpret tension-test data for a single crystal in terms of the motion of dislocations, we need an expression for the resolved shear strain analogous to Eq. (4.1) for the resolved shear stress. Consider an element of initial length $l_0$, Fig. 4.8(a), which is sheared to length $l_1$, Fig. 4.8(b). If $\lambda_0$ is the angle between the crystal axis and slip direction before shearing and $\lambda_1$ the corresponding angle after shearing, then as shown in Problem 6,

$$l_1/l_0 = \sin \lambda_0/\sin \lambda_1 = \sin \chi_0/\sin \chi_1. \tag{4.2}$$

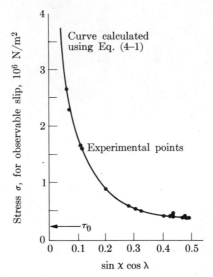

**Fig. 4.7** The effect of the geometrical factors on the stress necessary to initiate plastic deformation in an HCP crystal. The data are for a crystal of zinc tested at room temperature. (Data of D. C. Jillson.)

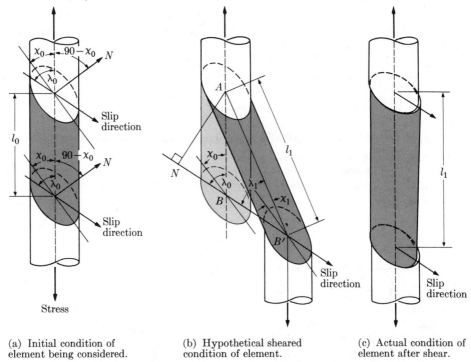

(a) Initial condition of element being considered.

(b) Hypothetical sheared condition of element.

(c) Actual condition of element after shear.

**Fig. 4.8** Schematic illustration of the geometrical parameters needed for calculation of the resolved shear strain.

By definition the shear strain is

$$\gamma = BB'/AN \tag{4.3}$$

and substitution then gives

$$\gamma = \frac{1}{\sin \chi_0} \left\{ \left[ (l_1/l_0)^2 - \sin \lambda_0 \right]^{1/2} - \cos \lambda_0 \right\}. \tag{4.4}$$

Only the *initial* orientation need be known to permit use of Eq. (4.4). If both initial and final orientations are known, an equivalent expression is

$$\gamma = \cos \lambda_1 / \sin \chi_1 - \cos \lambda_0 / \sin \chi_0. \tag{4.5}$$

The calculation in Problem 7 illustrates the use of Eq. (4.5). The corresponding equation for deformation in compression is the subject of Problem 8.

The stress-strain curves of metal single crystals can often be interpreted in terms of three stages of deformation, I, II, and III. The character and extent of these stages varies greatly from one metal to another and with change in the conditions of deformation, but the following general descriptions are widely applicable. Once the critical resolved shear stress has been exceeded, stage I deformation begins. It is characterized by slip on a single set of slip planes, and therefore the stress increases only slightly during stage I (hence the alternative designation, easy glide). Stage II is characterized by slip on more than one set of slip planes and therefore by a pronounced increase in stress as deformation proceeds. The increase is relatively uniform with increasing strain, so stage II is also known as the region of linear hardening. Stage III is said to begin when the rate of strengthening begins to fall off. The process of deformation is most complex during stage III and involves important contributions from new types of interactions of dislocations.

**Other modes of plastic deformation.** Single crystals may deform in an in-homogeneous manner for various reasons. For example, the end portions (which are gripped by the tension tester) restrain neighboring parts of the crystal from slipping in the normal manner. Also, the geometrical effects of large strains in tension or compression can influence the mode of deformation. This latter behavior is especially common in HCP metals or other crystals having few possible slip systems. Consider an HCP crystal tested in compression in the orientation shown in Fig. 4.9(a). Since $\cos \varphi = 0$, the resolved shear stress on the (0001) slip planes is zero, so deformation by slip cannot occur. The crystal can "kink," however, if two arrays of dislocations cause local bending in the manner illustrated. The formation of kink bands has two effects on the crystal. First, the crystal becomes shorter and thus tends to accommodate the applied force. For this reason the term *accommodation kink* is sometimes used. Second, the kinked region assumes a different orientation, for which $\cos \varphi$ has a finite value. Slip can then occur in this region provided that $\tau_0$ is attained. The formation of a twin, described in Chapter 3, can cause a similar reorientation, Fig. 4.10. In spite of the deformation produced by the twin, the grips of the tension tester force the ends of the

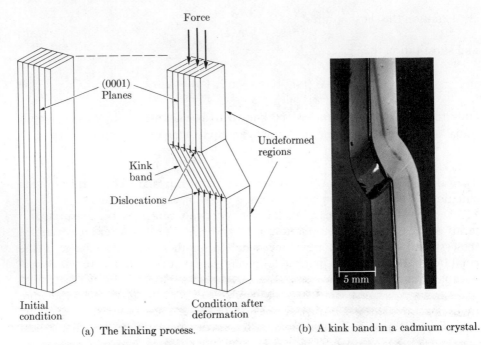

(a)  The kinking process.

(b)  A kink band in a cadmium crystal.

**Fig. 4.9** The formation of kink bands in hexagonal single crystals.  (Photomicrograph courtesy J. J. Gilman.)

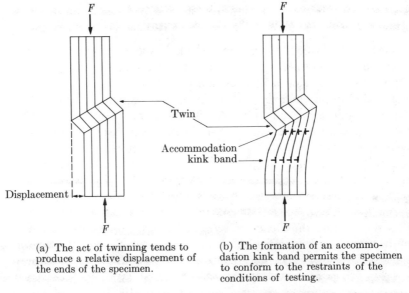

(a) The act of twinning tends to produce a relative displacement of the ends of the specimen.

(b) The formation of an accommodation kink band permits the specimen to conform to the restraints of the conditions of testing.

**Fig. 4.10** Schematic illustration of the nature of accommodation kink bands accompanying the formation of deformation twins.

Twins    Slip lines

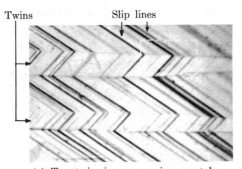

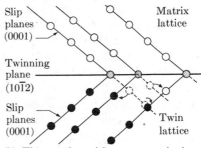

(a) Two twins in a magnesium crystal crossed by slip bands on (0001) planes. Area enclosed by dashed line is represented schematically in part (b).

(b) The atomic positions near a twinning plane in (a). (After C. S. Barrett.)

Accommodation kinks    Slip lines

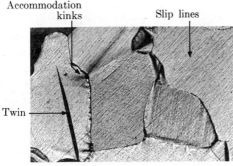

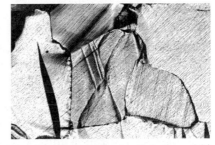

(c) Twins produced in zinc by tensile testing

(d) A later stage in the tensile testing of (c).

**Fig. 4.11** Illustrations of several phenomena accompanying twinning in metals. (Photomicrograph in (a) courtesy C. T. Haller; in (c) and (d) courtesy J. E. Burke.)

crystal to maintain their initial orientation. The transition is often accomplished by means of accommodation kink bands adjacent to one or both sides of the twin.

Several conditions determine whether deformation occurs by twinning. (1) The process of slip, which generally occurs preferentially, must be impeded for some reason; for example, because of geometrical factors. (2) The stress must be such that the twinning process tends to accommodate the stress. As can be deduced from Fig. 3.32, any vector to the left of $K_2$ (the second undistorted plane) is shortened during twinning, whereas any vector to the right of $K_2$ is lengthened. Therefore, a compressive stress along a direction to the left of $K_2$ can be accommodated, whereas a tensile stress must be along a direction to the right of $K_2$. (3) The $c/a$ ratio is a significant factor in HCP crystals. Consider twinning in zinc ($c/a = 1.86$) and in magnesium ($c/a = 1.62$), both of which have $(10\bar{1}2)$ as the twinning plane $K_1$. The angle between $K_1$ and the basal plane is 47° for zinc and 43° for magnesium; consequently, a zinc crystal increases in length parallel to the basal plane during twinning, whereas magnesium decreases in length. Therefore, a *tensile* stress parallel to the basal plane causes zinc to twin, whereas a *compressive* stress is needed for magnesium.

The various mechanisms of plastic deformation frequently occur jointly. Figure 4.11(a) is a photomicrograph of a crystal of magnesium that twinned after having experienced prior deformation by slip. The schematic drawing, Fig. 4.11(b), shows the relative orientations of the slip planes and the twinning plane. In a polycrystalline metal the stress state within an individual crystal may be especially complex and may lead to deformation of many kinds. Tensile testing of a poly-crystalline zinc specimen, Fig. 4.11(c), produces slip in each grain, twinning, and kink bands. Additional tensile straining, Fig. 4.11(d), causes existing twins to increase in size and additional twins to form.

## DEFORMATION OF POLYCRYSTALLINE ALLOYS

For purposes of engineering design, the deformation of polycrystalline alloys is treated quite differently from that of a single crystal of a pure metal. The entire bar or sheet of alloy is considered to be an isotropic continuum, the effects of grain structure and of the anisotropic behavior of individual crystals being averaged in the empirical constants employed. Although this continuum approach must be augmented in some instances (for example, if deformation occurs preferentially at grain boundaries, or if the grains of the alloy acquire a preferred crystallographic orientation during deformation), generally it permits a satisfactory, quantitative treatment of plastic deformation of commercial alloys under technically important conditions of fabrication and use.

**Definitions of stress and strain.** Plastic deformation may involve large changes in length and cross section of metal specimens. Therefore it is necessary to examine critically the definition of strain that is customarily used in the analysis of small, elastic deformations, before applying it to the study of plastic deformation. Consideration of the meaning of *strain* leads to the conclusion that any small increase in length $dl$ must be divided by the length $l$ *that the specimen has at that time*. That is, the small strain $d\varepsilon$ corresponding to the small deformation $dl$ is given by the equation

$$d\varepsilon = \frac{dl}{l}. \tag{4.6}$$

If the total deformation $\Delta l$ is small, say $\Delta l = 0.001 l_0$, then to a good approximation $l$ can be considered constant at its initial value $l_0$. The total strain $\varepsilon$ in this case is

$$\varepsilon = \int d\varepsilon = \frac{1}{l_0} \int_{l_0}^{l_0 + \Delta l} dl = \frac{\Delta l}{l_0} = 0.001. \tag{4.7}$$

This method of calculating strain is used in problems involving elastic deformation.

Evidently the approximation $l = l_0$ is not a good one if the total deformation $\Delta l$ is large. Fortunately, it is not difficult to calculate the *true* (exact) value of the strain, using Eq. (4.6). If a bar has been plastically deformed *uniformly* to a length $l$ from an initial length $l_0$, the *true strain* $\varepsilon$ is given by

$$\varepsilon = \int d\varepsilon = \int_{l_0}^{l} \frac{dl}{l} = [\ln l]_{l_0}^{l} = \ln \frac{l}{l_0}. \tag{4.8}$$

The strain, $\varepsilon$, is given correctly for small deformations by $\Delta l / l_0$, Eq. (4.7), but for larger deformations the expression $\ln (l/l_0)$ must be used to obtain the true strain. For a large total deformation such as $\Delta l = l_0$, the true strain calculated by Eq. (4.8), namely,

$$\varepsilon = \ln \frac{l}{l_0} = \ln \frac{2l_0}{l_0} = \ln 2 = 0.693, \tag{4.9}$$

is significantly different from the strain given by the approximate equation, Eq. (4.7):

$$\varepsilon' = \frac{\Delta l}{l_0} = \frac{l_0}{l_0} = 1.000. \tag{4.10}$$

The *nominal strain*, $\varepsilon'$ (rather than the more accurate $\varepsilon$) is used in the nominal stress-strain curves widely used in industry.

Equation (4.8) is valid only if the deformation is uniform over the length considered. If the deformation is nonuniform, that is, if *necking* of the specimen occurs (see Fig. 4.14), the plastic strain has different values in various parts of the bar. Since the *maximum* strain is usually of principal interest, it is sufficient to determine the strain at the necked section. The strain at this position is (see Problem 10)

$$\varepsilon = 2 \ln \frac{d_0}{d}, \tag{4.11}$$

where $d_0$ is the initial diameter and $d$ is the neck diameter of the bar.

As in the case of strain, nominal and exact definitions of *stress* are also in use. For small elastic deformations the stress is always very nearly

$$\sigma = \frac{F}{A_0}, \tag{4.12}$$

where $F$ is the load and $A_0$ is the original cross-sectional area of the specimen. This is also the definition of the nominal stress used in approximate stress-strain curves. However, since the actual area $A$ may be only a fraction of $A_0$ after severe plastic deformation, the exact definition

$$\sigma = \frac{F}{A} \tag{4.13}$$

must be used to calculate the *true stress*.

**Flow stress.**  A striking characteristic of the relation between stress and plastic strain is that *below a certain stress* (the *flow stress $S_0$*), essentially *no plastic strain is produced*.  Thus for elastic strain the stress-strain relation can be represented by the equation $\varepsilon = \sigma/E$, where $\varepsilon$ is the normal strain in the direction in which the stress $\sigma$ is acting.  However, for plastic strain a pair of equations is needed, such as*

$$\varepsilon \simeq 0 \qquad \text{for} \quad \sigma < S_0$$
$$\varepsilon = f(\sigma) \quad \text{for} \quad \sigma > S_0. \tag{4.14}$$

That is, plastic strain occurs as some function $f$ of the stress only when the stress is greater than the critical value $S_0$.  Before we consider the nature of the function that relates stress and plastic strain, we should learn something of the stress $S_0$ at which plastic deformation begins.

The complex nature of *small* plastic strains makes it difficult to define precisely the stress at which plastic deformation begins, even under the simple conditions of a tension test.  Fortunately, there is general agreement among engineers that a useful, conventional definition of the beginning of plastic deformation is the 0.2 *percent offset yield strength.*†  The method for determining this stress value is shown in Fig. 4.12.  From the point representing a strain of 0.2 percent on the zero stress axis a line is drawn parallel to the elastic portion of the stress-strain curve.  This line will intercept the stress-strain curve at the point at which the material being tested departs from elastic behavior by 0.2 percent.  The stress at this point is the (0.2 percent offset) *yield strength*.  Thus, for a metal plastically deformed in simple tension, the magnitude of the stress at which plastic deformation begins is the yield strength.  It is shown below that this is also the value of $S_0$.

*Combined stresses.*  Actual engineering structures often are subjected not merely to a single tensile stress but to a combination of tensile, compressive, and torsional stresses.  The question then arises: what is the condition for flow under such conditions of *combined stresses*?  Extensive investigations of metal specimens subjected to complex stress combinations have shown that for isotropic, polycrystalline metals plastic deformation begins when the elastic shear-strain energy reaches a critical value characteristic of the material.  This is the von Mises criterion of plastic flow, named after the man who first suggested it.  The corresponding

---

* These equations are strictly valid only for a simple stress state, such as that in a tension test.  More generally, not the *single* stress component $\sigma$, but rather the stress *function* $\bar{S}$ is related to $S_0$, and the pair of equations is then written

$$\bar{D} = 0 \qquad \text{for} \quad \bar{S} < S_0$$
$$\bar{D} = f(\bar{S}) \quad \text{for} \quad \bar{S} > S_0,$$

where $\bar{S}$ and $\bar{D}$ are defined by Eqs. (4.21) and (4.22).

† Although 0.2% offset is the most widely used definition of yield strength, 0.1% is often employed for steels, and as much as 0.5% for cast iron.

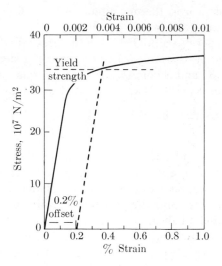

**Fig. 4.12** Illustration of the method for determining the 0.2 % offset yield strength from the stress-strain curve in tension.

equation is

$$\bar{S} = \left[ \frac{(S_1 - S_2)^2 + (S_2 - S_3)^2 + (S_3 - S_1)^2}{2} \right]^{1/2} = S_0, \qquad (4.15)$$

where $S_1$, $S_2$, and $S_3$ are the *principal stresses*,* and algebraically $S_1 > S_2 > S_3$. Thus the flow stress $S_0$ is not represented by a single physical stress in the most general case. Rather, flow stress *is the critical value of the stress function $\bar{S}$ at which plastic deformation begins.*

It is easy to determine the magnitude of $S_0$ by means of a tension test. When a given material is tested in tension, $S_1 = \sigma_1$ and $S_2 = S_3 = 0$. When these values are substituted in Eq. (4.15), this equation reduces to

$$\left[ \frac{(\sigma_1)^2 + 0 + (\sigma_1)^2}{2} \right]^{1/2} = \sigma_1 = S_0. \qquad (4.16)$$

That is, plastic flow begins in simple tension when the tensile stress $\sigma_1$ becomes

---

* The use of *principal axes* (see Chapter 2) permits the reduction of any stress state to one involving only three normal (tensile or compressive) stresses. The three stresses referred to the principal axes are the *principal stresses*. Stress problems that are described without the use of shear stresses are already referred to principal axes, for example in a tension test. That is, the stress $\sigma_1$ in a tension test is also the principal stress $S_1$. A torsion test, on the other hand, is usually described in terms of a torque (or the corresponding shear stresses), and a suitable conversion must be made to obtain the principal stresses. In this case the principal stresses consist of a tensile and a compressive stress.

equal to $S_0$. In other words, the magnitude of $S_0$ in Eq. (4.15) is just the yield strength determined in a tension test.

This information permits a prediction of the stress at which flow begins under any combination of applied stresses. For example, if a steel plate is subjected to tension in one direction and is compressed with an equal stress at right angles to the first direction, then $S_3 = -S_1$ and $S_2 = 0$. Equation (4.15) then becomes

$$\left[\frac{(S_1)^2 + (S_1)^2 + 4(S_1)^2}{2}\right]^{1/2} = S_0, \qquad S_1 = \frac{S_0}{\sqrt{3}} = 0.58S_0. \qquad (4.17)$$

That is, the steel plate will plastically deform at a tensile stress only about 58% as large as that required to produce plastic deformation in a tension test. This pronounced effect of the compressive stress can be explained qualitatively in terms of the maximum shear stress. Slip, the principal mechanism of plastic deformation, occurs at a critical value of shear stress. Therefore we might expect plastic flow to begin when the maximum shear stress reaches a certain value, regardless of the magnitudes of the individual principal stresses. In the case of equal tensile and compressive stresses at right angles, the maximum shear stress is twice as great at a given value of the tensile stress as it is in simple tension. This approach gives the relation $S_1 = 0.50S_0$, rather than the more exact shear-strain energy criterion of Eq. (4.17). In spite of its lack of accuracy, the maximum shear-stress criterion of the beginning of plastic deformation is widely used.

In contrast to the *decrease* in flow resistance described above, a condition of biaxial tension generally leads to an *increase* in flow resistance compared with that found in a tension test. For example, if $S_2 = \frac{1}{2}S_1$ and $S_3 = 0$, the value of $S_1$ at which plastic deformation begins can be shown by substitution in Eq. (4.15) to be $S_1 = 2S_0/\sqrt{3} = 1.15S_0$. A state of combined stresses such as this can reduce the ductility of a metal in comparison with the ductility that the metal exhibits in a tension test. Since a larger tensile stress $S_1$ must be exerted to produce initial plastic deformation, the stress for fracture will usually be reached after a smaller total amount of deformation has occurred.

Inasmuch as the flow stress (yield strength) is the practical limit on which design stresses are based, it is worth while to consider the more important factors that influence its value. Although the quantitative values vary greatly from one alloy to another, the following general principles govern the effect of alloying, of temperature, and of rate of straining on the magnitude of the flow stress.

*Influence of Alloying.* The flow stress of a solid solution is usually higher than that of the pure solvent metal. The effect of a given concentration of alloying element, say, 1 atomic percent, tends to be larger in proportion to the dissimilarity between the solute and solvent atoms. When applied in commercial alloys, solid-solution strengthening can increase the yield strength severalfold.

*Influence of Temperature.* The flow stress decreases with increasing temperature $T$. The quantitative relationship can be approximated in many cases by the

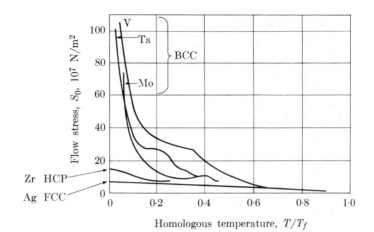

**Fig. 4.13** Variation of the flow stress of several metals as a function of homologous temperature, $T/T_f$. (Adapted from W. J. M. Tegart.)

exponential equation

$$S_0 = Ae^{B/T},\qquad(4.18)$$

where $A$ and $B$ are constants for a specific material having a given metallurgical structure. Comparison of the effect of temperature on various metals is facilitated by the use of their *homologous temperatures*, defined as the fraction $T/T_f$, where $T_f$ is the melting point of a given metal. The flow stress, $S_0$, changes quite differently as a function of $T/T_f$, Fig. 4.13, for BCC metals (V, Ta, and Mo) and for close-packed metals (Ag, FCC; Zr, HCP).

*Influence of Strain Rate.* If the stress $S(\varepsilon, T)$, is measured at a given strain, $\varepsilon$, and temperature, $T$, the measured value is found to vary with the rate of straining, $d\varepsilon/dt \equiv \dot{\varepsilon}$, according to the following empirical equation,

$$S(\varepsilon, T) = C\dot{\varepsilon}^m,\qquad(4.19)$$

$C$ and $m$ are constants; $m$ is called the strain-rate sensitivity. Although $m$ is a decisive factor under certain special conditions, for example in connection with superplasticity, usually the effect of strain rate is small. An increase in flow stress corresponding to a 100-degree *decrease* in temperature would be produced by about a $10^5$-fold *increase* in strain rate. BCC metals are more strongly affected by a given change in strain rate than are FCC metals.

*Influence of Microstructure.* The yield strength of an alloy depends to an important degree on the manner in which the constituent atoms are assembled; that is, it depends on the structure of the alloy on a microscopic scale. For example, suitable heat treatment can change the microstructure in such a manner that the

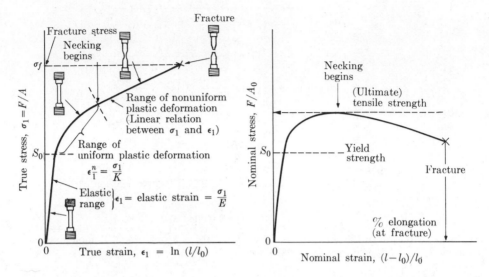

(a) The true-stress/true-strain curve, showing three regions of deformation behavior. For a tension test, $\overline{D} = \epsilon_1$ and $\overline{S} = \sigma_1$.

(b) The nominal-stress/nominal-strain curve (or load-elongation curve), showing the practical importance of the ultimate tensile strength.

**Fig. 4.14** The principal features of tension-test curves for a typical metal. Two different methods for plotting the same test data are illustrated in (a) and (b). The amount of elastic strain is shown greatly exaggerated.

yield strength is increased by a factor of ten. The microstructural features in question include phase constitution, grain size, and dislocation substructure, topics that are treated in later chapters.

**Stress-strain curves.** The data obtained from a tension test, Fig. 4.14, give much information about a particular metal or alloy at the temperature (usually room temperature) and strain rate (usually $10^{-3}$ sec$^{-1}$; that is, 0.1 percent per second) employed in the test. When the data are plotted as true stress ($\sigma_1 = F/A$) versus true strain [$\varepsilon_1 = \ln(l/l_0)$], Fig. 4.14(a), three regions of deformation behavior can be conveniently distinguished. (1) As the stress is gradually increased from zero, the metal can be considered to deform *only* elastically up to the yield strength, $S_0$. Elastic deformation continues to increase with increasing stress up to the fracture stress $\sigma_f$, but in the later stages of a tension test the amount of elastic deformation is small compared to the amount of plastic deformation. (2) In the range of stresses between $S_0$ and the beginning of necking, plastic deformation occurs uniformly, and the test data can often be approximated by the effective-stress/effective-strain curve described below. (3) When a "neck" begins to form in the test specimen, signaling the beginning of nonuniform deformation, the load, $F$, begins to decrease, Fig. 4.14(b).  Because of rapid decrease in the area, $A$, at the neck, however, the true stress, $F/A$, continues to increase (roughly linearly with increase in strain) until the specimen finally breaks at $\sigma_f$.

Because the raw data obtained from a tension test are load, $F$, and elongation, $l - l_0$, the results are usually plotted as the (engineering) stress-strain curve shown in Fig. 4.14(b).  The yield strength is identical to that in Fig. 4.14(a), but the point corresponding to the maximum load now appears as a prominent feature of the curve, the (ultimate) tensile strength.  The percent elongation at fracture and the percent reduction in area at the neck can be determined from measurements on the reassembled specimen after fracture (Problem 13).

*Effective Stress-Strain Curve.*  The behavior in the range of uniform plastic deformation in Fig. 4.14(a) can often be approximated by the relation,

$$\bar{S} = K\bar{D}^n. \tag{4.20}$$

$K$ is a constant called the *strength coefficient*; $n$ is a constant* called the *strain-hardening exponent*, since it measures the rate at which a metal becomes strengthened (hardened) as a result of plastic straining.

$\bar{S}$ and $\bar{D}$, the effective stress and effective strain, are defined in such a manner that Eq. (4.20) is valid for any arbitrary combination of principal stresses and strains:

$$\bar{S} = \left[ \frac{(S_1 - S_2)^2 + (S_2 - S_3)^2 + (S_3 - S_1)^2}{2} \right]^{1/2}, \tag{4.21}$$

$$\bar{D} = \left[ \frac{2\{(D_1)^2 + (D_2)^2 + (D_3)^2\}}{3} \right]^{1/2}, \tag{4.22}$$

where $S_1$, $S_2$, and $S_3$ are the principal stresses and $D_1$, $D_2$, and $D_3$ are the principal strains.  It is seen that for a simple tension test $\sigma_1 = S_1 = \bar{S}$, since $S_2 = S_3 = 0$. It can also be shown that $\varepsilon_1 = D_1 = \bar{D}$, since there is no appreciable volume change during plastic flow in metals and therefore $-D_1 = D_2 + D_3$.  Thus, the strength coefficient $K$ and the strain-hardening exponent $n$ can be determined for a given alloy by means of a tension test.  Table 4.2 gives representative values of $K$ and $n$ determined in this manner.  Also, if a stress-strain curve such as Fig. 4.14(a) is determined for a given alloy by means of a tension test, the same curve describes the behavior of the alloy when it is subjected to a complex combination of stresses. This is true because $\sigma_1$ and $\varepsilon_1$ in a tension test are equal to $\bar{S}$ and $\bar{D}$, and the relations between $\bar{S}$ and $\bar{D}$, such as Eq. (4.20), are valid for any arbitrary stress state.

*Yield-Point Phenomena.*  In some alloys the initial stage of plastic deformation is quite different from that shown in Fig. 4.14 because of the occurrence of *yielding*.

---

* More refined analyses indicate that the value of $n$ decreases at a stage in the deformation of a polycrystalline metal corresponding to the transition from stage II to stage III in a single crystal.

**TABLE 4.2**

Approximate Values of $S_0$, $K$, and $n$ for Several Metals and Alloys*

| Material | Flow stress $S_0$, N/m$^2$ | Strength coefficient $K$, N/m$^2$ | Strain-hardening exponent $n$ |
|---|---|---|---|
| Annealed low-carbon steel | $21 \times 10^7$ | $45 \times 10^7$ | 0.3 |
| Quenched and tempered 0.6% steel | $52 \times 10^7$ | $127 \times 10^7$ | 0.15 |
| Annealed copper | $5.5 \times 10^7$ | $32 \times 10^7$ | 0.54 |
| Annealed cartridge brass | $8 \times 10^7$ | $90 \times 10^7$ | 0.49 |
| Annealed 24S aluminum alloy | $11 \times 10^7$ | $34 \times 10^7$ | 0.21 |
| Precipitation-hardened 24S aluminum alloy | $31 \times 10^7$ | $69 \times 10^7$ | 0.16 |

* To convert from N/m$^2$ to lb/in$^2$, multiply by $1.450 \times 10^{-4}$.

When deformation begins in these alloys, Fig. 4.15, plastic deformation occurs nonuniformly, with the resulting production of markings on the surface called *Luder's lines*. This nonuniform yielding continues at nearly constant stress until the entire specimen has been deformed several percent. The overall deformation is then uniform, and further plastic deformation occurs in a uniform manner until necking begins. The yielding phenomenon occurs because certain impurity elements (nitrogen and carbon in steel) locate themselves preferentially adjacent to dislocations and make the dislocations more difficult to move. However, once the necessary high stress, called the upper yield stress, has torn some of the dislocations away from their restraining impurity atoms, the dislocations can then continue to move under the usual stress, the (lower) yield stress. When yielding has begun in one portion of the specimen, stress concentration in adjacent portions causes yielding to spread gradually throughout the metal at the yield stress. This yielding phenomenon is the exception rather than the rule in plastic deformation behavior of metals.

*Necking and Superplasticity.* The occurrence of a neck during plastic deformation detracts from the usefulness of an alloy in two respects: First, it localizes the deformation and causes eventual fracture in that location; second, it produces unevenness in the surface and cross section of metals severely deformed in tension, as in the deep-drawing of a cup. Certain alloys under special conditions of deformation resist the tendency to form a neck and exhibit the behavior of *superplasticity*, in which the alloy experiences uniform elongations 10 to 1000 times greater than the normal amount. This behavior can be understood by considering the variation

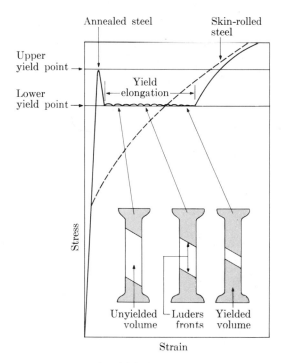

**Fig. 4.15** Schematic illustration of the yield phenomenon in an annealed low-carbon steel, caused by the propagation of Luder's bands. Slight cold-rolling (skin rolling) prevents yielding and the accompanying unevenness of the surface of the steel. (After D. J. Blickwede.)

with rate of straining of the stress $S$ required to continue a process of deformation, Eq. (4.19). The coefficient $m$, the strain-rate sensitivity, determines the increase in stress that results from an increase in rate of straining. The value for common steel, $m = 0.02$, is typical of most alloys; it is too low to effectively impede the development of a neck. In those special alloys for which $m$ is in the range above about 0.3, the formation of a neck becomes increasingly difficult: Necking involves a local high rate of straining (see Problem 15), but the local stress for flow is then strongly increased by the large value of $m$. Consequently, uniform deformation continues during extensive elongation of the alloy.

The achievement of adequately high values of the strain-rate coefficient depends on structural aspects of the alloy as well as on special conditions of deformation. A crucial structural feature is the presence of a large amount of internal boundary, either grain boundary or interphase boundary. Deformation can then occur largely by special processes in or near the boundary (for example, shear in boundaries and migration of vacancies along boundaries), and with a minimum of the usual motion of dislocations within the volume of the grains. The boundary processes are enhanced by performing the deformation under

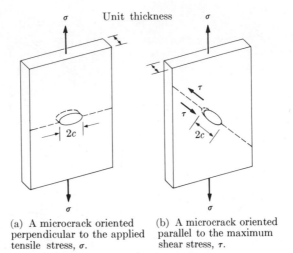

(a) A microcrack oriented perpendicular to the applied tensile stress, $\sigma$.

(b) A microcrack oriented parallel to the maximum shear stress, $\tau$.

**Fig. 4.16** Schematic illustrations for use in analyzing the growth of cracks by (a) cleavage and (b) shear rupture. A microcrack is assumed to have an elliptical cross section and to extend through the sheet of metal.

conditions of high temperature and low rates of straining. Because grain growth tends to occur at high temperatures (as discussed in Chapter 9), special procedures may be needed to maintain the required grain size, which is as small as 1 µm. Typical procedures are: cycling the alloy through a phase transformation during deformation; or subjecting the alloy to ordinary deformation at a low temperature prior to superplastic deformation at a high temperature. Superplasticity is of potential value for the forming of metals, but its present limitations have restricted its application.

## FRACTURE

The process of fracture is usefully pictured as consisting of two stages: the occurrence of the initial crack (which may be pre-existing or may require a nucleation step); and the growth of an existing crack. We postpone to a later section a discussion of the initiation of cracks and begin with the theory of the growth of cracks.

**Fracture strength.** For ease of mathematical analysis, a crack can be assumed to have an elliptical cross section, Fig. 4.16, and to extend through the entire thickness of the specimen. As the tensile stress, $\sigma$, on the specimen is gradually increased, the crack will begin to increase in size when a value $\sigma_f$ is reached. If $G$ is the energy necessary to create unit area of crack, the energy represented by a crack of length $2c$ and unit thickness is

$$W_c = 2cG. \tag{4.23}$$

We begin with a treatment of *brittle* fracture; that is, fracture in which plastic deformation has little effect. The energy available to cause growth of the crack is the portion of the elastic strain energy that is released as a result of the formation of the crack. In the case of a crack oriented perpendicular to the applied tensile stress, Fig. 4.16(a), the energy released, $W_e$, is (see Problem 17)

$$W_e = \pi c^2 \sigma^2 / E \tag{4.24}$$

where $E$ is Young's modulus. The existing microcrack will grow under the action of a given stress if the change in $W_e$ with increase in $c$, $dW_e/dc$, is at least equal to the corresponding change in $W_c$, $dW_c/dc$. Differentiation of Eqs. (4.23) and (4.24) leads to the equality,

$$2\pi c \sigma^2 / E = 2G. \tag{4.25}$$

The value of $\sigma$ given by this equation is the critical value required to cause existing cracks to grow and eventually to cause fracture,

$$\sigma_f = \left(\frac{GE}{\pi c}\right)^{1/2}. \tag{4.26}$$

$\sigma_f$, the fracture stress, is the magnitude of the externally applied tensile stress at which fracture occurs.

By considering the effect of stress concentration caused by the crack, we can determine the *fracture strength* $\sigma_{max}$ of a brittle metal. To simplify the calculation we consider that the specimen is a thin sheet; the stresses then satisfy the condition of *plane stress* in which the stress perpendicular to the sheet is zero. If the length of the crack, $2c$, is large compared to the radius, $\rho$, of its ends, the stress concentration factor, Eq. (2.39), then has the value

$$\sigma_{max}/\sigma_f = (4c/\rho)^{1/2}. \tag{4.27}$$

This means that the metal can withstand a stress up to $\sigma_{max}$ (not merely $\sigma_f$) at the ends of the crack, where the process of fracture is occurring. Substitution of Eq. (4.27) in Eq. (4.26) gives the expression,

$$\sigma_{max} \simeq (EG/\rho)^{1/2}, \tag{4.28}$$

for the fracture strength.

If the mechanism of fracture is the "pulling apart" or cleavage of adjacent planes in the metal, corresponding to Fig. 4.16(a), a theoretical estimate of $G$ is $Ed/10$, where $d$ is the spacing of adjacent planes of atoms. Substitution of this value in Eq. (4.28) gives for the fracture strength (assuming that $\rho \simeq 2.5d$),

$$\sigma_{max} = \left(\frac{E^2 d}{10\rho}\right)^{1/2} \approx E/5. \tag{4.29}$$

The high values corresponding to this estimate have been found experimentally for fibers and rods of $SiO_2$ (silica), but are not observed for metals and alloys.

**TABLE 4.3**

Theoretical Estimates of the Stress to Produce Fracture by Cleavage and by Shear in Metals*

| Metal | Young's modulus, $E$, N/m$^2$ | Stress for cleavage, N/m$^2$ | Stress for shear, N/m$^2$ |
|---|---|---|---|
| Copper (FCC) | $12 \times 10^{10}$ | $2 \times 10^{10}$ | $0.14 \times 10^{10}$ |
| Silver (FCC) | $12 \times 10^{10}$ | $2 \times 10^{10}$ | $0.07 \times 10^{10}$ |
| Iron (BCC) | $21 \times 10^{10}$ | $4 \times 10^{10}$ | $1 \quad \times 10^{10}$ |
| Tungsten (BCC) | $35 \times 10^{10}$ | $7 \times 10^{10}$ | $2 \quad \times 10^{10}$ |

* To convert from N/m$^2$ to lb/in$^2$, multiply by $1.450 \times 10^{-4}$.

The reason is that metals require less energy to fracture by a mechanism involving shear than by cleavage alone. The theoretical estimates in Table 4.3 show that the stress to cause fracture by shear in FCC metals is only about 5 percent of the stress needed to cause cleavage. In BCC metals the corresponding value is about 25 percent. We will see later that a BCC metal often fractures by a shear mechanism, but unfavorable circumstances can cause fracture by cleavage, Fig. 4.17. FCC metals fracture only by shear.

We now consider briefly the analysis of fracture in ductile materials, a more complex topic than brittle fracture. The principal phenomenon is now shear rupture, Fig. 4.16(b); its analysis involves two additional features. First, the shear

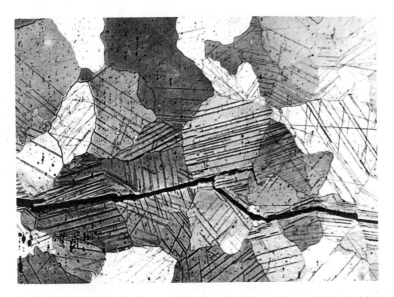

**Fig. 4.17** Photomicrograph of a cleavage fracture in a low-carbon steel tested in tension at $-195°C$. The slip lines show that much plastic deformation preceded fracture. (Courtesy J. R. Low, Jr., Carnegie-Mellon University.)

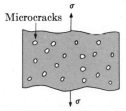

(a) Microcracks form as the specimen is being deformed.

Two microcracks coalesce

(b) Some neighboring micro-cracks coalesce by shear rupture, thereby creating visible cracks.

Neck region

(c) This process continues over the interior of the specimen on a plane roughly perpendicular to the applied stress.

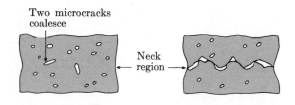

Smooth fibrous ———— Rough fibrous

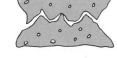

(d) The final stage of fracture occurs by shear rupture at about 45° to the applied stress.

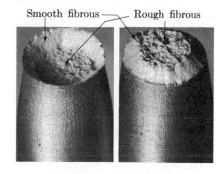

(e) Photograph of the two halves (the cup and the cone) of a specimen after fracture in tension.

**Fig. 4.18** Schematic illustration of the stages of the fracture process in ductile fracture (cup-and-cone).

stress causes plastic deformation (as well as rupture of atomic bonds). This deformation decreases the sharpness at the ends of the microcracks, thereby decreasing the stress-concentration factor, and therefore decreasing the maximum stress in the metal. Second, since the maximum shear stress is on a plane at 45° to the applied tensile stress, $\sigma$, shear cracks tend to occur at about this angle. Only a qualitative treatment of shear rupture is possible at present.

**Modes of fracture in steel.** When an ordinary low-carbon steel is tested in tension at room temperature, Fig. 4.14, it is ductile and absorbs much energy before it fractures. Fracture in this case develops through the stages shown in Fig. 4.18. In the course of uniform plastic deformation, many microcracks form throughout the body of the specimen, Fig. 4.18(a). The microcracks form preferentially at small particles of $Fe_3C$ or of impurities (oxides or sulfides), but even in the absence of second-phase particles a microcrack can form by dislocation mechanisms of the kind illustrated in Fig. 4.19. As deformation proceeds, the region between certain neighboring microcracks fractures in shear; in other words, the two voids

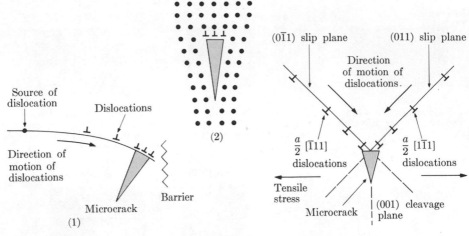

(a) Edge dislocations produced at a source during deformation pile up at a barrier (1). Several piled up dislocations form a microcrack (2) within the crystal lattice. (After A. N. Stroh.)

(b) Dislocations in a BCC metal can react at an intersection and produce sessile dislocations, the latter then forming a microcrack. (After A. H. Cottrell.)

**Fig. 4.19** Schematic illustrations of mechanisms for forming microcracks (a) at a barrier, such as a grain boundary, and (b) at the intersection of two slip planes.

coalesce and form a macroscopic crack, Fig. 4.18(b). The average stress is now higher in the partially cracked region, so deformation and additional coalescence occur here, accompanied by the formation of a local neck. In the interior of the specimen, restraints imposed by the surrounding regions cause the average plane of fracture to lie perpendicular to the tensile stress, Fig. 4.18(c), although the unit steps of shear rupture occur on planes at about 45°. These restraints diminish near the surface, so the final stage of fracture occurs by coalescence of microcracks lying on a 45° plane, Fig. 4.18(d). The two main stages of fracture can be seen on the final fracture surface, Fig. 4.18(e), as a rough, planar interior region surrounded by a smooth cup-and-cone rim. Both regions are fibrous (because of shear rupture) rather than smooth (cleavage fracture, Fig. 4.17), or granular (fracture along grain boundaries, Fig. 4.21).

Although fracture of any kind is undesirable in a structural metal, the ductile fracture described above is far preferable to brittle fracture. If tension tests are carried out on low-carbon steels at various temperatures, Fig. 4.20(a), the reduction in area at the neck is found to decrease from 60 percent at room temperature to essentially zero at $-195°C$. Although the microstructure, Fig. 4.17, reveals that a small amount of deformation, $\varepsilon$, occurs even at $-195°C$, the amount of energy, $\frac{1}{2}\sigma\varepsilon$, expended in producing fracture is only about 1 percent of that required at room temperature. This difference in behavior greatly influences the response of

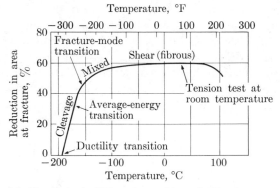

(a) Change in ductility, measured by reduction in area.
The types of fracture mode are shown, and also the
transition from large absorption of energy (at high
temperatures) to small absorption (at low
temperatures).

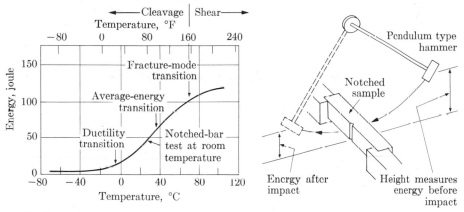

(b) Change in energy absorbed during fracture,
measured by notched-bar (Charpy type) tests.
Plastic deformation is essentially absent below
the ductility transition.

(c) Schematic illustration of the
nature of a notched-bar (or impact)
test.

**Fig. 4.20** The change in fracture behavior of hot-rolled 0.2% carbon steel with change in
temperature of testing for (a) tension tests and (b) notched-bar tests. (After E. R. Parker.)
To convert from J to ft-lb, multiply by 0.737.

a steel member to a stress, $\sigma$, in excess of the value allowed for in the design.
Ductile behavior permits the redistribution of such a stress by a small yielding
of the steel, thus absorbing the energy associated with the overstress. In brittle
behavior, on the contrary, the same overstress can completely cleave the steel
member and render it useless for supporting the load for which it was originally
designed.

Other factors besides temperature can significantly affect the fracture behavior of a given steel: among them are conditions of stressing (stress concentrators, and also combined stresses), rate of straining, and environmental effects (for example, a corrosive liquid on the surface). The data of Fig. 4.20(b) show that a steel may act in a relatively brittle manner in a notched-bar test at room temperature, even though it is ductile in a tension test. The explanation lies in the ratio of maximum tensile stress $\sigma$ to maximum shear stress $\tau$ for various types of testing.

| $\sigma/\tau$ | Type of test |
| --- | --- |
| 1 | Torsion |
| 2 | Tension |
| $\sim 4$ | Notched bar |

Tensile stresses promote cleavage fracture, whereas shear stresses encourage the competing process of plastic deformation. The tendency for brittle fracture to occur is greater at high rates of straining, because there is less time for plastic flow to occur near the microcracks and thus relieve the stress concentration. The deleterious effects of a corrosive environment will be considered in Chapter 11.

A special type of brittle fracture is the *intercrystalline* mode, in which the path of fracture goes along the grain boundaries, Fig. 4.21. This type requires even less energy than cleavage, which is *transgranular* (goes through the grains). Comparison of Figs. 4.17 and 4.21 reveals that considerably more plastic deformation occurs during cleavage than during intercrystalline fracture. A principal cause of intercrystalline fracture is the segregation of an impurity or alloying element at the grain boundaries. In extreme cases a second, brittle phase forms at the boundaries and promotes intercrystalline fracture. More commonly the segregation is within the original phase, but is sufficient to lower the energy for fracture along the grain boundaries.

**Fracture mechanics.** This term denotes a procedure for determining the fracture toughness $K_c$ for a given alloy. This quantity is then used to predict the applied stress, $\sigma_f$, at which fracture will occur. Certain conditions must, of course, be held constant; in particular, the temperature (that for which the test data are obtained), and the conditions of loading (specified by a subscript roman numeral). Let us consider the determination of $K_{Ic}$ for a given high-strength steel to be used at room temperature in the form of a thick plate under tensile loading. The steel is then in a condition of plane strain [$\sigma_3 = (\sigma_1 + \sigma_2)/2$ and $d\varepsilon_3/dt = 0$], denoted by the subscript I. In the presence of an elliptical crack of length $2c$, the fracture toughness is defined as

$$K_{Ic} = \sigma_f(\pi c)^{1/2}. \tag{4.30}$$

In view of the theoretical expression for $\sigma_f$ given by Eq. (4.26), $K_{Ic}$ is a material

**Fig. 4.21** Photomicrograph of intercrystalline fracture in a low-carbon steel tested in tension at $-195°C$. The embrittling impurity is oxygen. (Courtesy J. R. Low, Jr., Carnegie Mellon University.)

constant,

$$K_{Ic} = (GE)^{1/2}, \tag{4.31}$$

and depends on Young's modulus $E$ and on the energy $G$ needed to create unit area of crack. If the quantitative value for fracture toughness is known, the magnitude of the applied tensile stress, $\sigma_f$, that will cause the steel to fracture, is obtained by dividing the fracture strength, $K_{Ic}$, of the steel by the stress-concentrating effect, $(\pi c)^{1/2}$, of the most severe crack that is present in the structure. To determine the $K_{Ic}$ value for the steel in question, several thick plates, each containing an artificially produced crack of known length, are tested to fracture. A plot of the several values of $\sigma_f$ against the corresponding value of $1/(\pi c)^{1/2}$ gives a straight line whose slope is $K_{Ic}$, a constant for the steel. When $K_{Ic}$ values for various steels are plotted against the yield strength of each steel, Fig. 4.22, an inverse relationship is observed. Consequently, if a strong steel (high yield strength) is used for a structural member, its fracture toughness $(K_{Ic})$ is low, and care must be exercised to avoid fracture. The "TL" $K_{Ic}$ values shown in Fig. 4.22 are the highest that can be obtained with present technological practice; somewhat lower values must be allowed for in routine production.

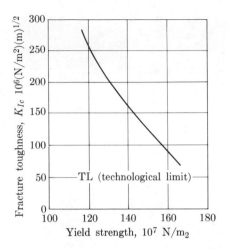

**Fig. 4.22** Data for various medium- and high-strength steels showing the inverse relation between $K_{Ic}$ and yield strength. (Data of W. S. Pellini.) To convert from $(N/m^2)(m^{1/2})$ to $(lb/in^2)(in^{1/2})$ multiply by $0.910 \times 10^{-3}$.

The procedure for using $K_{Ic}$ values in design involves careful inspection of the steel to determine the geometry of the most severe stress concentrators; for example, of the worst cracks or second-phase particles. If no cracks are visible with the best inspection techniques available, it is assumed that the largest crack is just below the limit of detection. Equation (4.30) can then be used to calculate the fracture stress, $\sigma_f$, from the values of $K_{Ic}$ and $c$. The design stress must, of course, be lower than $\sigma_f$ by a suitable factor of safety, typically about 0.7. The simple expression for fracture toughness given by Eq. (4.30) can be modified to adapt it to various design requirements. For example, if the cracks (rather than being very long compared to their width $2c$) have a disc shape, then

$$K_{Ic} = \sigma_f \left( \frac{\pi c}{1 - v^2} \right)^{1/2}, \tag{4.32}$$

where $v$ is Poisson's ratio. Also, other expressions denoted by $K_{II}$ and $K_{III}$ have been developed for other states of strain.

**Fatigue.** Yield strength and fracture strength are useful in designing for static loads, but a different parameter, *fatigue strength*, must be employed in applications involving repeated loading. An example of such loading is the axle of an automobile, the outer fibers of which are subjected to a cycle of tensile-compressive-tensile stressing about 10 times per second at 100 km/hr (60 miles/hr). During the life of an automobile, the axle is subjected to a total number $N \approx 10^7$ cycles of stressing. If fracture is to be avoided during this period, the design stress must

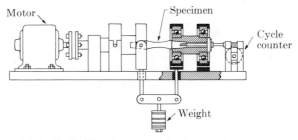

(a) An R. R. Moore-type rotating beam endurance testing machine.

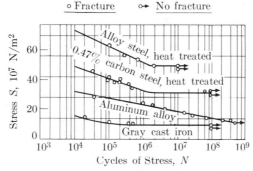

(b) Typical S-N curves, showing the experimental points, for several commercial alloys.

**Fig. 4.23** Illustration of a common type of endurance test and of results obtained by its use. (After H. F. Moore.)

be well below the fracture strength and even somewhat below the yield strength. Let us see how the appropriate design value can be determined.

*Laboratory Fatigue Tests.* Laboratory tests simulate the conditions of loading that a given machine member experiences in use. Loading can be in tension and/or compression, in torsion, or in bending. A rotating-beam tester, Fig. 4.23(a), subjects a highly polished specimen to cycles of reversed bending (as in an axle), the maximum tensile (and compressive) stress, S, being determined by the applied weight. The test data are plotted as an $S$–$N$ curve, Fig. 4.23(b), which shows the number of cycles, $N$, at which fatigue fracture occurs for a given stress, $S$. If $S$ has the value of the yield strength, $N$ is on the order of $10^4$. Two different types of behavior are found at lower stresses. (1) A few alloys including steel have an *endurance limit,* which is about one-half of the tensile strength. Below this value of stress the alloy can withstand an unlimited number of cycles of stressing without fracture. (2) Most alloys, such as the aluminum alloy in Fig. 4.23(b), fail to exhibit

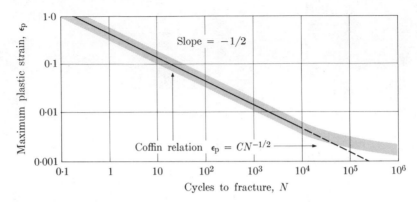

**Fig. 4.24** Typical behavior of aluminum alloys in the range of low-cycle fatigue. The strain, $\varepsilon_p$, is the maximum plastic strain in the outer fibers during cyclic bending.

an endurance limit. Consequently, their fatigue strength is specified for a given number of cycles. For example, this aluminum alloy has a fatigue strength of $14 \times 10^7 \, \text{N/m}^2$ for $N = 10^8$.

Some structural members are subjected to only a limited number of stress cycles during their useful life; the landing carriage of an aircraft is an example. Low-cycle fatigue (below about $N = 10^4$) is significantly different from ordinary (or high-cycle) fatigue. If the test data are plotted as log (plastic strain, $\varepsilon_p$) versus log (cycles to fracture, $N$), Fig. 4.24, a straight line is obtained from $N = \frac{1}{4}$ to about $N = 10^4$. Above $10^4$ the data deviate in the manner shown; this is the beginning of the region of high-cycle fatigue in which stress is a more convenient index of fatigue life, Fig. 4.23(b).

*Mechanisms of Fatigue.* Fracture in fatigue differs greatly from fracture under a static load. Plastic deformation encourages fatigue fracture, whereas it tends to inhibit brittle failure. Since plastic deformation occurs more easily at the surface (rather than in the constrained interior) of a metal, and especially when the stresses are higher at the surface, fatigue cracks usually initiate at the surface. In tensile fracture, on the other hand, crack growth occurs preferentially for microcracks near the center of the necked region of the specimen.

The initiation and growth of a fatigue crack generally occurs in the following stages. (a) Slip occurs on favorably oriented planes, Fig. 4.25(a), and causes roughening of the surface. (b) At the stress raisers created by this roughness, a crack develops and propagates by stepwise shearing on a plane at about 45° to the axis of stressing; this is called stage I growth, Fig. 4.25(b). (c) After a preparatory process, such as formation of microcracks, stage II growth eventually begins in a direction that, on the average, is perpendicular to the axis of stressing, though the actual path is somewhat irregular. The crack advances a finite amount during

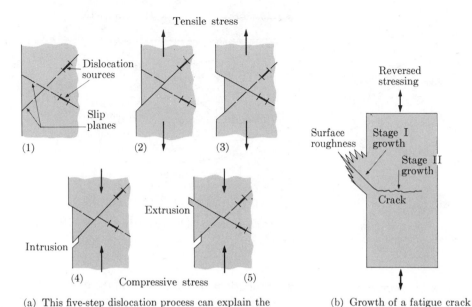

(a) This five-step dislocation process can explain the production of extrusions and intrusions at the surface. (After A. H. Cottrell and D. Hull.)

(b) Growth of a fatigue crack occurs in two distinctive stages. (After A. S. Tetelman and A. J. McEvily, Jr.)

**Fig. 4.25** Schematic illustrations of the stages in a typical process of fatigue.

**Fig. 4.26** Electron fractograph of the fracture surface in a specimen of a hardened aluminum alloy (2024-T4) that failed at a critical stress intensity of $14 \times 10^6$ $(N/m^2)m^{1/2}$. The yield strength of this alloy is $32 \times 10^7 N/m^2$ and its tensile strength is $47 \times 10^7 N/m^2$. (Courtesy D. A. Meyn, U.S. Naval Research Laboratory.)

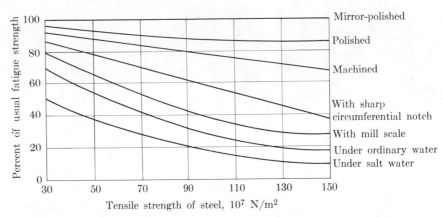

**Fig. 4.27** Effect of surface conditions in reducing the usual fatigue strength of steel, which is determined on mirror-polished specimens.

each cycle of stressing, and the pattern of advance can be seen in the final fracture surface, Fig. 4.26. (d) At a certain stage in the growth of a fatigue crack, the stress in the uncracked cross section reaches the level at which ordinary fracture occurs in a single application of tensile stress. Depending on the alloy and the conditions of service, this fracture may be cleavage or a type of shear rupture.

*Factors Affecting Fatigue Strength.* Although it would appear that fatigue-strength values could be used directly as the permissible stresses in rapidly rotating shafts, the situation is quite otherwise. Actual operating conditions differ significantly from those used in the standard fatigue tests, and the performance of an alloy may be greatly affected by the changed conditions. The influence of some of these service factors can be estimated roughly as follows.

1.  *Effect of surface conditions.* Rarely is an industrial machine component given the high polish that is used on a laboratory endurance test specimen. Figure 4.27 shows the effect of surface condition in reducing the actual endurance limit of steel. It is noteworthy that the endurance limit of a low-strength steel is only slightly affected by a given surface imperfection, compared with the decrease suffered by a high-strength steel. This behavior is a result of the higher notch sensitivity of stronger, less ductile materials. Since the ratio of the endurance limit to the tensile strength, the *endurance ratio,* is about 0.5 for steels, it is easy to show (Fig. 4.27) that there is no advantage in using a high-strength steel for an application requiring endurance strength in a corroding environment.

2.  *Effect of range of stress.* Frequently a part is subjected in service to a range of stresses in which the ratio $r$ of the minimum stress to the maximum stress is not $-1$, as in the case of the reverse bending employed in the usual endurance

test. For example, the ratio may be $r = 0$ if the stress range is from zero to a maximum tensile stress. It has been determined that the endurance limit $S_r$ corresponding to a certain value of $r$ is given by an equation of the form

$$S_r = S_e\left(\frac{3}{2 - r}\right), \tag{4.33}$$

where $S_e$ is the endurance limit determined in reverse bending.

3. *Effect of complex stresses.* In brittle alloys, such as gray cast iron, the behavior in an endurance test is determined only by the largest tensile stress. The significant stress in ductile materials, on the other hand, is determined by the combination of all the principal stresses that act on the specimen, Eq. (4.15). In particular, if permanent compressive stresses can be produced in the surface layer by such a means as shot-peening, the endurance limit can be increased by about 25%.

4. *Effect of section size.* Larger specimens of a given material have a lower endurance limit than the small laboratory specimen. If the diameter of a machine part is more than about 6 mm, it is customary to assume that the endurance limit will be about 20% below that usually determined.

5. *Effect of temperature and frequency.* Endurance strength and tensile strength are affected in approximately the same manner by changes in temperature or rate of straining. Therefore changes in frequency have little effect on the endurance limit at ordinary temperatures, but increasing the temperatures causes a pronounced decrease in endurance strength compared with the room-temperature value.

## OTHER TESTS OF MECHANICAL PROPERTIES

Many different procedures are employed to determine whether a given alloy is suitable for a certain application. The alloy may be tested for its ability to deform satisfactorily during a forming operation, or perhaps for its ability to operate under stress at a high temperature. For technological purposes, economy and ease of testing are important factors.

**Hardness tests.** It is possible in many cases to substitute for the relatively slow and expensive tensile test a more convenient test of the plastic deformation behavior of metals, a *hardness test.* Hardness is usually defined as *resistance to penetration,* and the majority of commercial hardness testers force a small sphere, pyramid, or cone into the body of the metal by means of an applied load. A definite number is obtained as the *hardness* of the metal, and this number can be related to the yield strength or tensile strength of the given metal as determined from a tension test. The tensile strength of a steel (in $N/m^2$), for example, is roughly equal to $3 \times 10^6$ times its Brinell hardness number. Hardness tests are essentially of a practical control nature, but they are of great value both in the plant and in the laboratory.

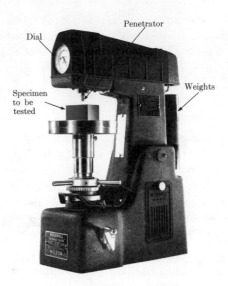

**Fig. 4.28** A Rockwell hardness testing machine. A few of its essential parts are indicated. (Courtesy Wilson Mechanical Instrument Company.)

Figure 4.28 shows the manual *Rockwell* hardness tester, an especially convenient machine. The following procedure is used in obtaining a hardness number with this machine. The specimen is brought into contact with the penetrator, the penetrator is then slowly forced into the specimen surface by weights acting through a system of levers, and when the load is released the dial pointer indicates the hardness number, which is determined by the depth reached by the penetrator. In automated production lines it is possible to measure the hardness of each piece at rates up to 1000 tests per hour, using a fully automatic *Rockwell* machine. Specimens are fed to the testing unit, and on the basis of the hardness reading they are automatically sorted so that only acceptable pieces continue in the production line.

The characteristics of several types of hardness testers are summarized in Table 4.4. The *Brinell* machine is widely used in production-line testing of large parts such as crankshafts, since a minimum of preliminary preparation of the test surface is needed. *Microhardness* tests can be made on individual particles that constitute the microstructure of an alloy. For example, the hardness of particles less than 0.2 mm in diameter can be determined by this means.

**Impact tests.** A metal may be very hard (and therefore presumably very strong, i.e., have a high tensile strength) and yet be unsuitable for applications in which it is subjected to sharp blows. *Impact resistance* is the capacity of a metal to withstand such blows without fracturing. Many different procedures are used to evaluate the impact resistance of metals, depending on the brittleness of the

**TABLE 4.4**

Characteristics of Several Hardness Testers

| Type | | Penetrator | Usual range of loads, kg | Usual range of hardnesses covered | Usual surface preparation before testing | Typical applications |
|---|---|---|---|---|---|---|
| Rockwell | C scale | Diamond cone | 150 | Medium to very hard | Fine grinding | Production testing of finished parts |
| | B scale | $\frac{1}{16}$ inch steel ball | 100 | Soft to medium | Fine grinding | Production testing of finished parts |
| Brinell | | 10 mm steel ball | 500–3000 | Soft to hard | Coarse grinding | Production testing of unfinished parts |
| Vickers | | Diamond pyramid | 5–100 | Very soft to very hard | Fine grinding | Laboratory investigations |
| Microhardness | | Diamond pyramid | 0.01–50 | Very soft to very hard | Fine polishing | Testing of micro-constituents of alloys |

metal and on the nature of the application. Figure 4.29 shows five of the principal tests. The tension impact test can be used on the most ductile metals, while the torsion impact test is suitable for extremely brittle materials such as hardened tool steels.

The term *notch-bend test* is sometimes applied to those tests (Izod and Charpy) involving a notched specimen, to emphasize that the effect of the notch is more important than the speed of testing in reducing the ductility of the metal. In the neighborhood of the base of the notch, bending of the specimen causes the appearance of radial and transverse tensile stresses as well as of the primary longitudinal tensile stress. This condition of triaxial tension raises the stress at which plastic deformation begins and tends to cause fracture with little plastic flow. For example, if the radial and transverve stresses are assumed to be one-half as large as the longitudinal stress, then from Eq. (4.15) $S_1 = 2S_0$; that is, the longitudinal stress at which flow begins is equal to twice the stress $S_0$ necessary to cause flow in a simple tension test. In some metals local deformation can occur (without fracture) in the highly stressed area at the base of the notch. The sharpness of the

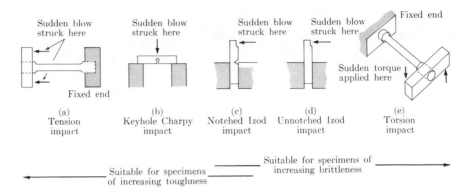

**Fig. 4.29** Impact tests suitable for testing metals of various degrees of brittleness.

notch is reduced by such deformation, and further ductile behavior is encouraged. Metals that behave in this manner are said to have *low* notch sensitivity. *High* notch sensitivity is characteristic of normally ductile metals that fracture in a brittle fashion in the triaxially stressed notch area. Initial cracking intensifies the stress concentration and causes rapid propagation of the crack across the entire specimen.

**High-temperature strength.** The need for materials that can operate for long periods at the high temperatures encountered in internal combustion engines, steam-power machines, and gas turbines has led to extensive study of the properties of metals at these temperatures. Although knowledge of many properties is necessary for designing a complex industrial machine, only the high-temperature strength is considered here. The effect of temperature on other properties, such as Young's modulus and corrosion resistance, is treated in other chapters. At moderate temperatures the ordinary yield strength is the property that limits the application of engineering alloys. For steels these temperatures are in the range below about 400°C (750°F). The rate of straining or the duration of loading has little effect on the yield strength or tensile strength, and it is therefore permissible to use a (rapid) tension test to determine the strength that a metal will exhibit on being stressed for many years in such a part as a bridge member or an automobile frame. At very high temperatures the situation is completely different, and it is necessary to determine the precise dependence both of tensile strength (rupture strength) and of yield strength (creep strength) on the time of application of the stress. The procedure and equipment for both creep and stress-rupture testing are similar, Fig. 4.30, but the treatment of the test data is sufficiently different that each of the tests will be described separately.

A *stress-rupture test* determines the constant approximate-stress value $(F/A_0)$ that causes a metal to break in a definite time at a given temperature. The specimen is loaded as shown in Fig. 4.30, and the elongation is automatically recorded from the time of the beginning of the test until the time of fracture. A typical set of three

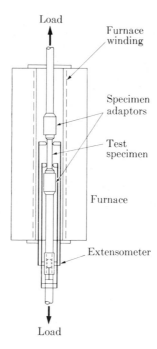

**Fig. 4.30**  A typical test method for obtaining creep and stress-rupture data.

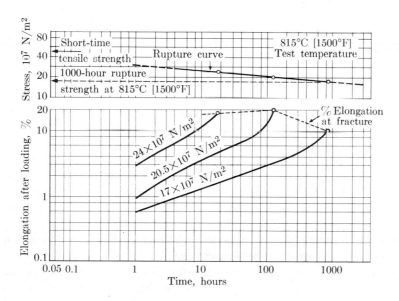

**Fig. 4.31**  Typical stress-rupture curves, and a plot of rupture stress vs. time to rupture. (After E. Epremian.)

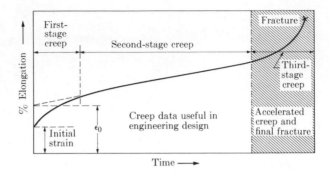

**Fig. 4.32** Typical creep behavior showing the relation of the useful portion of a creep curve to the eventual fracture of the alloy.

curves for tests at 815°C (1500°F) is shown in the lower part of Fig. 4.31 plotted as log (percent elongation) versus log (time). To estimate closely the load that would cause rupture to occur in a certain time, say 1000 hours, it is convenient to plot the experimentally determined pairs of stress-rupture time values in the manner shown in the upper curve of Fig. 4.31. The points in such a plot form a straight line that is useful both for interpolation and extrapolation. An ordinary tensile test at a high temperature is evidently similar to a very short time rupture test. The good correlation between extrapolated rupture strength and this *short-time tensile strength* is also shown in the figure.

At high temperatures plastic deformation (creep) appears to occur at all stresses, but the *rate* at which it occurs increases with increasing stress at a given temperature. Therefore the quantity that is used in design, the *creep strength*, is chosen so that plastic deformation occurs at a suitably low rate. Thus an objectionable amount of deformation will not occur during the expected life of the machine.

Careful measurements are required to determine the small rates of plastic deformation involved in creep. The principle of measuring creep strength is the same as that shown in Fig. 4.30, but special means are used to increase the precision of the strain measurement. Figure 4.32 shows the characteristics of the usual creep curve in which the applied stress causes a constant rate of creep to continue for the time of interest in the engineering design being considered. The relatively constant second-stage creep is preceded by a brief first stage of creep, and it would be followed eventually by accelerated creep and final fracture in the third stage of creep. Creep strength values represent the rate of elongation during second-stage creep at a given temperature and under a given stress.

It can be seen from Fig. 4.32 that the *total* elongation experienced by a specimen during creep of an engineering structure is the sum of the initial elastic and plastic strains plus the rapid first-stage creep deformation and the elongation calculated from the creep rate. Design data based on total elongation are given

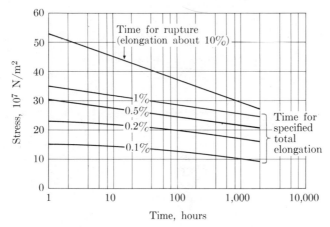

**Fig. 4.33** Design data for high-temperature alloy N-155* at 650°C (1200°F). (After Grant and Buchlin.)

in Fig. 4.33 for a high-temperature alloy. The lower curves in this figure show the time required for 0.1–1.0% elongation to occur when the alloy is subjected to given tensile stresses at 650°C (1200°F). Data from stress-rupture tests on this alloy are summarized in the top curve, which shows the time for rupture to occur under various tensile stresses. For applications in which only small deformations are permissible, creep resistance determines the maximum design stress. However, when large deformations can be tolerated, the rupture strength may be the limiting factor.

The alloying principles and structural considerations that apply to strength at normal temperatures are also generally valid for high-temperature strength. An important qualification, however, is that the microstructure (grain size, dislocation structure, second-phase particles, etc.) remain relatively stable. This aspect of strength is discussed in several of the following chapters.

**Low-temperature properties.** In designing equipment to operate at low temperatures, it is sometimes necessary to consider the decrease in ductility produced in some alloys by these low temperatures. Practically all alloys show an increase in yield strength and tensile strength at low temperatures. This behavior is shown for nickel and for two steels in the lower part of Fig. 4.34. In addition, all metals and alloys that have a *face-centered cubic lattice* appear to maintain their ductility substantially unimpaired at very low temperatures: −240°C (−400°F). However, nonface-centered cubic metals and alloys, such as ordinary pearlitic steels, usually suffer a marked decrease in ductility and impact strength at even moderately low

---

* The composition of alloy N–155 is 20% Cr, 20% Ni, 20% Co, 3% Mo, 2% W, 1% Nb, 0.1% C, and the balance iron.

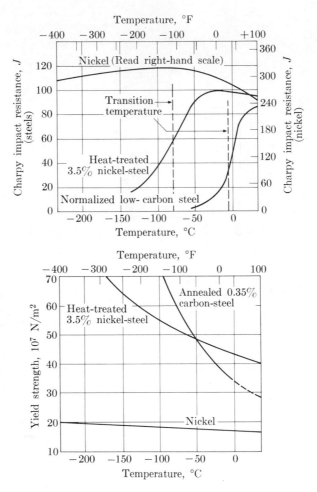

**Fig. 4.34** The effect of low temperatures on the impact resistance and yield strength of a face-centered-cubic metal (nickel) and two essentially body-centered-cubic alloys. [To convert from J to ft-lb multiply by 0.737.]

temperatures. Figure 4.34 shows how the impact strength curve for steels falls off at low testing temperatures. The impact strength of nickel, on the contrary, remains essentially unchanged down to $-240°C$ ($-400°F$). The *transition temperature* used here in describing the change from ductile to brittle behavior in steels will be recognized as the "average energy transition" of Fig. 4.20.

Since ductile behavior is essential for the satisfactory performance of many alloys in service, the high transition temperature of ordinary carbon steel makes it unsuitable for use under unfavorable conditions of stress even at only moderately low temperatures. Fortunately, it is possible to use steels at subzero temperatures provided their transition temperatures are lowered by the addition of about 4% nickel, by hardening and tempering, by complete deoxidation, or by refinement of grain size. Figure 4.34 shows the pronounced lowering of the transition temperature produced by the combination of heat treatment and the addition of nickel. The transition temperature of the heat-treated nickel steel is more than 60°C (100°F) lower than that of the normalized carbon steel.


## APPLICATIONS OF PLASTIC DEFORMATION

As indicated by the examples discussed above, plastic deformation influences many important uses of metals and alloys. An especially direct application of the principles of plastic deformation is found in the various processes of hot-working, which are commonly used in large-scale production. Wear resistance and machinability are strongly influenced by the plasticity of an alloy; therefore, these topics are also conveniently discussed in the present section. Other aspects of deformation will be considered in later chapters, especially Chapters 9 and 10.

**Hot-working.** One of the most common and commercially important techniques for fabricating metals and alloys is hot-working. When metals are deformed at high temperatures, the strained grains instantly reform (recrystallize*) during hot-working and essentially return to their initial condition. One important consequence of this constancy of structure is the fact that the stress required to deform the metal, the flow stress $S_0$, also remains essentially constant during a hot-working operation. This fact greatly simplifies the estimation of the forces required in commercial forming processes such as forging, extrusion, and rolling. A first approximation to the force $F$ required in a simple forming operation (such as press forging) for a block of metal of cross-sectional area $A$ is simply

$$F = S_0 A. \tag{4.34}$$

The corresponding work required to reduce the height of the block from the initial value $h_0$ to a final height $h_f$ is conveniently written

$$\text{Work} = \int_{h_0}^{h_f} F \times dh = S_0 A h \int_{h_0}^{h_f} \frac{dh}{h} = S_0 V \ln \frac{h_f}{h_0}, \tag{4.35}$$

where use has been made of the fact that the volume $V = Ah$ remains constant during plastic deformation. Since the quantity $\ln(h_f/h_0)$ is the true strain $\varepsilon$, according to Eq. (4.8), the *ideal work* of plastic deformation is

$$\text{Work} = S_0 V \varepsilon. \tag{4.36}$$

---

* The nature of the process of recrystallization is discussed in Chapter 9.

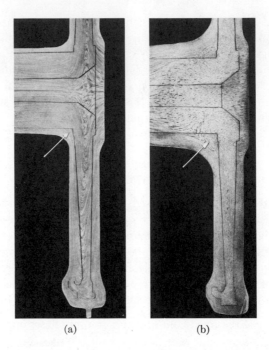

(a)                              (b)

**Fig. 4.35** Illustration of good (a) and of poor (b) grain flow lines in a forged blank for the compressor wheel of a jet aircraft.  Note that in (a) the grain flow parallels the finished part outline much more closely than in (b).  The arrow points to an especially critical area. (Courtesy S. M. Jablonski, Wyman-Gordon Company.)

Negative values of $\varepsilon$ are produced by compressive stresses (negative $S_0$), so that the work is always a positive quantity.

The actual work required in hot-forming operations is always greater than the ideal value because of several factors that were neglected in the above treatment.  The most important of these are (1) friction between the metal and the forming device, and (2) nonuniform flow of the metal.  Both phenomena are useful in many processes, for example in producing desirable grain flow lines in forgings, Fig. 4.35.  However, they cause the work of deformation to be larger than the ideal value by a factor $\alpha$ that ranges from about 2 for press forging to 10 or more for extrusion.

Some of the general characteristics of hot-working processes that contribute to their wide use are:

1. The energy necessary to form a metal to a desired shape decreases rapidly as the temperature of forming increases.

2. The ability of an alloy to flow without cracking usually reaches a maximum value at a temperature near its solidus temperature.

3. The compositional inhomogeneities associated with the cast, ingot structure of an alloy are more quickly removed at high temperatures, where atomic diffusion occurs rapidly.

This third factor is involved in the distinction between *wrought* (i.e., formed by deformation) and *cast* alloys. The tensile strength of the cast material is usually equal to that of the wrought alloy. However, the impact strength of a casting is relatively low, because of inhomogeneities in the grain boundaries and elsewhere. After sufficient hot-working of a casting (about 50% reduction) the new set of grains produced by recrystallization usually has much higher impact strength and somewhat better ductility.

**Wear resistance.** In *wear* of a metal or alloy, small particles are torn from its surface by materials moving against it. Although the mechanism of wear is not completely understood, it is likely that the following processes occur to some extent in the wearing of metals:

1. *Seizing or penetration of the metal surface.* Depending on the nature of the wearing agent, the first step in removing a metal particle may be digging into the metal surface or local welding because of intimate contact on an atomic scale. High hardness tends to prevent both penetration of the surface and the deformation necessary to permit relatively large contact areas. There is usually an excellent correlation between hardness and wear resistance.

2. *Deformation of the metal.* Small pieces of the metal must be deformed and torn loose if the process of wear is to take place. A tough metal tends to resist this action and so exhibit good wear resistance. In practice, toughness is often sacrificed in favor of high hardness.

3. *Corrosion of the metal surface.* A corrosion product, such as an oxide or a sulfide, is usually quickly removed from the metal surface by the wearing agent. Corrosion-resistant metals, therefore, show superior wear resistance in many applications.

4. *Heating of the metal surface.* Although actual melting of the metal occurs only in rare cases, less severe heating can reduce wear resistance by speeding corrosion reactions and by decreasing mechanical properties. Metals with a high melting point tend to have good wearing properties.

Although the qualitative significance of these four processes can often be estimated in a given instance, a quantitative theory of wear resistance is lacking. In fact, the conditions under which wear occurs in practice are so varied that it is necessary to divide the subject into more or less isolated sections for effective study.

The following are important types of wear:

| Type of Wear | Example |
|---|---|
| Metal-to-metal, lubricated | Shaft in bearing |
| Metal-to-metal, not lubricated | Wheel on track |
| Metal-to-nonmetal, dry friction | Belt on pulley |
| Metal-to-nonmetal, particle impact | Sandblast nozzle |
| Metal-to-fluid, liquid particle | Wet steam in turbine |

Wear testing is a difficult problem. A useful device for rating materials for a given application must closely duplicate service conditions, and the results must be verified by checking them against actual performance. Although most wear-testing machines are custom-made for a given purpose, the *Amsler machine* is a versatile device for testing metal-to-metal wear, and the *Brinell machine* is useful for testing metal-to-nonmetal wear.

*Hard-Surfaced Metals.* The rate of wear under many industrially important conditions can be reduced to an acceptable value by the use of high hardness at the wearing surface. Hardening an entire cross section has the disadvantage of decreasing the over-all ductility, and frequently limits the surface hardness to relatively low values. It is therefore common practice to harden just the surface layer, often by means of an extremely hard, brittle phase embedded in a more ductile matrix. Most methods of surface hardening involve atomic diffusion (discussed in Chapter 8). The essential features of some representative methods are summarized in Table 4.5.

*Bearing Metals.* A *bearing* supports a moving part of a machine. Since motion is essential to the proper working of almost all machines, the problem of ensuring dependable action with minimum frictional losses and at low cost is a most important one for engineers. This problem is complex, and there are almost as many solutions to it as there are conditions of service (conditions that range from the slow-moving, heavily loaded boom of a fifty-ton power shovel to the delicately balanced shaft of a small, high-speed electric motor). The solution of bearing problems lies largely in mechanical design, but intimately associated with this design are the characteristics of the available bearing materials.

Bearings are usually classified as (1) rolling contact bearings and (2) plain bearings. The self-lubricating bearing made by the powder-metallurgy process is a type of plain bearing. Rolling contact bearings have a number of advantages over plain bearings: (1) the starting friction is lower; (2) the shaft is held more precisely in the desired position; (3) maintenance costs are lower: (4) a combination of radial and thrust loads can be carried. The principal advantage of plain bearings is their lower cost, but in addition they may be chosen for conditions involving shock loading, or occasionally for more quiet operation.

**TABLE 4.5**    Methods of Producing Hard Surfaces on Metals

| Method | General nature of surface-hardening process | Metals usually hardened by this method | Process used | Characteristics of the process | Typical uses |
|---|---|---|---|---|---|
| Carburizing | A high-carbon surface is produced on a low-carbon steel and is hardened by quenching. | Plain carbon or alloy steels containing about 0.20% carbon. | Low-carbon steel is heated at 870–930°C (1600–1700°F) in contact with gaseous, solid, or liquid carbon-containing substances for several hours. The high-carbon steel surface is hardened by quenching. | Case depth is about 1.2 mm. Hardness after heat-treatment is $R_c$ 65. Negligible dimension change caused by carburizing. Distortion may occur during heat-treatment. | Gears Camshafts Bearings |
| Nitriding | A very hard nitride-containing surface is produced on the surface of a strong, tough steel. | Nitralloy steels containing aluminum; Nitralloy 135 contains 0.35% C, 1.1% Cr, 0.2% Mo, and 1.0% Al. The steel is hardened and then tempered at 540–700°C (1000–1300°F) before being nitrided. | The steel is heated at 500–540°C (930–1000°F) in an atmosphere of ammonia gas for about 50 hr. No further heat treatment is necessary. | Case depth is about 0.4 mm. Extreme hardness (Vickers 1100). Growth of 25–50 μm occurs during nitriding. Case is not softened by heating for long times at 425°C (800°F). Case has improved corrosion resistance. | Valve seats Guides Gears |
| Cyaniding | A carbon-and-nitride-containing surface is produced on a low-carbon steel and is hardened by quenching. | Plain carbon or alloy steel containing about 0.20% carbon. | Low-carbon steel is heated at 870°C (1600°F) in a molten 30 percent sodium cyanide bath for about one hour. Quenching in oil or water from this bath hardens the surface of the steel. | Case depth is about 0.3 mm. Hardness is about $R_c$ 65. Negligible dimension change is caused by cyaniding. Distortion may occur during heat-treatment. | Screws Nuts and bolts Small gears |

**TABLE 4.5** Methods of Producing Hard Surfaces on Metals (*continued*)

| Method | General nature of surface-hardening process | Metals usually hardened by this method | Process used | Characteristics of the process | Typical uses |
|---|---|---|---|---|---|
| Carbonitriding | Carbon and nitrogen are added to the surface of a low-carbon steel and permit hardening by an oil quench. | Plain carbon steels containing about 0.20% carbon. | Low-carbon steel is heated at 700–800°C (1300–1600°F) for several hours in a gaseous hydrocarbon and ammonia atmosphere. Nitrogen in the surface layer increases hardenability and permits hardening by an oil quench: | Case depth is about 0.5 mm. Hardness after heat treatment is $R_c$ 65. Negligible dimensional change occurs. Distortion is less than in carburizing or cyaniding. | Gears Nuts Bolts |
| Siliconizing (Ihrigizing) | A moderately hard, corrosion-resistant surface containing 14% silicon is produced on low-carbon steels. | Plain carbon steel containing 0.1–0.2% carbon. | The steel parts are heated at 925–1000°C (1700–1850°F) in contact with silicon carbide and chlorine gas for two hours. No further heat treatment is required. | Case depth is about 0.6 mm. Hardness is $R_B$ 85. Case has good corrosion resistance. Growth of 25–50 μm occurs during siliconizing. | Valves Tubing Shafts |
| Hard chromium plating | A hard chromium plate is applied directly to the metal surface | Generally a steel, low or high carbon, hardened or soft | The steel parts are plated in the usual plating bath, but without the usual undercoat of nickel. The plating is a hundred times thicker than decorative chromium plating. | Plating thickness is about 0.1 mm. Extreme hardness (Vickers 900). Plating has good corrosion resistance. Plating has a low coefficient of friction. | Dies Gages Tools |

TABLE 4.5    Methods of Producing Hard Surfaces on Metals(*continued*)

| Method | General nature of surface-hardening process | Metals usually hardened by this method | Process used | Characteristics of the process | Typical uses |
|---|---|---|---|---|---|
| Hard surfacing | A hard, high-alloy layer is welded to the surface of a steel or cast iron. | Almost any steel or cast iron, but usually a medium carbon steel. | A fusion welding process is used to melt the hard surfacing alloy and weld it to the base metal. | Typical hard surfacing alloy is composed of chromium, tungsten and cobalt. Surface layer is about 3 mm thick. Variety of hardness and toughness given by various alloys. | Building up worn or broken parts. Excavating equipment Metal-working dies |
| Flame hardening | The surface of a hardenable steel or iron is heated by a gas torch and quenched. | Steel containing 0.4–0.5% carbon; cast iron containing 0.4–0.8% combined carbon. | An oxygen-acetylene flame quickly heats the surface layer of the steel, and a water spray or other type of quench hardens the surface. | Hardened layer is about 3 mm thick. Hardness is $R_c$ 50 to 60. Distortion can often be minimized. | Gear teeth Sliding ways Bearing surfaces |
| Induction hardening | The surface of a hardenable steel or iron is heated by a high-frequency electromagnetic field and quenched. | Steel containing 0.4–0.5% carbon; cast iron containing 0.4–0.8% combined carbon. | The section of steel to be hardened is placed inside an induction coil. A heavy induced current heats the steel surface in a few seconds. A quenching operation, such as spraying with water, hardens the heated section. | Hardened layer is 3 mm thick or more. Hardness is $R_c$ 50 to 60. Distortion can often be minimized. Surface remains clean. | Bearing surfaces Gear teeth Hubs |

**TABLE 4.5**    Methods of Producing Hard Surfaces on Metals (*continued*)

| Method | General nature of surface-hardening process | Metals usually hardened by this method | Process used | Characteristics of the process | Typical uses |
|---|---|---|---|---|---|
| Austenitic (Hadfield) manganese steel | A tough, strong austenitic structure is converted to a hard martensitic structure by surface deformation. | This steel contains 1.2% carbon and 12–13% manganese. | The cast or rolled steel is toughened by an austenitizing heat treatment of water quenching from 1000°C (1850°F). Deformation in service causes a martensitic structure to form in the surface layer. | Hardened layer is of variable thickness, depending on severity of deformation. Hardness is $R_c$ 50. | Dredge buckets Railroad frogs and switches Rock crushers |

Ball bearings and roller bearings are almost invariably made of steel that can be hardened after machining. SAE 1090,* a plain carbon steel containing $0.90\%$ carbon, is an inexpensive and frequently used material, but alloy steels are required for many applications. Greater toughness at the high hardness needed for wear resistance is obtained by using a few percent of one or more of the alloying elements chromium, nickel, and molybdenum, and by lowering the carbon content to about $0.50\%$. A carburized, low-carbon alloy steel can be used to achieve a combination of hard, wear-resistant surfaces and a tough core. For applications in which corrosion resistance is essential, a $17\%$ chromium stainless steel is often chosen.

An extremely wide range of materials can be used for plain bearings; in fact, the subject of *bearing metals* usually deals only with alloys for use in plain bearings. The correct application of a bearing metal requires the matching of such bearing properties as cost, fatigue strength, compressive strength, ease of fabrication, antiseizure characteristics, and corrosion resistance, with such service conditions as type of load, amount of load, shaft speed, conditions of temperature, dirt and corrosion, oil supply, and shaft hardness. Table 4.6 lists a number of representative bearing metals and some of their properties. The minimum value of $ZN/p$ at which complete film lubrication exists is known as the bearing modulus. This modulus depends on such factors as the smoothness of the journal and of the bearing surface, and it may decrease by a factor of ten during the running-in period of a bearing. The values given in the table are approximate steady-state values. The bearing load factor $pV$ is also only approximate and depends on such variables as the type of lubrication and the conditions of loading.

**Machinability.** Another commercially important and many-sided property is *machinability*. An industrial machining operation has many objectives, such as rapid cutting, good surface finish, and long tool life, but an adequate definition of the machinability of a metal is the following: *The most machinable metal is the one which permits the removal of material with a satisfactory finish at lowest cost.* However, the actual value of lowest cost is determined (as well as by the quality of finish required) by such factors as: (1) the metal being cut (the work piece), (2) the metal of the cutting tool (the tool material), (3) the size and shape of the tool, (4) the kind of machining operation (drilling, planing, etc.), (5) the size, shape, and velocity of the cut, (6) the type and quality of the machine (lathe, shaper, etc.), (7) the conditions of lubrication. Although these factors are closely interrelated, attention is given here mostly to the work piece and to the tool material.

*Chip Formation.* Figure 4.36(a) shows the formation of discontinuous chips in the machining of a work piece by a tool. In every such machining process the following steps occur:

1. The metal is severely stressed just ahead of the cutting edge of the tool.

---

* The SAE and AISI classifications of steel are described in Chapter 10.

**TABLE 4.6**  Properties of Representative Bearing Metals

| Alloy | Tin-base babbit 91% Sn 4.5% Cu 4.5% Sb | Lead-base babbit 75% Pb 15% Sb 10% Sn | Bearing bronze 80% Cu 10% Sn 10% Pb | Copper-lead 70% Cu 30% Pb | Cadmium-silver 98.5% Cd 1.0% Ag 0.5% Cu | Silver with a thin coating of lead and indium | Aluminum-base 91.5% Al 6.5% Al 1% Cu 1% Ni | Self-lubricating bearing 89% Cu 10% Sn 1% graphite |
|---|---|---|---|---|---|---|---|---|
| **0.1% offset compressive yield strength, N/m²** | | | | | | | | |
| 20°C (70°F) | $3.0 \times 10^7$ | $2.4 \times 10^7$ | $10 \times 10^7$ | $2.1 \times 10^7$ | $10 \times 10^7$ | | $11 \times 10^7$ | $9 \times 10^7$ |
| 100°C (210°F) | $1.9 \times 10^7$ | $1.1 \times 10^7$ | | | | | $10 \times 10^7$ | |
| **Fatigue strength** Usual test specimen values, N/m² | $2.1 \times 10^7$ | $3.0 \times 10^7$ ($2 \times 10^7$ cycles) | | | | | $6.2 \times 10^7$ ($5 \times 10^8$ cycles) | |
| Approximate order of merit in a bearing; No. 1 is lowest | 1 | 2 | 8 | 5 | 4 | 7 | 6 | 3 |
| Thermal conductivity W/(m·°C) | 52 | 30 | | 127 | 75 | 375 | 180 | |
| Coefficient of thermal expansion per °C | $23 \times 10^{-6}$ | $25 \times 10^{-6}$ | $20 \times 10^{-6}$ | $22 \times 10^{-6}$ | $29 \times 10^{-6}$ | $20 \times 10^{-6}$ | $25 \times 10^{-6}$ | $20 \times 10^{-6}$ |
| Minimum $ZN/p$* | 10 | 10 | 5 | 5 | 5 | 2 | 5 | |

| | | | | | | | | |
|---|---|---|---|---|---|---|---|---|
| Maximum $pV$* | 40,000 | 40,000 | | 90,000+ | 90,000+ | 90,000+ | 100,000 | 900+ |
| Corrosion resistance (using ordinary lubricants) | very good | good | fair | poor | poor | fair | excellent | good |
| Solidus temperature | 223°C (433°F) | 240°C (464°F) | 946°C (1735°F) | 328°C (622°F) | 314°C (598°F) | Max. useful temp. 204°C (400°F) | 227°C (440°F) | |
| Conformability to journal | good | good | poor | fair | fair | fair | fair | poor |
| Ability to embed dirt | good | good | poor | fair | fair | fair | good | poor |
| Seizure resistance | excellent | very good | fair | good | good | good | good | good |
| Journal hardness | not critical | not critical | hardened steel | 300 Brinell minimum | 250 Brinell minimum | hardened steel | 300 Brinell minimum | 250 Brinell minimum |
| Cost | medium | lowest | medium | high | high | highest | medium | medium |
| Remarks | | | | | | Used only in pre-fit bearings | Generally used in the cold-worked temper with a thin coating of pure tin | Usually saturated with oil before use |

* $Z$ = viscosity in centipoises; $N$ = revolutions per minute; $p$ = average unit pressure, lb/in²; $V$ = peripheral speed, ft/s.

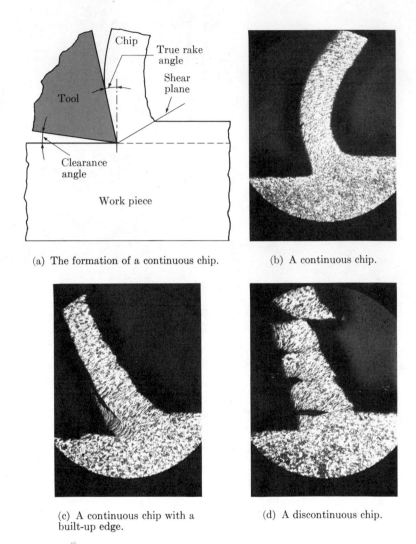

(a) The formation of a continuous chip.    (b) A continuous chip.

(c) A continuous chip with a built-up edge.    (d) A discontinuous chip.

**Fig. 4.36** Illustration of chip formation in machining operations. The photomicrographs show the character of three types of chip. (Courtesy M. E. Merchant and Hans Ernst, Cincinnati Milling Machine Company.)

2. The metal (a) fractures approximately perpendicular to the tool face and forms *discontinuous* chips, or (b) plastically deforms sufficiently so that fracture does not occur; that is, a *continuous* chip is produced. Continuous chips may form with or without a built-up edge.

3. The metal flows up the face of the tool.

**TABLE 4.7**

Characteristics of Chip Types

| Characteristic | | Type of Chip | | |
|---|---|---|---|---|
| | | | Continuous | |
| | | Discontinuous | Without built-up edge | With built-up edge |
| Results of having a given type of chip formation | Surface finish | good | excellent | poor |
| | Ease of chip disposal | good | fair | poor |
| | Type of tool failure | Wear of cutting edge | Wear of cutting edge and abrasion of tool face | "Cratering" of tool face and final spalling of cutting edge |
| | Tendency to heat tool | small | small | large |
| Conditions for obtaining a given type of chip | Optimum work-piece ductility | low | high | high |
| | Optimum chip thickness | large | small | large |
| | Optimum cutting speed | low | high | low |
| | Optimum rake angle | small | large | small |

Photomicrographs of discontinuous chips and of the two types of continuous chips are shown in Fig. 4.36(b–d). Some of the characteristics of the three kinds of chips are given in Table 4.7. Discontinuous chips are desirable for automatic screw machine work because of their easy disposal. Continuous chips with a minimum of built-up edge are preferred when the best finish is needed, and these chips may readily be broken artificially because they curl tightly in the absence of built-up edge. The tendency to form a built-up edge is decreased by using a

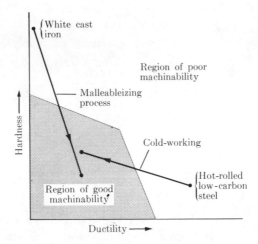

**Fig. 4.37** The improvement in machinability of alloys caused by changing their structures.

keen cutting edge, a highly polished tool face, and a tool material with a low coefficient of friction against the work piece. Suitable cutting fluids also aid materially in reducing the built-up edge.

*Work-Piece Characteristics.* Since the cutting edge of the tool must penetrate the work piece in a machining operation, it is usually desirable that the metal being cut have low resistance to penetration. Hardness is defined as resistance to penetration, and therefore low hardness is an almost universal requirement for easy machinability of a metal. The need for low hardness applies on a micro- as well as a macroscale, since hard, abrasive constituents in the microstructure will quite evidently produce rapid wear of the cutting tool. The second important characteristic of a work piece is low ductility, which promotes easy breaking of the chips in discontinuous chip formation. When continuous chips are formed, low ductility also increases the ease of cutting by decreasing the built-up edge and the deformation of the metal.

Some alloys, such as those of magnesium, naturally possess a desirable combination of low strength and low ductility. Other materials can have their properties changed by suitable treatment to bring them within the range of good machinability. Figure 4.37 is a schematic illustration of the improvement of the machinability of a low-carbon steel by *decreasing* its ductility (and incidentally *increasing* its hardness). This figure also shows the effect of malleableizing a white cast iron, a process that changes the hard, brittle carbide into softer constituents. Machinability can also be improved by suitable additions to alloys during manufacture. Lead is used for this purpose in a number of alloys, and steels can be treated with sulfur as well. The action of constituents such as these is to reduce

**TABLE 4.8**
Relative Machinabilities of Alloys

| Machinability | Ferrous | Nonferrous |
|---|---|---|
| Excellent | | Magnesium alloys |
| | | Aluminum alloys |
| | | Leaded brass |
| | | Zinc alloys |
| | Leaded steel screw stock | |
| | Low-carbon steel screw stock | Silicon-bronze |
| | Ferritic malleable cast iron | Leaded phosphor-bronze |
| | Pearlitic malleable cast iron | |
| | Ferritic gray cast iron | Yellow brass |
| | Free-cutting 12% Cr iron | Cast copper |
| | Low alloy steels | Nickel |
| | Ingot iron | |
| | Wrought iron | |
| | Free-cutting 18–8 stainless steel | |
| | | Monel metal |
| | Austenitic manganese steel | |
| | High-speed steel | |
| | 18–8 Stainless steel | |
| | White cast iron | |
| | | Stellite |
| Unmachinable | | Sintered carbides |

the rubbing friction between the chip and tool, since they act as a solid lubricant at that interface.

Since so many factors enter into a machinability rating, it is difficult to list the machinability of alloys in a definite order. The arrangement given in Table 4.8 can be considered only approximate.

*Tool-Material Characteristics.* The substances used as tool materials are conveniently divided into the five groups shown in Table 4.9. Because of their low cost, low-alloy steels are used when their properties are adequate for the application. However, for most production machining the use of the more expensive materials is well justified by the higher cutting rates they permit. For example, carbide-tipped tools maintain their cutting efficiency at 700°C (1300°F). The maximum temperature for high-speed steel is about 550°C (1000°F), and for carbon steel it is considerably lower. The nonsteel tool materials tend to produce a better finish in machining ferrous materials because of a lower friction coefficient. Sintered carbides, made by the powder-metallurgy process, consist of hard carbide

**TABLE 4.9**  Characteristics of Tool Materials

| Characteristic | Tool material | | | | |
| --- | --- | --- | --- | --- | --- |
| | Low-alloy steels | High-alloy steels | Cobalt-chromium-tungsten alloys | Sintered carbides | Aluminum oxide |
| Typical alloy | Plain carbon tool steel 1% C balance Fe* | High-speed tool steel 18% W 4% Cr 1% V 0.7% C balance Fe* | Stellite Star J-Metal 32% Cr 15% W 2.5% C balance Co* | Tungsten carbide alloy 94% WC 6% Co | 98% $Al_2O_3$ 2% binder |
| Relative cost | low | medium | high | higher | highest |
| Permissible metal cutting rate | low | medium | high | higher | highest |
| Resistance to softening by heat | poor | good | very good | very good | excellent |
| Toughness | good | fair | poor | poor | poorest |
| Wear resistance | poor | fair | good | very good | excellent |
| Typical uses | Pipe cutters Roll threading dies | Milling cutters Drills | Insert of cast alloy brazed to steel shank for machining cast iron, bronze, etc. | Insert brazed to steel shank for machining cast iron, plastics, etc. | Insert clamped in mechanical holder for finish machining |

* Plus small amounts of Mn, Si, etc.

particles bonded with 3 to 10% of cobalt metal. In addition to their extensive use as tips for cutting tools, sintered carbides are used in such devices as wire-drawing dies, in which very good wear resistance is needed. The new aluminum-oxide tool materials permit very high cutting speeds and suffer little wear. However, lack of toughness limits their application to the finer machining operations.

## PROBLEMS

**1.** Give two examples of metal parts in which plastic deformation behavior is important.

**2.** The total deformation of metals during creep is often less than 1%. Discuss the possibility of reducing (a) the elastic, (b) the anelastic, (c) the permanent plastic components of this total deformation by special treatment, such as periodic removal of the load.

**3.** Sketch a unit cell of the copper crystal structure (face-centered cubic). Show on it (a) the plane on which slip occurs, (b) the direction in which slip occurs. (c) What statement can be made about the density of atoms on the slip plane compared with other planes such as the {100} planes? (d) What statement can be made about the atomic spacing along the slip direction compared with other directions?

**4.** Consider the compression at room temperature of a single crystal of magnesium oriented so that the [0001] direction coincides with the axis of compression. Assuming that the stress for twinning on the $(10\bar{1}2)$ plane is 10 times $\tau_c$ for slip on the (0001) plane, determine whether the crystal will twin or slip.

**5.** A cylindrical FCC single crystal is stressed axially in tension. The initial orientation of the crystal is such that the [517] direction coincides with the cylindrical axis. (a) By analogy with Fig. 4.5, show the initial position of the axis of loading on a stereographic projection. (b) Which will be the primary slip system? (c) Determine the position of the axis of loading when over-shooting first begins. (d) At this stage of deformation, through what angle has the orientation of the crystal rotated?

**6.** Making use of triangles $ABB'$, $ABN$, and $AB'N$ in Fig. 4.8(b), verify the first equality in Eq. (4.2).

**7.** Using Eq. (4.5), calculate the shear strain $\gamma$ at which overshoot first begins in the crystal of Problem 5. Suitable manipulation of the stereographic projection gives the angles in question.

**8.** Derive an expression for the shear strain in compression, analogous to Eq. (4.5). [*Hint*: Make a sketch similar to Fig. 4.8(b) for an element that slips in the opposite direction.]

**9.** The approximate strain $\varepsilon'$ differs increasingly from the true strain $\varepsilon$ with increasing plastic deformation. Show that $\varepsilon'$ is 5% larger than $\varepsilon$ when $\varepsilon'$ has reached a value of about 0.1. [Recall the useful approximation for small values of $x$, $\ln(1 + x) = x - \frac{1}{2}x^2$.]

**10.** Consider a small cylindrical volume of metal in the necked region of a tension test specimen. Taking the diameter of the necked region as the diameter of the small volume, and assuming that no volume change occurs during plastic deformation, show that the equation

$$\varepsilon = 2 \ln \frac{d_0}{d},$$

gives the true strain at the neck of the specimen.

**11.** (a) Does the equation of Problem 10 also apply during the early stages of the tension test when uniform deformation is occurring? (b) Why?

**12.** For a given material the *yield strength* (which is determined by a tension test) and the *flow stress* have the same numerical value. What is the difference in the significance of these two terms with regard to mechanical design?

**13.** A typical specimen for a tension test has a gage length of 50 mm (recorded by two marks on the initial specimen) and a diameter of 13.0 mm. At the conclusion of the test the two parts of the fractured specimen, Fig. 4.14(a), can be reassembled to recreate the shape of the specimen at the moment of fracture. (a) If the final separation of the gage marks is 81 mm, what is the percent elongation at fracture? (b) If the minimum diameter in the neck region is 6.9 mm, what is the percent reduction in area at fracture?

**14.** The central portion of a long steel rod passes through a liquid that is under $7 \times 10^7$ N/m$^2$ pressure. What is the maximum tensile stress that the rod can sustain without plastic deformation? The yield strength of the steel is $20 \times 10^7$ N/m$^2$. [*Hints*: Make an analysis at a point in the central portion of the rod. Note that $S_1$ is along the axis of the rod. $S_2$ and $S_3$ are equal, are at right angles to each other, and lie in a plane that is perpendicular to the axis of the rod.]

**15.** When a potential neck attempts to develop during the deformation of a metal, two factors tend to increase the local stress and thus prevent the development of the neck : (1) strain hardening, and (2) strain-rate sensitivity. Using explanatory sketches and suitable equations: (a) explain the action of strain hardening (which is effective until the strain reaches a value about equal to the strain-hardening coefficient); (b) explain the additional effect due to strain-rate sensitivity. [*Hint*: Compared to the remainder of the specimen, the amount of deformation in the neck is greater; also, the strain rate is correspondingly greater.]

**16.** (a) What is the difference, if any, between the fracture stress and the tensile strength? (b) Compare the direct design significance of these two values.

**17.** (a) Apply the equation Work = $\int$ force $\times$ dx to show that $\sigma^2/2E$ is the work of elastic deformation. Consider that the tensile stress $\sigma$ is gradually applied to unit volume of a bar of metal whose Young's modulus is $E$. (b) Explain by analogy with gravitational potential energy why this calculation gives the strain energy per unit volume. (c) Show that the strain energy $W_e$ in Eq. (4.24) can be considered to come from a cylindrical region surrounding the crack, the diameter of the cylinder being slightly larger than the crack length $2c$.

**18.** Assume that the heat-treated alloy steel of Fig. 4.23(b) is to be used as a 40 mm shaft containing a keyway. (a) What is the (laboratory) endurance limit of this steel? (b) List the factors that would tend to decrease the endurance limit value in this case. (c) Make an estimate of the actual endurance limit of the steel under these conditions.

**19.** What are the service conditions for which (a) yield strength, (b) creep strength, (c) rupture strength is the most useful design value?

**20.** Consider the press forging of a block of 0.45% carbon steel at 980°C (1800°F), where the flow stress $S_0$ is $7 \times 10^7$ N/m$^2$. (a) Calculate the ideal work required to compress a block $100 \times 50$ mm in cross-sectional area from an initial height of 50 mm to a final height of 25 mm. (b) Describe with the aid of sketches the action of the two factors that would cause the *actual work* to be about twice the ideal work obtained in (a). (c) If a 4 kW motor drives the hydraulic press, and assuming an over-all mechanical efficiency of 70%, calculate the maximum number of pieces that could be formed per minute. Use $\alpha = 2$.

**21.** Discuss the interdependence of metallurgical and mechanical factors in bearing design. [For example, consider the fact that the *stress* in any bearing can be made smaller, for the same total load, by increasing the size of the bearing.]

**22.** List the steps that might be taken to make a low-carbon steel more suitable for machining in an automatic screw machine by (a) changes in the work pieces, and (b) changes in the machining procedure.

## REFERENCES

Backofen, W. A., *Deformation Processing*, Reading, Mass.: Addison-Wesley, 1971. Chapters 1–3, 6 and 8 treat the basic equations of plasticity in detail and apply them to several important practical processes.

Cottrell, A. H., *The Mechanical Properties of Matter*, New York: John Wiley, 1964. Chapter 10, "Plasticity" is a lucid summary of the continuum approach to plastic deformation.

Honeycombe, R. W. K., *The Plastic Deformation of Metals*, London: Edward Arnold, 1968. Chapter 14, "Fatigue" and Chapter 15, "Fracture" are especially recommended. The entire book is a detailed description of plasticity from the standpoint of mechanism.

McClintock, F. A. and A. S. Argon (Eds), *Mechanical Behavior of Materials*, Reading, Mass.: Addison-Wesley, 1966. Chapters 5, 8, 13, and 15 to 20 give a detailed treatment of most of the topics considered in this chapter.

McGregor Tegart, W. J., *Elements of Mechanical Metallurgy*, New York: Macmillan, 1966. Chapters 2 and 3 on compressive and torsional testing and applications are recommended supplementary reading.

CHAPTER 5

# METALLOGRAPHY

## INTRODUCTION

In the previous chapters we have studied some aspects of the phases that constitute metallic materials. We have also learned the importance of defects, especially dislocations, in determining the mechanical behavior of a given phase. Numerous methods permit quantitative, or at least qualitative, determination of the various characteristics of phases that are important for technical applications of metals and alloys. Among the methods treated here are microscopic (optical and electron), diffraction (x-ray, electron, and neutron), analytical (x-ray fluorescence, electron microprobe, Auger analysis), and also the powerful deductive technique of stereometric metallography. The information on phase structure obtained by these methods is the basis for the treatments of phase transformations (Chapter 9) and of strengthening mechanisms (Chapter 10).

The ability of the metallurgist to observe structure on an ever-finer scale has increased gradually with time. Even the ancient artisan was able to control the composition of brasses and bronzes by the color of their fractures. Other characteristics such as grain size, porosity, and segregation of a second phase were also observed and were recognized as influencing properties. Modern study of metallic structures (metallography) began in 1808 when Alois de Widmanstätten polished the surface of a meteorite and, etching it, discovered the beautiful geometric pattern that now bears his name, Fig. 5.1. As this meteorite was a nickel-iron alloy, scientists soon deduced that metals are actually crystalline, even though they lack the geometrical surfaces of typical nonmetallic crystals. More powerful methods for investigating the structure of metals were developed over the years: optical microscopy (1841), x-ray diffraction (1913), and electron microscopy (1932).

**Fig. 5.1** Widmanstätten structure on the polished and etched surface of a meteorite. (Courtesy E. P. Henderson, Smithsonian Institution.)

Present methods for investigating the phase structure of metallic materials, including their crystalline defects, can be classified in four categories: (1) Microscopic study of surfaces using the optical microscope, electron microscope, scanning electron microscope, and electron-emission microscope. (2) Microscopic study of internal structure, especially of dislocations, using the transmission electron microscope, radiography, x-ray topology, and the field-emission microscope. (3) Diffraction studies of crystal structure and related features using x-rays, electron beams, and neutron beams. (4) Chemical analyses of the various phases using, in addition to standard analytical techniques, such specialized procedures as x-ray fluorescence, Auger analysis, and the electron microprobe. An extensive literature is associated with each separate technique, so the following brief descriptions can only indicate the general nature of the technique and the major principles underlying its use.

## MICROSCOPIC OBSERVATION OF SURFACES

A microscope magnifies an image, such as a structural feature on the surface of a metal, by using lenses to suitably bend the light (or other radiation). But useful magnification eventually reaches a limit imposed by the nature of the radiation employed. As discussed below in connection with Eq. (5.2), a microscope using radiation of wavelength $\lambda$ cannot resolve details separated by a distance less than about $\lambda/2$. In particular an optical microscope operating with green light ($\lambda = 0.54\,\mu m$) can resolve only those details separated by $0.3\,\mu m$ or more. Since the human eye can resolve fine details separated by 0.1 mm, the *useful* magnification of such a microscope is only about $\times 400$ (although magnifications of twice this value are often employed for ease of viewing). Problem 1 shows that the use of ultraviolet light increases the useful magnification to $\times 800$ and justifies actual magnifications of about $\times 2000$. The limitation of resolving power by the wavelength of the radiation is removed when an electron beam ($\lambda \simeq 0.1\,\text{Å} = 0.1 \times 10^{-10}$ m) is employed. Various design aspects are then responsible for limiting resolution to the range from 2 to 100 Å.

**Optical techniques.** The fact that metals are opaque to light (in contrast to transparent crystals such as quartz) requires that they be examined by reflected light (rather than by transmitted light). Consequently, a metallurgical microscope has a distinctive lighting system, but in other respects its optical characteristics are similar to the more common biological microscope.

*Metallurgical Microscope.* The following procedure is commonly used to prepare a specimen for examination under a microscope, Fig. 5.2. A specimen about 20 mm on an edge is cut from the metal to be examined. (Alternatively, a small specimen can be mounted in a suitable plastic for grinding and polishing.) A mirror polish is produced on one face of the specimen by grinding on an abrasive wheel, polishing on successively finer emery papers, and lapping on revolving cloth-covered wheels

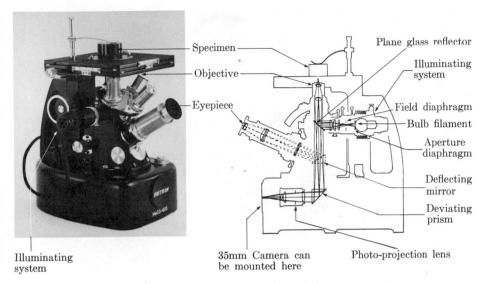

Specimen

Objective

Eyepiece

Plane glass reflector

Illuminating system

Field diaphragm

Bulb filament

Aperture diaphragm

Deflecting mirror

Deviating prism

Illuminating system

35mm Camera can be mounted here

Photo-projection lens

**Fig. 5.2** A metallurgical microscope and the light path through it. (Courtesy Unitron Instrument Company.)

with fine abrasives. To reveal the structural details (grain boundaries, a second phase, or inclusions), this polished surface is *etched* with a chemical solution (such as three percent nitric acid in ethyl alcohol for steel). The etchant attacks various parts of the specimen at different rates and reveals the structure. A metallograph, Fig. 5.3, permits visual observation of a specimen as well as the making of a *photomicrograph*. The photomicrograph in Fig. 5.3 shows a coarse pearlite structure (platelets of $Fe_3C$ in almost pure iron).

*Objective Lens and Resolution.* The most important component in an optical microscope is the objective lens, which gathers the light reflected from the specimen. The ability of a lens to gather light is determined largely by its numerical aperture (N.A.), defined as

$$\text{N.A.} = \mu \sin \theta. \qquad (5.1)$$

Here $\mu$ is the refractive index of the medium [air ($\mu = 1$), or perhaps cedar oil ($\mu = 1.5$)] between the objective lens and the specimen; $\theta$ is the half-angle subtended by the lens from a point on the specimen. A lens with a large value of $\theta$ must therefore operate at a small lens-to-specimen distance. The limit of resolution of closely spaced details is

$$\text{Limit of Resolution} = \frac{\lambda}{2 \times \text{N.A.}}, \qquad (5.2)$$

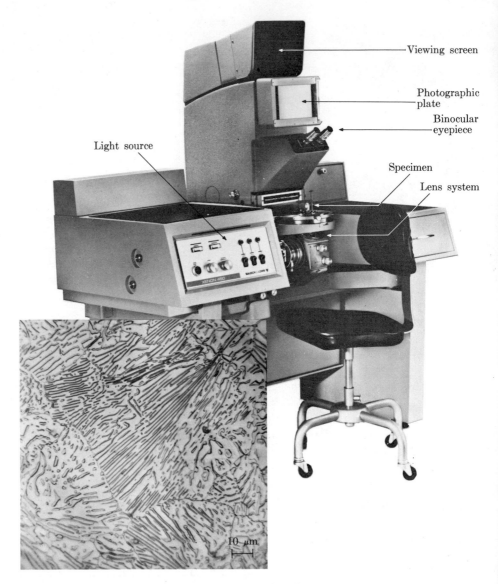

Viewing screen

Photographic plate

Binocular eyepiece

Light source

Specimen

Lens system

10 μm

**Fig. 5.3** A metallograph and a photomicrograph of pearlite in steel made using the metallograph. (Courtesy Bausch & Lomb.)

where $\lambda$ is the wavelength of the light employed. For the usual microscope this limit is about $0.3\,\mu m$ and increases to $1\,\mu m$ for a lens of low numerical aperature (N.A. = 0.3). Consequently, the *total* magnification of a microscope (given by the product of the magnification of the objective and the eyepiece) can easily be made

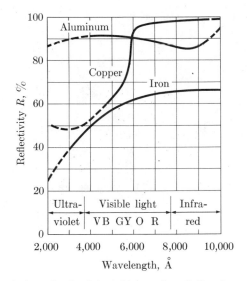

**Fig. 5.4** The variation of reflectivity with wavelength for clean metal surfaces.

equal to or greater than the *useful* magnification necessary to observe the details resolvable by the objective lens.

The *depth of field* of an objective is the distance within which the details are in focus. For low-power lenses (small N.A.) this factor is usually not critical, but with high-power lenses special care must be taken that the surface of the specimen is flat and precisely perpendicular to the lens. If the N.A. is 1.3, for example, the variation in lens-to-detail distance should not exceed 0.4 μm.

The *contrast* between features in the microstructure (for example, between two different phases) as viewed by the usual metallurgical microscope, arises from differences in the intensity of light reflected from various points. These differences have two causes; absorption and geometric effects. Differences in absorption (and therefore in reflectivity) can be traced to the nature of the inter- action of visible light with the surface of the specimen. Each phase has a charac- teristic spectrum of reflectivity, analogous to those shown in Fig. 5.4. Conse- quently, when viewed with light of a given wavelength (such as the filtered green light, $\lambda = 0.54\,\mu$m, commonly employed in photomicrography) the various phases differ in brightness. Other sources of contrast, geometrical effects, occur primarily because of differences in the level (topography) of the surface caused by the etching process. For example, etching usually produces grooves at grain boundaries; therefore, grain boundaries appear dark since the incident light is reflected outside of the lens by these grooves. The contrast in shading among grains of the same phase (such as in a pure metal) arises from tiny crystal facets produced in each grain by etching. Since these facets differ in orientation from one grain to

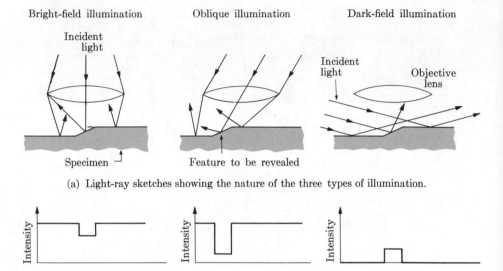

Bright-field illumination    Oblique illumination    Dark-field illumination

(a) Light-ray sketches showing the nature of the three types of illumination.

(b) Distribution of the intensity of light reflected back through the lens.

**Fig. 5.5** Schematic comparison of oblique and dark-field illumination with the usual bright-field illumination. (After D. G. Brandon.)

another, they produce different degrees of scattering of the light and hence different degrees of brightness.

Contrast can be enhanced by the use of oblique or dark-field illumination, Fig. 5.5, instead of the usual bright-field illumination. Oblique incidence of the light is easily produced by suitable displacement of an aperture. The resulting increase in contrast is obtained at the cost of a decrease in N.A. because only a portion of the lens acts to form the image. Dark-field illumination actually reverses the contrast, as shown schematically in Fig. 5.5. Because the incident light makes a small angle with the surface, only irregular features such as grain boundaries are able to reflect the light into the lens. These features then appear light against a dark background. The N.A. is not reduced for dark-field illumination, but the brightness of the image is low in comparison with bright-field illumination.

*Polarizing Microscope.* This microscope, also known as a petrographic microscope, is a standard instrument for examining transparent, anisotropic crystals with transmitted light. Adapted for use with reflected light, it is also a powerful tool for studying anisotropic opaque materials, such as metals and alloys. Two nicol prisms are used: one polarizes the incident light; the other analyzes the reflected light to determine the degree of rotation of the plane of polarization. The amount of rotation depends on the orientation of the anisotropic crystal that reflects the light. Consequently, good contrast is obtained among grains in a

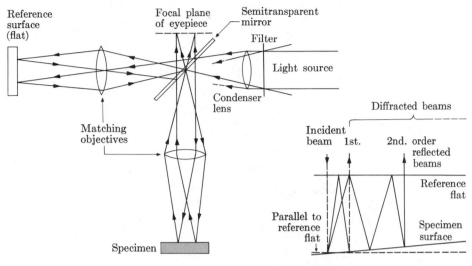

(a) Ray diagram for a two-beam interferometer.

(b) Interference system for multiple-beam interferometry.

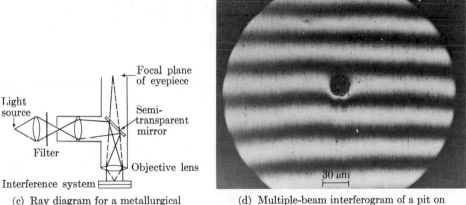

(c) Ray diagram for a metallurgical microscope adapted for multiple-beam interferometry.

(d) Multiple-beam interferogram of a pit on a polished surface of high purity aluminum.

**Fig. 5.6** The nature of interference techniques and a typical interferogram. (Adapted from D. G. Brandon, *Modern Techniques in Metallography*, Princeton, N.J.: Van Nostrand, 1966. Interferogram courtesy H. L. Craig, Jr., Reynolds Metal Company.)

noncubic phase (for example, the HCP form of titanium) by the use of polarized light. Even isotropic phases can be viewed to advantage in this manner provided that an anisotropic (or simply rough) surface film, such as an oxide, is first produced.

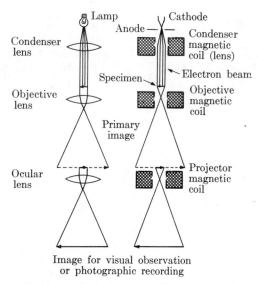

Lamp

Cathode

Anode

Condenser lens

Condenser magnetic coil (lens)

Objective lens

Specimen

Electron beam

Objective magnetic coil

Primary image

Ocular lens

Projector magnetic coil

Image for visual observation or photographic recording

(a) Diagrams illustrating the analogy between the light microscope and an electron microscope.

**Fig. 5.7** Illustration of the principal features of an electron microscope. (Courtesy Siemens Corporation.)

*Interference Microscope.* Precise details of the topography of a surface can be studied by the technique of *optical interferometry.* The simplest interferometer employs the interference between two beams of light, Fig. 5.6(a). The initial beam, from a monochromatic light source, is divided in two equal parts by a 50% transparent mirror. One beam is focused on the specimen and the second beam on an optically flat reference surface. The two reflected beams are then recombined by the beam splitter and pass through the eyepiece together. The two beams reinforce each other for those points on the specimen for which their path lengths are either the same or differ by an integral multiple of the wavelength, $n\lambda$. The beams cancel

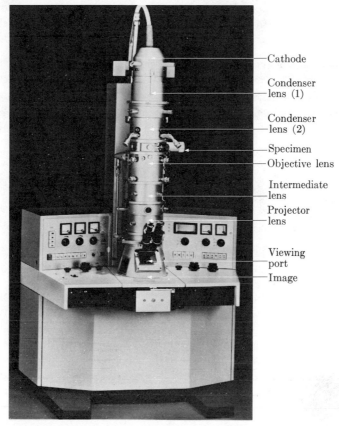

Cathode

Condenser
lens (1)

Condenser
lens (2)

Specimen
Objective lens

Intermediate
lens

Projector
lens

Viewing
port

Image

(b) An electron microscope suitable for both replica and
transmission techniques, Elmiskop 101.

for path differences of $n\lambda/2$. The relatively poor vertical resolution of a two-beam
interferometer, 250 Å, can be increased to 10 Å by use of the multiple-beam
principle, Fig. 5.6(b). The transparent reference surface is placed above the
specimen, and the two surfaces are made closely parallel. When a beam of normally
incident, monochromatic light is reflected from this interference system, the
reflected beam consists of a series of multiply reflected rays from a given point on
the surface. The greater the number of reflections, the greater is the difference in
phase. A metallurgical microscope adapted for multiple-beam interferometry
Fig. 5.6(c), can be used to measure slip steps (in plastically deformed crystals), the

perfection of surface finish or of surface films, and the thickness of vapor-deposited films. A typical interferogram, Fig. 5.6(d), demonstrates that the perfection of a polished surface is impaired only locally by the presence of an etch pit.

**Electron-optical techniques.** The electron beam that replaces the light beam in electron-optical instruments has two important characteristics: its wavelength and its ability to penetrate the specimen. The wavelength is determined by the voltage $V$ employed to accelerate the electrons composing the beam to a velocity $v$. From Eq. (1.2), $\lambda = h/p$. A second equation involving the momentum $p$ (neglecting relativistic effects) can be obtained from the relation,

$$\text{kinetic energy} = eV = \frac{mv^2}{2} = \frac{p^2}{2m}, \tag{5.3}$$

where $e$ is the electronic charge and $m$ is the electronic mass. Elimination of $p$ between these two equations gives the relation,

$$\lambda = \left(\frac{h^2}{2meV}\right)^{1/2} = \left(\frac{150}{V}\right)^{1/2} \text{Å}. \tag{5.4}$$

Thus, electrons accelerated by only 150 volts have, in theory, wavelengths small enough to permit resolution of atomic dimensions. Since the voltages employed in practice range from $10^3$ to $10^6$ volts, the limit of resolution should be on the order of 0.1 Å. Instrumental factors, however, limit the practical resolution to 2–100 Å.

Electrons have only limited ability to penetrate solid materials. The intensity $I$ of an electron beam is decreased by absorption according to the usual law,

$$dI = -\mu I \, dx, \tag{5.5}$$

where $\mu$ is the linear absorption coefficient and $dx$ is an increment of thickness of the absorbing material. The integrated form of Eq. (5.5) is

$$I = I_0 \, e^{-\mu l}, \tag{5.6}$$

where $I_0$ is the intensity of the initial beam and $I$ is the intensity remaining after penetration of a thickness $l$ of material. Although $\mu$ decreases with increase in energy of the electron beam, its value for most metals is so large that the penetration distance is generally in the range of 10–1000 Å (see Problem 2).

*Electron Microscope.* Although the "ray" optics of an electron microscope are analogous to those of a light microscope, Fig. 5.7, an electron beam requires quite a different design. The microscope column must be evacuated to at least $10^{-2}$ N/m² ($10^{-7}$ atm) to prevent appreciable absorption of the electron beam by the residual air. The electrons are supplied from a heated tungsten filament (the cathode) and are accelerated past the anode by a difference in potential of 50–1000 kV. The collimating, objective, and ocular "lenses" are magnetic coils employing soft magnetic pole pieces to concentrate the magnetic field. After

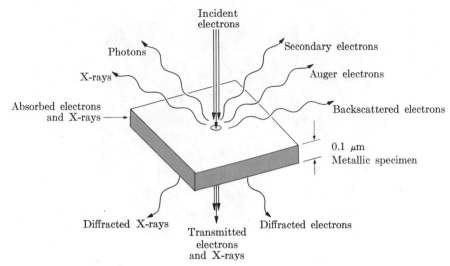

**Fig. 5.8**  Schematic illustration of the various ways in which an incident beam of electrons can interact with a metallic specimen.

collimation, the electron beam interacts with the atoms of the specimen, but an appreciable fraction of the beam succeeds in traversing the entire thickness. The magnetic objective lens then focuses the beam and forms an image magnified by as much as a factor of $10^2$. An additional lens (or lenses) produces further magnification to a total of as much as $\times 250,000$. The electrons finally strike a phosphorescent screen and produce the final image, which may be either viewed through a port or photographed.

Several phenomena occur during the interaction and transmission of an electron beam in a specimen, Fig. 5.8. Various techniques, described in later sections, employ one or more of these phenomena to obtain information about the specimen. The conventional electron microscope, however, uses only the transmitted electrons (replica technique) or the transmitted plus the diffracted electrons (transmission technique). The transmission technique uses a thin ($\sim 0.1\,\mu m$) section of the metal or alloy as the specimen and investigates the *internal* structure. The replica technique, on the other hand, uses only a thin layer of a material of low absorptivity, containing the pattern of *surface* features of the metal in question. Since the present section is devoted to microscopic observations of surfaces, discussion of transmission electron microscopy is deferred to page 223.

*Replica Technique.* The fine surface features produced by etching a metal or alloy can be revealed far more effectively by an electron microscope, Fig. 5.9(a), than by the best optical microscope, Fig. 5.9(b). Because electrons are so strongly absorbed by matter, a *replica* of the surface gives a more satisfactory image than

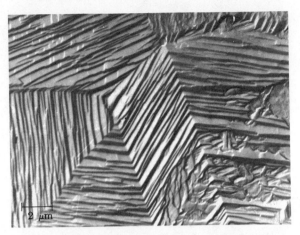

(a)  An electron micrograph of a shadowed carbon replica

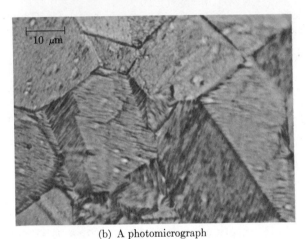

(b)  A photomicrograph

**Fig. 5.9**  Comparison of (a) an electron micrograph of a shadowed carbon replica, and (b) a photomicrograph, both of high-purity nickel. (Courtesy R. D. Heidenreich, Bell Telephone Laboratories.)

does a bulk specimen of the metal itself.  The replica, Fig. 5.10, can be produced by coating the etched surface of the metal with a thin ($\sim 0.1\ \mu m$) film of a suitable plastic or by evaporating a thin layer (of carbon, for example) onto the surface. The replica layer must faithfully mirror the details of the surface and must also remain intact after removal from the metal.  On the average, each electron transmitted through the replica undergoes one scattering event (that is, one interaction) with the atoms comprising the replica.   Since the linear absorption coefficient $\mu$ in Eq. (5.6) increases with increase in atomic number, light elements are chosen for making replicas.  Heavy elements, on the other hand, can be used

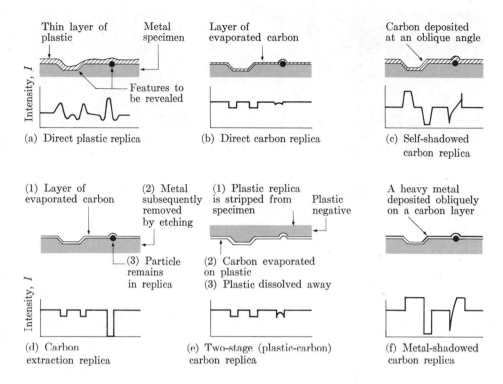

**Fig. 5.10** Schematic illustrations of the manner in which six replication techniques produce contrast; that is, differences in the intensity of the transmitted electron beam. (After D. G. Brandon.)

to enhance contrast by *shadowing* (evaporation at a small angle to the surface) as illustrated in Fig. 5.10(c) and (f). The simple plastic replica, Fig. 5.10(a), and the evaporated carbon replica, Fig. 5.10(b), produce contrast because of differences in thickness. Problem 3 illustrates the use of Eq. (5.6) in calculating contrast, including the effect of shadowing.

*Scanning Electron Microscope (SEM).* This type of microscope observes the surface of a specimen directly. Because its depth of focus is large, the SEM can be used for uneven surfaces such as the corroded alloy shown in Fig. 5.11. The instrument is similar to a conventional electron microscope except that the beam is extremely small and strikes only a tiny fraction of the specimen at any instant, Fig. 5.12. The beam is scanned over the surface of the specimen (in a manner similar to the operation of a television screen), and the resulting *backscattered* and/or *secondary electrons*, Fig. 5.8, form the image on a fluorescent screen.

(a) A large area of the surface viewed at moderate magnification. A feature of special interest is in the dashed box.

(b) Adjustment of the controls of the microscope brings the feature of interest to a high magnification.

**Fig. 5.11**  SEM micrographs of the corroded surface of a specimen of cartridge brass. Of special interest is the hillock of redeposited copper. (Courtesy S. R. Bates, University of Florida.)

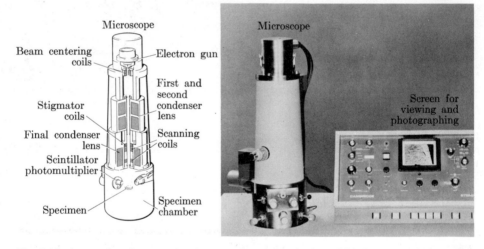

**Fig. 5.12**  A scanning electron microscope and a schematic view of the electron path through it. (Courtesy Kent Cambridge Scientific, Inc.)

Exposure times on the order of one minute may be required to photograph images at the higher magnifications, corresponding to the maximum resolution of about 100 Å.  The primary sources of contrast are differences in direction of scattering because of topography and differences in the efficiency of emission of secondary electrons as a function of composition. Tilting the specimen with respect to the beam gives contrast by shadowing and also permits stereoscopic photography by a three-dimensional effect.

*Electron-Emission Microscope.*  Electrons emitted directly from a specimen can be used to form an image of the surface.  The cause of emission may be thermal

(a)                          (b)                          (c)

**Fig. 5.13** Frames from a motion picture of the nucleation and growth of an annealing twin at the corner of an austenite grain at 980°C (1800°F) in a commercial steel, taken with a thermionic-emission microscope. The elapsed time from (a) to (c) was four minutes. (Courtesy W. L. Grube and S. R. Rouze, General Motors Research Laboratories. By permission from the *Canadian Metallurgical Quarterly*.)

energy, bombardment by ions, or the action of an extremely high electric field at the surface. *Thermionic-emission microscopes* rely mainly on thermal energy to permit the electrons to escape from the surface (treated with an activator, such as barium beryllate) and thus form the electron image. The electrons are then propelled to a fluorescent screen by an electric field through a suitable system of magnetic lenses. Figure 5.13 gives an example of magnified images produced by this technique. Grain contrast, similar to that obtained by etching, occurs because the ease of electron emission varies with orientation of the grains. The limit of resolution is about 200 Å. This microscope is especially useful for observing phase transformations as they actually occur at high temperatures.

In a sufficiently high electric field ($10^9$ V/m) electrons are "pulled" from the surface of the specimen, Fig. 5.14(a), and the corresponding image can be greatly magnified in a *field-emission microscope*. A practical difficulty with this technique is the necessity of having the specimen in the form of a needle with a tip radius of 1 μm or less (to achieve the required electric field). A field emission micrograph has a limit of resolution of 20 Å and reveals considerable detail about the configuration assumed by the crystal planes in a sharply curved surface; information on the electronic configurations in the surface is also obtained. Two sources of contrast are: (a) the variation in ease of electronic emission (or work function) from one crystal plane to another; and (2) the local differences in electric field associated with local differences in radius of curvature; for example, at the transition between adjacent crystal planes.

**Macrostructure.** As the use of a high-power microscope is almost always necessary to see the individual phases or microconstituents in alloys, the naked eye sees only other more obvious structural features. Such *macro*structural details include cracks, gas pockets, segregation of alloying elements, fracture characteristics, and forging flow lines. Three different kinds of procedures are used to observe the

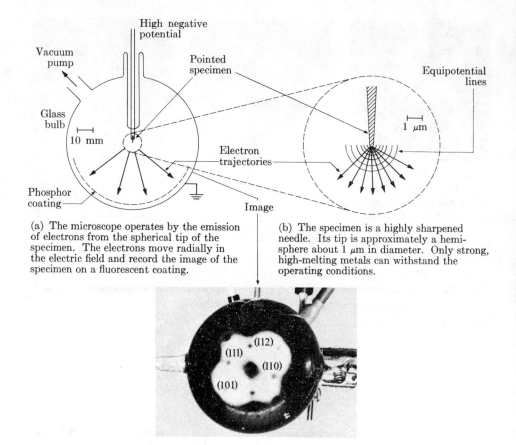

(a) The microscope operates by the emission of electrons from the spherical tip of the specimen. The electrons move radially in the electric field and record the image of the specimen on a fluorescent coating.

(b) The specimen is a highly sharpened needle. Its tip is approximately a hemisphere about 1 μm in diameter. Only strong, high-melting metals can withstand the operating conditions.

(c) A tungsten specimen produced this pattern on a fluorescent coating. The bright and dark regions arise because of different ease of emission of electrons from different crystal planes.

**Fig. 5.14**    Illustration of the principle of operation of a field-emission microscope; a typical FEM image is also shown (courtesy S. S. Brenner).

various types of macrostructure: examination of (1) the specimen surface, (2) a macroetched section, or (3) a fractured section.

*Examination of the Specimen Surface.*    Many defects (large cracks and surface irregularities in castings, for example) are immediately evident on looking at metal parts. However, in critical machine components, such as an aircraft engine crankshaft, even very small cracks may be dangerous. The location of small surface flaws is made easy by the use of magnetic particles for ferromagnetic alloys, and by the use of fluorescent dyes in penetrating oils for both magnetic and non-

  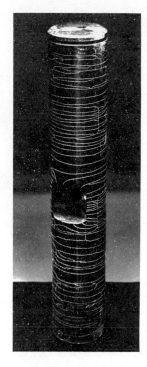

(a) Appearance of king
pin under visual
inspection.

(b) Previously unnoticed
cracks made visible by
the Magnaflux process.

(c) The same cracks
made visible by the
Magnaglo process.

**Fig. 5.15** The use of the Magnaflux and Magnaglo processes to locate cracks in a king pin for a truck. (Courtesy Magnaflux Corp.)

magnetic alloys. Figure 5.15(a) shows a metal surface on which no cracks are visible. The Magnaflux process reveals many networks of cracks, Fig. 5.15(b), as the small magnetic particles collect preferentially at these discontinuities in the surface. In the Magnaglo process, Fig. 5.15(c), a dye-containing oil is allowed to penetrate into the cracks, and they are then seen clearly when the dye is caused to fluoresce in ultraviolet light.

*Examination of a Macroetched Section.* Much information on the quality of hot-rolled steel bars and similar metal products can be obtained by deeply etching (usually in hot hydrochloric acid) a specimen representing a cross section of the part. This technique discloses the following types of macrostructure.

1. Coarse grain structural features such as columnar and dendritic patterns not completely removed by the hot-working process.

**Fig. 5.16** Typical endurance (fatigue) failure in a motorboat shaft containing a keyway. (Courtesy International Nickel Company.)

2. Internal cracks produced in the working process or during heating or cooling.

3. Center porosity, "pipe," or segregation derived from initial ingot defects.

4. Surface defects such as laps, seams, or decarburization produced by improper hot-working practice.

5. Flow lines in a forging; Fig. 4.35 shows both a desirable and an undesirable pattern of flow lines.

The segregation of certain elements such as sulfur and phosphorus can also be determined on a cross-sectional specimen. Photographic paper soaked in suitable reagents is placed on the specimen, and chemical reaction of the reagents with the segregated element produces a permanent record on the paper.

*Examination of a Fractured Section.* When a metal part fails in service, observation of the broken surfaces frequently reveals the nature of the failure. If a service failure is a result of the presence of internal cracks or gas pockets, these imperfections are easily detected in the fractured section. *Fatigue failures*, discussed previously in Chapter 4, have a characteristic appearance. In the fractured shaft shown in Fig. 5.16 the crack began at the keyway. During the slow spreading of the crack into the body of the shaft, rubbing of the two surfaces of the cracked portion produced a smooth area. The "clamshell" markings indicate successive boundaries of the crack. Final brittle failure of the reduced cross section produced the granular portion of the fracture surface.

## DIFFRACTION METHODS

In contrast to microscopic techniques, which study the surface of a specimen, diffraction techniques give information about the internal, crystalline structure

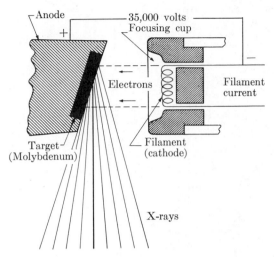

**Fig. 5.17**  The operation of an x-ray tube with a molybdenum target at 35,000 V.

of metals and alloys. As the atomic spacings in crystals are about 3 Å, the most generally useful tools for studying crystal structure make use of radiation (or particles) having a similar, very short wavelength. X-rays are most widely used for this purpose, but beams of electrons or neutrons are also used in special cases. In view of the great practical importance of x-ray diffraction, the theory of diffraction will be presented in connection with the description of this technique. The same principles apply, however, for electron diffraction and for neutron diffraction, which are treated in the following section.

**X-ray diffraction.**   One of the results of the interaction of electrons with a metal, Fig. 5.8, is the production of x-rays. To obtain a beam of x-rays suitable for diffraction techniques, the arrangement shown in Fig. 5.17 is employed. As discussed previously (see Problem 8 in Chapter 1), the electrons must have a relatively high energy (corresponding to a high accelerating voltage). These electrons produce x-rays by two distinct types of interaction with the atoms (ions) in the metal. The first type leads to a weak general (*white*) radiation, Fig. 5.18, characterized by a minimum wavelength $\lambda_{min}$ at one end (see Problem 4) and a gradual tailing off at the high-wavelength end. White x-rays originate from the stepwise dissipation of the energy of the incident electrons within the metal. The rare occurrence of dissipation in a single step corresponds to an x-ray with $\lambda_{min}$. The second type of interaction leads to *characteristic* x-rays, consisting of large peaks of intensity at a few wavelengths characteristic of the target metal. The mechanism in this case involves the actual ejection of an electron from an inner shell; the 1s shell leads to the K-type x-ray spectrum, for example. The refilling of the empty inner shell by an electron from one of various outer shells is accompanied by the emission of energy in the form of an x-ray photon ($K\alpha$ or $K\beta$,

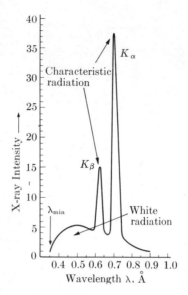

**Fig. 5.18** The range of wavelengths produced when molybdenum metal is the target of an x-ray tube operating at 35,000 V.

for example). An x-ray beam striking a metal or alloy can in turn generate (secondary) x-rays; this process is usually termed fluorescence, and the resulting radiation is called *fluorescent x-rays*. We will now consider some techniques that employ either characteristic or white x-rays.

*Bragg's Law.* Since the wavelength of x-rays is about equal to the distance separating the atoms in solids, diffraction effects are produced when a beam of x-rays strikes a crystalline substance. Although the basic explanation of such a diffraction effect involves the concept of each atom acting as a center of wave propagation, it is more convenient for many purposes to consider entire planes of atoms as mirrors reflecting the incident beam. This simple view assumes that the atoms form a perfect, infinite lattice, but it is satisfactory for many applications. In Fig. 5.19 the horizontal lines represent the cross sections of crystal planes of the set of parallel planes having, in general, Miller indices (*hkl*). When a beam of x-rays with wavelength $\lambda$ strikes this set of crystal planes at some arbitrary angle, there will usually be no reflected beam because the rays reflected from the various crystal planes of the set must travel paths of different lengths. Therefore, although the incident rays are in phase, the reflected rays are out of phase and cancel one another, Fig. 5.19(a). There is one particular degree of "out-of-phaseness" that does not have this destructive effect. If each ray is out of phase with the preceding one by exactly one wavelength (or exactly two, etc.), then there will be a reflected beam consisting of rays that are effectively in phase again, Fig. 5.19(b).

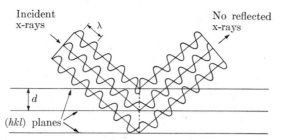

(a) No reflected beam is produced at an arbitrary angle of incidence.

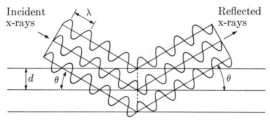

(b) At the Bragg angle, θ, the reflected rays are in phase and reinforce one another.

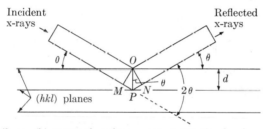

(c) Similar to (b) except that the wave representation has been omitted.

**Fig. 5.19** The reflection of an x-ray beam by the (*hkl*) planes of a crystal.

The angle at which this reflection occurs is known as the Bragg angle, $\theta$.

Figure 5.19(c) shows that the condition for the existence of the reflected beam is that the distance $MPN$ be equal to an integral number of wavelengths; that is,

$$n\lambda = MPN, \tag{5.7}$$

where $\lambda$ is the wavelength and $n$ can have the values 1, 2, 3, etc. Both $MP$ and $PN$ are equal to $d \sin \theta$, where $d$ is the interplanar spacing of the (*hkl*) planes. Therefore the condition for the production of a reflected x-ray beam is

$$n\lambda = 2d \sin \theta. \tag{5.8}$$

This equation is known as *Bragg's law*. Although for a given substance $n$ and $d$ ordinarily have only a few different values, both $\lambda$ and $\theta$ can be varied continuously

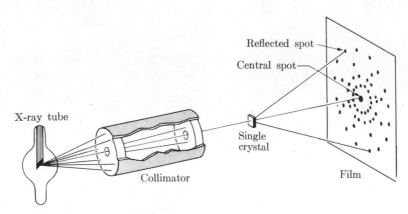

**Fig. 5.20** The Laue method of x-ray diffraction. The reflected spots are visible only after the film has been developed in the usual manner.

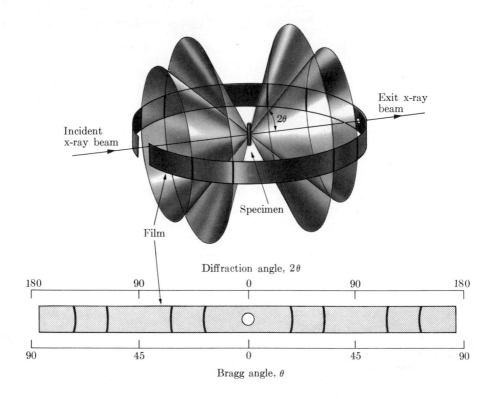

**Fig. 5.21** The production of a powder-method diffraction pattern. Each cone of reflected radiation is recorded in two places on the circular strip of film. (After B. D. Cullity.)

over a wide range. The various x-ray diffraction techniques differ according to whether $\lambda$ or $\theta$ is varied. The *Laue method* is shown in Fig. 5.20. A narrow (collimated) beam of white x-rays is caused to strike a stationary single crystal of the metal to be studied. Each of the planes in the crystal can select the wavelength that satisfies Bragg's law and can reflect this portion of the x-ray beam. Each of the reflected spots indicated on the film in Fig. 5.20 is produced by a different set of planes. The principal metallurgical uses of the Laue method are the determination of crystal orientation and the detection of imperfections in metal crystals.

The *rotating crystal method* uses characteristic (constant wavelength) x-radiation but the diffracting single crystal is rotated, and so values of $\theta$ are obtained that cause reflections to be produced. This method is used mostly for crystal structure determinations.

*The Powder Method*. This method uses an x-ray beam of constant wavelength and a specimen consisting of thousands of tiny crystals. The random orientations of the individual crystals ensure the required variation in angle, and hundreds of reflected beams are produced. As the name *powder method* suggests, the multi-crystal sample is often prepared by reducing larger crystals to powder, but an ordinary piece of metal wire can also be used, since it consists of a large number of very small crystals. Figure 5.21 shows the procedure used in obtaining a typical diffraction pattern from a powder specimen. A narrow x-ray beam strikes the specimen and is diffracted as cones of radiation. The reason for this behavior can be understood by first considering the reflections produced by the crystal planes that have the largest interplanar spacing $d$. Bragg's law shows that these planes will yield reflections at the smallest angle possible for this substance (using the given x-ray wavelength $\lambda$). The Bragg angle $\theta$ of these reflections will have some definite value; it is about $15°$ in Fig. 5.21. However, the angle between the x-ray beam and the reflections is $2\theta$, as shown in Fig. 5.19(c). The crystal planes in question will produce reflections for any orientation around the x-ray beam provided that the beam strikes their surfaces at the Bragg angle. Since many of the tiny crystals in a powder specimen are suitably oriented, the cone of reflection having a semi-apex angle of $2\theta$ is generated as shown in Fig. 5.21. Similar reasoning can be used to explain the production of the remaining cones of reflected radiation by additional sets of crystal planes having progressively smaller spacings. However, since $\theta$.cannot be greater than $90°$, it can be concluded from Eq. (5.8) that crystal planes having an interplanar spacing less than $\lambda/2$ cannot yield a reflection. Each cone of reflected radiation is made up of many reflections from individual crystals, and it is necessary to record only a representative portion of each cone on the x-ray film. Ordinarily two portions of each cone are recorded in order to make it easier to obtain the desired measurements.

Two principal kinds of information can be obtained from a powder diffraction pattern: the *positions* of the reflected beams, which are determined by the size and shape of the unit cell, and the *intensities* of the reflected beams, which are determined by the distribution and kinds of atoms in the unit cell. For simplicity, the present treatment will be restricted to diffraction patterns of pure metals that crystallize in the cubic system. It will then be necessary to consider only the positions of the reflected beams to determine the space lattice and the lattice constant that correspond to the diffraction pattern. This analysis is made easier by use of the following relation between $d$, the spacing of the $(hkl)$ plane, and $a$, the lattice constant of the cubic crystal (see Problem 6):

$$d = \frac{a}{\sqrt{h^2 + k^2 + l^2}}. \tag{5.9}$$

It is conventional to incorporate the order of reflection $n$ in the Miller indices, so that second-order reflection from the (100) plane is considered as first-order reflection from the (200) plane. Such higher-order planes may be fictitious in the sense that they need not have physical existence, but they are used conveniently and correctly in most x-ray analyses.

Bragg's law, Eq. (5.8), can be put into a form useful for the analysis of powder diffraction patterns of cubic crystals by substituting Eq. (5.9) for $d$ and including the order of reflection $n$ in the indices of the reflecting plane. The law then becomes

$$\lambda = \frac{2a}{\sqrt{h^2 + k^2 + l^2}} \sin \theta. \tag{5.10}$$

The square root can be eliminated by squaring both sides of this equation. Letting $Q^2 = h^2 + k^2 + l^2$ and rearranging the factors, the significant result is

$$Q^2 \left(\frac{\lambda^2}{4a^2}\right) = Q^2 C = \sin^2 \theta, \tag{5.11}$$

where $C$ is a constant. That is, the squares of the sines of the angles at which reflections occur are in the ratio of certain whole numbers. This relation among the $\theta$ values can be used to determine the crystal lattice of the metal that produces a given diffraction pattern, since the possible $Q^2$ values are different for the three cubic space lattices (Table 5.1). Each reflecting plane $(hkl)$ has a corresponding value of $Q$. However, not all values of $Q$ can occur in the analysis of diffraction patterns from most space lattices, since certain reflections cannot appear because of destructive interference produced by body- or face-centered atoms. For example, reflection from the (100) plane in the body-centered-cubic lattice cannot occur because the body-centered atoms form intermediate planes that produce reflected waves exactly out of phase with those from the corner atoms.

To illustrate the use of Eq. (5.11) in solving a diffraction pattern, film (d) in Fig. 5.22 will be analyzed. The first step, that of determining the diffraction angle $\theta$ associated with each line in the pattern, can be carried out using an equation derived with the aid of Fig. 5.23. The angle that the reflected beam makes with the exit beam is $2\theta$. Expressed in radians, $2\theta = s/r$, where $s$ is measured on the flattened film and is one-half the distance between the two arcs of the cone of reflected radiation; $r$ is the radius of the camera. It is more convenient to express $\theta$ in degrees and to measure $2s$ directly as the distance between the two arcs of a

**TABLE 5.1**

Reflecting Planes for the Three Cubic Space Lattices

| Reflecting plane (hkl) | $Q^2$ $(h^2 + k^2 + l^2)$ | Simple cubic | Body-centered cubic | Face-centered cubic |
|---|---|---|---|---|
| (100) | 1 | * | | |
| (110) | 2 | * | * | |
| (111) | 3 | * | | * |
| (200) | 4 | * | * | * |
| (210) | 5 | * | | |
| (211) | 6 | * | * | |
| | | | | |
| (220) | 8 | * | * | * |
| (300)(221) | 9 | * | | |
| (310) | 10 | * | * | |
| (311) | 11 | * | | * |
| (222) | 12 | * | * | * |
| (320) | 13 | * | | |
| (321) | 14 | * | * | |
| | | | | |
| (400) | 16 | * | * | * |
| (410)(322) | 17 | * | | |
| (411)(330) | 18 | * | * | |
| (331) | 19 | * | | * |
| (420) | 20 | * | * | * |
| (421) | 21 | * | | |
| (332) | 22 | * | * | |

* Indicates the occurrence of reflection.

given cone of reflected radiation. Therefore the expression for $\theta$ is usually written

$$\theta = \frac{s}{2r} \times \frac{360}{2\pi} = 14.32 \frac{2s}{r} \text{ degrees.} \qquad (5.12)$$

The value of $2s$ measured for each line of the pattern in Fig. 5.22(d) and the corresponding $\theta$, $\sin \theta$, and $\sin^2 \theta$ can be tabulated conveniently as follows:

| Line | 2s, mm | $\theta$, degrees | $\sin \theta$ | $\sin^2 \theta$ |
|---|---|---|---|---|
| 1 | 41.9 | 23.6 | 0.400 | 0.1600 |
| 2 | 60.9 | 34.4 | 0.565 | 0.3192 |
| 3 | 77.7 | 43.8 | 0.692 | 0.479 |
| 4 | 94.2 | 53.1 | 0.800 | 0.640 |
| 5 | 112.3 | 63.3 | 0.893 | 0.797 |
| 6 | 139.0 | 78.4 | 0.980 | 0.960 |

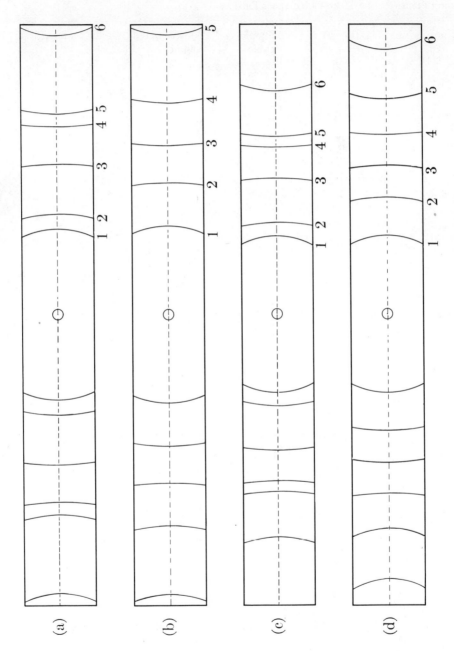

**Fig. 5.22** Simplified sketches of x-ray diffraction patterns obtained from wires of four different metals using a 50.8 mm diameter camera and cobalt $K\alpha$ radiation, $\lambda = 1.786$ Å.

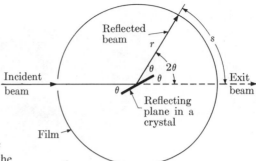

**Fig. 5.23** The relation between the angle of reflection, $\theta$, from a crystal plane and the distance $s$ measured on the film.

These six lines result from the first six planes that can cause reflection (Table 5.1). Depending on the space lattice of the unknown metal that produces the diffraction pattern, these first six planes may have $Q^2$ values of 1, 2, 3, 4, 5, 6 (simple cubic), 2, 4, 6, 8, 10, 12 (body-centered cubic), or 3, 4, 8, 11, 12, 16 (face-centered cubic). But Eq. (5.11) shows that the ratio of the $Q^2$ values is the same as the ratio of the $\sin^2 \theta$ values. Therefore, the space lattice of the unknown metal can be determined from the ratio of $\sin^2 \theta$ values. In the pattern being considered the $\sin^2 \theta$ values are seen to be very nearly in the ratio $1:2:3:4:5:6$. However, the diffraction pattern is that of a pure metal, and the true ratio of $Q^2$ values is almost certainly $2:4:6:8:10:12$, inasmuch as a simple cubic lattice is almost unknown in pure metals. From this ratio it can be concluded that the metal has a body-centered-cubic space lattice.

The next step is to determine the lattice constant $a$. Writing each value of $\sin^2 \theta$ in the form of Eq. (5.11) gives

| Line | $\sin^2 \theta$ | = | $Q^2 \times C$ |
|------|------|------|------|
| 1 | 0.1600 | | $2 \times 0.0800$ |
| 2 | 0.3192 | | $4 \times 0.0798$ |
| 3 | 0.479 | | $6 \times 0.0798$ |
| 4 | 0.640 | | $8 \times 0.0800$ |
| 5 | 0.797 | | $10 \times 0.0797$ |
| 6 | 0.960 | | $12 \times 0.0800$ |
| | | | 0.0799 average |

Since $C$ is equal to $\lambda^2/4a^2$ (Eq. (5.11)), and since in this case the wavelength used is $\lambda = 1.786$ Å, the value of $a$ is given by

$$a = \frac{\lambda}{2\sqrt{C}}, \qquad a = \frac{1.786}{2 \cdot 0.2826} = 3.16 \text{ Å}. \tag{5.13}$$

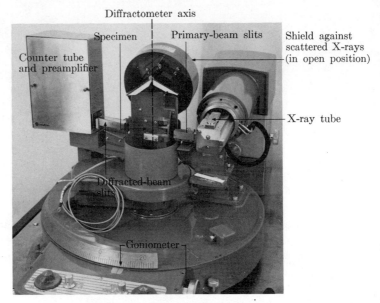

Diffractometer axis

Specimen    Primary-beam slits    Shield against
scattered X-rays
(in open position)

Counter tube
and preamplifier

X-ray tube

Diffracted-beam
slits

Goniometer

(a)  An x-ray diffractometer.

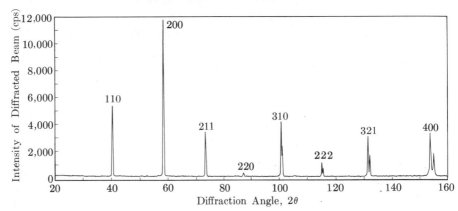

(b) Record of diffraction angles for a tungsten sample obtained by use of a diffractometer with copper radiation.

**Fig. 5.24**   The diffractometer method of x-ray diffraction.  (Courtesy General Electric Co.)

Table 1.2 shows that tungsten is the body-centered-cubic metal that has a lattice constant of 3.16 Å, and therefore tungsten is the pure metal that produces the diffraction pattern of Fig. 5.22(d).

Analyses of this type are used for many purposes, for example, to determine what kinds of solid phases (what kinds of crystal structures) are present in a given

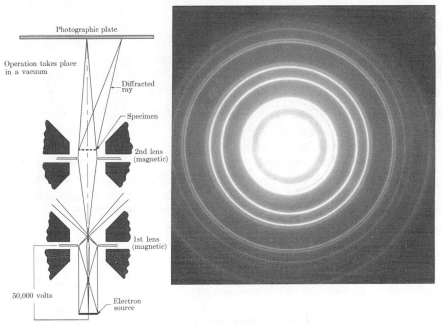

(a) Schematic illustration of the principle of transmission electron diffraction.

(b) The diffraction pattern obtained from a thin, polycrystalline specimen of gold.

**Fig. 5.25** The technique of transmission electron diffraction and a typical diffraction pattern. Specimens must be only 0.1 µm in thickness. (Courtesy James Hillier, RCA Laboratories.)

alloy. If two different kinds of crystals coexist, as in many industrial materials, each diffraction pattern appears on the film independently of the other. Inasmuch as almost all inorganic chemical compounds are crystalline substances, chemical analyses can be made by means of x-ray diffraction. Another important application of x-ray techniques is the investigation of reactions in solid metals that lead to softening, hardening, or to other property changes. The magnitude and condition of stress in metal grains can also be determined by this means.

In addition to photographic techniques, such as the powder method discussed above, there are methods of analysis that detect x-rays by the use of counter tubes. An x-ray *diffractometer*, Fig. 5.24(a), measures the intensity of the beam (counts per second) diffracted from a specimen over a range of angles and can be arranged for automatic recording of the data on a chart like that shown in Fig. 5.24(b) for a powder specimen. The $\theta$ values so obtained can then be analyzed in the same manner as in the photographic method.

**Electron and neutron diffraction.** These two techniques, although basically similar to x-ray diffraction, complement it in two important ways. Because electrons are strongly absorbed by metals, the corresponding diffraction pattern can give useful information about the surface layers of a metal. The technique of electron diffraction is about as simple as x-ray diffraction and is widely employed. Neutron diffraction, on the other hand, is a highly specialized technique requiring the use of a nuclear reactor. It has the advantage of relying on nuclear properties for a major fraction of the intensity of the diffracted beam. Consequently, unusual features of metallic structure can be studied by this technique, as described below.

*Electron Diffraction.* Figure 5.25(a) shows a typical arrangement for obtaining characteristic patterns of crystalline substances by the use of a beam of electrons. Electrons from a heated filament are given the desired velocity by being accelerated by a potential of 50,000 V. After the characteristics of the electron beam have been improved through the action of two magnetic lenses, the beam is then allowed to strike the specimen and produce a diffraction pattern on the photographic plate. The wavelength associated with electrons accelerated by 50,000 V is about one-twentieth that of x-rays. Thus, although Bragg's law also holds for electron diffraction, the diffraction angle is very much smaller than in the case of x-rays. A transmission pattern for gold is shown in Fig. 5.25(b). Electron diffraction can be used like x-rays to identify substances and to obtain information on the grain size, the state of stress, and the orientation of crystal grains.

In connection with transmission electron microscopy (see page 223), the technique of selected-area diffraction (SAD) has great importance. While a specimen is being viewed in the microscope, Fig. 5.26(a), a small area can be selected and studied by electron diffraction. For example, Fig. 5.26(a) shows how a bright-field image is obtained and Fig. 5.26(b) is an example of the detail that can be observed in a steel. (The martensite structure in steel is discussed in Chapter 10.) If the microscope is adjusted as shown in Fig. 5.26(e), then a diffraction pattern can be obtained of the same field of view, Fig. 5.26(f). Through the use of dark-field technique, Fig. 5.26(c), each diffraction spot can be assigned to the corresponding microstructural feature, and its structure can be determined.

Instead of an electron microscope, special instruments are sometimes used to obtain electron-diffraction patterns under unusual conditions such as ultra-high vacuum or with improved possibilities for analysis. HEED instruments (high-energy electron diffraction) are used in transmission with thin specimens; Fig. 5.25 is an example of this technique. Similar instruments designed for the so-called *reflection* technique are termed RHEED. In this case the incident electrons make a very small angle with the surface: They penetrate only the outermost layer where they are diffracted (not reflected). The resulting diffraction pattern is characteristic of the material in the surface layers only. The LEED instrument (low-energy electron diffraction) is another means for studying the atoms at the surface, in this case, because the low energies of the electrons (10–1000 V) do not permit appreciable penetration into the specimen. The backscattered electrons (Fig. 5.8) employed in this technique form a pattern that can be interpreted similarly to HEED patterns. In the usual experimental arrangement, Fig. 5.27(a), a collimated

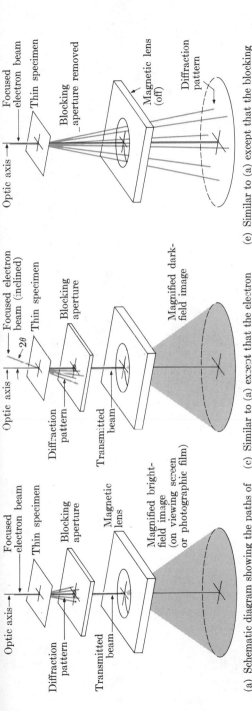

**Diagram (a) labels:** Optic axis · Focused electron beam · Thin specimen · Blocking aperture · Magnetic lens · Diffraction pattern · Transmitted beam · Magnified bright-field image (on viewing screen or photographic film)

**Diagram (c) labels:** Optic axis · Focused electron beam (inclined) · $2\theta$ · Thin specimen · Blocking aperture · Diffraction pattern · Transmitted beam · Magnified dark-field image

**Diagram (e) labels:** Optic axis · Focused electron beam · Thin specimen · Blocking aperture removed · Magnetic lens (off) · Diffraction pattern

(a) Schematic diagram showing the paths of the electrons during the formation of a magnified bright-field image. All diffracted beams are blocked at the aperture plane; only the direct beam can pass.

(c) Similar to (a) except that the electron beam is tilted. Consequently, one of the diffracted beams can be magnified to form the final (dark-field) image. The diffracted beam passes undistorted down the optic axis.

(e) Similar to (a) except that the blocking aperture has been removed. The magnetic (intermediate) lens is not used; therefore, the entire diffraction pattern is seen in the viewing screen.

(b) A bright-field micrograph of steel containing plates of martensite with internal twins. Carbides particles have precipitated at the matrix/twin interface.

(d) This micrograph shows the same area as in (b). Use of the dark-field image of the $(0\bar{2}0)$ spot in the twin causes the contrast of the twins to reverse.

(f) A selected-area-diffraction (SAD) pattern of the area shown in (b).

**Fig. 5.26** Illustration of the three modes of operation of an electron microscope: to produce a magnified bright-field image ((a) and (b)), a magnified dark-field image ((c) and (d)), and a diffraction pattern ((e) and (f)). (Micrographs courtesy D.-H. Huang and G. Thomas. By permission from *Metallurgical Transactions*.)

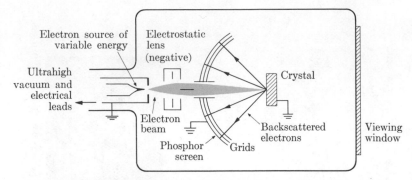

(a) Schematic illustration of the principle of operation of a LEED apparatus, permitting the observation of backscattered electrons. These electrons reveal the crystal structure of the surface of the specimen.

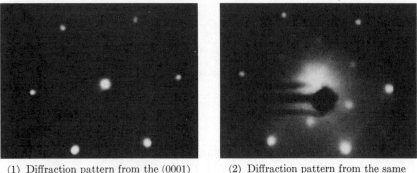

(1) Diffraction pattern from the (0001) surface of cleaned titanium.

(2) Diffraction pattern from the same surface after formation of the $TiO_3$ structure.

(b) LEED patterns are determined by the atomic arrangements in the surface layers. Pattern 1 reflects the hexagonal arrangement in titanium metal. Pattern 2 reveals the presence of an ordered distribution of oxygen atoms. (Courtesy G. W. Simmons and E. J. Scheibner.)

**Fig. 5.27** A description of the technique of low-energy electron diffraction (LEED) and an example of its application to study surface structure.

beam of electrons is backscattered from the specimen. Two or more grids between the specimen and the fluorescent screen pass only those electrons having a desired range of energies. Typical LEED patterns, Fig. 5.27(b), are those obtained from a cleaned metal surface (titanium) and from the same metal containing a large fraction of impurity atoms (oxygen) on its surface.

*Neutron Diffraction.* In contrast to x-ray and electron diffraction, both of which are important industrial techniques, neutron diffraction is still in the laboratory stage. Some of the practical problems in the use of this method become apparent in the sketch of a typical experimental arrangement, Fig. 5.28. A nuclear reactor

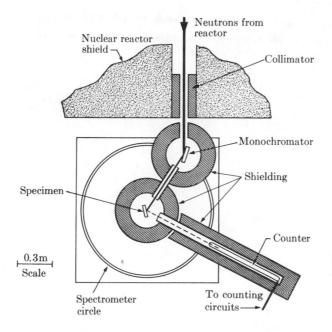

**Fig. 5.28**  An apparatus for neutron diffraction.  (After G. E. Bacon.)

is needed for the intense initial source of neutrons.  However, these neutrons are white radiation containing many wavelengths and for most purposes the characteristic radiation needed is obtained by reflection from a monochromator, often a single crystal of lead.  The intensity of the initial neutron beam is reduced by a factor of about $10^3$ by this operation, and after further diffraction by the sample the final beam is quite weak.  An efficient, well-shielded counter is required for adequate analysis of this beam using the diffractometer technique that was described for x-rays.

Neutron diffraction has advantages over the other two techniques because of differences in the source of diffraction.  The nucleus causes most of the diffraction of neutrons, while electrons are primarily responsible for diffraction of x-rays and electron beams.  Since the number of electrons increases uniformly with atomic number, the diffracting power of atoms for x-rays and electrons exhibits a similar increase.  Two consequences of this fact are especially noteworthy.  Light elements, such as hydrogen and carbon, have little diffracting power compared with heavier elements and so it is difficult to study structures such as metal hydrides containing both heavy and light atoms.  A second problem arises in studying structures made up of atoms whose atomic numbers are nearly the same.  For example, iron and cobalt have nearly identical scattering powers for x-rays, so it is difficult to determine by x-rays the positions in the crystal structure occupied by each kind of atom.

In contrast to the regularity in scattering power for x-rays, the various elements show an irregular variation in neutron-scattering power. The following representative values are from the reference by Bacon.

| Element | Relative neutron-scattering power | Relative x-ray-scattering power |
|---|---|---|
| Hydrogen | 1.8 | 1 |
| Deuterium | 5.4 | 1 |
| Beryllium | 7.7 | 4 |
| Carbon | 5.4 | 6 |
| Magnesium | 3.6 | 12 |
| Aluminum | 1.5 | 13 |
| Titanium | 1.8 | 22 |
| Iron | 11.4 | 26 |
| Cobalt | 1.0 | 27 |
| Molybdenum | 5.5 | 42 |
| Tungsten | 2.7 | 74 |

These values show clearly that iron and cobalt atoms can be distinguished readily by neutron diffraction. Also, since light atoms such as hydrogen and beryllium have neutron-scattering powers that are comparable to those of heavy atoms, neutron diffraction is a useful tool for studying metal hydrides, beryllides, and the like. In addition to the effect of the nucleus, an important contribution to the neutron-diffraction pattern is made by the electrons in unfilled shells in magnetic substances.

## DETERMINATION OF CHEMICAL COMPOSITION

The standard methods for determining the composition of metals and alloys are the usual "wet" chemical procedures and various instrumental methods such as spectroscopy. These methods are convenient for bulk materials and for determining *average* compositions (for example, of a steel consisting of both ferrite and carbide phases). Special methods, however, are able to provide such valuable information as the *individual* compositions of the carbide phase and of the ferrite phase as they exist in intimate contact in the fine-scale microstructure of the steel. Several such special analytical tools are described in this section.

**Electron-beam microprobe.** The essential idea of the microprobe, Fig. 5.29(a), is that an ordinary metallographic specimen can be viewed under a microscope and a selected point can be subjected to an extremely fine beam of electrons. Electrons striking a given kind of atom cause it to emit characteristic x-rays of definite wavelength. Also, the more atoms of a given kind are present, the greater is the intensity of the corresponding x-rays. Thus, by setting the analyzing crystal at the

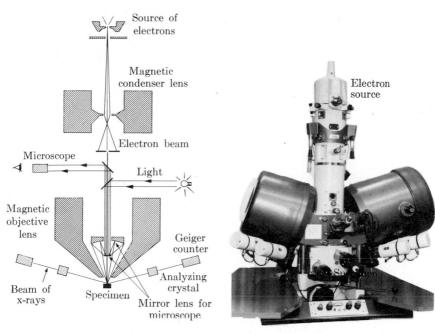

(a) The principle of operation.

(b) A commercial instrument, the Cameca Electron Microprobe. (Courtesy Cameca Instruments, Inc.)

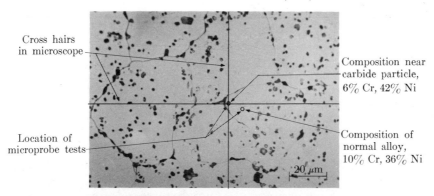

(c) An example of chemical analysis at two points in the micro-structure of a nickel-chromium steel. (Courtesy J. Philibert.)

**Fig. 5.29** Use of the electron-beam microprobe as a metallographic tool.

Bragg angle for a given element, say chromium, and then measuring with the Geiger counter the intensity of the emitted x-rays, it is possible to determine the quantity of any element at the selected point in the alloy.

When a specimen is viewed through the microscope, it is seen as an ordinary microstructure (Fig. 5.29(c)). In this case a special heat treatment, simulating injurious conditions of use, has caused the precipitation of carbide particles within the otherwise seemingly homogeneous alloy. Using cross hairs to locate a point of interest for testing (near a carbide particle in a grain boundary in our case), one can then cause the electron beam to fall on an area of the size shown by the small dot of diameter one micron. By setting the analyzing crystal at the x-ray wavelength for chromium and then for nickel, the composition values at the given point can be determined. In the example illustrated, it was found that there are zones of heterogeneity around the carbide particles. These zones are impoverished in chromium but enriched in nickel compared to the normal alloy located at some distance from carbide particles. The diffusion process underlying this phenomena will be discussed in Chapter 8.

**X-ray fluorescence analyzer.** The principle of operation of this method is similar to that of the electron-beam microprobe; the essential difference is that the average composition of a macroscopic specimen is determined. A large, powerful beam of x-rays strikes an area about $10\,mm^2$ and generates the characteristic radiations of the various kinds of atoms in the sample. Elements down to atomic number 9 (fluorine) can be analyzed, and an amount as small as $10^{-14}$ g can be detected in the sample. As with the microprobe, quantitative analysis of a given element requires a measurement of the intensity of one of the characteristic x-ray wavelengths of this element. Because many factors determine the measured intensity (including absorption of the x-rays by several causes) comparative measurements on reference standards of known composition are commonly employed as the basis for quantitative analyses.

**Auger analysis.** One of the results of electron bombardment of a metal, Fig. 5.8, is the ejection of *Auger electrons*. A beam of high-energy electrons first ejects electrons from an inner shell of an atom in the specimen. This excited atom may then return to its normal state by a two-step process; not only is a characteristic x-ray photon emitted, but also a weakly bound (Auger) electron is ejected from an outer shell. The Auger electrons have such low energy that they can escape with unchanged energy only if they are within a few atomic distances of the surface of the metal. Since each element produces Auger electrons of definite energies, observation of the spectrum of these energies is the basis of a method of chemical analysis. Auger spectroscopy has the basic advantage of analyzing only those atoms at the surface of a specimen. Thus, if an impurity atom is adsorbed only on the surface, it can be detected by this method even if it occupies as little as 1% of the surface (1% of a monolayer). Light elements such as carbon, oxygen, and nitrogen are common contaminants on metal surfaces, but their detection

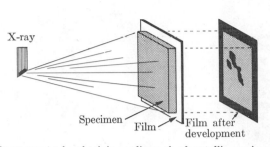

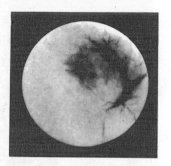

(a) Arrangement for obtaining radiograph of metallic specimen.

(b) Radiograph showing shrink-age in brass casting (courtesy General Electric X-ray Corporation).

**Fig. 5.30** Illustration of the radiographic technique. Irregular dark areas in both (a) and (b) indicate voids in the metal.

is almost impossible by the usual x-ray techniques. Fortunately, Auger spectoscopy becomes more sensitive as the energy of the incident x-rays is decreased; only relatively weak x-rays are sufficient to excite the lighter elements. This method was able, for example, to detect traces of carbon, oxygen, and sulfur atoms on the surface of the titanium specimen of Fig. 5.27 even after special efforts at producing an ultra-clean surface

## OBSERVATION OF INTERNAL STRUCTURE

In contrast to the techniques described previously, which study internal structure either indirectly or by deduction from surface patterns, several methods permit direct observation of internal structure in bulk metallic specimens. X-ray methods have the advantage of relatively high penetrating power, but a serious disadvantage is the fact the x-rays, unlike light or an electron beam, cannot be appreciably bent by a system of lenses. Consequently, only moderate magnifications are attainable. In spite of the low penetrating power of electrons, transmission electron microscopy (TEM) is the most widely employed technique and is capable of high magnifications. Still higher magnifications are attainable with field-ion micro-scopy (FIM), but only the surface layer of atoms can be examined. Information on the three-dimensional structure can be obtained, however, by stripping layer after layer off the surface. These direct methods for observing internal structure will now be described.

**Radiography.** One of the outstanding characteristics of x-rays is their ability to pass through objects opaque to ordinary light. Fortunately this ability is limited and varies in degree from one substance to another; therefore it is possible to obtain x-ray photographs showing the degree of uniformity of such articles as metal castings and welded assemblies. Figure 5.30(a) illustrates one method of obtaining such a *radiograph*. In the example shown, certain areas of the metallic

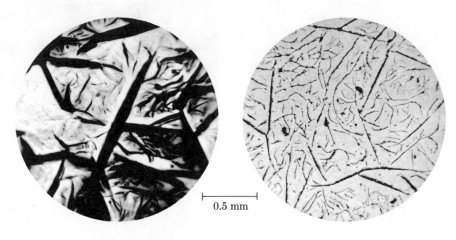

(a) A microradiograph.                    (b) The corresponding photomicrograph

**Fig. 5.31** Comparison of (a) a microradiograph and (b) a photomicrograph of an unetched gray cast iron. (Courtesy Naval Research Laboratory.)

specimen transmit the x-rays more readily than others, resulting in nonuniform darkening of the photographic film. It can be inferred that the more "transparent" areas of the metal contain hidden nonuniformities, such as holes produced in the casting process. An actual radiograph of a brass casting is shown in Fig. 5.30(b).

Radiography permits the examination of large metal specimens for gross defects that make up more than about 2% of the thickness, but this method fails to reveal the finer details of metal structure. For this purpose a refined version of the same technique, called *microradiography*, is used successfully. Only a small area of a specimen that has been reduced to a thickness of about 0.1 mm is examined, usually with x-rays of long wavelength. When the resulting microradiograph is magnified about one hundred diameters, the distribution of the different constituents of the alloy can be seen, as shown in Fig. 5.31.

Although a microradiograph gives a picture similar to a photomicrograph, it differs from the latter in depending on the whole volume of the specimen rather than merely on the surface. Also, contrast between the constituents of an alloy is produced, not as a result of etching, but because of differences in absorption coefficients. In Fig. 5.31(a) the graphite flakes appear dark because graphite absorbs x-rays comparatively weakly. Even if two constituents have generally similar absorption coefficients, good contrast can be obtained if the x-ray wavelength is chosen to lie between their absorption edges. Consideration of Table 1.2 shows that copper $K_\alpha$ radiation ($\lambda = 1.54\,\text{Å}$) would be suitable in this respect for a microradiograph of an iron-copper alloy.

A related metallographic technique, *autoradiography*, involves radioactive isotopes rather than x-rays. The term "auto" (that is, "self") is used because the

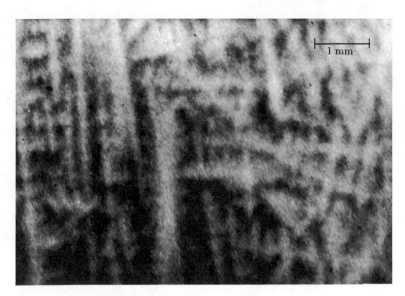

**Fig. 5.32** Autoradiograph showing interdendritic segregation of manganese in a steel casting. (Courtesy R. G. Ward. Used by permission from *The Physical Examination of Metals*, 1960, Edward Arnold, London.)

decomposing isotopes give directly recordable evidence of the distribution of the given element in the microstructure. In this technique the first step is the introduction of the radioactive atoms of the element to be studied; for example, manganese in steel. Direct addition of radioactive manganese to the liquid steel is the usual procedure in this case, but often it is possible to accomplish the same purpose by irradiating a solid alloy in a nuclear reactor and then selecting conditions that give a photographic record of the desired element. If the activated sample is then placed in close contact with a photographic film, electrons produced during disintegration of the radioactive atoms make a record of the location in the alloy of the element under study. Figure 5.32 shows an autoradiograph obtained in this way in a study of the distribution of manganese in a steel casting. Since the photographic darkening occurs in the spaces between the dendrites, it can be concluded that the manganese segregates into these interdendritic spaces.

**Nondestructive testing.**    There are a number of techniques in addition to radiography for determining whether a metal part contains internal flaws (techniques that do not require internal access). Magnetic and electrical methods have been developed for a number of applications, but the use of high-frequency sound waves has proved most generally useful. Figure 5.33 illustrates the principle of this *ultrasonic* inspection. A combination generator and receiver of ultrasonic waves obtains a wave pattern from the metal specimen being tested. This pattern is caused to appear on the screen of a cathode-ray tube, and the presence of a flaw is indicated by the appearance of extra reflections on the screen.

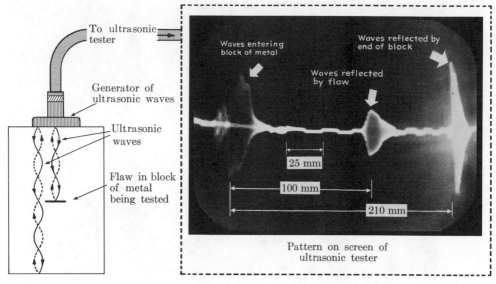

To ultrasonic
tester

Generator of
ultrasonic waves

Ultrasonic
waves

Flaw in block
of metal
being tested

Waves entering
block of metal

Waves reflected by
end of block

Waves reflected
by flaw

25 mm

100 mm

210 mm

Pattern on screen of
ultrasonic tester

**Fig. 5.33** The principle of ultrasonic flaw detection. (Ultrasonic Reflectoscope pattern, courtesy Sperry Products, Inc.)

**X-ray-diffraction microscopy.** This technique and TEM (transmission electron microscopy) permit a study of linear defects (dislocations) and planar defects in the interior of a metal. Since these defects determine most of the technically important properties of alloys, their microscopic study receives much attention. The operating principle of "defect microscopy" is the same for both techniques, but it will be described here using an x-ray beam as an example. The essential difference in the case of an electron beam is the fact that the beam can be bent by a system of lenses, both before and after passage through the specimen.

If a monochromatic beam of x-rays of intensity $I_0'$ strikes a thin crystal, the beam will generally pass through the crystal and emerge as the transmitted (or *direct*) beam of intensity $I_0$. (The effect of absorption in decreasing the intensity is not important for the present discussion.) An unusual behavior, however, occurs if a set of crystal planes, denoted by the general indices $(hkl)$, is at the Bragg angle $\theta$ for reflection. In this case the initial beam is so strongly diffracted that the intensity of the direct beam is only $I_0 - I_D$, the balance of the intensity now being in the diffracted beam, $I_D$. Depending on which beam is employed, two types of microscopic images can be formed (see Fig. 5.26): *Bright-field image* employs the direct beam; *dark-field image* employs the diffracted beam. The following discussion uses the dark-field image as the primary example, but the bright-field image is correspondingly affected by the conditions in question. In TEM, bright-field images are actually employed more frequently.

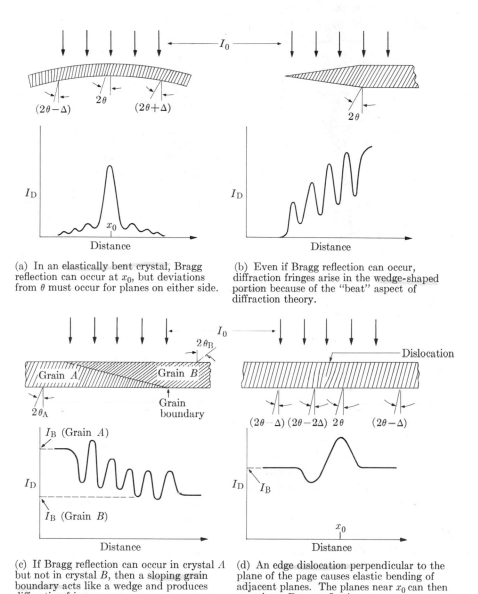

(a) In an elastically bent crystal, Bragg reflection can occur at $x_0$, but deviations from $\theta$ must occur for planes on either side.

(b) Even if Bragg reflection can occur, diffraction fringes arise in the wedge-shaped portion because of the "beat" aspect of diffraction theory.

(c) If Bragg reflection can occur in crystal $A$ but not in crystal $B$, then a sloping grain boundary acts like a wedge and produces diffraction fringes.

(d) An edge dislocation perpendicular to the plane of the page causes elastic bending of adjacent planes. The planes near $x_0$ can then experience Bragg reflection.

**Fig. 5.34** Schematic illustrations of the dark-field images produced in several important cases encountered in x-ray-diffraction microscopy and in TEM. $I_0$ is the incident intensity, $I_D$ is the diffracted intensity, and $I_B$ is the background intensity.

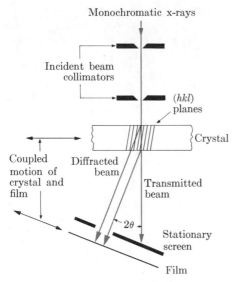

Monochromatic x-rays

Incident beam
collimators

(hkl)
planes

Crystal

Coupled     Diffracted
motion of     beam
crystal and
film

Transmitted
beam

2θ

Stationary
screen

Film

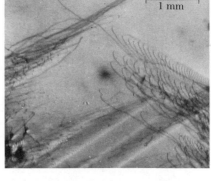

1 mm

(a) Schematic illustration of the technique, showing the method for obtaining a serial image of the diffracted beam for the entire crystal. (After L. V. Azároff.)

(b) Micrograph showing a number of dislocations in a single crystal of copper. (Courtesy F. W. Young, Jr., Oak Ridge National Laboratory.)

**Fig. 5.35**  The Lang technique of x-ray diffraction and a micrograph produced by its use.

The reason for the production of images of defects by *diffraction contrast* can be understood by considering a few important examples. If a crystal is bent elastically, Fig. 5.34(a), then even if the planes near $x = x_0$ are in the correct position for diffraction of the incident beam, neighboring planes are *not* in position for diffraction. The simple Bragg-law analysis then predicts a sharp maximum at $x_0$, which is actually observed. The additional oscillations of intensity, however, can be explained only by a more complex analysis of the diffraction phenomena. Often these oscillations are of secondary importance, but in the case of a wedge-shaped crystal, Fig. 5.34(b), or the closely analogous sloping grain boundary, Fig. 5.34(c), the entire diffraction effect arises from this source. If an edge dislocation is present in a crystal which is almost in position for Bragg reflection, Fig. 5.34(d), then the elastically bent planes lying on one side of the dislocation (at $x_0$ in the figure) may come into position for reflection. Thus, the image in this case is not precisely at the location of the dislocation but slightly to one side.

A useful technique of x-ray diffraction microscopy is the *Lang method*, Fig. 5.35(a), in which the mechanisms of diffraction contrast described in Fig. 5.34 are utilized to reveal such structural features as dislocations and grain boundaries. Dislocations in copper revealed by this technique are illustrated in Fig. 5.35(b); the cause of diffraction contrast is essentially that described in Fig. 5.34(d). The width of the image associated with a given dislocation is several microns because the x-ray beam is in position for appreciable diffraction over a very large number of lattice planes. Since the image of each dislocation is so wide and since the specimen is relatively thick (about 100 μm), only low dislocation densities can be examined in detail. A further disadvantage is the absence of any means to magnify the image.

**Transmission electron microscopy.** Because TEM has several advantages over the corresponding x-ray technique, it is the most generally used method of defect microscopy. A principal advantage is the possibility of magnifying the image through the use of magnetic lenses, thus taking advantage of the inherent resolving power of the short-wavelength radiation. Also, through selected-area diffraction (Fig. 5.26) any small area of the specimen can be selected for examination. Because of the short wavelength of the electron beam, even small changes in orientation of the crystal tend to produce the condition for Bragg reflection and therefore to produce images of defects (see Problem 9). Only a few seconds are needed to record an image in TEM, whereas hours may be required in the x-ray method. To illustrate the capabilities of TEM, two important applications will now be discussed.

*Analysis of Dislocation Images.* In view of the importance of dislocations for many properties of metals, methods have been developed for identifying the type and orientation of dislocations observed in TEM patterns. A basic result from diffraction theory states that when

$$\mathbf{g} \cdot \mathbf{b} = 0 \tag{5.14}$$

diffraction contrast will be essentially zero. Here $\mathbf{b}$ is the Burgers* vector of the dislocation and $\mathbf{g}$ is the vector perpendicular to a given plane $(hkl)$, and has the magnitude $1/d$, where $d$ is the interplanar spacing. As an example of the use of Eq. (5.14), consider the screw dislocation of Fig. 2.6(b) with Burgers vector in the $z$-direction; that is, $\mathbf{b} = [00l]$. This dislocation causes a displacement in the $z$-direction, given by Eq. (2.26), but no displacement in either the $x$- or $y$-direction. Consequently, any set of lattice planes $(hk0)$ parallel to the $z$-direction will not be bent by the presence of this screw dislocation. The normal to any of these planes has the form $[mn0]$, since the component in the $z$-direction is zero. Equation (5.14) can be written,

$$\mathbf{g} \cdot \mathbf{b} = [mn0] \cdot [00l] = m \cdot 0 + n \cdot 0 + 0 \cdot l = 0. \tag{5.15}$$

On the other hand any plane, such as (002), that does not satisfy Eq. (5.14) can give an image due to the presence of this screw dislocation. The procedure for making use of Eq. (5.14) involves tilting the specimen until only one set of planes (character-ized by the diffraction vector $\mathbf{g}_1$, for example) is diffracting strongly. If the dis-location is then invisible, or nearly so, $\mathbf{g}_1 \cdot \mathbf{b} = 0$. The procedure is repeated

---

* The lattice constant can often be omitted from the expression for the Burgers vector. In the present case $\mathbf{b} = a[00l]$ can be replaced by $\mathbf{b} = [00l]$ with a corresponding change in the notation for the $\mathbf{g}$ vector.

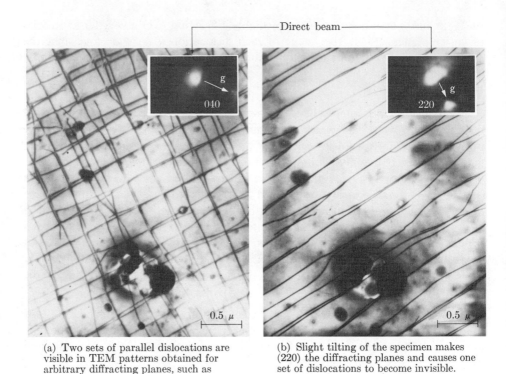

(a) Two sets of parallel dislocations are visible in TEM patterns obtained for arbitrary diffracting planes, such as (040) shown here.

(b) Slight tilting of the specimen makes (220) the diffracting planes and causes one set of dislocations to become invisible. A Burgers vector of $\pm\frac{1}{2}[1\bar{1}0]$ satisfies Eq. (5–14).

**Fig. 5.36** Example of the determination by TEM of the Burgers vectors in a network of dislocations in a crystal of silicon. The [001] direction is approximately perpendicular to the plane of the specimen. The diffraction pattern inset in each TEM pattern identifies the strongly diffracting planes and therefore the vector **g**. (Courtesy G. Thomas. By permission from ASM Seminar, *Thin Films*, Metals Park, Ohio, 1964.)

until a second set of planes, characterized by $\mathbf{g}_2$, also gives no contrast; that is. $\mathbf{g}_2 \cdot \mathbf{b} = 0$. These two conditions can be solved for $\pm\mathbf{b}$, but the sign of the Burgers vector can be determined only from other aspects of the geometry of diffraction. A specific example of the determination of Burgers vectors is given in Fig. 5.36 for a network of dislocations in silicon. If some arbitrary family of planes is diffracting, such as the (040) in Fig. 5.36(a), then two parallel sets of dislocations are visible because Eq. (5.14) is not satisfied for either set of dislocations. For certain reflecting planes, however, such as the (220) planes in Fig. 5.36(b), Eq. (5.14) is satisfied for one set of dislocations, so this set is invisible. Its Burgers vector can then be determined.

*Identification of a Second Phase.* If a second phase is present in an alloy (as discussed in Chapters 6 and 7), this phase can be identified and its orientation

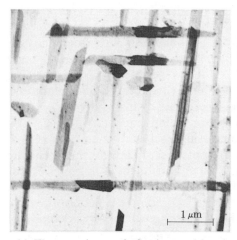

(a) Electron micrograph showing particles of the second phase in the matrix.

(b) Electron-diffraction pattern showing spots from both phases. The spots from the matrix are indicated by a cross.

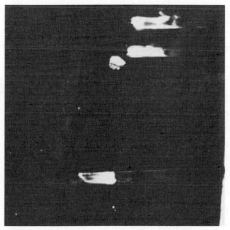

(c) Dark-field image from a set of reflecting planes in the second phase, labelled $P_1$ in (b).

(d) Dark-field image from a set of reflecting planes in the second phase, labelled $P_2$ in (b).

**Fig. 5.37** Illustration of the procedure for identifying a second phase and determining its orientation relative to the matrix. An aluminum alloy containing 6 at. % silver is used in this example.

relationship to the matrix phase can be determined by TEM. Consider, for example, the alloy A1-6 a/o Ag after a treatment that has produced the particles of second phase shown in the micrograph of Fig. 5.37(a). An electron-diffraction pattern, Fig. 5.37(b), viewed in the electron microscope shows diffraction spots from both phases. The diffraction spots from the matrix are more intense. Since the matrix is FCC, its orientation is readily determined (see Problem 11). The spots of the precipitate can·be analyzed in a similar manner; in the present case the precipitate is found to be HCP and can be identified as the $\gamma$ phase (see Fig. 7.28) from its lattice parameters. Selected-area diffraction (SAD), Fig. 5.26, permits the examination of an image from a single set of diffracting planes. If the image is dark-field and if the planes are those of the second phase, the precipitate particles that cause a particular diffraction spot appear bright as in Fig. 5.37(c) or (d). The precipitate may have several distinct orientations, and the diffraction patterns from all precipitate particles are superimposed on the pattern from the matrix. Particular diffraction spots can be ascribed to given orientations of the precipitate by examination of suitable dark-field images, as shown in Fig. 5.37(c) and (d). Analysis of the data of the diffraction pattern by stereographic projection permits a determination of the orientation relationships of the precipitate to the matrix.

**Field-ion microscopy (FIM).** The relatively simple technique of FIM produces the highest resolution available by any method. The technique is similar to that of field-emission microscopy, Fig. 5.14, except that the sharply pointed specimen is given a *positive* electrical potential in the presence of helium at a low pressure $(0.1 \text{ N/m}^2)$. Helium ions (or similar "imaging gas," such as neon) produce an image of the specimen in the manner shown schematically in Fig. 5.38(a). When a helium atom approaches an atom in the specimen, an electron from the helium may be transferred to the ionic core of the atom since the positive electrical potential pulls back the electron cloud surrounding the atom and exposes the ion. The positively charged helium ion is then strongly propelled toward a fluorescent screen, where it produces a visible pulse. Countless repetitions of this process create the desired image. The magnification is approximately $L/R$, where $R$ is the radius of the tip and $L$ is the distance from the tip to the screen, typically 0.05 m. Since $R$ must be in the range $10^{-7}$ to $10^{-8}$ m to achieve the required electric field, typical magnifications are about $10^6$, sufficient for resolution of atomic dimensions.

The principal features of a typical FIM pattern, Fig. 5.38(b), reflect the atomic configuration at the surface of the specimen. Although the tip is almost a hemisphere, it is actually faceted and consists of a series of circular terraces. Each terrace corresponds to a definite crystallographic plane. The atoms at the edge of each plane produce the brightest image because the electric fields are highest there. In some high-index planes (for example, $46\bar{4}$) in crystal $A$) all the atoms are visible. Examination of the grain boundary between crystals $A$ and $B$ shows that it produces little distortion of the atomic distributions.

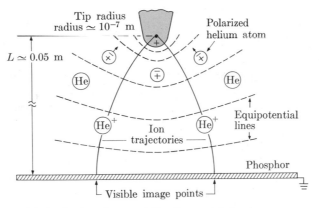

(a) At the sharply curved tip, the crystal structure leads to a terrace structure. Ionization of the helium occurs at the steps in the terrace.

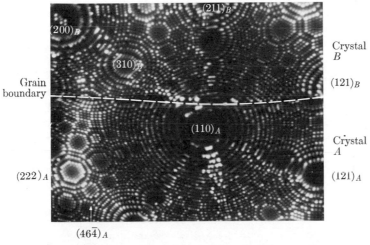

(b) Field-ion micrograph of a grain boundary in tungsten. The indices of major planes in crystals $A$ and $B$ are shown. The misorientation across the boundary is about $35°$.

**Fig. 5.38** Schematic illustration of the principle of operation of the field-ion microscope and a typical micrograph obtained by its use.

Although each FIM pattern reveals only the structure at the surface, a three-dimensional picture of the atomic structure can be constructed from a series of such patterns for successive positions within the bulk of the specimen. For this purpose, the electric field is increased sufficiently to cause desorption of the atoms of the specimen, a process referred to as *field evaporation*. A single atom can be removed by the superposition of a short pulse of voltage on the precisely regulated imaging voltage. More commonly, an entire atomic layer is removed by this means, and then an FIM pattern of the newly exposed structure can be obtained. By repetition of this procedure detailed studies can be made of such features as dislocations, grain boundaries, or special aggregates of atoms. In another mode of operation the FIM can serve as an *atom probe* to identify individual atoms in the pattern being observed. A short pulse of field evaporation liberates the atom in question, and the time required for the atom to reach a detector is then determined. This information permits a calculation of the mass/charge ratio of the iron, thus identifying it.

## STEREOMETRIC METALLOGRAPHY

As the term *stereo* (meaning "space") suggests, the goal of stereometric metallography is to obtain quantitative information about the three-dimensional microstructure of an alloy. (This discipline is also known as quantitative metallography.) An obvious procedure would be to use experimental methods of various kinds (radiography, preferential dissolution of phases, serial sectioning, etc.) to study the spatial configurations in question. In fact, the principal procedures in stereometric metallography involve the deduction of three-dimensional features from detailed studies of ordinary (two-dimensional) images or micrographs. For example, quantitative study of micrographs of a two-phase alloy permits an exact determination of the amount of second phase present in the bulk alloy. Other structural features that can be described quantitatively include: size distribution of second-phase particles; shape of particles or grains; grain size of matrix; and density of dislocations. The analytical procedures that permit these useful deductions are described in the following sections.

Brief mention should first be made, however, of two bases of the entire framework of stereometric metallography: statistical analysis and limit of resolution. An obvious aspect of microstructures is the variability of size, shape, and distribution of the various structural features as seen in two-dimensional micrographs. Since the corresponding variability exists in the spatial features, adequate statistical procedures offer the only practical means for deducing reliable information on three-dimensional structures from two-dimensional micrographs. Some of the mathematical principles are discussed in a later section, but an essential part of statistical analysis is the necessity for obtaining a sufficiently large number of measurements for reasonable precision. Fortunately, such measurements can often be done automatically by microscopes employing television screens and computing devices.

The limit of resolution of the original micrograph (or microscopic image) is clearly important and has two possible origins. In optical micrographs the resolution is limited by the wavelength of the light, as discussed on page 183. In electron-microscopic images (or field-ion patterns, etc.) the limiting factor may

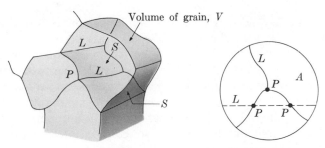

Volume of grain, $V$

(a) The character of $V$, $S$, $L$, and $P$ in a three-dimensional structure.

(b) The related quantities in a two-dimensional section through the structure.

**Fig. 5.3 9** A schematic comparison of the character of volumes ($V$), surfaces ($S$), lines ($L$), and points ($P$) in three dimensions and in two dimensions. (Adapted from E. E. Underwood, *Quantitative Stereology*, Reading, Mass.: Addison-Wesley, 1970.)

be the flux $N$ of particles (electrons, ions, or photons), where $N$ is the number of particles per square meter per second striking the fluorescent screen.. The limit of resolution, corresponding to Eq. (5.2) for the light microscope, is $N^{-1/2}$ m. An adequate image is created only if about $10^3$ particles contribute to the formation of a point in the image during 0.1 second, which is the span of time over which the human eye integrates. Consequently, the flux must be about $10^4$ particles per second, and the statistical limit of resolution is

$$\delta \simeq (10^4/N)^{1/2}. \tag{5.16}$$

Spatial structures (and also two-dimensional sections through these structures) are conveniently described in terms of volumes ($V$), surface areas ($S$ or $A$), length of lines ($L$), and number of points ($P$). These are, of course, three-, two-, one-, and zero-dimensional entities, respectively. Two special aspects of this type of description can be explained with the aid of Fig. 5.39. A three-dimensional feature in a spatial structure (such as the volume $V$ of a grain), Fig. 5.39(a), becomes a two-dimensional feature $A$ in the usual (planar) micrograph of a section through the structure, Fig. 5.39(b). Similarly, a (generally *nonplanar*) surface $S$ in space, a two-dimensional feature, appears as a one-dimensional line $L$ in a section. Thus, the character of a feature of interest, say a grain boundary, is different depending on whether one visualizes it in space or studies it in a micrograph. This distinction is unimportant, however, for use of the relations of stereometric metallography, since they apply regardless of the origin of the features in question. The second special aspect is the use of two alternative symbols to describe two-dimensional features, already illustrated above. The symbol $A$ refers to a planar surface. The symbol $S$ generally refers to a curved surface.

The relations of stereometric metallography involve quantities expressed in terms of two letters; for example, $S_V$. This symbol is read: "(amount of) surface area per unit volume," and may be a measure of the amount (in square meters) of grain-boundary area per cubic meter of spatial structure. A related quantity is $L_A$, the amount (in meters) of linear grain boundary per square meter of planar section through the spatial structure. In every case, the main letter represents an amount of a certain microstructural feature, and the subscript denotes unit amount of "tested region." Table 5.2 gives definitions of the principal symbols employed in stereometric metallography. The experimental determination of quantities such as $S_V$ is a major goal, but this aspect of stereometric metallography is greatly simplified by the existence of relations of the type, $S_V = (4/\pi)L_A$. These two topics will now be discussed.

**Measurement techniques.** Consider the problem of determining the fraction $A_A$ of the area of a microstructure occupied by particles of a second phase. Of the various possible methods that might be employed, the preferred one is *point counting*, Fig. 5.40(a), on a statistically adequate number of micrographs that uniformly sample the alloy in question. A test grid, which forms a uniform array of points, is placed in a random position on a micrograph and the fraction $P_P$ is recorded: $P_P$ is the fraction of the total points that lie in the second phase. Repetition of this procedure on this micrograph and also on other micrographs representative of the structure gives a statistically correct value of $P_P$ (which is equal to the desired quantity $A_A$, as discussed below). Point counting can also be done directly with an optical microscope if a test grid is mounted in the eyepiece.

Quantities such as the amount of grain-boundary area are conveniently determined by measurements of $P_L$, the number of points (intersections of the boundary) per unit length of test line. A circular pattern, Fig. 5.40(b), is convenient for nonuniform structures. The number $P$ of intersections is counted for many random positions of the pattern on numerous micrographs of the structure in question. Division by the total length $L$ of line in the pattern gives $P_L$. A related quantity is $P_A$, the number of points per unit area. Examples are the number of triple points (junctions of three grains), Fig. 5.40(d), or the number of etch pits (intersections of dislocations with the surface). The number of *spatial* objects per unit area is denoted by $N_A$; for example, the number of grains in a micrograph of a single-phase alloy. In this case, the total number of grains, $N_t$, can be adequately approximated by the sum

$$N_t = N_w + \tfrac{1}{2}N_i, \tag{5.17}$$

where $N_w$ is the number of grains completely within the field of view outlined by the device used for counting, and $N_i$ is the number intercepted by the border. If $N_i$ is sufficiently large, the average intercepted grain is divided in half. If $A_t$ is the total area viewed, then

$$N_A = \frac{N_t}{A_t} = \frac{1}{\bar{A}}, \tag{5.18}$$

**TABLE 5.2***

List of Basic Symbols and their Definitions

| Symbol | Dimensions | Definition |
|---|---|---|
| $P$ | | Number of point elements, or test points |
| $P_P$ | | Point fraction. Number of points (in areal features) per test point |
| $P_L$ | $m^{-1}$ | Number of point intersections per unit length of test line |
| $P_A$ | $m^{-2}$ | Number of points per unit test area |
| $P_V$ | $m^{-3}$ | Number of points per unit test volume |
| $L$ | $m$ | Length of lineal elements, or test line length |
| $L_L$ | $m/m$ | Lineal fraction. Length of lineal intercepts per unit length of test line |
| $L_A$ | $m/m^2$ | Length of lineal elements per unit test area |
| $L_V$ | $m/m^3$ | Length of lineal elements per unit test volume |
| $A$ | $m^2$ | Planar area of intercepted features, or test area |
| $S$ | $m^2$ | Surface or interface area (not necessarily planar) |
| $A_A$ | $m^2/m^2$ | Area fraction. Area of intercepted features per unit test area |
| $S_V$ | $m^2/m^3$ | Surface area per unit test volume |
| $V$ | $m^3$ | Volume of three-dimensional features, or test volume |
| $V_V$ | $m^3/m^3$ | Volume fraction. Volume of features per unit test volume |
| $N$ | | Number of features (as opposed to points) |
| $N_L$ | $m^{-1}$ | Number of segments of test line that lie within the phase of interest per unit length of test line |
| $N_A$ | $m^{-2}$ | Number of interceptions of features per unit test area |
| $N_V$ | $m^{-3}$ | Number of features per unit test volume |
| $\bar{L}$ | $m$ | Average lineal intercept, $L_L/N_L$ |
| $\bar{A}$ | $m^2$ | Average areal intercept, $A_A/N_A$ |
| $\bar{S}$ | $m^2$ | Average surface area, $S_V/N_V$ |
| $\bar{V}$ | $m^3$ | Average volume, $V_V/N_V$ |

* Adapted from E. E. Underwood, *Quantitative Stereology*, Reading, Mass.: Addison-Wesley, 1970.

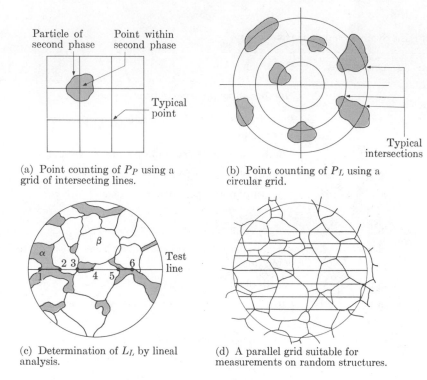

(a) Point counting of $P_P$ using a grid of intersecting lines.

(b) Point counting of $P_L$ using a circular grid.

(c) Determination of $L_L$ by lineal analysis.

(d) A parallel grid suitable for measurements on random structures.

**Fig. 5.40** Schematic illustrations of several methods for making quantitative evaluations of microstructural features on planar sections.

where $\bar{A}$ is the mean area of the grains. If the object being counted is a second phase for which $V_V$ is small compared to unity, the corresponding relation is $N_A = A_A/\bar{A}$. Problem 14 is an example of the use of Eq. (5.18).

*Lineal analysis* is often an alternative procedure to point counting and may be used to study quantities that are difficult to measure by other means. This method may be preferred in such cases as the use of an electronic "sweep" for mechanized operations. The quantity determined in this case is $L_L$, the fractional length of line within the feature of interest. Fig. 5.40(c) shows the application of this method for determining the amount of a second phase. The test line is placed in a random position on the micrograph; the portions (shown as a heavy line) within the second phase are measured and their sum is expressed as a fraction $L_L$ of the total length. Alternatively, the number of intercepts $N_L$ per unit length of line may be easily counted.

**Basic relations.** The determination of a physically significant quantity (for example, the fraction $V_V$ of second phase in a three-dimensional alloy microstructure) is greatly facilitated by the existence of the following relations among the various quantities.

$$V_V = A_A = L_L = P_P, \qquad (5.19)$$

$$S_V = (4/\pi)L_A = 2P_L, \qquad (5.20)$$

$$L_V = 2P_A, \qquad (5.21)$$

$$P_V = \tfrac{1}{2}L_V S_V = 2P_A P_L. \qquad (5.22)$$

The quantity on the left of each of these relations is the spatial quantity, which is usually of major interest. The remaining quantities are those that can be measured on two-dimensional sections (micrographs), and therefore are the means for determining the spatial quantity. A typical derivation of these basic relations is the subject of Problem 15. Other derivations can be found in Chapter 2 of *Quantitative Stereology*, listed among the references at the end of the chapter. A graphical summary of the relations of Eqs. (5.19)–(5.22) is given in Table 5.3. The quantities of major interest (shown inside squares) are not measured directly; instead, they are calculated from one (or more) of the measured quantities (shown inside circles) by means of the basic relations.

TABLE 5.3*
Relationship of Measured (○) to Calculated (□) Quantities

| Microstructural feature | Dimensions of symbols $m^0$ | $m^{-1}$ | $m^{-2}$ | $m^{-3}$ |
|---|---|---|---|---|
| Points | $(P_P)$ | $(P_L)$ → | $(P_A)$ → | $[P_V]$ |
| Lines | $(L_L)$ | $(L_A)$ | $[L_V]$ | — |
| Surfaces | $(A_A)$ | $[S_V]$ | — | — |
| Volumes | $[V_V]$ | — | — | — |

* Adapted from E. E. Underwood, *Quantitative Stereology*, Reading, Mass.: Addison-Wesley, 1970.

The geometrical characteristics of grains and of second-phase particles (their size, shape, spacing, and distribution) strongly influence the mechanical properties of metals and alloys. A well-known example is the effect of a fine grain size in increasing the strength of a pure metal, a topic discussed quantitatively in Chapter 10. Although the grain size of a pure metal can usually be determined fairly easily, the structural characteristics of a complex alloy may be difficult to describe in a useful manner. Two approaches that are often applicable will now be described.

*Average Size.* The mean *intercept length*, $\bar{L}_3$, is a general measure of the average dimension in a collection of grains or particles. It is independent of shape and is defined as simply

$$\bar{L}_3 = \frac{L_L}{N_L} = \frac{1}{N} \sum_{i=1}^{N} (L_3)_i \qquad (5.23)$$

The subscript 3 identifies this quantity as pertaining to three-dimensional features, such as grains or phases.

   If many arbitrary lines are passed through an irregularly shaped particle, the lengths intercepted by the particle vary from zero to some maximum value. $\bar{L}_3$ is the average length of all possible intercepts. In the case of the grains of a single-phase alloy, $\bar{L}_3$ gives a basic measure of grain size. The appropriate equation in this case can be obtained by substitution of Eq. (5.20) into Eq. (5.23) and use of the relations $N_L = P_L$ and $L_L = V_V$, giving

$$\bar{L}_3 = \frac{2V_V}{S_V} = \frac{2}{S_V} \qquad (5.24)$$

since $V_V = 1$ for a one-phase alloy. In contrast, for a typical two-phase ($\alpha$ and $\beta$) structure,

$$(P_L)_{\alpha\beta} = (2N_L)_{\alpha\beta} \qquad (5.25)$$

since the number of points of intersection with the $\alpha$–$\beta$ interface is twice as large as the number of line segments within particles of the $\alpha$ phase. The analog of Eq. (5.24) then becomes,

$$(\bar{L}_3)_\alpha = \frac{4(V_V)_\alpha}{(S_V)_{\alpha\beta}}. \qquad (5.26)$$

In structures having comparable amounts of the $\alpha$ and $\beta$ phases, interfaces of the $\alpha\alpha$, $\alpha\beta$, and $\beta\beta$ types must be considered (see Problem 16).

*Average Spacing.*  In Chapter 10 we will find that a crucial factor in the strengthening effect of a dispersion of particles of the $\alpha$ phase is the mean-free spacing $\lambda$ of the particles. This quantity can be determined by lineal analysis. Since a test line has the same number $N_L$ of intercepts with the matrix and with the dispersed particles, the lineal fraction for the matrix is given by

$$\lambda(N_L)_{\alpha\beta} = (V_V)_\beta = 1 - (V_V)_\alpha. \qquad (5.27)$$

Combination of this result with Eqs. (5.20), (5.25), and (5.26) gives the useful relation,

$$\lambda = \bar{L}_3 \frac{1 - (V_V)_\alpha}{(V_V)_\alpha}. \qquad (5.28)$$

Thus, the mean-free spacing $\lambda$ varies directly with the average diameter of the particles, $L_3$, with a proportionality constant determined by the volume fraction of the dispersed phase.

**Statistical analysis.**  The reliability of quantities determined by the techniques of stereometric metallography depend on two factors: first, control of systematic errors (such as incorrect sampling, malfunction of instruments, etc.); and second, adequate statistical control. The following discussion of statistical analysis assumes that systematic errors are under control and that the measured values $x_1, x_2, \ldots, x_n$ given by $n$ experimental observations are affected only by random variations in the physical phenomenon (plus the uncertainty in measurement, if any). The basic statistical measure of a metallographic feature is its average or *arithmetic mean* value, $\bar{x}$, defined as

$$\bar{x} = \frac{1}{n} \sum_{i=1}^{n} x_i. \tag{5.29}$$

$\bar{x}$ is usually taken as the value of the feature being measured.

Since $\bar{x}$ is simply the average of many, perhaps widely scattered values, a procedure is needed for quantitatively assessing the confidence that one can place in the value of $\bar{x}$. Although this confidence is decreased by large values of the difference $(x_i - \bar{x})$ between individual measurements $x_i$ and the average value $\bar{x}$, this factor can be compensated by the use of a sufficiently large number $n$ of measurements. The basic statistical measure in this case is the *standard deviation*, $S(x)$, conveniently evaluated as its square, the *variance*, $S^2(x)$. The latter is defined as

$$S^2(x) = \left(\frac{1}{n-1}\right) \sum_{i=1}^{n} (x_i - \bar{x})^2 \approx \frac{1}{n} \sum_{i=1}^{n} x_i^2 - (\bar{x})^2. \tag{5.30}$$

In terms of the standard deviation, the correct value of the measured quantity has a 95% probability of lying in the range $\bar{x} \pm 2S(\bar{x})$. To decrease the uncertainty of the average value $\bar{x}$, one need only increase the number of measurements. In practice, limitations on a feasible time for completing the measurements often determine the precision achieved.

## PROBLEMS

**1.** The use of special lenses in an optical microscope permits the use of ultraviolet light ($\lambda = 0.365 \, \mu\text{m}$, for example). If the numerical aperture is 1.5, determine:

a) the limit of resolution, and

b) the useful magnification.

**2.** The permissible thickness of a specimen through which electrons must be *transmitted* depends strongly on the absorption coefficient $\mu$ of the material of the specimen for the electron energies in question.

a) For a beryllium specimen, calculate the thickness for which the intensity of the transmitted beam is 50% that of the initial beam. Assume that $\mu$ is $5 \times 10^{-6} \, \text{m}^{-1}$.

b) Repeat the calculation for uranium, for which $\mu$ is $165 \times 10^{-6} \, \text{m}^{-1}$.

**3.** The carbon replica shown in Fig. 5.41(a) is used to detect steps of height $h$ on the surface of a specimen.

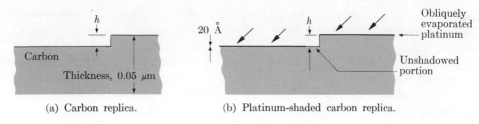

(a) Carbon replica.            (b) Platinum-shaded carbon replica.

**Fig. 5.41** Illustration for Problem 3.

a) If the absorption coefficient $\mu$ is $5 \times 10^{-6}\ \mathrm{m}^{-1}$ for carbon, calculate the minimum vertical step height that can be detected. Assume that a 5% difference in intensity is required for detection.

b) Consider the effect of oblique shadowing with platinum, Fig. 5.41(b), and calculate the minimum detectable step height in this case. Assume that $\mu$ is $65 \times 10^{-6}\ \mathrm{m}^{-1}$ for platinum.

**4.** a) Explain why an incident electron beam, accelerated by a given voltage $V$, cannot produce x-rays having wavelengths below a certain minimum value $\lambda_{min}$.

b) Using Eq. (1.3), show that $\lambda_{min} = (12.4 \times 10^3)/V\text{Å}$.

**5.** a) Use Bragg's law, Eq. (5.8), to determine whether the (001) plane of the grain of iron (Fig. 5.42) will produce an x-ray reflection when copper $K\alpha$ radiation is used. Since iron is body-centered cubic, reflections from $\{100\}$ type planes can occur only for even values of $n$, Table 5.1. The crystal axes of the grain of iron coincide with the $x$-, $y$-, and $z$-axes in the figure.

b) Make similar analyses for the (100) and (010) planes.

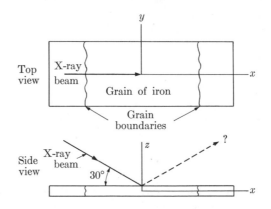

**Fig. 5.42** Illustration for Problem 5.

**6.** The value of $d$, given by Eq. (5.9), can be determined as the perpendicular distance between the origin and the first plane in the set of $(hkl)$ planes. Sketch the first $(hkl)$ plane on a set of coordinate axes; recall that its intercepts are $a/h$, $a/k$, and $a/l$. Erect the perpendicular from the origin to this plane; its length is $d$ and it makes angles $\alpha_1, \alpha_2$, and $\alpha_3$ with the three coordinate axes.   Now, show that Eq. (5.9) follows from the geometrical relation $\cos^2 \alpha_1 + \cos^2 \alpha_2 + \cos^2 \alpha_3 = 1$.

**7.** From an analysis of the diffraction patterns, identify the metals that produced patterns (a), (b), and (c) of Fig. 5.22.

**8.** Show by means of *qualitative* sketches the essential difference between diffraction patterns that would be produced by

a) a mechanical mixture of 50% copper powder with 50% nickel powder,

b) an alloy of 50% copper and 50% nickel. [*Note*: This alloy is discussed in Chapter 7 and is shown to consist of only one crystal structure, a solid solution in which copper and nickel atoms occupy lattice points at random.]

**9.** a) Using 0.04 Å as a typical wavelength of an electron beam for TEM, show that the Bragg angle for the reflection from the (200) plane of a FCC metal with a lattice spacing of 4 Å is $10^{-2}$ radians.

b) Explain why a small change in orientation of a crystal is likely to produce the condition for Bragg reflection from some (*hkl*) plane.

**10.** a) Calculate the values of $d_{hkl}$ for the spots from the matrix in the diffraction pattern of Fig. 5.37(b).  Use Bragg's law with $n = 1$ and $\lambda = 0.04$ Å.  The film-to-specimen distance is 0.663 m and $a$ for aluminum is 4.04 Å; assume $\sin \theta \simeq \theta \simeq \tan \theta$.

b) Using the relation of $d_{hkl}$ to the Miller indices of the diffracting plane, Eq. (5.9), determine the indices of the spots.

**11.** Using a procedure similar to that of Problem 10, analyze the diffraction spots from the matrix in Fig. 5.37(b) and determine the crystallographic orientation of the matrix.

**12.** Using sketches similar to those of Fig. 5.34, explain why the dark-field image of the second phase is bright in Fig. 5.26(d).

**13.** a) Explain why only some of the precipitate particles appear bright in Fig. 5.37(c) and (d).

b) Describe how the orientation relationship between the HCP precipitate and the FCC matrix can be deduced by the dark-field method.

**14.** Calculate $N_A$ and $\bar{L}_3$ for the grains in Fig. 5.40(d). Assume that the magnification is ×100.

**15.** Consider a total surface area, $A_t$, containing a single particle of the $\alpha$ phase in a matrix. Let $A_\alpha$ represent the area of the particle in the plane of polish.

a) If a single test point is placed at random on the surface, show that the probability is $A_\alpha/A_t$ that the point will fall within the $\alpha$ phase.

b) If a total of $P_t$ test points are used, what fraction of them will fall in the $\alpha$ phase?

c) Show that this result is equivalent to $A_A = P_P$.

d) Repeat the argument in three dimensions and show that $A_A = V_V$.

**16.** a) Use a number of test lines to determine the mean diameter $(\bar{L}_3)_\alpha$ of the particles of $\alpha$ phase in Fig. 5.40(c).  Assume that $(V_V)_\alpha = 0.2$. [*Hint*: Tabulate values of $(P_L)_{\alpha\alpha}$ and $(P_L)_{\alpha\beta}$. Use these values to computer the total number of line segments that lie within the $\alpha$ phase; $(N_L)_{total} = (P_L)_{\alpha\beta}/2 + (P_L)_{\alpha\alpha}.$]

b) Use a procedure analogous to that of part (a) to compute $(\bar{L}_3)_\beta$.

**17.** Show by a direct expansion of the first expression for $S^2(x)$ in Eq. (5.30) that the second (approximate) expression is valid for large values of $n$.

## REFERENCES

*Atlas of Microstructures of Industrial Alloys*, Vol. 7 of *Metals Handbook*, 8th edn., Metals Park, Ohio: American Society for Metals, 1972. A comprehensive collection of microstructures of a wide variety of metals and alloys.

Bacon, G. E., *Neutron Diffraction*, 2nd edn, Oxford: Clarendon Press, 1962. A full treatment of the procedures and results of neutron diffraction.

Biggs, W. D., "Elements of Metallography," in *Physical Metallurgy*, Ed. R. W. Cahn, 2nd edn, p. 655, Amsterdam: North Holland Publishing Company, 1970. A brief, well-illustrated survey of the various types of metallographic instruments.

Brandon, D. G., *Modern Techniques in Metallography*, Princeton, N.J.: Van Nostrand, 1966. A quantitative treatment of metallographic techniques at an intermediate level.

DeHoff, R. T. and F. N. Rhines (Eds), *Quantitative Microscopy*, New York: McGraw-Hill, 1968. Several authors present thorough treatments of the principal aspects of this subject.

Nutting, J. and R. G. Baker, *Microstructure of Metals*, London: Institute of Metals, 1965. An excellent collection of micrographs representing the wide range of structures exhibited by metals and alloys.

Rack, H. J. and R. W. Newman, "Microstructures," in *Physical Metallurgy*, Ed. R. W. Cahn, 2nd edn, p. 705, Amsterdam: North Holland Publishing Company, 1970. An advanced, quantitative treatment of microstructures emphasizing the geometrical and stereometrical aspects.

Rostoker, W. and J. R. Dvorak, *Interpretation of Metallographic Structures*, New York: Academic Press, 1965. Integrates the study of microstructure with other aspects of physical metallurgy such as crystallization, transformations and diffusion.

Underwood, E. E., *Quantitative Stereology*, Reading, Mass.: Addison-Wesley, 1970. This is a textbook on stereometric metallography and covers the subject comprehensively at an intermediate level.

CHAPTER 6

# PHASES IN METAL SYSTEMS

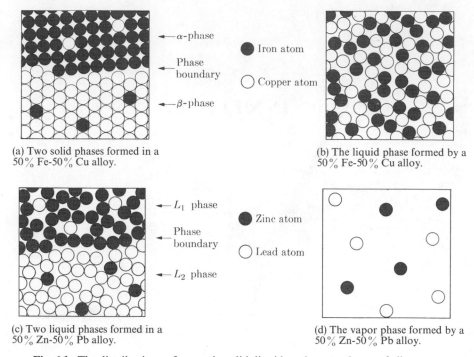

(a) Two solid phases formed in a
50% Fe-50% Cu alloy.

(b) The liquid phase formed by a
50% Fe-50% Cu alloy.

(c) Two liquid phases formed in a
50% Zn-50% Pb alloy.

(d) The vapor phase formed by a
50% Zn-50% Pb alloy.

**Fig. 6.1**   The distributions of atoms in solid, liquid, and vapor phases of alloys.

## INTRODUCTION

The aggregates of atoms involved in typical metallurgical reactions are often more complex than simply the solid metallic elements, to which most of the preceding treatments have been limited. For example, an aggregate may be in the vapor or liquid state, rather than in the solid state, or it may be a mixture of two or of all three of these states. Also, the aggregate may consist of two or more different kinds of atoms. In this case, although the vapor state continues to be unique, the solid state of aggregation often consists of two or more types (*phases*), each of which is homogeneous and has characteristic physical and chemical properties. Two different phases are usually separated by a bounding surface, the phase boundary. Occasionally a liquid alloy also consists of two (or more) phases.

Some typical examples of phases are shown schematically in Fig. 6.1. A solid alloy consisting of 50% Fe atoms and 50% Cu atoms does not exist as a homogeneous solid under equilibrium conditions. Instead, the atoms form two different solids, one rich in iron and the other rich in copper, Fig. 6.1(a). To distinguish between these two solids, it is conventional to label them with the first two letters of the Greek alphabet, $\alpha$ and $\beta$. One then speaks of the $\alpha$-phase and the $\beta$-phase. The nature of the phase boundary is considered in detail in a later section.

Two or more different kinds of metal atoms will usually form a homogeneous solution in the liquid state. This behavior is shown for a 50% Fe-50% Cu alloy in Fig. 6.1(b). Here also it is conventional to use the word *phase* rather than *state* and to speak of the liquid phase. In a few instances, such as that of the 50% Zn-50% Pb alloy shown in Fig. 6.1(c), two liquid phases exist at equilibrium. The usual nomenclature in this case is simply to use $L_1$ (liquid 1) and $L_2$ (liquid 2). All gases are completely miscible in one another and metal vapors are no exception to this rule. Figure 6.1(d) shows the vapor phase in which the 50% Pb-50% Zn alloy can exist under suitable conditions of temperature and pressure.

**One-component phase diagrams.** Before considering in some detail the phases that constitute metal systems, it will be helpful to understand the nature of the equilibrium relations among these phases. The relations are relatively simple for a one-component system; that is, a system consisting of a single, pure metal. It is convenient to consider only this simple system now and to postpone the treatment of more complex equilibria to Chapter 7.

Thermodynamics is a useful tool for describing the conditions under which a given phase will exist at equilibrium. Thermodynamics is largely concerned with energy, and specifies that a metal or alloy tends to consist of that phase (or mixture of phases) which results in minimum energy under the thermodynamic conditions in question. These conditions are most usefully expressed for metallic systems through the thermodynamic variables, pressure ($P$), temperature ($T$), and composition* ($N$). For a pure metal the composition is fixed (the mole fraction is $N = 1$), and therefore if we specify the pressure and the temperature, we determine the phase that exists at equilibrium. Sufficient time must, of course, be allowed for the attainment of equilibrium. Also, the present treatment assumes that other external forces—mechanical, electrical, gravitational, etc.—are negligibly small. Usually this assumption is valid.

The tendency of a substance to change its phase state can be described by one of the three terms, unstable, metastable, or stable. If the substance is at thermodynamic equilibrium it has no tendency to change: It is *stable*. The meanings of unstable and metastable can be understood by the mechanical analogy of an oblong block of wood. If one balances the block on one of its edges, the block is in an *unstable* state and quickly falls to a more stable position. If one stands the block vertically, it is in a *metastable* state and may remain there indefinitely. A suitably large force, however, can cause the block to fall to the horizontal position, in which it is stable with respect to the type of changes being considered. A familiar

---

* The chemical potential, $\mu$, is a generally more useful measure of concentration for thermodynamic purposes. It is related to the concentration in mole fraction, $N$, by the equation,

$$\mu = \mu_0 + RT \ln \gamma N, \tag{6.1}$$

where $\mu_0$ is a constant at a given temperature and pressure, and $\gamma$ is the activity coefficient. See also Eq. (6.33).

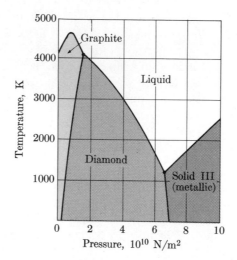

**Fig. 6.2** The one-component phase diagram for carbon. Each region of the diagram shows the range of temperature and pressure within which the indicated phase is thermodynamically stable. (Data of C. G. Suits.) (To convert from $N/m^2$ to atm multiply by $0.987 \times 10^{-5}$.)

example of metastability is diamond. Laboratory investigations at high temperatures and pressures have shown that carbon exists at equilibrium as the liquid phase or as one of three solid phases, Fig. 6.2. For example, at 2000 K and $2 \times 10^{10}$ $N/m^2$ ($\sim 2 \times 10^5$ atm) the equilibrium form of carbon is diamond according to the one-component phase diagram. If carbon in the form of diamond is cooled to room temperature under pressure and then taken down to one atmosphere, it is metastable relative to its equilibrium state (graphite) and retains the crystal structure of diamond for an indefinitely long time (see Problem 1).

As the subject for a more detailed study, we consider the one-component phase diagram of magnesium, Fig. 6.3. In this case the three phases that can exist correspond to the three states of matter—vapor, liquid, and solid. The vapor phase consists of single atoms and is as transparent as air, whereas the liquid phase is similar in appearance to mercury (quicksilver). Under the microscope the solid phase, Fig. 6.4, is seen to consist of many grains, made visible under polarized light because each differently oriented grain rotates the plane of polarization differently (as explained in Chapter 5).

The interpretation of equilibrium diagrams is made easy by the following rule: *Only points in the diagram have physical significance.* Since the variables are temperature and pressure, evidently a *point* is determined by a temperature and a pressure. In Fig. 6.3 the point $A$ is at pressure $P_A$ and at temperature $T$. Since $A$ is in the field of the diagram labeled *liquid*, magnesium metal held long enough at pressure $P_A$ and temperature $T$ for equilibrium to be established is entirely liquid. Similarly, magnesium metal at pressure $P_B$ and temperature $T$ is a mixture

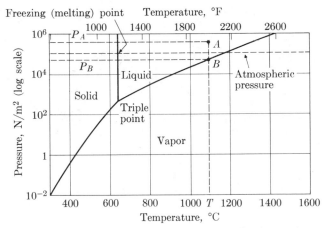

**Fig. 6.3** The one-component phase diagram for magnesium.

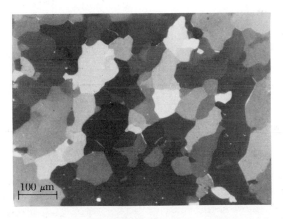

**Fig. 6.4** A typical microstructure of magnesium metal, the solid phase in the one-component diagram, Fig. 6.3. Photographed using polarized light to produce contrast between grains. (Courtesy H. A. Diehl, Dow Chemical Company.)

of liquid and vapor. The temperature and pressure of the triple point will cause solid, liquid, and vapor to be in equilibrium. When two or more phases coexist, the relative amount of each phase cannot be determined from this diagram.

An elementary principle of chemistry states that every liquid or solid tends to be in equilibrium with a particular pressure of its vapor. How then, it may be asked, can there be a completely liquid region in the one-component diagram? Will not vapor exist in equilibrium with the liquid at point $A$ in Fig. 6.3? The explanation lies in the manner in which the one-component system is determined,

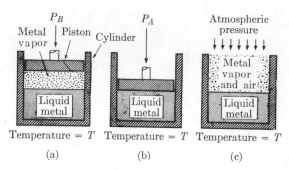

**Fig. 6.5** The difference between the conditions under which one-component diagrams are valid ((a) and (b)), and (c) the usual conditions of heating a metal.

Fig. 6.5. Only the metal being investigated is contained in the cylinder that exerts pressure on the system; even air is excluded. If the external pressure is equal to the vapor pressure of the liquid metal at the given temperature, both liquid and vapor exist in equilibrium in the cylinder. This is the condition at point $B$, Fig. 6.3, and is shown schematically in Fig. 6.5(a). However, if the external pressure is greater than the vapor pressure of the liquid metal, the piston is forced down, the vapor condenses, and only the liquid phase remains. This is the condition at point $A$, Fig. 6.3, and is shown schematically in Fig. 6.5(b). An entirely different situation exists if magnesium metal is held at temperature $T$ in contact with air at one atmosphere pressure ($10^5$ N/m$^2$), Fig. 6.5(c). Of the total pressure of one atmosphere, only the partial pressure $p_B$ is supplied by magnesium vapor. In this instance the significant pressure for most purposes is that of the vapor, not that of vapor and air combined.

Just as the line separating the liquid and vapor areas of the diagram in Fig. 6.3 gives the vapor pressure of the liquid metal as a function of temperature, so the corresponding line for the solid gives the vapor pressure of the solid. It is significant that a solid metal need not melt on being heated. For example, if magnesium is held in a vacuum of $10^2$ N/m$^2$ and is heated to almost 600°C. the metal will completely vaporize without the formation of any liquid. Moderate pressures above that of the triple point have little effect on the melting point of a metal, and the line separating the solid and liquid regions is almost vertical.

**The phase rule.** The thermodynamic conditions for equilibrium are easily extended to an alloy consisting of $C$ components. The components are usually the pure elements: If $C = 2$, we have a *binary* alloy; if $C = 3$, we have a *ternary* alloy, etc. Equilibrium now requires, not only that the pressure and temperature are uniform throughout the system, but also that each component has the same chemical potential in each of the phases that exist. J. Willard Gibbs showed that these requirements for equilibrium limit the freedom an investigator has to adjust the

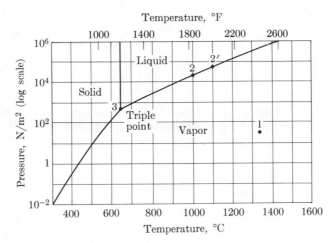

**Fig. 6.6** The one-component phase diagram for magnesium with points in one-, two-, and three-phase regions of the diagram.

thermodynamic variables (pressure, temperature, and compositions of the phases). A principle of mathematics states that if $x$ variables are related by $n$ equations, the number

$$F = x - n \qquad (6.2)$$

of the variables are undetermined and may be freely adjusted. In an alloy with $C$ components, only $C - 1$ compositions are independent since the sum of the mole fractions is a constant; namely, unity. Therefore, the total number of variables is $x = (C - 1) + 2$, the 2 representing pressure and temperature. The number of equations is $n = P - 1$, where $P$ is the number of coexisting phases. If only one phase is present, the investigator can adjust all of the composition variables. If two phases coexist, however, the requirement that all their chemical potentials are equal reduces by one the number of adjustable compositions. When these expressions for $x$ and $n$ are substituted in Eq. (6.2), the result is the *phase rule,*

$$P + F = C + 2, \qquad (6.3)$$

where $P$, $F$, and $C$ represent the *number* of phases, of degrees of freedom, and of components in the given system.* (The *degrees of freedom* are the number of such variables as temperature, pressure, or concentration that can be changed independently without changing the number of phases present.)

The application of the phase rule to the one-component diagram for magnesium, Fig. 6.6, will show the significance of its terms. In this case there is only

---

*A useful though irreverent mnemonic version of this equation is

$$\text{Police Force} = \text{Cops} + 2.$$

one component, magnesium. At point 1 in Fig. 6.6 there is only one phase present, gaseous magnesium; therefore at this point the phase rule becomes

$$P + F = C + 2,$$
$$1 + F = 1 + 2,$$
$$F = 2.$$

That is, there are two degrees of freedom. Two of the variables can be changed *independently* without causing a change in the number of phases. Thus, the temperature and the pressure may be varied separately without losing the vapor phase.

The situation at point 2 is significantly different. Since two phases, liquid and vapor, exist in equilibrium at this point, the phase rule yields

$$P + F = C + 2,$$
$$2 + F = 1 + 2,$$
$$F = 1.$$

One may ask why only one degree of freedom exists when clearly *both* temperature and pressure must change in going from point 2 to point 2′, where the two phases continue to exist. The answer is that only *one* of the variables can be changed *independently.* An increase in pressure must be accompanied by a proper change in temperature if neither of the two phases is to be lost. Similarly, the temperature can be changed arbitrarily but the two phases will continue to exist only if the pressure is also varied by a specific, *not an independent*, amount. Only one of the variables, therefore, can be independently changed at point 2 if two phases are to remain. A similar analysis shows that there are zero degrees of freedom at point 3, the triple point, in Fig. 6.6.

The phase rule and much of the other information on one-component systems can be applied to the multicomponent alloys that are of greater commercial interest. Of course, the presence of a second component metal makes it possible to vary the chemical compositions of the phases as well as the temperature and pressure.

## THERMODYNAMIC BACKGROUND

Because thermodynamic concepts are useful for treating many topics in physical metallurgy, the essential background in this subject will now be presented. In a thermodynamic analysis one chooses a definite *system* for treatment; a typical system is one mole of a particular alloy, say, 0.7 mole fraction copper and 0.3 mole fraction zinc. The remainder of the universe is the surroundings of the system. An *isolated* system is separated from its surroundings so completely that no exchange of any kind occurs (heat, work, or mass). A *closed* system does not exchange mass, while an *adiabatic* process does not involve exchange of heat although the other types of exchanges may occur. For our purposes, the *state* of a system is the phase condition in which it exists; for example, α-phase, or β-phase, or α- plus β-phases (each of specified composition, pressure, and temperature). *State functions* are those quantities whose values are determined simply by the state of the system (but not by the process or *path* employed in reaching this state). Internal energy, $U$, enthalpy, $H$, entropy, $S$, and free energy, $G$, are examples of important state functions. The work, $W$, done by the system on its surroundings is an example of a nonstate function (see Problem 5).

**First law of thermodynamics.** "The energy of the world is constant" is a well-known statement of the first law of thermodynamics. The usual mathematical statement is made in terms of the *internal energy*, $U$,

$$dU = \delta Q - \delta W. \tag{6.4}$$

The change in the state function, $U$, is the difference between the quantity of heat $\delta Q$ *added to* the system and the quantity of work $\delta W$ *done by* the system. The use of $\delta$ rather than $d$ reminds us that $Q$ and $W$ are nonstate functions. A familiar type of work is $\delta W = P\,dV$, the work done by the system against the pressure $P$ of the surroundings when the volume of the system increases by the amount $dV$. Other types of work are electrical, magnetic, and gravitational. The essential difference between a state function and a nonstate function can be seen by considering a mole of magnesium vapor at a pressure of $10^3$ N/m$^2$ ($10^{-2}$ atm), first at 800°C (where the internal energy has a value $U_1$) and then at 810°C (where the internal energy is $U_2$). Provided that the magnesium is at equilibrium, the values $U_1$ and $U_2$ are each characteristic of the temperature, regardless of the process employed in changing the temperature. One method of raising the temperature is to add heat to the system. Since the specific heat, $C_P$, is 20.8 J/(mol · K) for magnesium vapor,

$$\delta Q = 20.8 \times 10 = 208 \text{ joules} \tag{6.5}$$

will raise the temperature of a mole of vapor by 10 K. An alternative method for raising the temperature is to add the same amount of energy as work. The required work of compression of the magnesium vapor is

$$\text{work} = -\int \delta W = -\int P\,dV = \int_{10^3}^{P_f} \frac{RT}{P}\,dP = 208; \tag{6.6}$$

$R$ is the gas constant, 8.32 J/(mol · K). After integration (assuming that $T$ is constant), Eq. (6.6) can be solved for the final pressure $P_f$,

$$\ln P_f = \frac{208 + RT \ln 10^3}{RT} = 6.931$$

$$P_f = 1.023 \times 10^3 \text{ N/m}^2. \tag{6.7}$$

Thus, a two percent increase in pressure is sufficient to raise the internal energy from $U_1$ to $U_2$.

Since most processes involving metals and alloys occur at constant pressure (usually atmospheric), many useful relations are derived subject to this restriction. For example, since $V\,dP$ is then zero, the first law of thermodynamics can be written,

$$d(U + PV) = dH = \delta Q, \tag{6.8}$$

where the *enthalpy*, $H$, is defined,

$$H = U + PV. \tag{6.9}$$

Since $dH = \delta Q$ at constant pressure, the term *heat content* is also used for the enthalpy. The enthalpy can be determined experimentally as the heat absorbed by the system during a process (such as a chemical reaction) at constant pressure (see Problem 6). The change, $\Delta H$, is positive if heat is absorbed and the reaction is then called *endothermic*. An exothermic reaction evolves heat; $\Delta H$ is then negative. In either case, part of the heat comes from a change in internal energy $U$ and part from the work against the surroundings.

**Entropy and the second law of thermodynamics.** "The entropy of the world strives toward a maximum" is one of the statements of the second law of thermodynamics. The concept of entropy, $S$ (which is a state function), can be approached in two ways. First, from a general thermodynamic argument one can show that this useful quantity must have the definition,

$$dS = \frac{\delta Q_{\text{rev}}}{T}, \tag{6.10}$$

where $\delta Q_{\text{rev}}$ is the amount of heat added to the system during an (ideal) reversible process at temperature $T$. The entropy $S_2$ in state 2 can thus be expressed in relation to the entropy in state 1,

$$S_2 = S_1 + \int_1^2 \frac{\delta Q_{\text{rev}}}{T}. \tag{6.11}$$

From such practical observations as the fact that a perpetual-motion machine cannot be devised, one can conclude that the change in entropy in any real process taking place within the system must be positive (or, in an ideal process, zero); that is,

$$dS \geq 0. \tag{6.12}$$

*For an isolated system*, Fig. 6.7(a), the condition $\Delta S = 0$ can be used to define an equilibrium state of a system. A typical system, however, is not isolated, so a definition in terms of a different state function, $G$, is more generally useful and is discussed below.

A real process (such as a chemical reaction) can occur only if the system is not at equilibrium. Consequently, $dS > 0$, and

$$\delta Q > \delta Q_{\text{rev}}. \tag{6.13}$$

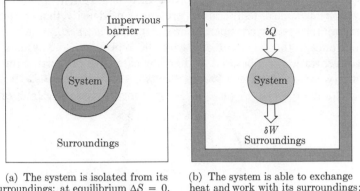

(a) The system is isolated from its surroundings: at equilibrium $\Delta S = 0$.

(b) The system is able to exchange heat and work with its surroundings; at equilibrium $\Delta G = 0$.

**Fig. 6.7** Schematic illustrations of two different relationships of a system to its surroundings.

As a simple example, consider an isolated system consisting of two bars of copper, one at temperature $T_1$ and the other at some higher temperature $T_2$. The initial increment in entropy, $dS$, accompanies the transfer of an infinitesimal amount of heat, $\delta Q$, from the bar at $T_2$ to the bar at $T_1$ (the latter term is positive because the heat enters the bar at $T_1$).

$$dS = \frac{\delta Q}{T_1} - \frac{\delta Q}{T_2}. \tag{6.14}$$

The sum of these two terms is greater than zero and remains so as $T_1$ and $T_2$ gradually approach a common temperature, $T_e$, at which the system is in equilibrium.

The second approach to the concept of entropy is through statistical mechanics and has the advantage of giving a physical picture; namely, that entropy is a measure of the randomness of a system. For convenience, randomness is divided into two types: (1) the randomness in arrangement (configuration) of the atoms in a solution of metals 1 and 2, giving rise to $\Delta S^{id}$, the *ideal entropy* of mixing; and (2) all other types of randomness, primarily associated with the vibration of the lattice in solids, giving rise to $\Delta S^{xs}$, the *excess entropy* of mixing. Therefore, the total change in entropy, $\Delta S$, accompanying a given process is the sum,

$$\Delta S = \Delta S^{id} + \Delta S^{xs}. \tag{6.15}$$

As shown below, the molar entropy of mixing is

$$\Delta S^{id} = -R(N_1 \ln N_1 + N_2 \ln N_2), \tag{6.16}$$

where $N_i$ is the mole fraction of component $i$. Typical uses of $\Delta S^{id}$ are for treating the diffusion of radioactive atoms (which form an ideal solution with the normal atoms) and for deriving the expression for the concentration of vacancies.

**Thermodynamics of solutions.\*** The properties of a system are of two kinds; *intensive properties* such as temperature, pressure, and density, that are independent of the amount of system considered, and *extensive properties* such as volume, weight, and energy that vary directly with the amount of the system. Various units of concentration can be used to describe extensive properties, but the most widely employed is the mole fraction, $N_i$, of each component $i$. If a uniform solution contains $n_i$ moles of each of $C$ components, the mole fraction of component 1 is

$$N_1 = \frac{n_1}{\sum\limits_{i=1}^{C} n_i}.$$

(6.17)

Corresponding equations apply for $N_2$, $N_3$, etc. This equation takes an especially simple form if

$$\sum\limits_{i=1}^{C} n_i = 1,$$

(6.18)

that is, if the system contains a total of 1 mole of atoms. In this case $N_1 = n_1$, $N_2 = n_2$, etc. The extensive quantities then refer to one mole of solution and are simply written $V$ (molar volume), $H$ (molar enthalpy), $S$ (molar entropy), etc. The addition of a prime (') indicates an extensive quantity for an arbitrary amount of solution; thus, $V'$, $H'$, and $S'$.

In a solution containing two components, 1 and 2, how is a given extensive property to be divided between the two kinds of atoms? Volume is a property that can be readily visualized in this respect (see Problem 7), but the same question applies to any extensive property†, $G$. The solution to this dilemma lies in the concept of partial molal quantity, $G_1$, defined as the partial derivative

$$G_1 = \left(\frac{\partial G'}{\partial n_1}\right)_{P,T,n_2},$$

(6.19)

that is, $G_1$ is the change in $G'$ (per mole of component 1 added) when the pressure, temperature, and the number of moles of component 2 are held constant. In practice, only a very small amount of component 1 is added in order that the value of $G_1$ will be determined at essentially constant composition. The importance of this definition is the fact that it leads to the following basic relation for the thermodynamics of solutions,

$$G = N_1 G_1 + N_2 G_2 + \cdots + N_C G_C$$

(6.20)

---

\*This treatment is adapted from L. S. Darken and R. W. Gurry, *Physical Chemistry of Metals*, New York: McGraw-Hill, 1953. Derivations and additional information can be found there.
† Here the symbol $G$ represents any extensive thermodynamic property. This same symbol is used later for Gibbs free energy.

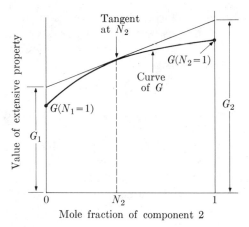

**Fig. 6.8** Graphical method for determining the partial molal properties, $G_1$ and $G_2$, from the curve of the molar property, $G$. (Adapted from L. S. Darken and R. W. Gurry.)

for a solution containing $C$ components. A second useful relation, derived from Eq. (6.20), is the Gibbs–Duhem equation,

$$N_1\, dG_1 + N_2\, dG_2 + \cdots + N_C\, dG_C = 0. \tag{6.21}$$

In a binary solution, for example, this equation gives the values of $G_2$ if the curve of $G_1$ versus $N_2$ is known. A graphical method for determining $G_1$ and $G_2$ at any concentration $N_2$ is shown in Fig. 6.8. The equation of the tangent line is

$$G = G_1 + (G_2 - G_1)N_2. \tag{6.22}$$

Since $N_1 - (1 - N_2)$, the value of $G$ at the concentration $N_2$ can be written,

$$G = N_1 G_1 + N_2 G_2. \tag{6.23}$$

Thus, the intercepts of the tangent at $N_1 - 1$ and at $N_2 = 1$ give the values of $G_1$ and $G_2$. In general these values are different from the molar quantities $G(N_1 = 1)$ and $G(N_2 = 1)$ for the pure components, as shown in Fig. 6.8.

As previously discussed for entropy, any state function can be divided into two portions, $G^{id}$ and $G^{xs}$. This concept is mostly used to describe the change, $\Delta G$, that accompanies the process of mixing of two pure substances. A partial molal quantity can also be similarly divided,

$$\Delta G_1 = \Delta G_1^{id} + \Delta G_1^{xs}. \tag{6.24}$$

The availability of tabulated values of the various thermodynamic functions for binary metal systems* makes it possible to use thermodynamics in the solution of many problems.

---

* R. Hultgren, *et al.*, *Selected Values of Thermodynamic Properties of Binary Alloys*, Metals Park, Ohio: American Society for Metals, 1973.

**Free energy and equilibrium.** During a typical reaction, such as the solidification of an alloy, the system exchanges both heat and work with its surroundings, Fig. 6.7(b). The condition for equilibrium in such cases is described in terms of a new state function, the Gibbs free energy $G$,

$$G = H - TS,$$
(6.25)

defined in terms of the enthalpy $H$ and entropy $S$ of the system, but not dependent on the parameters of the surroundings.

To understand how Eq. (6.25) arises, imagine that the system plus the surroundings (affected by the reaction in question) are both enclosed within an impervious barrier, Fig. 6.7(b). By the second law of thermodynamics, the change in entropy during an infinitesimal amount of reaction is

$$dS_{total} = dS_{system} + dS_{surroundings} \geq 0.$$
(6.26)

The condition $dS_{total} = 0$ corresponds to a state of equilibrium. If the reaction occurs at constant temperature and $\delta Q$ is the amount of heat entering the system from the surroundings,

$$dS_{system} - \frac{\delta Q}{T} \geq 0.$$
(6.27)

If the pressure is constant and if the only work is that done against this pressure, $\delta W = P\,dV$; then Eq. (6.4) can be substituted for $\delta Q$, giving

$$dU + P\,dV - T\,dS \leq 0.$$
(6.28)

Since all of these quantities refer to the system, the subscript on $dS$ has been omitted. In view of the definition $H = U + PV$, Eq. (6.9), the three terms in Eq. (6.28) are equal to the differential, $dG$, of the function $G$ defined by Eq. (6.25). Consequently, for a reaction at constant temperature and pressure,

$$dG \leq 0,$$
(6.29)

that is, the Gibbs free energy must either decrease (for a real, irreversible process) or be zero (for a reversible process occurring under equilibrium conditions).

A related function, the Helmholtz free energy, $F^*$, defined as

$$F = U - TS,$$
(6.30)

can be derived in a similar manner for a process occurring at constant volume and constant temperature. Since in most reactions involving condensed phases the term $P\,dV$ in Eq. (6.28) is negligibly small, $dG$ and $dF$ are essentially identical. $F$ is often more convenient in derivations, but the function listed in tables is

---

* A possible source of confusion is the use of the symbol $F$ in the older literature for the Gibbs free energy.

invariably $G$.  Consequently, both functions are employed in thermodynamic treatments of metallic systems.

In view of the fact that all of the quantities entering into the definition of free energy refer to the system (rather than to the surroundings), it is instructive to consider how the two parts of $G$, Eq. (6.25), interact during a real process, for which $\Delta G$ must be negative.  The process of solidification of a pure metal is used here as an example.  The atoms in the crystal lattice are at a lower energy than are the atoms in the liquid state.  Therefore, the enthalpy decreases during crystallization.  On the other hand, the atoms become more orderly, so the entropy decreases, and the term $(-TS)$ increases in value.  At the equilibrium freezing temperature, $T_f$, these two factors exactly balance, so $\Delta G = 0$.  At temperatures below $T_f$, however, the decrease in enthalpy accompanying crystallization is greater than the increase in $(-TS)$, and $\Delta G$ is negative.  As in the analogous case of a mechanical system, the system can be pictured as moving in the direction that reduces the total energy.

Treatments involving the partial molar Gibbs free energy $G_i$ are often made in terms of the chemical potential $\mu_i$ or the thermodynamic activity $a_i$.  The chemical potential is simply an alternative name and symbol, since $G_i = \mu_i$.  The thermodynamic activity is related to $G_i$ through the defining equation,

$$G_i = \mu_i = G_i^0 + RT \ln a_i, \qquad (6.31)$$

where $G_i^0$ is the value in a reference state and $R$ is the gas constant, 8.31 J/(mol · K).  The activity $a_i$ is usefully pictured as representing the concentration $N_i$ corrected for nonideal behavior by the activity coefficient $\gamma_i$; that is,

$$a_i = \gamma_i N_i. \qquad (6.32)$$

If the pure component in question is chosen as the reference state, $G_i^0$ can usually be set equal to zero.  Equation (6.31) then becomes

$$G_i = \mu_i = RT \ln \gamma_i N_i. \qquad (6.33)$$

An example of the use of Eq. (6.33) is the derivation of the expression for $\Delta S^{id}$, the molar entropy of mixing of an ideal solution.  The mixing process is described by the reaction

$$N_1 \text{ (pure component 1)} + N_2 \text{ (pure component 2)} \rightarrow \alpha, \qquad (6.34)$$

where $\alpha$ represents one mole of the solid solution of components 1 and 2.  Since the initial value of the free energy $[N_1 G(N_1 = 1) + N_2 G(N_2 = 1)]$ can be taken as zero by setting $G(N_1 = 1)$ and $G(N_2 = 1)$ equal to zero, the total change in free energy is given by Eq. (6.33) as,

$$\Delta G = \Delta H^{id} - T \Delta S^{id} = N_1 RT \ln N_1 + N_2 RT \ln N_2. \qquad (6.35)$$

The activity coefficients do not appear because $\gamma = 1$ for an ideal solution.  By definition $\Delta H^{id}$ is zero, since the atoms of the various components of an ideal solution are chemically identical. Therefore, Eq. (6.35) reduces to the final expression for $\Delta S^{id}$, Eq. (6.16).

## THE VAPOR PHASE

Of the three states of aggregation of metal atoms, the vapor state is the simplest to treat. As was pointed out above, even in many-component systems there is only a single vapor phase. Also, the vapors formed by metal atoms are essentially the same as the more common gases; they have low density, are transparent, and obey the gas laws, although under most conditions they are monatomic in contrast to such diatomic gases as oxygen and nitrogen. The large interatomic distances in the vapor phase, Fig. 6.1(d), effectively isolate each atom so far as interchange of electrons is concerned.

**Vapor pressure.** An important aspect of the vapor phase in many commercial processes is the metal vapor that tends to be in equilibrium with every solid or liquid metal. The method for determining vapor pressures for a pure metal from its one-component diagram was described above, and it can be seen from the trend of the solid–vapor line in Fig. 6.3 that vapor pressures tend to be very low at room temperature. At higher temperatures, however, the vapor pressure may be large enough that significant evaporation can occur.

If the value of the heat of evaporation, $\Delta H_v$, is known, a good estimate can be made of the vapor pressure, $P$, of a pure metal as a function of temperature. By manipulation of the thermodynamic expression for the free energy (see Problem 8), one can show that

$$\frac{dP}{dT} = \frac{\Delta S_v}{\Delta V_v},$$

(6.36)

where $\Delta S$ and $\Delta V$ are the changes in entropy and in volume accompanying the evaporation of one mole of metal. Substitution of the expression in Eq. (6.10) for $\Delta S$ gives a useful form of the *Clausius–Clapeyron equation*,

$$\frac{dP}{dT} = \frac{\Delta H_v}{T \Delta V_v}.$$

(6.37)

As the volume of the initial liquid is usually negligible in comparison with the volume of the final vapor, and since the vapor obeys the ideal gas law, $\Delta V_v \approx V_v = RT/P$. Therefore, Eq. (6.37) can be written

$$\frac{d \ln P}{d(1/T)} = -\frac{\Delta H_v}{R}.$$

(6.38)

Thus, on a plot of $\ln P$ versus $1/T$, a straight line with slope $-\Delta H_v/R$ describes the variation of vapor pressure with temperature of the liquid phase. If some reference value of $P$ is known (such as the pressure at the triple point, Table 6.1), then the curve of vapor pressure versus temperature is completely determined. Problem 9 considers the analogous case of the vapor pressure above the solid metal.

More precise values of vapor pressure than those given by the approximation of Eq. (6.38) can be obtained from the following empirical equation,

$$\log P = \frac{A}{T} + B \log T + CT + D. \tag{6.39}$$

Here log is the logarithm to the base 10, $P$ is the vapor pressure in torr, $T$ is the temperature in Kelvin, and $A$, $B$, $C$, and $D$ are constants for a given liquid (or solid) metal, Table 6.1. The terms $B \log T$ and $CT$ represent small corrections to the integrated form of Eq. (6.38). Equation (6.39) gives the line separating the liquid (or solid) area from the vapor area in Fig. 6.3. The line separating the solid and liquid areas is essentially vertical. Few one-component diagrams are given in the literature, but from the triple-point temperatures and pressures listed in Table 6.1 and from vapor pressure data it is easy to construct a usable diagram. Frequently the triple-point pressure alone gives enough information for many purposes, for example, determining the possibility of melting chromium in a good vacuum (see Problem 4).

**Kinetic theory.** The picture of the vapor phase as an assembly of individual atoms, Fig. 6 1(d), can easily be developed into a quantitative description of important aspects of the behavior of the gas phase. A simplified treatment of kinetic theory is used here for that purpose. Consider that there are $n$ atoms, each of mass $m$, in a cubic container $l$ meters on an edge. It is convenient to assume that all the atoms are moving with the average velocity $u$, that one-third of them are moving parallel to each of the coordinate axes, and that collisions of the atoms with one another can be neglected. The pressure exerted by the gas on a wall of the container is created by collisions of the atoms with the wall. Since an atom moves $u$ meters each second, it will cover the distance $l$ a total of $u/l$ times per second and it will strike a given wall $u/2l$ times a second. A total of $n/3$ atoms are assumed to move in the direction perpendicular to this wall, so that the total number of collisions per second is $nu/6l$.

To determine the force $F$ exerted on the wall by these collisions, it is convenient to use Newton's law $F = ma$ in the form $F = (d/dt)(mu)$, where $mu$ is the momentum. At each collision the momentum of an atom is changed from $+mu$ to $-mu$, since the direction of its velocity is assumed to be completely reversed in an elastic impact with the wall. Therefore the total momentum change per second is $nmu^2/3l$. But this is $(d/dt)(mu)$, the rate of change of momentum, and it follows that $F = nmu^2/3l$. Since the pressure $P$ is the force per unit area,

$$P = \frac{nmu^2}{3l^3}. \tag{6.40}$$

**TABLE 6.1**

Data on the Vapor Pressures and Triple Points of Some of the Elements*

| Phase | $A$ | $B$ | $C \cdot 10^3$ | $D$ | Triple point Temperature, °C | Triple point Pressure, torr |
|---|---|---|---|---|---|---|
| Ag(s) | $-14{,}710$ | $-0.755$ | — | 11.66 | 961 | $2.65 \times 10^{-3}$ |
| Ag(l) | $-14{,}260$ | $-1.055$ | — | 12.23 | | |
| Al(l) | $-16{,}450$ | $-1.023$ | — | 12.36 | 660 | $1.88 \times 10^{-8}$ |
| Ca(s) | $-10{,}300$ | $-1.76$ | — | 14.97 | 838 | 1.97 |
| Ca(l) | $-9{,}600$ | $-1.21$ | — | 12.55 | | |
| Cd(s) | $-5{,}908$ | $-0.232$ | $-0.284$ | 9.717 | 321 | $1.15 \times 10^{-1}$ |
| Cd(l) | $-5{,}819$ | $-1.257$ | — | 12.287 | | |
| Cr(s) | $-21{,}200$ | $-1.56$ | — | 15.75 | 1875 | 7.70 |
| Cr(l) | $-20{,}400$ | $-1.82$ | — | 16.23 | | |
| Cu(s) | $-17{,}770$ | $-0.86$ | — | 12.29 | 1083 | $3.91 \times 10^{-4}$ |
| Cu(l) | $-17{,}520$ | $-1.21$ | — | 13.21 | | |
| Fe(s) | $-21{,}080$ | $-2.14$ | — | 16.89 | 1537 | $5.46 \times 10^{-2}$ |
| Fe(l) | $-19{,}710$ | $-1.27$ | — | 13.27 | | |
| Ge(s) | $-20{,}150$ | $-0.91$ | — | 13.28 | 937 | $5.78 \times 10^{-7}$ |
| Ge(l) | $-18{,}700$ | $-1.16$ | — | 12.87 | | |
| Hg(l) | $-3{,}305$ | $-0.795$ | — | 10.355 | $-38$ | $1.55 \times 10^{-6}$ |
| Mg(s) | $-7{,}780$ | $-0.855$ | — | 11.41 | 650 | 2.80 |
| Mg(l) | $-7{,}550$ | $-1.41$ | — | 12.79 | | |
| Mo(s) | $-31{,}060$ | $-0.2$ | — | 9.41 | 2610 | $2.69 \times 10^{-2}$ |
| Ni(s) | $-22{,}100$ | — | $-0.131$ | 10.75 | 1453 | 1.84 |
| Ni(l) | $-18{,}000$ | — | — | 8.17 | | |
| Pb(l) | $-10{,}130$ | $-0.985$ | — | 11.16 | 327 | $3.26 \times 10^{-9}$ |
| Pt(s) | $-28{,}460$ | $-1.27$ | — | 14.33 | 1769 | $2.42 \times 10^{-4}$ |
| Pt(l) | $-27{,}890$ | $-1.77$ | — | 15.71 | | |
| Si(s) | $-18{,}000$ | $-1.022$ | — | 12.83 | 1410 | $3.69 \times 10^{-2}$ |
| Si(l) | $-17{,}100$ | $-1.022$ | — | 12.31 | | |
| Ti(s) | $-24{,}275$ | — | $-0.23$ | .10.66 | 1668 | $3.80 \times 10^{-3}$ |
| Ti(l) | $-22{,}110$ | — | — | 9.135 | | |
| W(s) | $-42{,}000$ | $+0.146$ | $-0.164$ | 9.84 | 3410 | $3.31 \times 10^{-2}$ |
| Zn(s) | $-6{,}850$ | $-0.755$ | — | 11.24 | 420 | $1.49 \times 10^{-1}$ |
| Zn(l) | $-6{,}620$ | $-1.255$ | — | 12.34 | | |

To convert from torr to $N/m^2$, multiply by $1.333 \times 10^2$.

* Adapted from O. Kubaschewski and E. LL. Evans, *Metallurgical Thermochemistry*, Oxford: Pergamon Press, 1958; and A. N. Nesmeyanov, *Vapor Pressure of the Chemical Elements*, Amsterdam: Elsevier, 1963.

It is convenient to work with a kilogram atomic weight of the gas, $M = Nm$, where $N$ is Avogadro's number, since the volume $l^3$ is then the kilogram-molar volume $V$, 22.4 m$^3$. Equation (6.40) then becomes

$$PV = \frac{Mu^2}{3}. \tag{6.41}$$

Metal vapors can be assumed to obey the ideal gas law, so that $PV = RT$ and Eq. (6.41) can be written

$$u = \sqrt{3RT/M}. \tag{6.42}$$

Examples of the use of this equation are given later in connection with nucleation and rates of evaporation.

Because of its basic importance for reactions in the liquid and solid states as well, let us consider the assumption made in using an average velocity, $u$, in the above treatment. In a system (solid, liquid, or gas) containing one mole of atoms, the molar free energy $G$ is distributed among the $N$ atoms; the free energy, $g$, per atom ranges from essentially zero upwards toward very high values. By analogy with the Boltzmann distribution, the probability, $p$, that a given atom has free energy $g$ is

$$p = Ce^{-g/kT}, \tag{6.43}$$

where $C$ is a proportionality constant and $k$ is Boltzmann's constant. Since the probability is unity that an atom has an energy between zero and infinity,

$$1 = \int_0^\infty Ce^{-g/kT}\, dg = kTC. \tag{6.44}$$

Therefore, Eq. (6.43) becomes

$$p = \frac{1}{kT}e^{-g/kT}. \tag{6.45}$$

In terms of a mole of atoms, Fig. 6.9(a), this equation means that there are most atoms (per increment of energy, $dg$) at $g = 0$, and that the number decreases exponentially with increasing energy. Thus, the majority of the atoms have small amounts of energy; almost two-thirds of them having an energy less than $kT$.

Consider now the amount of energy, $(gpN)\, dg$, possessed by the atoms within a given increment of energy $dg$, Fig. 6.9(b). Near $g = 0$ the atoms have little energy; consequently, the two-thirds of the total atoms with energies below $kT$ possess only about one-third of the total energy of the system. The relatively small number of atoms having high energies account for the majority of the energy of the system.

For example, the extremely small number of atoms having energies greater than $5kT$ possess about 4% of the total energy, Problem 10. The arithmetic mean

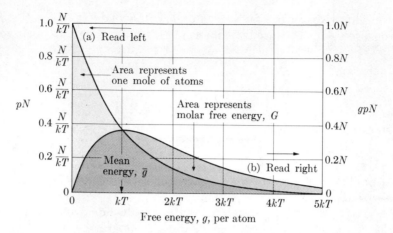

**Fig. 6.9** Curves showing the general character of the distribution of energy among a mole $N$ of atoms. (a) The relative number $pN$ (or probability) of atoms having a given free energy, $g$. (b) The amount of energy $(gpN)\,dg$ possessed by those atoms within an increment of energy $dg$.

value of the energy, $\bar{g}$, is determined as,

$$\bar{g} = \frac{G}{N} = \frac{\int_0^\infty gpN\,dg}{N} = \int_0^\infty \frac{g}{kT}\,e^{-g/kT}\,dg = kT. \qquad (6.46)$$

In the case of a monatomic gas, the energy is simply $g = mu^2/2$, and a velocity corresponding to $\bar{g}$ is the average velocity $u$.

To treat problems such as diffusion in gases, one must consider the collisions among the atoms. The cross-sectional area of an atom is $\pi r^2$ where $r$ is the radius. If $n$ is the number of atoms per unit volume, then the *mean free path* $l$ between collisions is given roughly by

$$\pi r^2 l = \frac{1}{n} \quad \text{or} \quad l = \frac{1}{\pi r^2 n}, \qquad (6.47)$$

since $\pi r^2 l$ can be pictured as the portion of the total volume available to a given atom. The collision frequency, $\Gamma$, is then

$$\Gamma = \frac{u}{l}, \qquad (6.48)$$

which is a value on the order of $10^{10}$ collisions per second. From Eq. (8.4), derived in Chapter 8, the relation between the diffusion coefficient and $\Gamma$ is

$$D = \frac{\Gamma \alpha^2}{6} = \frac{\Gamma l^2}{6} = \frac{ul}{6} = \frac{u}{6\pi r^2 n}. \qquad (6.49)$$

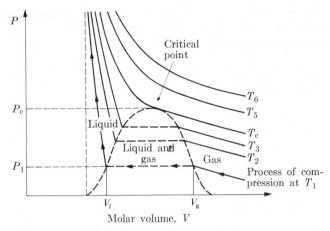

**Fig. 6.10** A pressure-volume phase diagram for one mole of a pure component showing the character of the gas → liquid transition at successively higher temperatures, $T_1$ to $T_6$. Below the critical temperature, $T_c$, both phases coexist at constant pressure, over a range of (total) volume.

Substitution of typical values in this equation (such as $u = 4 \times 10^2$ m/s for oxygen at 0 K) shows that $D$ is about $10^{-3}$ m$^2$/s at atmospheric pressure. Because diffusion occurs relatively slowly, convection is a more important mechanism of transport in gases.

## THE LIQUID PHASE

Liquid metals are of commercial interest for two reasons. First, most metals are produced and refined in the liquid state and are later subjected to a casting process as a step in their fabrication. Second, there have been a number of direct engineering applications of molten metals; their use as cooling media in nuclear power production is an example. Since the properties that must be considered in working with metal liquids are similar to those of other liquids (for example, hydrodynamic properties and surface tension), we will look briefly at the essential character of the liquid state. We begin from the viewpoint of the gas phase, but will see later that a close relationship exists with the solid phase as well.

**Gas → liquid transition.** When a one-component gas is severely compressed at a temperature $T_1$ slightly above the melting temperature, Fig. 6.10, eventually a second phase forms (the liquid). Application of the phase rule, Eq. (6.3)

$$P + F = C + 2$$

$$2 + 1 = 1 + 2$$

shows that there is only one degree of freedom, which has already been exercised in fixing the temperature at $T_1$. Consequently, both the pressure $P_1$ and the molar volume ($V_g$ and $V_l$, respectively) must remain constant for the gas phase and for the liquid phase during the gas → liquid transition. The amount of the gas phase gradually decreases and the amount of the liquid phase increases until at the end of the transition the system is entirely liquid. During this period the (total) molar volume $V$ obeys Eq. (6.20),

$$V = N_g V_g + N_l V_l, \tag{6.50}$$

and therefore decreases linearly from $V_g$ to $V_l$ as $N_g$ decreases from unity to zero.

At sufficiently high temperatures (2000 K for mercury vapor, for example) compression of progressively increasing degree causes a gradual transition from the gas phase to the liquid phase, curve $T_6$ in Fig. 6.10. The density of the "fluid" simply increases gradually from the low values normally associated with a gas to the high values characteristic of a liquid. At an intermediate temperature $T_c$, the *critical temperature*, when the pressure reaches the critical pressure $P_c$, the system is just on the verge of forming two phases. This *critical phenomenon* is of practical importance in such applications as steam turbines and is also of great theoretical interest. Most metals, however, are used at temperatures far below their critical temperatures, and therefore a clear difference exists between the gas phase and the liquid phase with which it is in equilibrium.

**Structure of liquids.** As in the case of solid crystalline phases, the structure of a liquid can be studied by diffraction techniques, such as x-ray diffraction. As discussed previously, Chapter 5, crystalline phases give diffraction patterns consisting of sharp lines; in contrast, liquids give only a few broad diffuse bands of diffraction. This type of pattern indicates that, although the atoms of the liquid do not have the long-range order of a crystalline substance, there is a tendency toward packing of the neighboring atoms around a given atom in the liquid. A quantitative interpretation of a diffraction pattern is made by considering that diffraction is caused by the number of atoms $(4\pi r^2 \rho)\,dr$ in each spherical shell of thickness $dr$ around the reference atom. If the number of atoms per cubic meter, $\rho$, were constant in the liquid, the quantity $4\pi r^2 \rho$ would simply plot as a parabola with increasing $r$, as shown by the dashed line in Fig. 6.11. The actual variation of the density of atoms determined from the x-ray pattern is shown by the solid line for liquid sodium. The density is essentially zero for distances out to the radius of the reference atom. Beyond this point it rises to a maximum value at the center of the first shell of packed neighboring atoms and then falls to a minimum before again rising to a second, less distinct maximum.

If the number of nearest neighbors (the coordination number) of the reference atom is determined from the area under the curve out to the minimum point, it is found to be 10.6. Since solid sodium is body-centered cubic, the coordination number of the solid is only 8, although the vertical lines in Fig. 6.11 show that there are six additional atoms only slightly beyond the eight nearest neighbors.

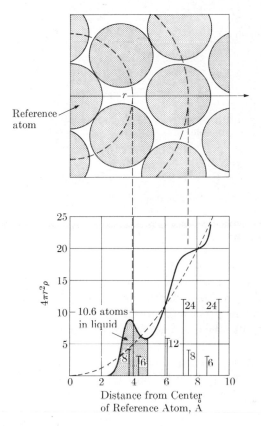

**Fig. 6.11** Atomic distribution curve for liquid sodium calculated from x-ray diffraction data. The upper drawing explains the two maxima in the curve in terms of "shells" of neighboring atoms around the reference atom. The dashed curve represents a uniform distribution of atoms, and the vertical lines show the positions and numbers of neighboring atoms in solid sodium.

Face-centred-cubic and close-packed-hexagonal metals have a higher coordination number (12) in the solid state, but their liquid structures are about the same as that of liquid sodium. Thus, the structures of various liquid metals show more similarity than do the structures of solid metals. This similarity is reflected in the properties of liquid metals, where much greater uniformity is exhibited than in the case of solid metals.

**Solid → liquid transition.** Changes from the liquid to the solid state are important commercially in connection with the solidification of liquid alloys; these changes also help give an insight into the relation between the nature of the solid and liquid states. In terms of thermodynamics, the free energy of a solid metal *below* its melting point $T_f$ must be lower than the free energy of the (perhaps hypothetical)

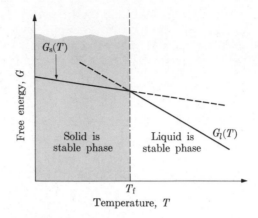

**Fig. 6.12** Relation of the molar free energies of the solid and liquid phases to the melting point $T_f$ of a pure metal.

liquid metal, Fig. 6.12. Therefore, if a liquid metal is cooled to below $T_f$, the change in free energy accompanying the process of solidification is $(G_s - G_l) < 0$. Since $\Delta G$ is negative, the process can occur and produce the (stable) solid phase. Similar reasoning relates the free energies to the process of melting of a solid at a temperature above $T_f$.

An analysis of the equilibrium between the solid and liquid phases at $T_f$ can give information on the change in entropy, $\Delta S_f$, that accompanies melting. Since $\Delta G = 0$,

$$H_l - H_s = T_f(S_l - S_s)$$

or

$$\Delta H_f = T_f \Delta S_f. \tag{6.51}$$

Since $\Delta H_f$ is positive (that is, heat must be added to the system to cause the process of melting to proceed), $\Delta S_f$ is also positive. This means that the liquid phase has a higher entropy (more randomness) than the solid phase. This conclusion, of course, agrees with the usual picture of the relative disorder of a liquid. A widely accepted theory of the structure of liquids, however, stresses the importance of "clumps of ordered atoms," made mobile relative to one another by the presence of the 10% of vacant space or "holes" in the liquid structure.

Some of the changes in properties on melting are especially interesting. Since most metals have their atoms in a close-packed arrangement in the solid, their molar volumes increase by several percent on melting. Antimony, bismuth, and gallium, on the other hand, exhibit a decrease in molar volume because their crystal structures pack the atoms quite inefficiently; consequently their liquid structures are more dense, in spite of the appreciable fraction of holes. Melting causes a pronounced change in the mobility of the metal. The mobility of a given atom relative to its immediate neighbors (diffusivity) increases by about a factor

**TABLE 6.2**

Viscosities of Several Liquid Metals and Alloys

| Material | Temperature, °C | Viscosity,* $N \cdot s/m^2 \times 10^{-3}$ |
|---|---|---|
| Aluminum | 700 | 3.0 |
| Copper | 1200 | 3.2 |
| Iron | 1600 | 6.2 |
| Iron—3.4% carbon | 1300 | 8.9 |
| White cast iron—3.4% carbon | 1300 | 2.4 |
| Steel | (above melting range) | approx. 3 |
| Magnesium | 680 | 1.2 |
| Zinc | 500 | 3.7 |

\* To convert from $N \cdot s/m^2$ to poise, multiply by 10.

of $10^5$. The overall flow (viscosity) of one portion of the metal relative to another macroscopic portion increases by a factor of about $10^{15}$. These two phenomena will now be discussed more fully.

**Diffusivity and viscosity.** Compared to diffusion in solids and gases, diffusion in liquids is poorly understood. In the theory of random walk (discussed in Chapter 8), for example, the jump distance $\alpha$ and the jump frequency $\Gamma$ have definite physical significance for both solids and gases and can be evaluated for important systems. The mechanism of diffusion in liquids is uncertain at present, but it probably is different for the ordered regions and for the surrounding portions rich in holes. Consequently, average values of $\alpha$ and $\Gamma$ have little relation to the atomic processes occurring in liquids. Nevertheless, the methods described in Chapter 8 permit the application of Fick's second law, Eq. (8.13), to experimental data to determine empirical values of the interdiffusion coefficient $\tilde{D}$ for a pair of liquid metals. The values so determined are uniformly about $10^{-9}$ $m^2/s$, and thus are intermediate between those for gases ($10^{-3}$) and solids ($10^{-11}$ to $10^{-20}$). As in the case of gases, under normal conditions convection in liquids is a far more important mechanism of transport than is diffusion.

When a shear stress $\tau$ acts on a portion of liquid to produce a steady rate of shear strain $d\gamma/dt$, the coefficient $\eta$ in Newton's law of viscosity

$$\tau = \eta \frac{d\gamma}{dt} \tag{6.52}$$

is the dynamic viscosity of the liquid. It is the shear stress required to maintain unit rate of shearing, $1 \ m/(m \cdot s)$. Typical values for liquid metals, Table 6.2, are on the order of $10^{-3} \ N \cdot s/m^2$ and thus are extremely small compared to the shear strength of solid metals somewhat below the melting point.

*Fluidity*, a property related to viscosity, is of great interest in the casting of metals. It is usually studied in terms of the distance in inches a liquid alloy will flow in a standard groove mold before solidifying. This flow distance depends on several factors in addition to viscosity, and therefore it is sometimes called *casting fluidity*. Although oxide films and dissolved gases are complicating variables, usually fluidity increases linearly with temperature and decreases with initial additions of alloying elements. The data for the aluminum-silicon system given in Fig. 7.15 illustrate a typical variation of casting fluidity with alloy composition.

The flow of liquid alloys during casting operations or in engineering equipment is governed by the usual laws of hydrodynamics. For example, hydraulic pressure can be obtained from a head of liquid metal for use in forcing the metal to fill difficult mold sections. The Reynolds number concept also applies and can be used to predict whether flow will be laminar or turbulent under given conditions. Actually, the flow of metal in castings is usually turbulent, but there appears to be an upper limit where increasing turbulence is detrimental. An unusual aspect of metal casting is the fact that the flow of metal in the initial stages of filling the mold is largely dependent on the momentum of the liquid stream entering the mold.

## SOLID PHASES

Many kinds and distributions of solid phases are found in metallurgical systems. The underlying cause of this variety is the crystalline nature of these phases, which was discussed in Chapter 1. However, several additional aspects of crystal structure will be studied here that profoundly affect the actual microstructures found in commercial metals and alloys.

**Allotropy or polymorphism.** Many of the metallic elements exist in alternative crystalline forms depending on the external conditions of temperature and pressure. This phenomenon is called *allotropy* or *polymorphism*. Under the usual condition of atmospheric pressure each allotropic form of a metal exists at equilibrium over a range of temperatures, as shown for iron in Fig. 6.13. The iron atoms are at the points of a body-centered-cubic space lattice at temperatures below 910°C. If this piece of iron is heated above 910°C the atoms proceed to form a different *solid* phase, a phase in which the iron atoms are at the points of a face-centered-cubic lattice. A second allotropic change in iron occurs on heating the iron above 1400°C. The face-centered-cubic phase then becomes unstable and changes to a body-centered-cubic phase. Almost all the properties of a metal change when the metal transforms from one allotropic form to another, but the change of principal commercial importance is the apparently trivial one of a decrease in the solubility of solid iron for carbon when the iron changes from the gamma to the alpha form. This effect forms the basis of steel hardening and will be studied in detail in Chapter 10.

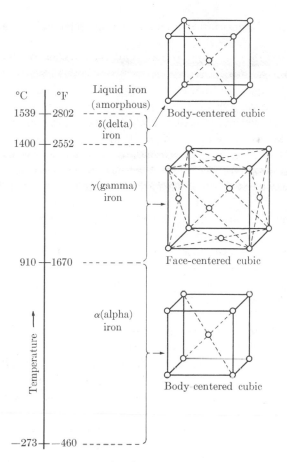

°C  °F

1539 — 2802

1400 — 2552

910 — 1670

−273 — −460

Liquid iron
(amorphous)

Body-centered cubic

δ(delta)
iron

γ(gamma)
iron

Face-centered cubic

α(alpha)
iron

Body-centered cubic

Temperature →

**Fig. 6.13** The temperature ranges in which the allotropic forms of iron exist under equilibrium conditions at atmospheric pressure.

An allotropic change is a phase change. The solid metal changes from one crystalline form to another, from one phase to another phase. In a more easily recognized phase transformation, the change from the *liquid* to a *solid* phase, two phenomena are observed: (1) supercooling and (2) liberation of the heat of reaction. Both these effects are also associated with phase transformations in the solid state.

*Supercooling*, the delay of a phase change to a temperature below the equilibrium temperature for such change, is especially pronounced in solid-state transformations. Whereas liquids usually supercool a few degrees before solidifying, solids may be cooled to tens or hundreds of degrees below the equilibrium temperature before transformation to the new phase occurs. The consideration of phase diagrams in Chapter 7 will not be complicated by constant reference to supercooling, but its importance in many commercial heat treatments will be discussed later.

The heat effects that accompany phase transformations in the solid state are small compared with those associated with the solidification of liquid metal. For example, 271 joules are liberated when a gram of liquid iron solidifies, while only 15 joules are liberated by a gram of iron when it changes from the gamma to the alpha form. However, even this small heat of transformation is detectable, and *thermal analysis* is often used to study solid-state transformations as well as solid-liquid reactions. This technique consists basically in detecting departures from the normal cooling (or heating) curve of a substance and interpreting these as evidence for absorption or evolution of heat during phase transformations.

Although the allotropic change in iron is of most importance industrially, there are two other examples of allotropy that deserve attention. One of these is the change in titanium from close-packed-hexagonal to the body-centered-cubic structure at temperatures above 885°C. This transformation vitally affects the technology of titanium alloys. A second important example of allotropy is that in uranium. The low-temperature phase α is orthorhombic and exists up to 668°C. From 668 to 774°C uranium exists as the complex, tetragonal β-structure containing 30 atoms per unit cell, while above 774°C a body-centered-cubic lattice is the stable phase. When uranium is forged or rolled in the α-phase condition there is a strong tendency toward the development of preferred orientation of crystal axes similar to that sketched in Fig. 2.9. Crystals of α-uranium behave in an unusual manner on being heated in that they contract in one direction and expand strongly in the other two directions. The combined effects of preferred orientation and nonuniform thermal expansion can cause such severe distortion of a uranium specimen that it may grow to many times its original size. This difficulty is largely overcome if the preferred orientation is eliminated by heating the uranium into the temperature range where the β-phase forms and then rapidly cooling to re-form randomly oriented grains of the α-phase.

**Types of solid phases.** Pure metals are the simplest examples of solid phases, but several other kinds of phases are commonly found in metallurgical systems. As discussed in Chapter 3, when a second element dissolves in a solid metal the resulting phase is called a *solid solution*. The solute atom may occupy two alternative types of position in the lattice of the matrix (solvent) metal, as shown in Fig. 6.14. If the two atoms are of comparable size, the solute atom will substitute at random for one of the matrix atoms in the crystal lattice, Fig. 6.14(a). This kind of structure is called a *substitutional* solid solution. There are a few relatively small atoms that can be accommodated in the interstices between the matrix atoms (Fig. 6.14(b)) to form an *interstitial* solid solution. Examples of these two kinds of solid solution will be given in Chapter 7, and it will be shown that no sharp dividing line exists between a pure metal and its solid solution.

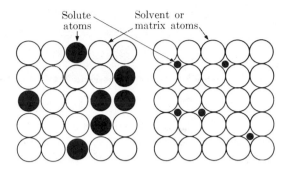

(a) Substitutional solid solution.     (b) Interstitial solid solution.

**Fig. 6.14** The two different types of solid-solution phases.

*Vacancies in Solid Phases.* As mentioned in Chapter 3, all solid phases at temperatures above absolute zero contain point defects of various kinds. Vacancies are especially important and are discussed here for the special case of a pure metal, but a closely similar analysis applies to any alloy phase. Equation (6.16) can be written for a solution of $N_V$ mole fraction of vacancies in pure metal 1,

$$\Delta S^{id} = - R(N_V \ln N_V + N_1 \ln N_1). \qquad (6.53)$$

For the hypothetical reaction involving the addition of *one* vacancy to a mole of perfect metal 1 (with no vacancies initially), one can show that the molar entropy change is $56R$ (see Problem 11). If the metal is at 1000 K, and neglecting $\Delta S^{xs}$ for the moment, the change in free energy $G$, (Eq. (6.25), is negative (and therefore the hypothetical reaction can occur) if $\Delta H$ is less than $T \Delta S$ or $1000 \times 56R$. Since $\Delta H$ is typically about $10^4 R$, the change in entropy of the system more than compensates for the energy $\Delta H$ needed to form the first vacancy.

The equilibrium concentration of vacancies, $N_V^0$, exists when the creation of one *additional* vacancy causes no change in free energy, Eq. (6.29). From Eqs. (6.25), (6.15), and (6.53),

$$\Delta G = \Delta H - T \Delta S = \Delta H - T(\Delta S^{id} + \Delta S^{xs})$$

$$= \Delta H - T \Delta S^{xs} + RT[N_V \ln N_V + (1 - N_V) \ln (1 - N_V)]. \qquad (6.54)$$

Therefore, at equilibrium (see Problem 12),

$$dG = \frac{d(\Delta G)}{dN_V} dN_V = 0 = (\Delta H - T \Delta S^{xs}) \, dN_V + RT \ln \frac{N_V^0}{(1 - N_V^0)} dN_V. \qquad (6.55)$$

Since the equilibrium concentration of vacancies, $N_V^0$, is very small compared to unity, Eq. (6.55) can be written

$$N_v^0 = \exp\left[-\frac{(\Delta H_f - T\,\Delta S_f^{xs})}{RT}\right] = \exp\left(-\frac{\Delta G_f}{RT}\right) = \exp\left(\frac{\Delta S_f^{xs}}{R}\right)\exp\left(-\frac{\Delta H_f}{RT}\right). \quad (6.56)$$

Here $\Delta G_f$ is known as the free energy of formation* of vacancies even though it lacks the term $T\,\Delta S^{id}$,

$$\Delta G_f = \Delta H_f - T\Delta S_f^{xs}. \quad (6.57)$$

A typical variation of $N_v^0$ with temperature is shown in a later chapter, Fig. 8.5, for gold. Near the melting point $N_v^0$ is almost $10^{-3}$ but the value drops to less than $10^{-6}$ at $0.5\ T_f$, where $T_f$ is the absolute melting temperature.

*Substitutional Solid Solutions.*  In the course of an alloy development it is frequently desirable to increase the strength of a given alloy by adding a metal that will form a solid solution.  Unfortunately, if an alloying element is chosen at random it is likely to form an objectionable intermediate phase instead of a solid solution.  Largely through the work of Hume-Rothery, a number of general rules governing the formation of substitutional solid solutions are available to aid in the proper choice of such alloying elements.  These rules can be summarized as follows.

1. *Relative size factor.*  If the sizes of two metallic atoms (given approximately by their lattice constants) differ by less than 15%, the metals are said to have a *favorable* size factor for solid-solution formation. So far as this factor is concerned, each of the metals will be able to dissolve appreciably (on the order of 10%) in the other metal.  If the size factor is greater than 15%, solid-solution formation tends to be severely limited and usually is only a fraction of 1%.

2. *Chemical affinity factor.*  The greater the chemical affinity of two metals, the more restricted is their solid solubility.  When their chemical affinity is great, two metals tend to form an intermediate phase rather than a solid solution.

3. *Relative valency factor.*  If the alloying element has a valence different from that of the base metal, the number of valence electrons per atom, called the *electron ratio*, will be changed by alloying.  Crystal structures are more sensitive to a decrease in the electron ratio than to an increase.  Therefore, a metal of high valence can dissolve only a small amount of a lower valence metal, whereas the lower valence metal may have good solubility for the higher valence metal.

4. *Lattice-type factor.*  Only metals that have the same type of lattice (face-centered cubic, for example) can form a complete series of solid solutions.  Also,

---

* The subscript "f" (for *formation*) is widely used in the literature.  Therefore, this notation is employed here even though the same subscript also refers to melting (fusion); for example, $T_f$ and $\Delta H_f$.

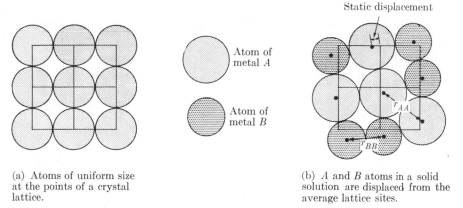

(a) Atoms of uniform size at the points of a crystal lattice.

(b) $A$ and $B$ atoms in a solid solution are displaced from the average lattice sites.

**Fig. 6.15** The static displacement of atoms in a solid solution. (After B. L. Averbach.)

for complete solid solubility the size factor must usually be less than 8%. Copper-nickel and silver-gold-platinum are examples of binary and ternary systems exhibiting complete solid solubility.

A qualitative estimate of the solid solubility of one metal in another can be obtained by considering these four factors. It should be noted that an *unfavorable relative size factor alone is sufficient to limit solid solubility to a low value*. If the relative size factor is *favorable*, then the other three factors should be considered in deciding on the probable degree of solid solubility. It must be emphasized that numerous exceptions to these Hume-Rothery rules are known.

X-ray studies have revealed several important features of the structure of actual solid solutions. In a solution of two atoms of different size, Fig. 6.15, the tendency toward close packing causes a typical atom to be somewhat displaced from the ideal lattice site. This distortion of the lattice can be evaluated quantitatively as the average (root mean square) static displacement of the atoms and has been found to be about 0.1 Å for systems in which the two atoms differ in size by 10% to 15%.

The size of an atom in a pure metal is ordinarily determined as the distance from the center of a given atom to the center of a nearest-neighbor atom (see Problem 13). The corresponding distance, $r_{AA}$ for example, can also be determined in a solid solution from suitable x-ray data. The results of a study of this kind are shown in Fig. 6.16(b) for gold and nickel, which can dissolve completely in each other over the entire range of compositions. It is seen that the size of the gold atom decreases from its size in pure gold as the fraction of nickel atoms in the alloy increases. The nickel atoms show a corresponding change. The size of the average atom in a solid solution determines the lattice parameter, which ideally varies linearly with composition according to *Vegard's law*. The gold-nickel solid solutions exhibit a positive deviation from Vegard's law.

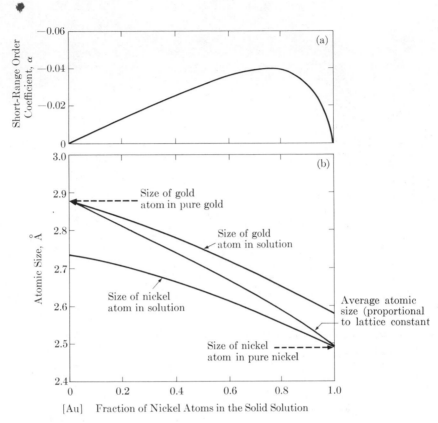

**Fig. 6.16** Results of x-ray studies of the gold-nickel solid solutions. (a) The short-range order coefficient. (b) The sizes of the atoms in solution. (After Flinn, Averbach, and Cohen.)

When the distribution of $A$ and $B$ atoms in a solid solution is completely random, the probability $p_A$ that a given neighbor of a $B$ atom is an $A$ atom is just the fraction $N_A$ of $A$ atoms in the alloy. That is, $p_A/N_A = 1$. The amount $\alpha$ by which this ratio differs from unity in a given case is a useful measure of the departure of a solid solution from randomness:

$$\alpha = 1 - (p_A/N_A), \tag{6.58}$$

where $\alpha$ is the short-range order coefficient. There are two possible departures from $\alpha = 0$, the value for a random solution. If $B$ atoms group themselves preferentially around other $B$ atoms, then $p_A$ is less than $N_A$ and $\alpha$ is positive. This behavior is called *clustering* in the solid solution. On the other hand, if a given $B$ atom is preferentially surrounded by $A$ atoms, $\alpha$ is negative and the solid solution is said to show *short-range ordering.* The curve of Fig. 6.16(a) shows that the degree of ordering varies with composition in the gold-nickel system and that in this case even the maximum departure from a random distribution is not large.

*Interstitial Solid Solutions.*  The elements hydrogen, boron, carbon, nitrogen, and oxygen have such small atomic diameters that they occupy the interstices in the crystal lattices of many metals, Fig. 6.14(b).   Such interstitial solid solutions usually have a limited range of composition and are most common in the more open BCC crystal structure.  They can affect the properties of some metals very significantly; for example, the interstitial solution of carbon in iron is the basis of the hardening of steel, discussed in Chapters 9 and 10.  Very small amounts of hydrogen introduced into steels during acid pickling (cleaning), plating, or welding operations cause a sharp decrease in ductility, known as *hydrogen embrittlement.* Interstitial nitrogen is useful not only in the nitriding process but also as an important factor in maintaining 18 Cr-8 Ni stainless steel in the austenitic condition (Chapter 11).

In some alloys both interstitial and substitutional solid solutions are formed to an appreciable extent.  For example, a chromium-nickel steel contains interstitially dissolved carbon and substitutionally dissolved chromium, nickel, and minor elements.

*Amorphous Solid Phases.*  Nonmetallic materials sometimes exist in a solid, but noncrystalline, form known as the glassy state.  Metals and alloys, on the other hand, have such a strong tendency to crystallize that an amorphous (glassy) solid can be obtained only under special circumstances.  During deposition from the vapor phase, for example, amorphous deposits tend to form if their temperatures are maintained below about 75 K.  A liquid alloy will solidify as an amorphous solid only if it is cooled at extremely high rates, on the order of $10^6$ to $10^8$ K/s. One of the techniques employed for this purpose is *splat cooling,* in which a drop of liquid is propelled at high velocity against a heat-conducting surface.  The solidified alloys resulting from such treatments may exhibit one or more of the following nonequilibrium structures: (1) amorphous phases; (2) nonequilibrium crystalline phases; or (3) crystalline solid solutions extending well beyond their equilibrium solubility.  These nonequilibrium structures tend to be highly unstable; often they must be maintained at low temperatures to prevent transformation to more stable configurations.  At present, amorphous alloys and the related nonequilibrium crystalline phases are of scientific, rather than industrial, interest.

**Intermediate phases.**  When an alloying element is added to a given metal in such amount that the limit of solid solubility is exceeded, a second phase appears with the solid solution.  This second phase may be the (primary) solid solution of the alloying element.  The aluminum-silicon system (Fig. 7.13) is an example of this behavior.  When the solubility of aluminum for silicon is exceeded, the silicon-rich $\beta$-phase appears with the aluminum-rich $\alpha$-phase.  More often the second phase that appears is an intermediate phase, such as the $Mg_2Pb$ compound in the lead-magnesium diagram, Fig. 7.26. It is convenient to classify the possible intermediate phases into categories determined by their structures.

*Interstitial Compounds.* When the solubility of an interstitially dissolved element is exceeded, an intermediate phase is produced in which these elements are again at interstitial positions. Compounds of this type, such as TaC, TiN, WC, $TiH_2$, $Mn_2N$, and $Fe_3C$, have high hardnesses and melting points, and they are useful in cemented carbides and for strengthening steels.

*Valency Compounds.* Two chemically dissimilar metals tend to form compounds showing ordinary chemical valence. As would be expected from the ionic or covalent binding characteristic of these compounds, their properties are essentially nonmetallic; moreover, they are brittle and have poor electrical conductivity. Examples of valency compounds are $Mg_2Si$, $Mg_2Pb$, and AlSb.

*Electron Phases.* The intermediate phases that appear at definite compositions in certain binary equilibrium diagrams have been shown to depend on the ratio of electrons to atoms at those compositions. Consider a monovalent base metal like copper, which is face-centered cubic. The initial *electron ratio* is then one electron per atom. If a divalent alloying metal, such as zinc, is added to the base metal, the electron ratio increases. For example, at 50 atomic percent zinc, the electron ratio is three electrons to two atoms, or 1.50. At this ratio a body-centered-cubic $\beta$-*phase* appears. A complex cubic $\gamma$-*phase* appears at a ratio of 1.62, and a close-packed-hexagonal $\varepsilon$-*phase* appears at 1.75. The essentially metallic nature of these phases is indicated by the electronic character of their binding and by the fact that each phase exists over a range of compositions. The copper-zinc diagram in Fig. 7.47 shows the electron phases $\beta$, $\gamma$, and $\varepsilon$ in addition to the primary solid solutions, $\alpha$ and $\eta$.

*Ordered Phases.* The extension of a primary solid solution over a large composition range is sometimes interrupted by the occurrence of an ordered form of that solid solution, Fig. 7.30. Such an ordered phase is roughly similar in mechanical and physical properties to the (disordered) solid solution from which it forms. A fuller treatment of order-disorder transformations is given in Chapter 10.

*Defect Phases.* A few intermetallic phases contain an unusually large number of vacant lattice sites in certain ranges of concentration and are known as defect phases. An example is NiAl, which has the CsCl structure at the equiatomic composition. If the aluminum content is increased beyond 50 a/o, the CsCl structure is maintained, but at the expense of producing vacancies on the sites normally occupied by nickel atoms. The maximum concentration of vacancies, 8%, occurs at about 54% aluminum. Nickel contents above 50 a/o, on the other hand, are accommodated in the usual manner; namely by nickel substituting for aluminum on sites normally occupied by aluminum atoms. Defect phases are interesting scientifically because of the unusual effects produced by the large vacancy concentrations, for example in diffusion.

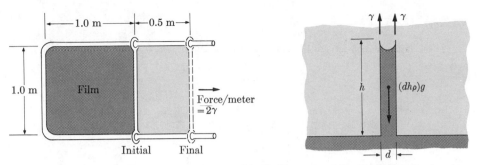

(a) Stretching a soap film on a wire frame.    (b) Capillary rise of liquid metal during brazing.

**Fig. 6.17**  Illustrations of the action of surface tension in liquids.

## SURFACES

The properties of a phase in its surface layers are usually quite different from those in the bulk of the phase. Because the overwhelming majority of the atoms are in the interior of a phase, its bulk properties determine the behavior of a phase under most conditions. Some phenomena (such as adsorption, adhesion, and catalysis), however, depend primarily on the properties of the surface layers. We will see that the microstructure of metals and alloys is also greatly influenced by the energy associated with surfaces of various types.

Two different terms, *surface* and *interface*, are commonly used to describe the surface region of a liquid or solid phase depending, respectively, on whether the adjacent phase is a gas or a second condensed phase. The surface energies are usually higher, $> 1 \, \text{J/m}^2$ compared to $< 0.5 \, \text{J/m}^2$ for interfaces. In both cases, however, the depth of the surface layer is only about two atom distances; below this depth the positions and energies of the atoms are little affected by the presence of the surface. Grain boundaries are a special type of interface since they are a boundary within a single phase. Nevertheless these boundaries (the structures of which were discussed in Chapter 3) are usually treated along with other types of interfaces.

**Surface energy.**  Since an additional energy $\gamma \, \text{J/m}^2$ is associated with each square meter of surface, a given system tends to minimize its surface area, thus minimizing this aspect of its energy. Droplets of water, for example, tend to be spherical. A familiar example of the action of surface energy is the surface tension of a soap film, Fig. 6.17(a). Since a soap film has two surfaces, 1 m² of new surface is created if the force/meter, $2\gamma$, acts over a distance of 0.5 meter,

$$2\gamma \times 0.5 = \gamma \, \text{J/m}^2. \tag{6.59}$$

Thus, the surface energy acts like a surface tension, $\gamma \, \text{N/m}$, in each surface of the soap film. Many problems in surface energy or interfacial energy are more conveniently visualized in terms of the corresponding surface tensions.

**TABLE 6.3**

Surface Tensions of Liquid Metals and Other Liquids

| Liquid | Temperature | | Surface tension $\gamma$, N/m* |
|---|---|---|---|
| | °C | °F | |
| Aluminum | 750 | 1380 | 0.520 |
| Aluminum and oxide film | 700 | 1290 | 0.840 |
| Copper | 1200 | 2200 | 1.160 |
| Iron | 1600 | 2900 | 1.360 |
| Lead | 350 | 660 | 0.453 |
| Magnesium | 681 | 1258 | 0.563 |
| Mercury | 20 | 68 | 0.465 |
| Zinc | 600 | 1100 | 0.770 |
| Fused salts | 400 to 930 | 750 to 1700 | 0.035 to 0.180 |
| Water | 20 | 68 | 0.074 |

* To convert from N/m to dyn/cm, multiply by $10^3$.

An important instance of surface tension is the capillary attraction that draws liquid metal (Fig. 6.17(b)) upward against the force of gravity into a narrow space of width $d$ between two pieces of metal. In a specimen of unit length perpendicular to the plane of the drawing, the surface tension exerts a force $2\gamma$ (assuming that the liquid "wets" the solid metal). This force is balanced by the force of gravity, $(dh\rho)\,g$, acting on the column of liquid of height $h$ and having density $\rho$. The liquid tends to rise to the height at which these two forces are in balance; therefore.

$$h = \frac{2\gamma}{g\rho d}. \tag{6.60}$$

This equation can be applied to the brazing process in which liquid copper is used to join two pieces of a higher-melting metal. At 1200°C the surface tension of liquid copper is 1.16 N/m, Table 6.3. If the spacing is $10^{-4}$ m, as in brazing practice, the copper tends to rise:

$$h = \frac{2 \times 1.16}{9.781 \times 8.9 \times 10^3 \times 10^{-4}} = 0.27 \text{ m}.$$

Often this effect is desirable, but in other cases it must be guarded against by avoiding long, narrow channels adjacent to areas that are to be brazed.

**Interfacial energy.** The two types of interfaces to be considered in this section, solid–liquid and solid–solid, differ significantly from a liquid–vapor interface.

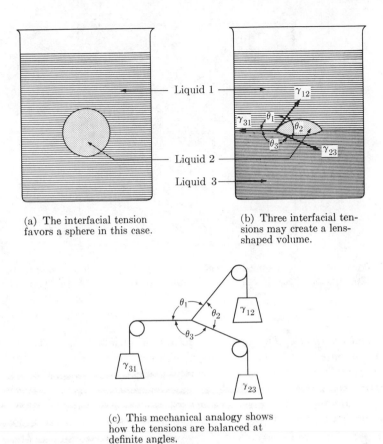

(a) The interfacial tension favors a sphere in this case.

(b) Three interfacial tensions may create a lens-shaped volume.

(c) This mechanical analogy shows how the tensions are balanced at definite angles.

**Fig. 6.18** Equilibrium of interfaces among immiscible liquid phases. (After C. S. Smith.)

First, a solid phase does not easily change its shape in response to a surface tension; consequently the effects of surface energy develop only gradually. Nevertheless, when solid–liquid or solid–solid systems are permitted adequate time to adjust their configurations, surface energy often exerts a decisive influence. A second factor is important in determining the shape of the interface separating one solid phase $\alpha$ from a different solid phase $\beta$. This factor is the strain energy associated with particles of $\beta$, for example, within a matrix of the $\alpha$-phase. If a particle of the $\beta$-phase occupies a larger volume than the $\alpha$-phase from which it forms, then the strain energy is lower if the particle assumes the form of a thin disk rather than a spherical shape. A third factor is the variation of surface energy with orientation of the surface relative to the crystal axes of a solid phase. This factor is especially significant for solid–solid interfaces; for example, in Chapter 3 we saw that a boundary in twin orientation between two grains has an unusually low interfacial energy.

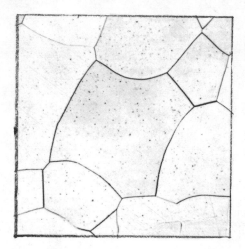

**Fig. 6.19**  Polygonal grains in annealed iron, showing the tendency for three grain boundaries to meet at 120° angles ($\times$ 250). (Courtesy C. S. Smith, Massachusetts Institute of Technology.)

*Solid–Liquid Interfaces.*  The mode of action of surface tension at boundaries between condensed phases is shown especially clearly in the equilibrium between two or more different liquid phases.  If one liquid phase is completely surrounded by another, Fig. 6.18(a), it will tend to assume a spherical shape, since this shape minimizes the interfacial area.  However, if the same liquid phase lies between two other liquids, Fig. 6.18(b), then the geometry of the interfaces is determined by the three interfacial tensions $\gamma_{12}$, $\gamma_{23}$, and $\gamma_{31}$ existing between the pairs of liquid phases.  These tensions can be pictured as exerting forces on the periphery of the lens-shaped volume of liquid 2.  The angles $\theta_1$, $\theta_2$, and $\theta_3$ between the three pairs of interfaces are those existing when the tensions are balanced vectorially in the manner shown by the mechanical analogy of Fig. 6.18(c).  The following relation then applies:

$$\frac{\gamma_{12}}{\sin \theta_3} = \frac{\gamma_{23}}{\sin \theta_1} = \frac{\gamma_{31}}{\sin \theta_2}. \qquad (6.61)$$

This equation can be applied when three grains meet to form a boundary in a solid metal.  If all the boundaries then have the same surface tension, the three angles should be equal when equilibrium has been established.  Statistical studies of grain geometry in pure metals have verified this conclusion.  Figure 6.19 shows the occurrence of 120° angles in annealed iron.

An important example of interfacial equilibrium is that occurring on a solid surface, for example, the equilibrium shown in Fig. 6.20 involving a liquid phase and a vapor phase.  Under the conditions shown the liquid tends to form a spherical cap with a contact angle $\theta$ determined by the equilibrium of the hori-

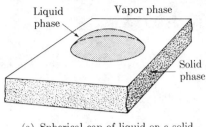
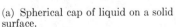

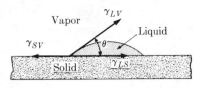

(a) Spherical cap of liquid on a solid surface.

(b) Section through the liquid cap showing the equilibrium of the interfacial tensions.

**Fig. 6.20** Equilibrium of interfacial tensions for a spherical cap of liquid phase in contact with a solid and a vapor.

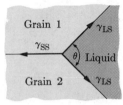

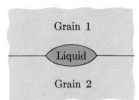

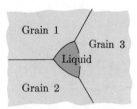

(a) Configuration employed in the derivation of Eq. (6-64).

(b) For moderate values of dihedral angle, particles of liquid assume this shape, along a grain boundary.

(c) For moderate values of dihedral angle, particles of liquid assume this shape if at a grain junction.

**Fig. 6.21** Important examples of equilibrium configurations involving solid–liquid interfaces.

zontal components of the three interfacial tensions,

$$\gamma_{SV} = \gamma_{LS} + \gamma_{LV} \cos \theta. \tag{6.62}$$

The vertical component of the liquid–vapor interfacial tension is balanced by elastic forces that it sets up in the solid. Surface-tension values for the interface between the liquid phase and an inert vapor phase were given in Table 6.3. Representative values of other types of interfacial tensions are listed in Table 6.4.

When a junction occurs involving two grains of a given solid metal and a region of liquid, Fig. 6.21(a), the angle $\theta$ formed by the liquid region at equilibrium is called the *dihedral angle*. Since the horizontal components of the surface tensions must balance,

$$\gamma_{SS} = 2\gamma_{LS} \cos \frac{\theta}{2}, \tag{6.63}$$

and therefore,

$$\cos \frac{\theta}{2} = \frac{\gamma_{SS}}{2\gamma_{LS}}. \tag{6.64}$$

**TABLE 6.4**

Surface and Interfacial Energies of Solid Metals

| Metal | Type of interface | Energy, J/m$^2$ |
|-------|-------------------|-----------------|
| | Surface energies | |
| Lead | Solid–vapor | 0.500 |
| Silver | Solid–vapor | 1.130 |
| Gold | Solid–vapor | 1.350 |
| Copper | Solid–vapor | 1.650 |
| | Interfacial energies | |
| Lead | Solid–liquid | 0.030 |
| Silver | Solid–liquid | 0.120 |
| Gold | Solid–liquid | 0.130 |
| Copper | Solid–liquid | 0.150 |
| Silver | Grain boundary | 0.500 |
| Copper | Grain boundary | 0.600 |
| Copper | Twin boundary | 0.025 |

If $\gamma_{LS}$ is about equal to $\gamma_{SS}$, then $\theta$ has a value in the vicinity of 120°. A small volume of liquid then assumes a lenticular shape along the grain boundary, Fig. 6.21(b), or an analogous shape at the junction of three grains of the solid, Fig. 6.21(c). If $\gamma_{LS}$ is several-fold larger than $\gamma_{SS}$, then $\theta$ approaches 180° and the liquid region approaches a spherical shape. In this case the energy of the system decreases relatively little by virtue of the location of the sphere of liquid on a grain boundary (compared to a location within a single grain), Problem 15.

An important phenomenon occurs if $\gamma_{LS}$ equals $\gamma_{SS}/2$ or less; Eq. (6.64) then gives $\theta = 0$. In Fig. 6.21(a) one can picture that the relatively large value of $\gamma_{SS}$ then "pulls" the liquid phase completely along the grain boundary. In fact, the liquid tends to spread as a thin layer throughout all the grain boundaries. This type of structure can cause difficulties of at least two kinds in the solidified alloy (the liquid then having become the solid phase $\beta$). If the $\beta$-phase is brittle, as in the case of small amounts of bismuth impurity in copper, the alloy behaves in a brittle manner because of fracture along the grain boundaries. The second difficulty occurs when such an alloy is heated for such a purpose as hot-working. The $\beta$-phase may melt far below the melting temperature of the $\alpha$-matrix phase, and thus cause the alloy to crack and break rather than to deform plastically. Such an alloy is said to be *hot short*; an example is steel containing a sulfide phase (sulfide inclusions).

Grain-boundary phases of this type can sometimes be avoided by eliminating the impurity in question, but often a more practical solution is to increase the value of $\gamma_{LS}$ by adding a suitable element that is soluble in the offending liquid.

For example, when about 0.5% manganese is added to molten steel, the sulfide phase that forms is now a manganese sulfide. This liquid phase has a sufficiently large value of $\gamma_{LS}$ that it forms rounded particles rather than a grain-boundary network. In some cases, such as in *liquid-phase sintering*, spreading of a liquid phase along the boundaries of the solid-phase particles is desirable. For example, if 90% powdered tungsten carbide (hard, high-melting) is mixed with 10% cobalt powder, the mixture can be pressed into the desired form, such as a tool bit. When the resulting compact is heated to the melting point of cobalt, the cobalt spreads between the grains of the tungsten carbide. On subsequent cooling, the relatively ductile and adequately strong cobalt phase bonds the tungsten carbide grains, which then serve as the main constituent of the tool bit in a lathe or other machine tool.

*Solid–Solid Interfaces.* The above equations involving interfacial tensions apply also to a mixture of solid phases, provided that opportunity is provided for equilibrium to be approached. During the initial formation of a second phase $\beta$ from a matrix phase $\alpha$, however, kinetic factors may produce a distribution of phases quite different from the distribution favored by interfacial energies. This subject is discussed in Chapter 9, but here we can point out an essential concept, the "trading" of interfacial energies. If a spherical particle of $\beta$-phase, analogous to the sphere in Fig. 6.18(a), forms inside a grain of $\alpha$-phase, the entire $\gamma_{\beta\alpha}$ interfacial energy must be supplied. In contrast, if the $\beta$-phase particle forms at a grain boundary, Fig. 6.21(b), the energy $\gamma_{\alpha\alpha}$ of the portion of the grain boundary that disappears is traded for (at least part of) the energy $\gamma_{\beta\alpha}$ of the boundary of the $\beta$-particle. An even better trade occurs at a grain junction, Fig. 6.21(c). The following orders of magnitude of interfacial energies (in terms of Young's modulus, $E$) permit a rough estimate of the result of trades of this type.

$$\gamma_{sv} = \frac{10^{-10}E}{10}$$ 
Fracture stress, Eq. (4.29), and separation of $10^{-10}$ m at fracture gives: work = $(E/5) \times 10^{-10}$ J for producing 2 m$^2$ of surface.

$$\gamma_{\alpha\alpha} = \frac{10^{-10}E}{25}$$ 
Estimated from theoretical predictions for (high-angle) grain boundaries, Fig. 3.27.

$$\gamma_{LS} = \frac{10^{-10}E}{50}$$ 
About half the grain-boundary energy $\gamma_{\alpha\alpha}$ since the liquid facilitates matching across the interface.

$$\gamma_{twin} = \frac{10^{-10}E}{100}$$ 
Only a small fraction of $\gamma_{\alpha\alpha}$ because of coincidence sites, where the atoms are common to both crystals.

This sequence shows that the interfacial energy of an atom at an interface decreases progressively as its environment approaches that of the bulk metal.

An example of the action of interfacial energies is the geometry of (coherent) twin boundaries in typical microstructures. Unlike the usual 120° triple grain junction, a twin boundary is almost perpendicular to the line formed by the

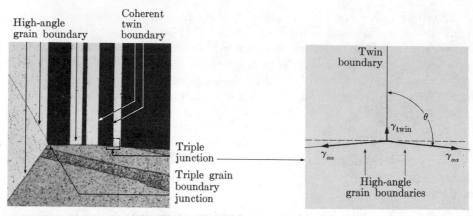

High-angle
grain boundary

Coherent
twin
boundary

Twin
boundary

$\gamma_{twin}$

$\theta$

$\gamma_{\alpha\alpha}$          $\gamma_{\alpha\alpha}$

High-angle
grain boundaries

Triple
junction

Triple grain
boundary
junction

(a) Twins in cartridge brass result in many
triple junctions. (Micrograph courtesy
R. A. Rummel, University of Florida.)

(b) Diagram of triple junction showing that $\theta$
near 90° balances the vertical forces.

**Fig. 6.22** Illustration of the effect of interfacial energies on a microstructure involving coherent twin boundaries.

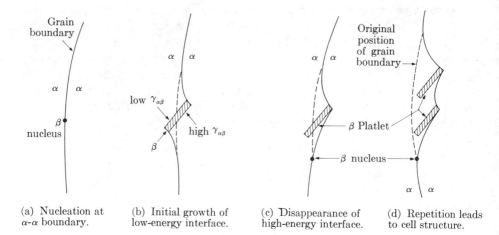

Grain
boundary

$\alpha$ | $\alpha$

$\alpha$ | $\alpha$

low $\gamma_{\alpha\beta}$

$\beta$
nucleus

high $\gamma_{\alpha\beta}$

$\beta$

Original
position
of grain
boundary

$\alpha$ | $\alpha$

$\beta$ Platlet

$\beta$ nucleus

$\alpha$ | $\alpha$

(a) Nucleation at
$\alpha$-$\alpha$ boundary.

(b) Initial growth of
low-energy interface.

(c) Disappearance of
high-energy interface.

(d) Repetition leads
to cell structure.

**Fig. 6.23** A sequence of sketches showing the nature of the nucleation and growth of a second phase ($\beta$) by the cellular transformation of a supersaturated solid solution ($\alpha$). Based on the results of K. N. Tu and D. Turnbull for a Pb-7 a/o Sn alloy.

remaining two (high-angle) boundaries, Fig. 6.22(a). A simple balance of forces, Fig. 6.22(b), explains this phenomenon, since $\gamma_{twin}$ is only a small fraction of $\gamma_{\alpha\alpha}$. Incoherent twin boundaries, usually produced during a process of deformation, have a range of energies near 0.5 $\gamma_{\alpha\alpha}$; therefore, these twins have more effect on triple-junction geometry.

Examples of the influence of interfacial energies on the kinetics of phase transformations are given in Chapter 10, but it will be helpful to consider a typical mechanism here, that of a *cellular transformation*. If a solid solution, $\alpha$, is supersaturated in solute, the $\alpha$–$\alpha$ grain boundary is a preferred site for the nucleation of a second phase, $\beta$, Fig. 6.23(a). Since the two $\alpha$-grains have different orientations, a particle of $\beta$-phase can have a low-energy interface with only one $\alpha$-grain; the left-hand grain in Fig. 6.23(b). In the case of the precipitation of a tin-rich $\beta$-phase from a Pb–7 a/o Sn solid solution ($\alpha$), the plates of $\beta$-phase form on the (111) plane (called the *habit plane*) of the $\alpha$-phase. The orientation relationship that makes this a plane of low energy is

$$(010)_\beta \parallel (111)_\alpha$$

$$[001]_\beta \parallel [110]_\alpha$$

As the high-energy interface is replaced by low-energy interface, Fig. 6.23(c), a platelet of $\beta$-phase is produced; this platelet grows into the right-hand grain of $\alpha$ using solute carried to it by diffusion along the grain boundary. The process of nucleation and growth is then repeated at neighboring sites, Fig. 6.23(d), and eventually leads to a "cell" composed of an array of $\beta$-platelets within the transformed $\alpha$-phase.

## PROBLEMS

1. a) Using Fig. 6.2, determine the stable phase of carbon at room temperature and atmospheric pressure.

   b) Explain how those diamonds now found in mines might have been produced in former geologic ages.

   c) Explain whether you would recommend that diamond be used as a tiny, critical part of a laboratory device operating at 3000 K in a protective gas at atmospheric pressure.

2. a) Apply the phase rule to the triple point in Fig. 6.6.

   b) What change in the condition of the metal corresponds to shifting the pressure-temperature point from the triple point to the liquid region of the diagram?

3. a) Sketch a qualitative one-component diagram for nickel, using the data of Table 6.1.

   b) Use the diagram of part (a) to determine the vapor pressure of nickel when it is held just above its melting point in an open crucible exposed to air.

4. Gaseous impurities are largely eliminated from some metals by melting them in a vacuum of $10 \text{ N/m}^2$. Would you suggest this procedure for (a) chromium? (b) nickel? (c) copper?

5. Picture one mole of magnesium metal, initially in the form of a cube at atmospheric pressure and 400°C. This system is in a definite thermodynamic state, Fig. 6.3. Imagine that the cube of magnesium is formed by extrusion into a long thin rod, with the expenditure of about $10^3$ joules of energy by the extrusion press.

a) What is the approximate quantity of work $\delta W$ in this case?

b) If the grain structure and atomic arrangement of the magnesium metal are the same after extrusion as before, explain why $W$ is not a state function.

c) What would be the change in internal energy $E$ as a result of extrusion under the conditions in (b)?

**6.** 100 grams of liquid magnesium at atmospheric pressure was slowly cooled past its melting point, 650°C, in a calorimeter. $37.2 \times 10^3$ joules were liberated by the magnesium during the time in which it solidified, essentially at constant temperature. Calculate the molar enthalpy, $\Delta H$, for the reaction, $Mg(l) \rightarrow Mg(s)$.

**7.** Imagine that the curve of Fig. 6.8 describes the variation of molar volume $V$ with mole fraction of component 2. Let $V_1 = 5 \times 10^{-6} \, m^3/mol$ and $V_2 = 9 \times 10^{-6} \, m^3/mol$.

a) A straight line connecting $V_1$ (at $N_2 = 0$) and $V_2$ (at $N_2 = 1$) is the curve of $V^{id}$ versus $N_2$. Explain why such a curve is obtained if the two types of atoms maintain their normal sizes even when in solution.

b) At $N_2 = 0.5$, what is the excess molar volume, $V^{xs}$?

c) Explain why it is impossible (other than by the graphical method of Fig. 6.8) to apportion this excess volume quantitatively between the two types of atoms.

**8.** Derive Eq. (6.36). The derivation involves the differential, $dG$, of the free energy $G = U + PV - TS$ and substitution of the appropriate expression for $dU$; namely, $(T dS - P dV)$.

**9.** Using an analysis similar to that employed in obtaining Eq. (6.38), show that the same equation applies for a solid metal except that the enthalpy is that for sublimation, $\Delta H_s$.

**10.** a) By a suitable integration of Eq. (6.45) determine the fraction of atoms having energies greater than $g = 5kT$.

b) By a similar integration of $gp \, dg$, determine the fraction of the total energy possessed by this small fraction of atoms.

**11.** Using the value $6.02 \times 10^{23}$ atoms per mole and the relation $\ln(1 - x) = -x$, which applies when $x$ is much smaller than unity, calculate $\Delta S^{id}$, Eq. (6.53), for the addition of one vacancy to a mole of initially vacancy-free metal. Show that the corresponding value for a mole of vacancies is $56R$.

**12.** Verify the correctness of Eq. (6.55) by separately justifying the simplification of the two terms on the right-hand side.

a) The first term was initially

$$\frac{d}{dN_V}(\Delta H - T \Delta S^{xs}) \, dN_V.$$

Since both $\Delta H$ and $\Delta S^{xs}$ are independent of $N_V$, show that this term is equivalent to $(\Delta H - T \Delta S^{xs}) dN_V$.

b) The second term was initially

$$\frac{d}{dN_V}\{RT[N_V^0 \ln N_V^0 + (1 - N_V^0) \ln(1 - N_V^0)]\} \, dN_V.$$

Perform this differentiation and show that the result is

$$RT \ln \frac{N_V^0}{1 - N_V^0} \, dN_V.$$

**13.** Use the data of Table 1.2 on the lattice constant of pure nickel to verify the value shown in Fig. 6.16 for the size of a nickel atom.

**14.** Difficulty was encountered with excessive rise of a silver braze at 1000°C in a $2 \times 10^{-4}$ m space between two copper plates. On the basis of Eq. (6.60), suggest possible changes that might be made to limit the rise to a reasonable amount, say, $10^{-2}$ m.

**15.** If $\gamma_{LS} = 2.0$ J/m$^2$ and $\gamma_{ss} = 0.4$ J/m$^2$, estimate the difference in energy of the system for

a) a $10^{-12}$ m$^3$ particle of liquid located at a grain boundary, and

b) located in the center of a grain. Use Eq. (6.64) to determine the shape of the particle.

## REFERENCES

Barrett, C. S. and T. B. Massalski, *Structure of Metals*, 3rd edn, New York: McGraw-Hill, 1966. Chapters 10 and 11 give detailed information on the structure of many types of metallic phases.

Cottrell, A. H., *An Introduction to Metallurgy*, London: Edward Arnold, 1967. Chapter 14 surveys many of the topics covered in the present treatment.

Darken, L. S. and R. W. Gurry, *Physical Chemistry of Metals*, New York: McGraw-Hill, 1953. This is the standard treatment of the thermodynamics of metallic solutions and of phase equilibria. Chapters 10 to 14 are especially pertinent.

Gordon, Paul, *Principles of Phase Diagrams in Materials Science*, New York: McGraw-Hill, 1968. Chapters 2 and 3 develop the thermodynamic basis and then show its application to one-component phase diagrams.

CHAPTER 7


# PHASE DIAGRAMS

# INTRODUCTION

Phase diagrams are concerned with the relations among the *phases* in metals and alloys. Chapter 6 showed that atoms may combine to form a gaseous phase, a liquid phase, or a solid phase; and of course the properties of these phases strongly influence the properties of a given alloy. However, the behavior of an alloy depends also on the manner in which these phases are related. An alloy composed of two solid phases, for example, can have a variety of properties, depending on the *structure* built up by these two phases. The commercial use of structural changes in steel will help illustrate this point.

Contrary to popular belief, steel is useful not because it is *hard* but because it is *hard and soft*. Numerous substances are as hard as or harder than steel, but as with the problem of storing the hypothetical universal solvent, it is difficult to form these deformation-resistant bodies into useful shapes. This problem is minimized in steels for the following reason. In one condition at room temperature steel is a mixture of crystals of ferrite and cementite (see p. 474), and the alloy is then soft enough to be machined readily and ductile enough to be formed by bending or pressing operations. However, after the necessary shaping has been accomplished, the soft steel can be caused to pass through changes of phase that render it extremely hard. These changes involve a third solid phase, produced by heating, and its later decomposition during quenching. The iron-carbon diagram, studied in this chapter, is the basis for heat treatments that soften and harden steel. This diagram will be shown to have other uses also.

Phase changes in the solid state are of paramount interest in physical metallurgy, as this example shows, but liquid-solid reactions like those that occur in the solidification of castings are also important. Because the physical picture of a liquid solidifying or of a solid melting is easily grasped, it is convenient to begin using liquid-solid reactions as examples of equilibrium diagrams. The nature of solid-solid reactions can then be understood by analogy. The gas phase does not occur in the usual equilibrium diagrams, since these diagrams are for use at atmospheric pressure, which is high compared with the vapor pressures of metals in the temperature range involved.

**Free energy-composition diagrams.** Just as the free energy determines the liquid-solid equilibrium in a pure metal (as described in connection with Fig. 6.12) it also predicts the equilibrium phases in a system consisting of two or more components. The free energy of a given phase is now a function of composition (as well as of temperature and pressure). We will consider here only systems at atmospheric pressure, and we begin with binary systems composed of quite similar components 1 and 2. At a temperature $T_0$ above both melting points, Fig. 7.1(a), the free energy of the liquid $G_l$ lies below that of the solid phase $G_\alpha$ at all compositions. Both curves of $G$ are concave downward because their shape is dominated by the entropy of mixing in the expression for the change, $\Delta G$, of the free energy on mixing, Eq. (6.35); see Problem 1.

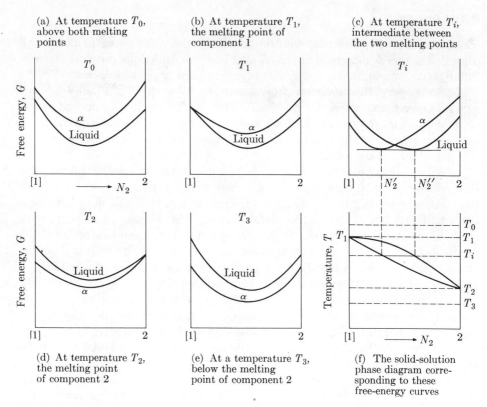

(a) At temperature $T_0$, above both melting points

(b) At temperature $T_1$, the melting point of component 1

(c) At temperature $T_i$, intermediate between the two melting points

(d) At temperature $T_2$, the melting point of component 2

(e) At a temperature $T_3$, below the melting point of component 2

(f) The solid-solution phase diagram corresponding to these free-energy curves

**Fig. 7.1** Curves of free energy versus composition for a binary system of the solid-solution type. (Adapted from A. Prince, *Alloy Phase Equilibria*)

At the melting temperature $T_1$ of component 1, Fig. 7.1(b), the liquid and solid phases are in equilibrium at $N_2 = 0$ (that is, at pure component 1); therefore, the two free energies are equal at this composition. At all other values of $N_2$, however, $G_l$ is lower than $G_\alpha$, and the liquid phase is stable compared to the solid phase. In the range of temperatures intermediate between $T_1$ and $T_2$, Fig. 7.1(c), a new aspect of the free energy causes a mixture of the liquid and solid phases to be stable over the range of compositions between $N_2'$ and $N_2''$. Whenever two free-energy curves cross, a common tangent can be constructed in the manner shown. The values of $G$ along this tangent represent the sum of $G_\alpha$ plus $G_l$ for various mixtures of the two phases, ranging from all solid phase at $N_2'$ to all liquid phase at $N_2''$. Since the tangent lies below both the curve for $G_\alpha$ and that for $G_l$, the stable condition of the system is a mixture of liquid and solid in the range of compositions between $N_2'$ and $N_2''$ at temperature $T_i$. At temperature $T_2$, Fig. 7.1(d), the solid phase is stable at all compositions up to $N_2 = 1$, at which point the solid and liquid phases are in equilibrium (the melting point of component 2). At any temperature $T_3$ below $T_2$, Fig. 7.1(e), the solid phase is stable at all compositions.

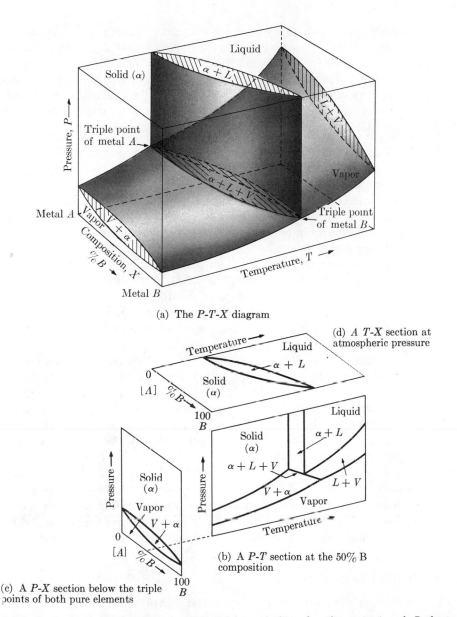

(a)  The *P-T-X* diagram

(d)  A *T-X* section at atmospheric pressure

(b)  A *P-T* section at the 50% B composition

(c)  A *P-X* section below the triple points of both pure elements

**Fig. 7.2**  Typical pressure-temperature-composition relations for elements *A* and *B* that dissolve completely in each other in the liquid and solid states.

The behavior of this relatively simple system can be summarized in a solid-solution phase diagram, Fig. 7.1(f). The data from the free-energy curves of parts (b) and (d) correspond to the melting points, $T_1$ and $T_2$, of the pure components. The two points, $N'_2$ and $N''_2$, from part (c) plot at $T_i$ as points along the two curves that enclosed the two-phase region in the diagram. The free-energy curves of part (c) show that the solid phase is stable between $N_2 = 0$ and $N_2 = N_2'$, whereas the liquid phase is stable for $N_2$ values greater than $N_2''$. In a later section we will see how to interpret a solid-solution diagram. If the components of a binary system interact in a more complex manner, the free-energy curves are correspondingly more complex and lead to phase diagrams of other types. An additional example is considered later in connection with the eutectic phase diagram.

**P-T-X diagrams.** A useful transition between the one-component diagram (pressure-temperature) studied in Chapter 6 and the usual two-component phase diagram (temperature-composition) considered in the present chapter is the three-dimensional diagram of pressure, temperature, and composition,* known as a *P-T-X* diagram. This type of diagram for most two-component alloys is quite complex and few diagrams have been completely investigated. However, when the two metals in a binary system dissolve completely in each other in both the liquid and solid states, then the *P-T-X* diagram is similar to a one-component diagram extended along the composition axis, Fig. 7.2(a). A single solid phase, the α solid solution, exists in the upper-left portion of the diagram and joins the solid-phase regions of the two one-component diagrams that form the front and back faces of the *P-T-X* model. The regions occupied by the liquid and vapor phases can be described in a similar manner.

A new feature in the *P-T-X* diagram is the fact that the two-phase and three-phase equilibria extend over a region rather than being limited to a line or point as in the one-component diagram. This behavior is shown in the *P-T* section (at constant composition), Fig. 7.2(b).† It is often useful to have a section at constant temperature through the *P-T-X* diagram. A *P-X* section of this kind is shown in Fig. 7.2(c) for a temperature below the triple points of both component metals. By far the most useful section through the *P-T-X* diagram is the one at constant pressure, at atmospheric pressure in particular. Figure 7.2(d) shows the diagram obtained in this instance. All the equilibrium diagrams considered in the balance of the chapter are *T-X* diagrams and unless otherwise indicated they refer to a fixed pressure of one atmosphere; the first to be discussed is similar to Fig. 7.2(d); namely, the solid-solution diagram previously derived from the free-energy curves.

## SOLID-SOLUTION SYSTEMS

Depending on the nature of the two metals involved, several types of binary equilibria can occur. Moreover, a number of different kinds of solidification

---

* The symbol *X* represents composition expressed in any units; examples are weight percent, volume percent, or $kg/m^3$. The symbol *N* refers specifically to composition in mole fraction, a unit that is especially convenient in thermodynamics.

† Sections of this type are not convenient for reasoning based on the phase rule, since the compositions of the phases in equilibrium are not shown.

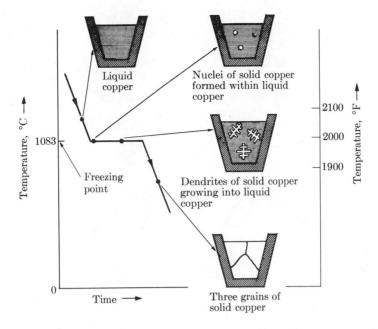

**Fig. 7.3**  Time–temperature curve for the solidification of a small crucible of liquid copper.

behavior and solid-state reactions may be combined in a single two-component diagram. The iron-carbon diagram is an important example of such complex reactions and will be considered later in the chapter. Fortunately, the interpretation of even the most complicated diagram is hardly more difficult than understanding each of the types of alloy behavior that it includes. This is true because every separate region of a complex diagram can be treated individually and independently. However, there remains the task of learning to analyze the typical reactions. For this purpose, especially simple two-component diagrams have been chosen for study. Each diagram contains only a single type of reaction; for example, we consider first only solid-solution freezing.

**Solid-solution freezing.**  We shall use the copper-nickel system as an example of a solid-solution equilibrium diagram. It has been found *experimentally* that alloys of these two metals form solid solutions when they are cooled from the liquid condition, and the experimental data have been conveniently summarized in the form of an equilibrium diagram (Fig. 7.5). An insight into the significance of the diagram can be obtained by considering the freezing of a number of compositions of copper-nickel alloys. When pure copper in the molten state is slowly cooled under equilibrium conditions, solidification occurs at the freezing point in the manner shown schematically in Fig. 7.3.

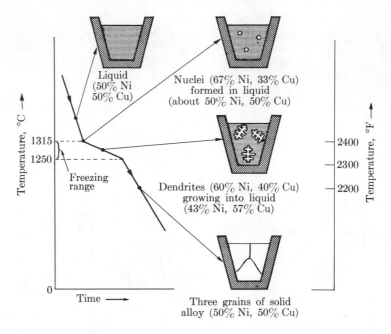

**Fig. 7.4** Time–temperature curve for the solidification of a small crucible of 50% copper, 50% nickel alloy.

The equilibrium solidification of an alloy of 50% (by weight) copper and 50% nickel is slightly more involved than the freezing of a pure metal. Complete solidification does not occur at a single temperature but progressively over a small temperature range; there is an accompanying continuous variation of the chemical compositions of the liquid and of the solid metal. The experimental facts in this case are indicated in the sketch of Fig. 7.4. When the liquid alloy reaches 1315°C, solidification begins with the formation of nuclei of the solid phase, composed of 67% nickel, 33% copper. As the temperature decreases, solidification continues with the growth of the solid phase, often in the form of dendrites (described later in connection with Fig. 9.9). When solidification is about half completed, the composition of the solid phase is 60% nickel, 40% copper, and the liquid phase has changed its composition to 43% nickel, 57% copper. Upon completion of solidification the liquid has disappeared, and the solid phase has the composition of the original liquid phase, namely 50% nickel, 50% copper. The composition of the solid phase is able to change during the course of equilibrium solidification because diffusion occurs relatively rapidly at high temperatures.

The equilibrium diagram for a solid-solution alloy system, Fig. 7.5, summarizes the essential information obtained from time-temperature curves. In this case, data are plotted for a series of alloys ranging in chemical composition from pure copper to pure nickel. For example, the data from the curve for the solidification

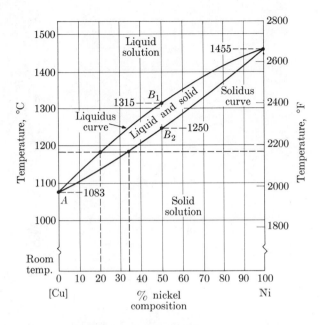

**Fig. 7.5** Copper-nickel equilibrium diagram. Points $A$, $B_1$, and $B_2$ show how the diagram is constructed from time-temperature curve data.

of pure copper, Fig. 7.3, are given adequately by point $A$ in the equilibrium diagram. This point indicates that when the composition is completely free of nickel (pure copper) the liquid phase changes to the solid phase *only at the freezing point of copper*. Similarly, the data of Fig. 7.4 are recorded by the points $B_1$ and $B_2$ plotted at the composition 50% nickel. These points fully record the information that the 50% nickel alloy begins to solidify at 1315°C and is completely solid at 1250°C. It will be shown that data on the chemical compositions of phases are also furnished by this diagram. The complete *liquidus* and *solidus* lines are constructed by the use of many pairs of points similar to $B_1$ and $B_2$ obtained for other alloy compositions. The liquidus and solidus lines divide the equilibrium diagram into regions where the liquid phase, liquid and solid phases, and solid phase exist. The microstructure of the equilibrium solid-solution phase is identical with that of a pure metal. It is noteworthy that such a structure is single phase, although many grains of the phase exist and are separated by grain boundaries.

**Uses of the solid-solution diagram.** Since an equilibrium diagram is merely a concise presentation of experimental data obtained on a given alloy system, it follows that the original data can be obtained again from the diagram. For a specified alloy composition and temperature it is possible, under equilibrium conditions, to tell (1) the phases that are present, (2) the chemical composition of each phase, and (3)

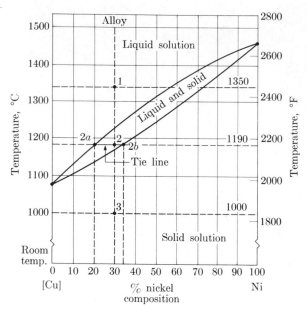

**Fig. 7.6** The equilibrium conditions of a 30% nickel alloy at three different temperatures.

the amount of each phase. The procedures for obtaining this information will be illustrated using the alloy of 30% nickel shown by the vertical dashed line in Fig. 7.6.

*Prediction of Phases.* The equilibrium diagram is plotted with temperature as the *ordinate* and composition as the *abscissa*. Therefore, specific information can be obtained from it only if a temperature and a composition are specified. Such a pair of values locates a *point* in the diagram. Points of this kind are used repeatedly in analyzing equilibrium diagrams. For example, the state of the alloy of *composition* 30% nickel can be determined only with reference to a certain *temperature*. Thus, when this alloy is at 1350°C point 1 is determined in Fig. 7.6, and if this alloy is at 1190°C point 2 is determined.

Once the *point* of interest is located in the diagram, it is easy to ascertain which phase or phases are present. *Those phases are present that correspond to the phase field in which the point lies.* For example, the 30% nickel alloy at 1350°C (point 1) consists of only one phase, the liquid solution. On the other hand, at 1190°C the same alloy consists of a mixture of liquid solution and solid solution, since point 2 lies in the liquid and solid field of the diagram. At 1000°C (point 3) only the solid-solution phase exists. Evidently a similar analysis can be made for any point (any alloy composition and temperature) in the diagram. For practical purposes, a point on the liquidus or solidus line is usually considered to be in the two-phase field.

*Prediction of Chemical Compositions of Phases.* When only a single phase is present, as at points 1 and 3 in Fig. 7.6, the chemical composition of the phase must be the same as the composition of the alloy. Since at point 1, for example, the 30% nickel alloy is completely liquid, the chemical composition of the liquid phase is obviously 30% nickel and 70% copper.

When two phases are present each phase has a different composition, and since neither of these compositions is the same as the composition of the alloy as a whole, it is necessary to consider two additional points in the diagram. If the two phases are to be in equilibrium, they must be at the same temperature; the points representing these phases, therefore, must lie at the temperature of the original point. For example, point 2 is determined by the alloy composition 30% nickel and the temperature 1190°C, but the two phases that exist in the alloy at this temperature are not at point 2 but only somewhere along the horizontal line at 1190°C. The exact points characterizing the liquid and solid phases are given by the intersections of the temperature horizontal with the liquidus and solidus lines, respectively.* Thus the composition (and temperature) of the liquid phase in the 30% nickel alloy at 1190°C is given by point 2a (20% nickel, 1190°C), while the solid phase is specified by point 2b (35% nickel, 1190°C). The chemical composition of a given phase is read on the composition axis as indicated by the vertical dashed lines from points 2a and 2b. The horizontal (constant temperature) line connecting the points that represent the two different phases is called a *tie line*.

In summary, the chemical compositions of the phases that occur in an alloy at a given temperature are determined as follows.

1. When only a single phase is present, its composition is the same as that of the alloy. (It is unnecessary to draw a tie line when the alloy-temperature point lies in a single-phase field.)

2. When two phases are present, a horizontal line is drawn through the alloy-temperature point, and their compositions are read at the intersections of this line with the boundary lines of the two-phase field.

*Prediction of Amounts of Phases.* If the total amount of alloy is known, it is possible to determine the amount of each phase that is present at equilibrium at a given temperature. When the total weight of alloy is 100 pounds or 100 grams the numerical relations are especially convenient, but the same convenience can be retained for any weight of alloy by expressing the amount of each phase as a percentage of the weight of the alloy.

If only a single phase is present, as at point 1 in Fig. 7.6, the weight of the phase must be equal to the weight of the alloy. There is 100% liquid phase present in this instance. If the weight of the alloy is 100 grams, the amount of liquid phase at point 1

---

* Considering the manner in which the liquidus and solidus curves are plotted from experimental data, it can be seen that this procedure for obtaining the compositions of the liquid and solid phases is equivalent merely to reobtaining the data that were used in plotting these curves.

is 100 grams. If the weight of the alloy is only 2.5 grams, there is still 100% liquid phase or 2.5 grams.

When two phases are present, their relative amounts are determined by the relation of their chemical compositions to the composition of the alloy. This is true because the total weight of one of the metals, say metal $A$, present in the alloy must be divided between the two phases. This division can be represented by the equation

$$W_0 \times \frac{\%A_0}{100} = W_L \times \frac{\%A_L}{100} + W_S \times \frac{\%A_S}{100}, \tag{7.1}$$

$$\begin{array}{ccc} \text{(weight of metal} & = & \text{(weight of metal} & + & \text{(weight of metal} \\ A \text{ in the alloy)} & & A \text{ in the liquid} & & A \text{ in the solid} \\ & & \text{phase)} & & \text{phase)} \end{array}$$

where $W_0$, $W_L$, and $W_S$ are the weights of the alloy, liquid phase, and solid phase, respectively, and $\%A_0$, $\%A_L$, and $\%A_S$ are the respective chemical compositions in terms of metal $A$. Since the weight of the alloy is the sum of the weight of the liquid phase and the weight of the solid phase, the following relation exists:

$$W_0 = W_L + W_S. \tag{7.2}$$

This equation can be used to eliminate $W_L$ from Eq. (7.1), and the resulting equation can be solved for $W_S$ to give the expression

$$W_S = W_0 \frac{(\%A_0 - \%A_L)}{(\%A_S - \%A_L)}. \tag{7.3}$$

Although a similar expression can be obtained for the weight of the liquid phase $W_L$, the weight of the second phase is more easily obtained by means of Eq. (7.2).

Since the weight of each phase is determined by chemical composition values according to Eq. (7.3), the tie line used in obtaining composition values is also significant in computing the weights of phases. In terms of lengths in the tie line (Fig. 7.6), Eq. (7.3) can be written

$$W_S = W_0 \frac{(\text{length of line } 2 - 2a)}{(\text{length of line } 2b - 2a)}, \tag{7.4}$$

where the lengths are expressed in terms of the numbers used for the concentration axis of the diagram. Another characteristic of the tie line with respect to the amounts of the two phases leads to the name *lever law* for Eqs. (7.3) and (7.4). The relative amount of a given phase is proportional to the length of the tie line on the *opposite* side of the alloy point of the tie line. Thus, the weights of the two phases are such that they would balance as in Fig. 7.7.

Using Eq. (7.3), we can find the *weight* of solid-solution phase at point 2 in Fig. 7.6 as follows:

$$W_S = W_0 \frac{(30 - 20)}{(35 - 20)},$$

$$W_S = W_0 \times 0.667.$$

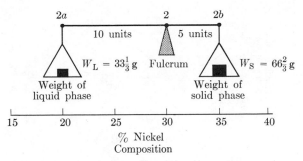

**Fig. 7.7** Schematic illustration of the lever law at point 2 in Fig. 7.6. The weight of the alloy is assumed to be 100 grams:

The *percentage* of solid-solution phase can be determined by use of the equation

$$\text{Percentage of solid} = \frac{W_S}{W_0} \times 100 = \frac{(\%A_0 - \%A_L)}{(\%A_S - \%A_L)} \times 100. \qquad (7.5)$$

At point 2 the percentage of solid phase is

$$\% \text{ solid} = \frac{(30 - 20)}{(35 - 20)} \times 100 = 66.7\%.$$

The percentage of liquid phase is the difference between 66.7 and 100%, that is, 33.3%.

**Copper-nickel: a typical solid-solution system.** The copper-nickel system is not unique in solidifying to produce a single, solid-solution phase. A number of other binary systems show a similar behavior; for example, the silver-gold system involves only solid-solution freezing (see Problem 5). Alloy systems of a given type have a similar pattern of solidification behavior, and characteristic variations in mechanical and physical properties appear with change in chemical composition. A knowledge of these regularities is useful in making rough predictions of the properties of alloys of various compositions even in the more complex equilibrium diagrams. The causes of the changes in properties accompanying the addition of an alloying element to a matrix metal are discussed in later chapters.

Figure 7.8 shows typical variations in properties of commercial copper-nickel alloys as the chemical composition of the solid solution* changes from 100%

---

* The solid solutions shown in Fig. 7.8 illustrate the difficulty of giving a solid solution a completely descriptive name. In this case is the complete series of solid solutions based on nickel or on copper? It is difficult to say. Also, the addition of a few percent of a third element would not change the essential nature of the solid solution. Because of these complications, it is customary to use the Greek letters to name the solid solutions in a given equilibrium diagram. In this case there is only one solid solution, and it is called by the first Greek letter, the alpha ($\alpha$) solid solution. In Fig. 7.12 there are two solid solutions, and they are arbitrarily called the alpha and beta ($\beta$) solid solutions.

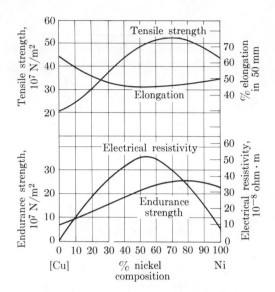

**Fig. 7.8** Typical property variations with change in composition in commercial copper-nickel alloys. (To convert from N/m$^2$ to lb/in$^2$, multiply by $1.45 \times 10^{-4}$. To convert from ohm · m to ohm · cm, multiply by $10^2$.)

copper to 100% nickel. Such strength properties as hardness and yield strength also pass through a maximum, while the ductility values show a minimum. Properties that are almost unaffected by atom interactions vary more linearly with composition. Examples of such properties of the solid solution are the lattice constant, thermal expansion, specific heat, and specific volume.

Figure 7.8 lists the properties characteristic of alloys under equilibrium conditions, but alloys that solidify at the cooling rates found in commercial casting operations are not in equilibrium. Figure 7.9(a) shows the microstructure of a cast alloy that is 30% nickel, 70% copper. The undesirable, inhomogeneous structure (called *coring*) that developed in the course of solidification is clearly visible and reveals the dendritic mode of solidification. The interior of the dendrite, which solidifies first, differs in chemical composition from that of the outer region, and the etching process discloses this difference as a slight contrast in appearance under the microscope. In the absence of sufficient time for diffusion in the growing particles of solid alloy, such an inhomogeneous, cored structure is an inevitable result of the mechanism of freezing in solid-solution systems. The equilibrium condition can be approached by prolonged heating of the solidified alloy at a temperature below the melting range. The same homogenization of the cast structure, however, can be produced more quickly by combining plastic deformation of the alloy with the heating, for example by hot-working the alloy, Fig. 7.9(b).

Although only a relatively small number of alloy systems, like copper-nickel, show complete solid solubility, almost all equilibrium diagrams exhibit certain

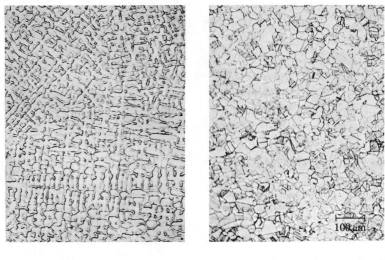

(a) Cored structure in the cast alloy    (b) Homogeneous structure produced
                                          by working and heating

**Fig. 7.9**  Microstructures of a 30% nickel, 70% copper alloy in the cast and in the homogenized conditions. (Courtesy Anaconda American Brass Company)

ranges of solid solubility in connection with other types of alloy formation. These limited regions of solid solubility can be analyzed in exactly the same manner as a complete solid-solution diagram. The variations in properties correspond to those found at low percentages of the solute metal. Since these limited ranges of solid solubility occur so widely in equilibrium diagrams, the strengthening effects and other property changes associated with them are of industrial importance.

## EUTECTIC SYSTEMS

Equilibrium diagrams merely record the experimentally observed behavior of alloy systems. While solid-solution formation represents one manner in which a liquid solution of two metals may solidify, the *eutectic* reaction is an equally prominent mechanism of solidification. A new feature is the coexistence of two solid phases, a phenomenon that is important for many other types of phase equilibria. Let us first consider the nature of the free-energy curves for this case.

**Free energy of alloy phases.**  The treatment used in connection with Fig. 7.1 must be modified to account for the existence of two solid phases: the $\alpha$ solid solution based on metal $A$, and the $\beta$ solid solution based on metal $B$. The change of free energy on mixing

$$\Delta G = \Delta H - T \, \Delta S \tag{7.6}$$

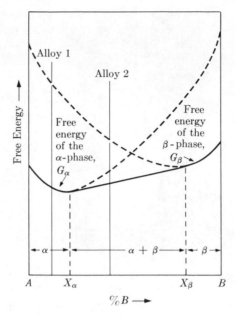

**Fig. 7.10** Free-energy curves for the $\alpha$ and $\beta$ phases. The line tangent to these curves determines the composition range in which a mixture of the two phases exists.

is still applied to each of these solid solutions; for example, $\alpha$. The entropy term still acts to produce a minimum in free energy. However, frequently the enthalpy becomes very high as large percentages of metal $B$ are (hypothetically) forced into solid solution with metal $A$. The free-energy curve for the $\alpha$ solid solution then has the form shown in Fig. 7.10, dropping to a minimum at small percentages of metal $B$ and then rising to a very high value as pure $B$ is approached. A similar free-energy curve exists for the $\beta$ phase. In Fig. 7.10 it is assumed that the crystal structure of metal $B$ is different from that of $A$. For the reason discussed previously in connection with Fig. 7.1(c), between compositions $X_\alpha$ and $X_\beta$ the alloy consists of a mixture of the $\alpha$ and $\beta$ phases. Problem 6 shows how the tangent line in this range of composition can be obtained from the lever law.

In the manner employed previously for a solid-solution alloy, the solidification of a liquid alloy in a eutectic system can be explained as a relative shifting of the free-energy curves for the various phases, Fig. 7.11. At the highest temperature, $T_0$, the curve for the liquid phase lies below the other two curves, so that the liquid phase is the stable structure for alloys of all compositions. At a somewhat lower temperature, $T_1$, the free-energy curve for the liquid has shifted upward relative to the other two curves and it lies above one of them at each end of the diagram. The argument of Fig. 7.10 then predicts the phase regions shown in the figure.

At a critical temperature $T_2$ in Fig. 7.11 all three free-energy curves have a common tangent and the three phases exist in equilibrium with one another. The

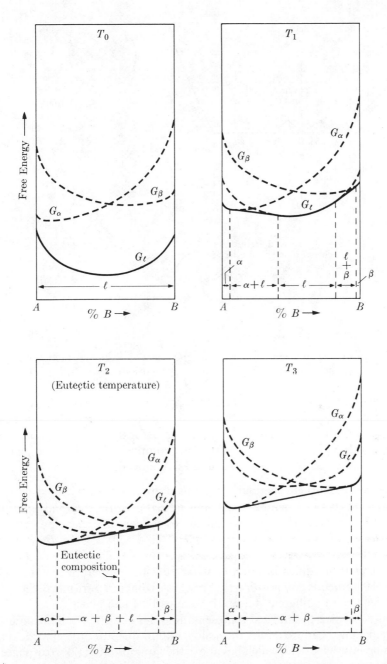

**Fig. 7.11** Explanation of phase changes in a eutectic system in terms of the free-energy curves for the liquid, $\alpha$ and $\beta$ phases. The temperatures decrease from $T_0$, which is the highest. (After J. B. Newkirk)

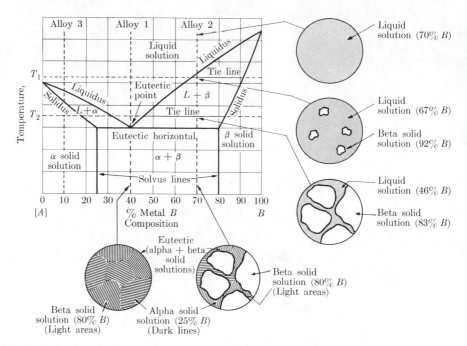

**Fig. 7.12** A generalized eutectic equilibrium diagram involving the hypothetical metals $A$ and $B$. The solidification processes of alloy 1 (eutectic composition) and of alloy 2 are indicated.

phenomena that occur at this (eutectic) temperature will be considered in the following section. Below $T_2$, for example at $T_3$, the free-energy curve for the liquid phase lies above the common tangent of the curves for the two solid phases. Therefore, here the liquid phase is not stable at any composition. The solid phases that exist under these conditions were discussed in connection with Fig. 7.10.

**The eutectic equilibrium diagram.** This diagram can be developed from time-temperature curves in a manner analogous to that used for the solid-solution diagram, but in this case the experimental curves show a different kind of behavior. *Over a wide range of compositions a portion of the solidification of the alloy occurs at a fixed temperature, the eutectic temperature.* In one alloy (the eutectic composition) complete solidification occurs at the eutectic temperature. Although the freezing of the eutectic composition thus resembles that of a pure metal, the resulting solid is significantly different, since it is composed of two phases.

Figure 7.12 is a generalized representation of a eutectic equilibrium diagram involving metals $A$ and $B$. Like the solid-solution diagram, the eutectic diagram has (1) a liquid-phase field, which lies above the liquidus lines, (2) solid-phase fields (the $\alpha$ phase and the $\beta$ phase), which lie below the solidus lines, and (3) phase fields containing both liquid and solid, which lie between the liquidus and solidus lines. Because of the presence of the eutectic-reaction horizontal, a phase field containing two solid phases in equilibrium is also introduced. This field is bounded on the sides by *solvus lines*.

Full information concerning the amounts and compositions of the phases that occur in a given alloy at a certain temperature can be obtained for the eutectic diagram by the same rules employed for solid-solution alloys. However, it is useful to have a general picture of the nature of the eutectic reaction before beginning detailed calculations. The basic equation of this reaction on cooling is

$$\text{Eutectic liquid} \rightarrow \text{alpha solid solution} + \text{beta solid solution}, \tag{7.7}$$

where the compositions of the $\alpha$ and $\beta$ phases are given by the end points of the eutectic horizontal. This reaction occurs at the temperature of the eutectic horizontal and involves liquid *of eutectic composition* (the composition of alloy 1 in Fig. 7.12).* When eutectic liquid is cooled to the eutectic temperature, the two solid phases (the alpha and beta solid solutions) begin to form in the liquid alloy. But since one of the solid phases is richer (in metal $B$, for example) than the liquid phase, and the other solid phase is poorer than the liquid phase, it is natural for these two solid phases to form side by side in a given portion of liquid. The additional amount of (metal $B$) component needed by one of the solid phases is then obtained from the other solid phase (which needs to lose some of this component). This picture of the solidification process of a eutectic liquid accounts for the fact that the two solid phases generally occur as an intimate mixture. This mixture is called the *eutectic structure* or the *eutectic microconstituent*. A schematic representation of a solidified alloy consisting entirely of eutectic structure is given by the sketch of alloy 1 in Fig. 7.12. A photomicrograph of a typical eutectic structure is shown in Fig. 7.13; the 12% silicon alloy is composed entirely of eutectic microconstituent. During equilibrium heating the reverse of Eq. (7.7) occurs, and eutectic liquid forms from the mixture of alpha and beta solid solutions.

Gibbs' phase rule, Eq. (6.3), can be used to demonstrate that the eutectic reaction must occur at constant temperature. There are two components (metals $A$ and $B$) in the system, and since three phases exist during the eutectic reaction, the phase rule becomes

$$P + F = C + 2,$$

$$3 + F = 2 + 2,$$

$$F = 1.$$

But this one degree of freedom has already been used in fixing the pressure at one atmosphere; therefore neither the temperature nor the composition of any of the phases can change during the course of the eutectic reaction under equilibrium conditions.

---

* If the liquid alloy is not of eutectic composition initially, it is essential that preliminary solidification of *only one* of the solid phases occur until the remaining liquid does attain the eutectic composition. This point is considered later.

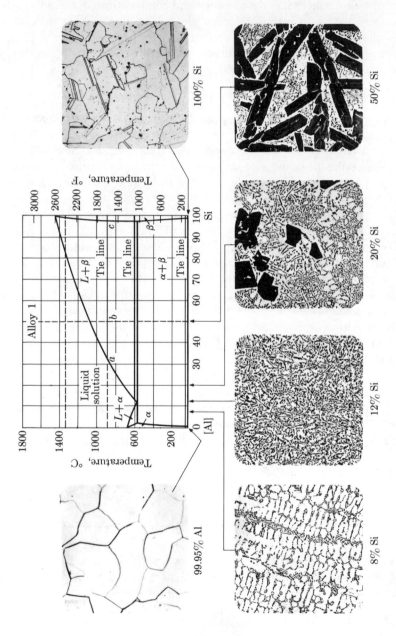

**Fig. 7.13**   The aluminum-silicon alloy system with some room-temperature microstructures. (Photomicrographs courtesy Aluminum Research Laboratories)

Most alloys that undergo the eutectic reaction as *part* of their melting or solidification process are not of eutectic composition (alloy 2 in Fig. 7.12 for example). Since only liquid of eutectic composition can decompose by the eutectic reaction, Eq. (7.7), it is usually necessary that the liquid phase change its composition. This is accomplished through the freezing of *primary crystals* of one of the solid-solution phases. Consider, for instance, that the liquid alloy initially contains a higher percentage of metal B than that corresponding to the eutectic composition. The primary solid phase that forms first is then the solid phase ($\beta$ phase) that is richer in metal B. However, the formation of this solid phase depletes the remaining liquid phase in metal B, and the composition of the liquid shifts toward the eutectic composition during cooling to the eutectic temperature. In general, then, as solidification of the primary crystals proceeds, the liquid phase gradually shifts its composition (and temperature) along the liquidus line toward the eutectic point (the eutectic composition and the eutectic temperature). This process is shown by sketches of the solidification of alloy 2. Here the possibility of dendritic growth is ignored. When the liquid reaches eutectic composition the remaining solidification occurs according to Eq. (7.7), and the primary crystals are surrounded by a finely divided mixture of the two solid phases in a eutectic structure. It is seen that the final structure of alloy 2 consists of two *microconstituents*, that is, primary beta crystals and the eutectic microconstituent.

If an alloy composition is such that the vertical line representing it does not cross the eutectic horizontal, no eutectic reaction occurs during the equilibrium solidification of that alloy. Alloy 3 in Fig. 7.12 is an example of such a composition, and trial will show that its solidification is entirely that characteristic of a solid-solution diagram. Moreover, even a general composition in this diagram, for example alloy 2, experiences solidification of the solid-solution type until the eutectic horizontal is reached. In fact there are helpful similarities of this sort in all the various types of equilibrium diagrams.

**Aluminum-silicon: a typical eutectic system.** The equilibrium diagram for aluminum-silicon alloys is given in Fig. 7.13, with photomicrographs showing representative structures at room temperature. The photomicrograph of 99.95% aluminum exhibits the typical equiaxed grain structure of a pure metal. The microstructure of the 8% silicon alloy shows long *dendrites* of primary alpha solid solution surrounded by the eutectic microconstituent. In contrast, when the primary solid phase is the silicon-rich beta phase (in the 20% and 50% silicon alloys), the primary phase crystals have *geometric* shapes. This dissimilar behavior reflects the pronounced difference between aluminum, a typical metal, and silicon, which has predominantly nonmetallic properties.

It is not quite exact to speak of the two solid phases in the aluminum-silicon system as though they were pure aluminum and pure silicon, respectively. There is some solid solubility in both cases. Solid aluminum can dissolve about 1% silicon. In the alpha solid solution both aluminum and silicon atoms in the ratio of about 99:1 are randomly placed on the crystal lattice of aluminum, the face-centered-cubic lattice. Similarly, the beta phase is not pure silicon but consists of silicon plus about 1% aluminum; both kinds of atoms are arranged on the diamond cubic lattice characteristic of silicon.

The following definition of a phase is a useful guide in interpreting photomicrographs. *A phase is a homogeneous portion of matter separated from an adjoining phase by a bounding surface.* In the photomicrograph of the 20% silicon alloy, for example, the dark primary beta crystals are easily recognized as a phase. The fact that beta is not 100% silicon is immaterial, since the aluminum solute atoms are homogeneously distributed and are dissolved on an atomic scale that does not permit mechanical separation. On the other hand, is the background microconstituent of this photomicrograph a phase? Evidently not, since a bounding surface separates the small dark particles from the lighter ones. This microconstituent is the eutectic structure and consists of two phases, the dark beta and the light alpha. The beta in the eutectic structure (called *secondary* beta) and the large crystals of *primary* beta are merely different-sized particles of the same beta phase.

*Prediction of Phases.* The eutectic diagram is used in exactly the same manner as the solid-solution diagram to obtain information on the chemical compositions and the amounts of the phases present. Therefore, only a single example, the 50% silicon alloy, will be considered, Fig. 7.13. At 1320°C the *point* determined by this temperature and the 50% silicon composition lies in the liquid-phase field. Only the liquid phase is present at this temperature. Its composition must be the same as that of the alloy, 50% silicon, and its amount is 100% of the weight of the alloy.

When the temperature of the molten alloy is lowered to about 1080° C, solidification begins with the nucleation and growth of primary crystals of beta solid solution. Typical of the phase analyses that can be made in this region of the diagram is the following at 870°C. The alloy composition-temperature point lies in the liquid-plus-beta phase field; therefore these two phases are present. The ends of the tie line drawn across this two-phase field determine the chemical compositions of the phases: the liquid phase is 34% silicon (66% aluminum); the beta solid-solution phase is 98% silicon. These composition values can be used in applying the lever law to determine the amounts of the two phases. In this instance Eq. (7.5) can be written

$$\% \ \beta\text{-phase} = \frac{(\% \ Si_0 - \% \ Si_L)}{(\% \ Si_\beta - \% \ Si_L)} \times 100,$$

or, in terms of lengths along the tie line,

$$\% \ \beta\text{-phase} = \frac{ab}{ac} \times 100.$$

In either case the numbers are

$$\% \ \beta\text{-phase} = \frac{50 - 34}{98 - 34} \times 100 = \frac{16}{64} \times 100 = 25\%.$$

Since the remaining portion of the alloy is liquid phase, there must be 75% liquid. A convenient form for recording this information is the following:

Point—50% silicon alloy at 870°C
Phases—*liquid*        and        *beta*
Compositions—34% Si                98% Si
Amounts—75%        $\dfrac{50 - 34}{98 - 34} \times 100 = 25\%$

This concise tabular presentation of phase data will be used frequently in the following pages.

As the cooling of this 50% silicon alloy is continued below 870°C, the amount of primary $\beta$ phase increases until the eutectic temperature is reached. At this temperature the liquid phase has reached the eutectic composition, 12% silicon. On cooling through the eutectic temperature, the liquid solidifies at constant temperature to form the eutectic structure consisting of an intimate mixture of small grains of the $\alpha$ and $\beta$ phases. Since the amounts of the liquid and solid phases change *at* the eutectic temperature, it is impossible to calculate the amounts of the phases present at exactly this temperature. A phase analysis at room temperature is representative of the condition of the alloy in the alpha-plus-beta region:

Point—50% silicon alloy at room temperature
Phases—*alpha*        and        *beta*
Compositions—1% Si                99% Si
Amounts—50%        $\dfrac{50 - 1}{99 - 1} \times 100 = 50\%$

The lever law is applied to two solid phases in exactly the same way as to a liquid and a solid phase. The validity of this procedure will be seen when the derivation of the lever law is recalled.

The amount of $\beta$ phase (50%) given by the above calculation is the sum of the primary and secondary beta. Frequently it is necessary to calculate not the amounts of the *phases* but the amounts of the *microconstituents*. Thus, in the photomicrograph of the 50% silicon alloy at room temperature the microconstituents are (1) the primary beta crystals and (2) the eutectic structure. Consideration of the manner of solidification of this alloy will show that a calculation of the amounts of these microconstituents can be made conveniently just above the eutectic temperature. At this temperature solidification of the primary $\beta$ crystals is complete, and all the liquid present must assume eutectic structure on cooling through the eutectic temperature. If the tie line shown in Fig. 7.13 is considered to practically coincide with the eutectic horizontal, the amount of primary beta can be calculated as

$$\% \text{ primary beta} = \frac{50 - 12}{98 - 12} \times 100 = 44\%,$$

and therefore the amount of eutectic liquid is 56%. However, since all this liquid

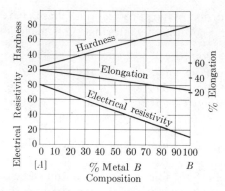

**Fig. 7.14**  Idealized variation of properties with composition in a simple eutectic alloy system.

forms eutectic structure on cooling through the eutectic temperature, there will also be 56% of eutectic structure.  Changes occurring between the eutectic temperature and room temperature in the amounts of the microconstituents may usually be neglected if the solvus lines are nearly vertical.

**Property variations in eutectic systems.**  In a series of alloys whose compositions extend across a eutectic diagram, as in Fig. 7.13, the physical and mechanical properties might be expected to show a linear variation, Fig. 7.14.  This ideal behavior, though it might be expected as the natural consequence of the linear variation in the amounts of the two phases present, is seldom found in practice. Irregularities are caused by the influence of the manner of distribution of the two phases on the properties of the alloy.  For example, as the composition changes from *hypoeutectic* (less than eutectic) to *hypereutectic* (more than eutectic) in terms of metal *B*, the primary crystals change from alpha phase to beta phase.  Since alpha and beta usually differ sharply in properties, a nonlinear change in properties across the eutectic composition would be expected.

Of still greater significance are changes in conformity with the *principle of the continuous phase*, which states that many important properties of two-phase alloys are primarily determined by the properties of the phase that forms a continuous path through the alloy.  This phase need not be present in the larger amount, since less than 1% bismuth in copper can form a continuous grain-boundary network that disastrously embrittles the copper.  Related to the principle of the continuous phase is the *principle of increasing ductility*.  If a two-phase alloy can be deformed or worked, there is a tendency for the continuity of the ductile phase to be increased at the expense of breakup of the brittle phase.  For example, certain cast alloys can be successfully worked only by slow compression at temperatures near the melting range.  After this initial deformation, however, these alloys can be worked by more conventional procedures.

As an example of property variations across an actual eutectic system, some properties of cast aluminum-silicon alloys containing up to 14% silicon are given

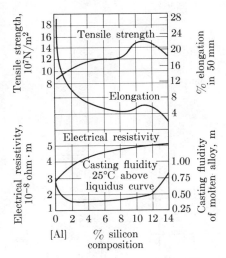

**Fig. 7.15** Variation of typical properties of cast aluminum-silicon alloys across a portion of this eutectic system, Fig. 7.13.

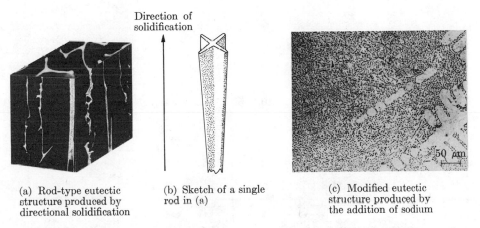

(a) Rod-type eutectic structure produced by directional solidification

(b) Sketch of a single rod in (a)

(c) Modified eutectic structure produced by the addition of sodium

**Fig. 7.16** Special eutectic microstructures produced in an aluminum-12% silicon alloy (Courtesy of James A. Bell and W. C. Winegard)

in Fig. 7.15. Anomalies in the tensile strength and elongation values near the eutectic composition are evident.

**Controlled eutectic microstructures.** The eutectic structure of the 12% silicon alloy in Fig. 7.13 can be described as acicular (needle-like) and is found in ordinary castings of this alloy. By causing solidification to occur unidirectionally, one can produce a rod-type eutectic structure, Fig. 7.16(a). The appearance in three dimensions of a single rod is sketched in Fig. 7.16(b). The whole array of seemingly

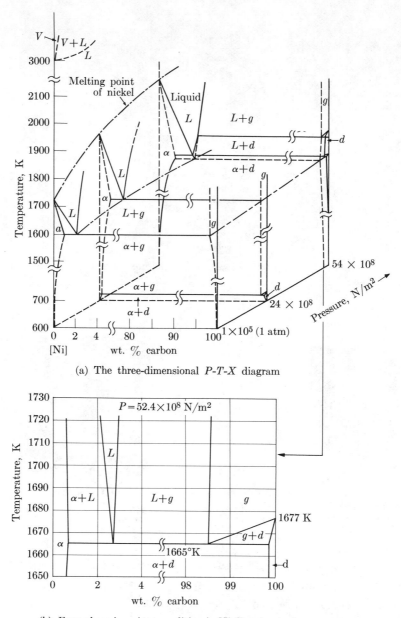

(a) The three-dimensional *P-T-X* diagram

(b) Four-phase invariant condition in Ni-C system

**Fig. 7.17** A suggested *P-T-X* diagram for the nickel-carbon system for pressure ranging upward from atmospheric pressure. Note the greatly expanded composition axis at the right-hand edge of the diagram. (Courtesy H. M. Strong and R. E. Hanneman, General Electric Company)

separate rods is actually interconnected and together constitutes a grain of the silicon phase. Similarly, entire groups of "needles" in the acicular structure are interconnected. A second method of controlling eutectic microstructures, *modification,* consists in treating the molten alloy with a suitable addition agent and then casting in the normal manner. Figure 7.16(c) illustrates the modified eutectic produced in aluminum-silicon alloys by the addition of sodium. This structure is actually the acicular-type eutectic on an extremely fine scale.

**P-T-X eutectic diagrams.** Compared to the solid-solution diagram, Fig. 7.2, the *P-T-X* diagram for eutectic systems shows some new features. We will use the nickel-carbon system, Fig. 7.17, as an example because of its importance for the synthesizing of diamonds. Since both nickel and graphite have high boiling points, the vapor phase enters only incidentally in this diagram. At atmospheric pressure graphite is the stable form of carbon, and the usual *T-X* diagram involves only a eutectic reaction between nickel and graphite. This diagram forms the front face of the three-dimensional *P-T-X* diagram; the top portion of the liquid-plus-graphite region has been omitted to give a freer view of the other two *T-X* sections. At a pressure of about $24 \times 10^8$ N/m² ($24 \times 10^3$ atm), an equilibrium involving the diamond phase of carbon can first be observed as a eutectoid* reaction in which the graphite solid, $g$, decomposes on cooling into the nickel solid solution, $\alpha$, and the diamond phase, $d$. This reaction occurs below the temperature of the eutectic reaction, $L \rightarrow \alpha + g$, but the usual type of phase analysis can be applied to the solidification and subsequent reaction of an alloy as it cools past these two reaction horizontals (see Problem 8). As the pressure increases, the eutectoid reaction moves to higher temperatures and at $52.4 \times 10^8$ N/m² coincides in temperature with the eutectic reaction. The resulting four-phase equilibrium is shown in the separate *T-X* diagram, Fig. 7.17(b). At a still higher pressure, for example, $54 \times 10^8$ N/m², the reaction, $g \rightarrow d + L$, lies at a higher temperature than the eutectic reaction. The metastable equilibria involving graphite, indicated by the dashed lines in this *T-X* section in Fig. 7.17(a), govern the behavior of a mixture of nickel and graphite being heated under a pressure of $54 \times 10^8$ N/m² in the process of synthesizing diamonds.

**Nonequilibrium solidification in eutectic systems.** It will be recalled that rapid cooling of a solid-solution alloy from the liquid state produces a cored, inhomogeneous structure. So also may cored grains and even additional phases be produced in rapidly cooled alloys of eutectic systems. The mechanism of such nonequilibrium solidification in an aluminum-4% copper alloy is shown in Fig. 7.18. The equilibrium diagram predicts that this alloy should solidify to form grains of alpha solid solution, Fig. 7.18(b). However, during the rapid cooling that takes place in most commercial casting operations there is not time for diffusion to occur in the

---

* This reaction is like the eutectic reaction except that a solid phase decomposes, rather than a liquid phase. The eutectoid reaction in steels is discussed on p. 320.

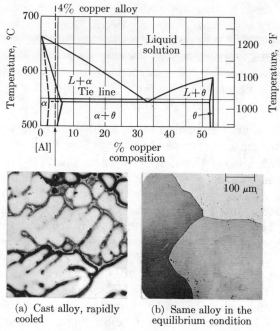

(a) Cast alloy, rapidly cooled

(b) Same alloy in the equilibrium condition

**Fig. 7.18** Qualitative explanation of the appearance of eutectic as an additional microconstituent in an aluminum-4% copper alloy. (Courtesy Aluminum Research Laboratories)

growing crystals of the alpha solid solution. As a result, the composition of a complete crystal varies from about 0.5% copper at the center, which solidifies at about 650°C, to 5.5% copper at the edge, which solidifies just above the eutectic temperature. This coring or variation in chemical composition of a solid-solution crystal can sometimes be seen under the microscope, as in the cast copper-nickel alloy, Fig. 7.9(a).

Not only coring, but also the formation of (nonequilibrium) eutectic structure can result from rapid cooling. At any temperature the *average* chemical composition of the cored α phase (Fig. 7.18) lies to the left of the equilibrium solidus line. A possible curve of *average* α phase composition versus temperature for the 4% copper alloy is shown by a dashed line, and the corresponding extension of the eutectic-reaction horizontal is also indicated. Since this dashed curve is the logical left-hand end of tie lines for the cored 4% copper alloy, it follows that solidification has not been completed when the eutectic temperature is reached during rapid cooling. The 5–10% of eutectic liquid that is still present solidifies to produce the alpha-plus-theta eutectic structure in the spaces between the growing dendritic primary crystals, Fig. 7.18(a). This 4% copper alloy consists of a single phase, alpha, at 540°C under equilibrium conditions. Therefore, heating the cast alloy at this temperature for sufficient time (to allow diffusion to occur) results in the production of the equilibrium structure, Fig. 7.18(b).

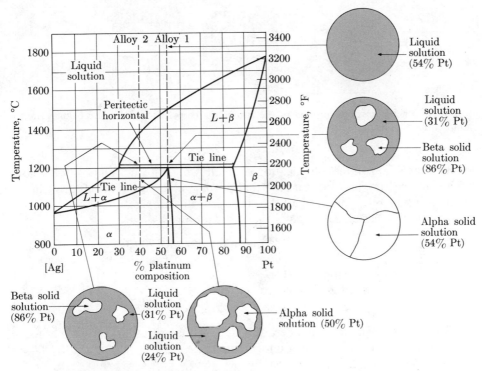

**Fig. 7.19** Simplified silver-platinum diagram showing the peritectic reaction.

## PERITECTIC SYSTEMS

Solid-solution formation and eutectic reaction are two important mechanisms of solidification shown by various metal systems. Another experimentally observed type of freezing, though less common, is the *peritectic reaction*. Although superficially the peritectic diagram with its peritectic-reaction horizontal may appear to resemble a eutectic diagram, their behaviors have little in common. There is no peritectic microconstituent corresponding to the eutectic structure and, in fact, the peritectic reaction consumes two phases to produce one different phase, behavior just the opposite of the eutectic reaction.

**The peritectic equilibrium diagram.** Although peritectic reactions are a part of such commercially significant diagrams as the iron-carbon (steel) and copper-zinc (brass) diagrams, there are only a few minor systems whose solidifications are characterized by a single peritectic reaction. The silver-platinum system, Fig. 7.19, is an example of such behavior. As usual the equilibrium diagram consists of one- and two-phase regions separated by phase-boundary lines. Of special interest is the peritectic horizontal at which the peritectic reaction occurs. The peritectic composition (54% platinum) is that of the central *solid* phase produced by the

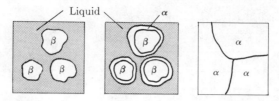

**Fig. 7.20** Schematic representation of the course of the peritectic reaction at the peritectic temperature.

peritectic reaction. If the solidification of an alloy of peritectic composition is considered (alloy 1 in Fig. 7.19), it is seen that primary beta-phase crystals grow in the liquid solution as the alloy is cooled to the peritectic temperature. Just above this temperature a phase analysis gives the following results.

$$
\begin{array}{ccc}
\text{Point}\text{---}54\% \text{ platinum alloy at } 1200°C \\
\text{Phases}\text{---}liquid & \text{and} & beta \\
\text{Compositions}\text{---}31\% \text{ Pt} & & 86\% \text{ Pt} \\
\text{Amounts}\text{---}58\% & & \dfrac{54-31}{86-31} \times 100 = 42\%
\end{array}
$$

Slightly below the peritectic temperature a phase analysis of this alloy of peritectic composition gives this surprising information:

$$
\begin{array}{c}
\text{Point}\text{---}54\% \text{ platinum alloy at } 1150°C \\
\text{Phase}\text{---}alpha \\
\text{Composition}\text{---}54\% \text{ Pt} \\
\text{Amount}\text{---}100\%
\end{array}
$$

How did the liquid and beta phases disappear and give way to homogeneous alpha grains? The peritectic reaction that accomplishes this change on cooling can be expressed by the equation

$$\text{Liquid} + \text{beta solid solution} \rightarrow \text{alpha solid solution,} \tag{7.8}$$

where the chemical compositions of the liquid and beta phases are given by the end points of the peritectic-reaction horizontal and the alpha solid solution has the peritectic composition (54% platinum). Like the eutectic reaction, the peritectic reaction occurs at constant temperature under ideal equilibrium conditions. The physical picture of this reaction is a peculiar one, since the alpha phase must begin to form at the surface of the primary beta crystals, where the liquid and beta are in contact, Fig. 7.20. However, further growth of the equilibrium alpha phase can occur only as a result of diffusion (through the existing alpha) of platinum atoms from the beta phase and of silver atoms from the liquid. If sufficient time is allowed at the peritectic temperature, the formation of homogeneous alpha grains will be completed.

For a peritectic reaction to occur, it is not necessary that an alloy be exactly of peritectic composition but only that its composition pass through the peritectic horizontal. Alloy 2 in Fig. 7.19 is an example of this more general composition. To show that equilibrium diagrams allow the prediction of phase changes that occur on heating an alloy as well as on cooling, the equilibrium *heating* of alloy 2 will be considered. At low temperatures this alloy consists only of alpha solid solution. On heating to about 1070°C, the liquid phase begins to form from the alpha, and a typical analysis in this two-phase region is the following.

$$
\begin{array}{lcc}
\text{Point—40\% platinum alloy at } 1150°\text{C} \\
\text{Phases—}liquid & \text{and} & alpha \\
\text{Compositions—24\% Pt} & & 50\% \text{ Pt} \\
\text{Amounts—39\%} & \dfrac{40-24}{50-24} \times 100 = 61\%
\end{array}
$$

As heating is continued to just below the peritectic temperature additional liquid phase forms at the expense of the alpha, and the compositions of both liquid and alpha change to those characteristic of the peritectic reaction. On heating through the peritectic temperature, the reverse reaction of Eq. (7.8) occurs, and the remaining alpha phase decomposes into beta solid solution and additional liquid. A phase analysis typical of the liquid and beta regions gives the amounts of these two phases just above the peritectic reaction horizontal.

$$
\begin{array}{lcc}
\text{Point —40\% platinum alloy at } 1200°\text{C} \\
\text{Phases—}liquid & \text{and} & beta \\
\text{Compositions—31\% Pt} & & 86\% \text{ Pt} \\
\text{Amounts—84\%} & \dfrac{40-31}{86-31} \times 100 = 16\%
\end{array}
$$

As heating is continued, the liquid phase increases in amount until the alloy is entirely liquid above about 1360°C.

**Nonequilibrium solidification in peritectic systems.** The use of peritectic equilibrium diagrams for the prediction of microstructures or of mechanical and physical properties is severely limited by the almost universal occurrence of nonequilibrium structures. For example, the 40% platinum alloy should consist of alpha phase grains at room temperature, according to the equilibrium diagram of Fig. 7.19, but the actual structure of the cast alloy is quite different, Fig. 7.21. The reason for the persistence of the dendritic beta phase is evident from the course of the peritectic reaction shown in Fig. 7.20. The alpha phase, in forming from the liquid and beta phases, surrounds or encases the beta-phase particles. This *surrounding* or *encasement* shields the beta phase from further reaction with the liquid, and diffusion in the solid phases is usually insufficient to allow equilibrium to be established during cooling.

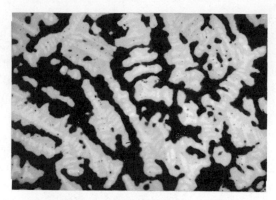

**Fig. 7.21** The microstructure of cast silver-40% platinum alloy showing the dendritic beta phase (light) encased by the alpha phase (light gray, gray, and black). (Courtesy F. N. Rhines and McGraw-Hill Book Company)

If the solidified alloy is subjected to a combination of plastic deformation and high temperature, as in the process of hot-working, the diffusion process is accelerated and equilibrium may be established. The physical and mechanical properties of the alloy can then be predicted roughly by means of the general rules given for solid-solution and for two-phase alloys.

## NONEQUILIBRIUM SOLIDIFICATION

In ideal solidification of alloys it is assumed that complete equilibrium is maintained between the solid and liquid phases. Extremely slow rates of cooling are required if this condition is to be approached; hence most alloys actually experience nonequilibrium solidification. Several undesirable results of nonequilibrium freezing have already been discussed for solid-solution, eutectic, and peritectic alloys. This phenomenon can also be used to purify metals, to even out nonuniformities in concentration, and to study properties such as diffusion in the liquid phase. As a background for understanding these useful aspects of nonequilibrium solidification, it is helpful to begin with a quantitative study of practical freezing conditions.

**Segregation during normal freezing.** Consider the freezing of a liquid alloy containing a concentration, $C'_L$, of solute. If the corresponding portion of the equilibrium diagram has the form shown in Fig. 7.22, the first solid to form (at temperature $T_1$) has the concentration $C'_S$, and the ratio $k_0 = C'_S/C'_L$ is called the *equilibrium distribution coefficient*. Because the liquidus and solidus curves are straight lines in the diagram of Fig. 7.22, at a later stage of equilibrium freezing at temperature $T_2$ this ratio would still have the same value, although the compositions of the phases would now be $C_L$ and $C_S$; that is,

$$k_0 = \frac{C_S}{C_L}. \tag{7.9}$$

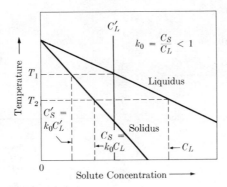

**Fig. 7.22** Equilibrium solidification of a liquid alloy of composition $C'_L$ in an equilibrium diagram for which $C_S < C_L$ and $k_0 < 1$.

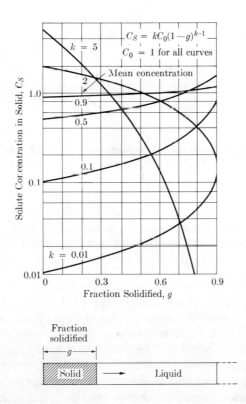

**Fig. 7.23** Segregation produced along a solidified bar during normal freezing for various values of $k$. (After W. G. Pfann)

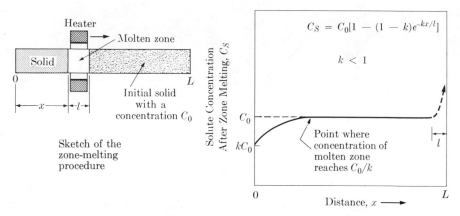

**Fig. 7.24**  The solute concentration produced by zone melting a bar having an initial, uniform concentration $C_0$.  (After W. G. Pfann)

During an actual solidification process the layer of solid being formed at any instant has the composition $C_S$ given by this equation, since equilibrium is closely approached at the liquid-solid interface. Usually it is assumed that the liquid phase has a uniform composition and that no diffusion occurs within the solidified alloy. Under these conditions the concentration of solute in the solid increases along the length of a solidifying bar in the manner shown in Fig. 7.23 for various $k$ values less than one. In Problem 11 it is shown that $k$ can also be larger than one, corresponding to the case in which the initial solid is richer in solute than the liquid from which it is forming. The concentration of solute then decreases along the length of the bar. The curves of Fig. 7.23 are based on the equation shown, in which $C_S$ is the concentration of solute in the layer of solid that solidified at the fractional length $g$, and $C_0$ is the mean solute concentration in the entire specimen. The *effective* distribution coefficient $k$ is used rather than $k_0$ to take account of the enriched (or impoverished) layer of liquid that tends to build up at the solid-liquid interface. In the definition analogous to Eq. (7.9), $k = C_S/C_L$, $C_S$ is the concentration of solute in the layer of solid being formed and $C_L$ is the mean concentration of all the remaining liquid.

**Zone melting.**  Figure 7.24 shows a convenient method for taking advantage of the difference in solute content of a liquid and the solid forming from it. A heater is used to form a molten zone of length $l$ in a bar of length $L$ and initial, uniform concentration $C_0$. The molten zone is initially at the left end of the bar. As the zone slowly progresses to the right the first solid that forms has the concentration $C_S = kC_L = kC_0$. Since this solid contains less solute than the solid that is melted by the advancing molten zone, the solute concentration of the molten zone increases, and therefore the concentration of the solidifying solid also increases. If the molten zone reaches the concentration $C_0/k$, then the solid that forms is $C_S = kC_L = k(C_0/k) = C_0$. That is, the concentration of solute is the same in the

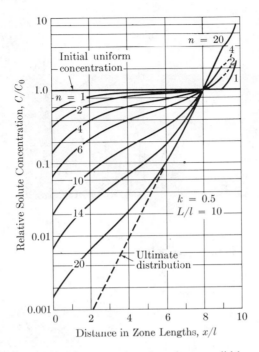

**Fig. 7.25** Progressive change in solute concentration along a solid bar when zone melting is repeated *n* times. (After W. G. Pfann)

solid being formed as in the solid being melted. Use is made of this behavior in the *zone-leveling* technique to produce a more uniform solute concentration in a bar that initially has nonuniformities in concentration. This and other topics are fully discussed in the reference by Pfann given at the end of the chapter. Figure 7.24 shows that in the last zone length the solute concentration of the solid rises as the enriched liquid freezes. The equation shown does not hold for this final portion of the curve.

Purification of a metal by the zone-melting technique is called *zone refining*. Usually the molten zone is passed through the bar a number of times to produce successive improvements in the purity of the solid, as shown in Fig. 7.25. After a large number of passes, an ultimate distribution is approached as the theoretical limit. Zone refining is widely used for producing the exact solute concentrations required in semiconductors.

## INTERMEDIATE PHASES

Most phase diagrams do not consist merely of *one* eutectic reaction or *one* peritectic reaction, but rather of a combination of the various fundamental reactions. For example, the iron-carbon diagram contains reactions of the solid-solution, peri-

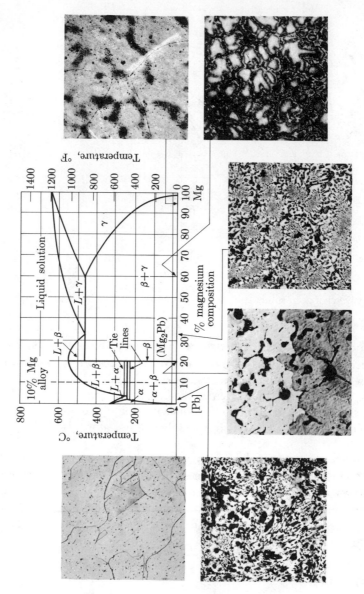

**Fig. 7.26**  The lead-magnesium diagram, illustrating the division of an equilibrium diagram into independent sections by a congruently melting intermediate compound ($Mg_2Pb$).  The cast structures of several alloys are shown by photomicrographs.  (Courtesy American Magnesium Corporation and Dow Chemical Company)

tectic, and eutectic types. In most cases the appearance of several reactions in a single binary diagram is the result of the presence of *intermediate phases*. These are phases whose chemical compositions are intermediate between the two pure metals, and whose crystal structures are different from those of the pure metals. This difference in crystal structure distinguishes intermediate phases from primary solid solutions, which are based on the pure metals. Some intermediate phases can accurately be called *intermetallic compounds* when, like $Mg_2Pb$, they have a fixed simple ratio of the two kinds of atoms. However, as pointed out in Chapter 6, many intermediate phases exist over a range of compositions and are considered *intermediate* or *secondary solid solutions*.

Intermediate phases are divided on the basis of their melting behavior into congruently melting and incongruently melting phases. An *incongruently melting phase* decomposes into two different phases, usually one solid and one liquid phase, instead of melting in the usual manner. A peritectic reaction is produced by this kind of decomposition (see Problem 10). The peritectic composition is represented by the composition of the intermediate phase at the peritectic temperature. A *congruently melting phase* melts in the same manner as a pure metal. In this case the equilibrium diagram is divided into essentially independent sections. In Fig. 7.26 the congruently melting beta phase divides the lead-magnesium diagram into two eutectic reactions, each of which can be analyzed separately.

To illustrate the solidification of a typical alloy in a many-reaction diagram, the lead-10% magnesium alloy will be considered. At high temperatures the alloy is liquid, for example,

$$\text{Point—10\% magnesium alloy at 550°C}$$
$$\text{Phase—}liquid$$
$$\text{Composition   10\% Mg}$$
$$\text{Amount—100\% liquid}$$

Solidification begins at 480°C with the nucleation and growth of beta-phase grains. A tie line is drawn only within the two-phase region and therefore terminates at the line that represents the composition of the beta phase. An analysis just above the eutectic temperature is typical of analyses in this two-phase region. The eutectic composition is 2.5% magnesium.

$$\text{Point—10\% magnesium alloy at 260°C}$$
$$\text{Phases—}liquid \quad \text{and} \quad beta$$
$$\text{Compositions—2.5\% Mg} \qquad 19\% \text{ Mg}$$
$$\text{Amounts—55\%} \qquad \frac{10 - 2.5}{19 - 2.5} \times 100 = 45\%$$

The liquid phase is of eutectic composition, and it forms eutectic structure as it cools through the eutectic temperature. The amount of eutectic structure is therefore equal to the amount of this liquid phase, namely 55%. Note that a *phase* calculation for the solidified alloy gives the amounts of the alpha and beta phases.

Point—10% magnesium alloy at 230°C
Phases—*alpha*        and        *beta*
Compositions—1% Mg              19% Mg
Amounts—50%        $\dfrac{10-1}{19-1} \times 100 = 50\%$

From the results of these phase analyses at 230 and 260°C it can be concluded that (1) the microstructure of the cast 10% magnesium alloy consists of 45% primary beta crystals and 55% eutectic structure, and (2) the eutectic structure is composed of 50/55 parts alpha phase and 5/55 parts beta phase. Sketches of similar solidification behavior are shown in Fig. 7.12 (alloy 2).

## SOLID-STATE REACTIONS

The diagrams in the preceding sections have described the decomposition of a *liquid* solution into a solid solution, eutectic, and so on. A *solid* solution can undergo exactly the same kinds of reactions, and such solid-state decompositions are of principal interest in industry. The theory and method of phase analysis are identical for liquid and solid transformations, and it is only necessary to use the slightly different words *eutectoid* and *peritectoid* in order to deal adequately with the phase relationships in the decomposition of a solid solution.

Although liquid- and solid-state reactions may be identical in phase theory, we might naturally doubt whether a solid body can decompose into one or more new phases in the same manner that a mobile, liquid solution can. This doubt is justified. Solid-state reactions differ in two important ways from liquid reactions in the manner in which they attain the equilibrium conditions predicted by the phase diagram.

1. Solid-state reactions occur much more slowly, are subject to greater hysteresis (e.g., supercooling), and rarely correspond in practice to true equilibrium conditions.

2. Solid phases consist of atoms arranged in certain crystal structures, and new solid phases forming from an existing solid phase tend to take definite positions with respect to the existing crystal structure. That is, the crystal structure of the new phase may have an *orientation relationship* to the crystal structure of the phase from which it formed. The geometric (Widmanstätten) pattern sometimes formed by the new phase in such cases is shown in Fig. 7.28(b).

However, while slow reaction rates and orientation relationships are of great practical importance, these factors need not be considered now in connection with the study of *equilibrium* solid-state reactions.

**Eutectoid reactions.** The fact that steel can be hardened is a direct result of a eutectoid reaction, as given in the iron-carbon equilibrium diagram. (The actual heat treatment of steel is studied in detail in Chapter 10; only equilibrium changes

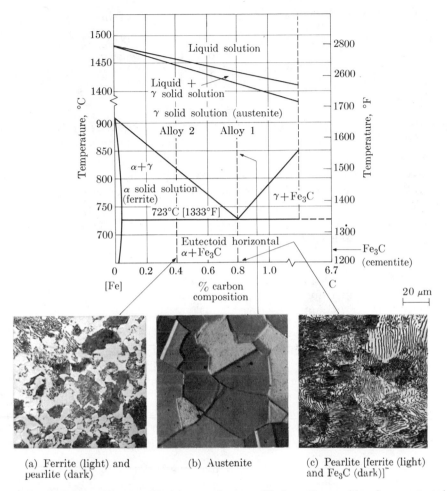

**Fig. 7.27** A portion of a simplified iron-carbon equilibrium diagram showing a eutectoid reaction with the gamma solid solution acting as the liquid solution in a eutectic reaction. (Photomicrographs courtesy J. R. Vilella, United States Steel Company)

in steel structures are considered here.) Just as a *eutectic* reaction involves the decomposition of a *liquid* solution, so does a *eutectoid* reaction involve the decomposition of a *solid* solution into two other solid phases. In Fig. 7.27 this solid solution is called *austenite* or the gamma solid solution. (Austenite forms from the liquid state by solid-solution freezing, which we have already studied.) The equilibrium decomposition of austenite will now be considered.

Austenite of eutectoid composition (which forms 100% eutectoid structure when it is slowly cooled through the eutectoid-reaction horizontal) has the simplest decomposition behavior. Alloy 1 in Fig. 7.27 has this composition, and in the austenite region of the diagram a typical phase analysis of this alloy is

Point—0.8% carbon alloy at 760°C
Phase—gamma solid solution (*austenite*)
Composition—0.8% C
Amount—100%

As alloy 1 is cooled through the eutectoid-reaction horizontal the austenite phase decomposes into the alpha solid solution (*ferrite*) and the iron-carbide phase ($Fe_3C$ or *cementite*). The nature of this eutectoid decomposition is strictly analogous to that of eutectic decomposition. The two new solid phases form side by side in a given region of the austenite to produce a *nodule* of pearlite, the eutectoid microconstituent. As the austenite phase is consumed by the growth of many pearlite nodules, the nodules meet one another to form the array that is seen in a photomicrograph, such as Fig. 7.27(c). A *phase* analysis just below the eutectoid-reaction horizontal gives the results

Point—0.8% carbon at 700°C
Phases—*alpha*        and        $Fe_3C$
Compositions—0.03% C                6.7% C
Amounts—88%        $\dfrac{0.80 - 0.03}{6.7 - 0.03} \times 100 = 12\%$

Since the corresponding microstructure, Fig. 7.27(c), is 100% pearlite, it follows that pearlite consists of 88% ferrite and 12% cementite.

Alloy 2 in Fig. 7.27 is a more general composition in this eutectoid system. As this alloy is cooled from the austenite region, primary ferrite crystals begin to form in the austenite grains starting at about 820°C. As cooling continues, more ferrite forms and the composition of the austenite approaches the eutectoid composition. Just above the eutectoid temperature the phase analysis is

Point—0·4% carbon alloy at 725°C
Phases—*alpha*        and        *gamma*
Compositions—0.03% C                (almost) 0.8% C
Amounts—52%        $\dfrac{0.40 - 0.03}{0.80 - 0.03} \times 100 = 48\%$

In cooling through the eutectoid temperature, the austenite (which is then of eutectoid composition) changes to pearlite, the eutectoid structure. Thus the *microstructure* of the alloy shows about 48% pearlite and 52% ferrite. Of course the percentage of *phases* is quite different, as shown by the following analysis:

Point—0.4% carbon at 700°C
Phases—*alpha*        and        $Fe_3C$
Compositions—0.03% C                6.7% C
Amounts—94%        $\dfrac{0.40 - 0.03}{6.7 - 0.03} \times 100 = 6\%$

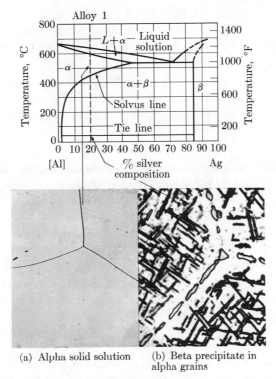

(a) Alpha solid solution     (b) Beta precipitate in alpha grains

**Fig. 7.28** Simplified aluminum-silver equilibrium diagram showing precipitation from a solid solution. (Courtesy A. H. Geisler)

In the photomicrograph of the 0.4% carbon alloy in Fig. 7.27 it is evident that the *microconstituents*, pearlite and (primary) ferrite, are often convenient quantities of reference. However, for other purposes, such as rating the machinability of a steel, the amount of the hard $Fe_3C$ *phase* may be more useful.

**Precipitation from solid solution.** A common type of solid-state reaction is the formation of a second phase within the grains of a solid-solution phase. The reason for such *precipitation* of the second phase is a decrease in solubility of the solute metal in the base metal. Figure 7.28 shows a portion of the aluminum-silver diagram in which the solubility of silver in aluminum varies from almost 50% at 540°C to only 2% near room temperature. Alloy 1, containing 20% silver, solidifies from the liquid state to form the alpha solid solution in the manner already discussed for solid-solution equilibrium diagrams. In the alpha-phase region the following analysis can be made, Fig. 7.28(a):

Point—20% silver alloy at 540°C
Phase—alpha solid solution
Composition—20% Ag
Amount—100%

As the solid alloy is further cooled, the solubility of the silver in aluminum drops to 20% at 450°C and precipitation of the silver-rich beta solid solution begins. As cooling continues, the amount of beta phase increases, until near room temperature a phase analysis gives the following results, Fig. 7.28(b):

Point—20% silver alloy at 40°C

| Phases—*alpha* | and | *beta* |
| Compositions—2% Ag | | 85% Ag |
| Amounts—78% | | $\dfrac{20-2}{85-2} \times 100 = 22\%$ |

The alpha and beta phases are *not* in the form of a eutectic structure, since the eutectic microconstituent can be produced only by the solidification of eutectic liquid. Rather, the beta crystals precipitate as plates on the {111} planes of alpha grains. The resulting geometric pattern on the surface of a polished specimen is called a *Widmanstätten structure*. Since the orientations of the various alpha grains differ, the character of the Widmanstätten pattern also varies from grain to grain.

**Order-disorder reactions.** A solid-state transformation that has no parallel in liquid-solid reactions is the *ordering* of a solid solution. It will be recalled that a solid solution consists of solvent and solute atoms distributed *at random* on the points of the solvent metal lattice. Figure 7.29(a) is a two-dimensional representation of a body-centered solid solution of equal numbers of atoms of metal $A$ and metal $B$. The *disordered* distribution of atoms in this solid solution is quite different from the regular arrangement in the *ordered* structure, Fig. 7.29(b). It is evident that perfect ordering of a solid solution can occur only if the two kinds of atoms are present in certain ratios, such as the 1:1 ratio shown in Fig. 7.29. Also, increased thermal vibration of the atoms at high temperatures tends to decrease the perfection of ordering. However, as shown in Fig. 7.30, the more or less perfectly ordered alpha prime ($\alpha'$) phase exists over a range of composition and temperature and can be distinguished from the disordered alpha phase.

The analysis of the solidification and cooling of an alloy in a diagram such as Fig. 7.30 offers no new problems. For example, the general composition, alloy 2, solidifies according to the solid-solution reaction, and at 600° the phase analysis is

Point—25% $B$ alloy at 600°
Phase—*alpha*
Composition—25% $B$
Amount—100%

As the temperature is lowered to about 420°, the ordered structure begins to form, and at 400° the following analysis can be made:

Point—25% $B$ alloy at 400°

| Phases—$\alpha$ | and | $\alpha'$ |
| Compositions—24% $B$ | | 27% $B$ |
| Amounts—67% | | $\dfrac{25-24}{27-24} \times 100 = 33\%$ |

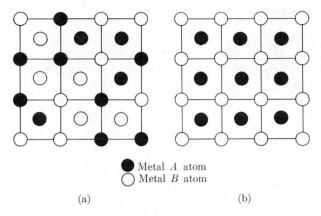

● Metal $A$ atom
○ Metal $B$ atom

(a)                                                    (b)

**Fig. 7.29**  The arrangements of equal numbers of atoms of metal $A$ and metal $B$ on the points of a two-dimensional "body-centered" space lattice.  (a) An ordinary solid solution.  The atoms are in a disordered position.  (b) An ordered structure.

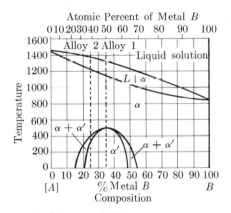

**Fig. 7.30**  Equilibrium diagram showing the formation of an ordered structure, $\alpha'$, from the disordered solid solution, $\alpha$.

On further cooling, the $\alpha'$ structure continues to form and completely replaces $\alpha$ at about 350°.  At room temperature the situation is simply

Point—25% $B$ alloy at room temperature
Phase—$\alpha'$
Composition—25% $B$
Amount—100%

The unusual properties associated with the formation of an ordered structure are considered in Chapter 10.

## TERNARY SYSTEMS

Since the properties of a pure metal may be improved by addition of *one* alloying element, why not add *several* and obtain a greater improvement? This is exactly what is done in practice, so that most commercial alloys are more complex than binary alloys. Metals for high-temperature service, for example, may contain ten important elements. At present there is no satisfactory method of representing phase relations in such multicomponent systems, and three-component diagrams are the most complex that are commonly encountered.

**Representation of ternary systems.** To plot adequately the variations of pressure, temperature, and the two concentration variables in a three-component system would require the use of four dimensions. By fixing the pressure at one atmosphere, three dimensions are sufficient. Since even this kind of plotting is inconvenient, the temperature also can be fixed to reduce the necessary diagram to two dimensions. Figure 7.31 shows the usual triangular method for plotting the two composition variables in a ternary system when the pressure and temperature are fixed. A vertex of the diagram represents 100% of one component, such as metal A. The base of the diagram opposite the metal A vertex represents 0% of metal A, and lines parallel to this base indicate varying percentages of metal A. Thus, the point representing alloy 1 lies 20% of the distance between the base and vertex A; therefore alloy 1 contains 20% A. Similar reasoning shows that alloy 1 contains 60% B and 20% C.

This simple composition plot at constant temperature (and pressure) is valuable in determining the phase constitution of alloys. Use of the three-dimensional ternary equilibrium diagram of Fig. 7.32(a) should clarify this fact. Temperature is plotted perpendicular to the composition triangles. As usual, *points* in the diagram are determined by composition and temperature. Phase regions are volumes rather than areas, but the phases present in a given instance are still determined by the phase region in which the *point* falls. Thus, during the solidification of alloy 1 the composition-temperature point passes through the following regions: (1) liquid, (2) liquid and beta, (3) liquid, beta, and alpha, and (4) beta and alpha.

**Phase analyses in ternary systems.** The space diagram is not convenient for making detailed phase analyses. However, a given analysis is made at a single temperature and therefore a constant temperature section through the space diagram is adequate for the analysis. Consider the analysis of alloy 1 in the liquid region, for example at temperature $T_1$. The section through the space diagram at this temperature is shown in Fig. 7.32(b), and it is evident that the *point* lies in the liquid-phase field. The phase analysis is simply

> Point—alloy 20% A, 60% B, (20% C) at temperature $T_1$
> Phase—*liquid*
> Composition—20% A, 60% B
> Amount—100%

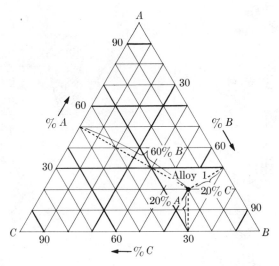

**Fig. 7.31** A method of plotting compositions in a ternary alloy system. The position of alloy 1, containing 20% $A$, 60% $B$, and 20% $C$, is shown.

The other phase fields that exist at temperature $T_1$ are not involved in this analysis.

When alloy 1 cools to the temperature of the liquidus surface for this composition, solid solution solidification begins. This solidification reaction is identical with that already studied in two-component systems, except that the beta solid solution formed in this case has both metal $A$ and metal $C$ atoms in solid solution on the space lattice of metal $B$. A typical phase analysis of alloy 1 in the liquid-plus-beta region can be made at temperature $T_2$ using the isothermal section of Fig. 7.32(c):

Point—alloy 20% $A$, 60% $B$ at temperature $T_2$
Phases—*liquid* and *beta*
Compositions—21% $A$, 58% $B$      10% $A$, 83% $B$
Amounts—92%      $\frac{2}{25} \times 100 = 8\%$

For convenience, the lengths needed in the lever law calculation were measured on the % $B$ composition scale. The tie lines in these isothermal sections are determined experimentally.

When a certain amount of the primary beta phase has formed, the liquid reaches a pseudoeutectic composition and a mixture of secondary beta and alpha phases begins to form in a structure similar to the eutectic microconstituent in a binary system. Temperature $T_3$ is within the range of temperatures over which the alpha and beta phases crystallize simultaneously from the liquid in alloy 1. The isothermal section at $T_3$ is shown in Fig. 7.32(d), and the following phase analysis applies. (It should be noted that the *compositions* of the three phases are read at the vertices of the triangular three-phase region.)

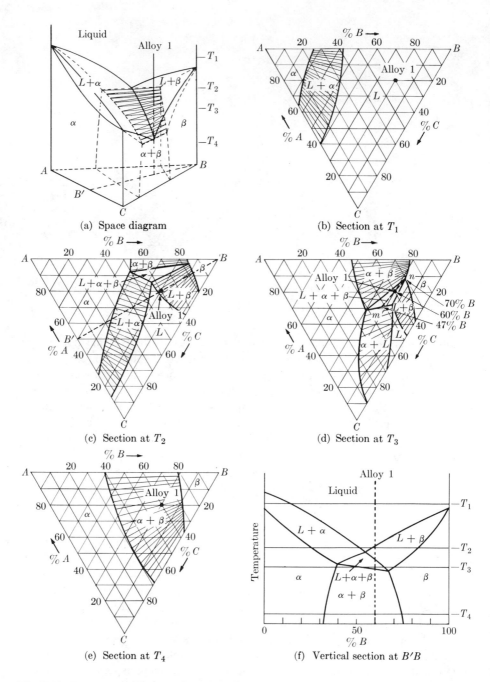

(a)  Space diagram

(b)  Section at $T_1$

(c)  Section at $T_2$

(d)  Section at $T_3$

(e)  Section at $T_4$

(f)  Vertical section at $B'B$

**Fig. 7.32**  Ternary equilibrium diagram and sections through it.  The lines in the two-phase regions represent experimentally determined tie lines.

Point—alloy 20% $A$, 60% $B$ at temperature $T_3$

| Phase—*liquid* | and | *alpha* | and | *beta* |
|---|---|---|---|---|
| Compositions—17% $A$, 55% $B$ | | 30% $A$, 38% $B$ | | 18% $A$, 70% $B$ |

$$\frac{ml}{mn} \times 100 =$$

| Amount—25% | | $\frac{5}{27} \times 100 = 19\%$ | | $\frac{13}{23} \times 100 = 56\%$ |
|---|---|---|---|---|

The construction used to determine the amount of beta phase is shown in Fig. 7.32(d), and is seen to be analogous to the usual lever law relation.* Similar levers can be set up to determine the amounts of the other phases. The total amount of beta phase includes both primary crystals and the secondary beta.

When the last liquid has disappeared, only the alpha and beta phases are present. A phase analysis at a temperature $T_4$ in the solid region of the diagram can easily be made with the aid of the isothermal section given in Fig. 7.32(e):

Point—alloy 20% $A$, 60% $B$ at temperature $T_4$

| Phases—*alpha* | and | *beta* |
|---|---|---|
| Compositions—41% $A$, 33% $B$ | | 11% $A$, 72% $B$ |
| Amounts—31% | | $\frac{27}{39} \times 100 = 69\%$ |

Although vertical sections through the space diagram, such as Fig. 7.32(f), bear a certain resemblance to binary diagrams, they lack an important characteristic. *Tie lines do not usually lie in the plane of a vertical section.* This can be seen by comparing the direction of the tie line through alloy 1 in Fig. 7.32(c) with the direction of the trace of section $B'B$ in this figure. This restriction on vertical sections of ternary diagrams must be borne in mind when these sections are used. Nevertheless, the vertical sections are quite valuable in showing the phases that are present in an alloy during equilibrium cooling and heating. The vertical sections also reveal the temperatures at which the various phase changes occur.

The ternary diagram of Fig. 7.32 is a relatively simple example of possible ternary systems. The construction and visualization of most technically important ternary diagrams is extremely difficult. However, once the necessary isothermal diagram sections are available, the principles illustrated in Fig. 7.32(a) are sufficient for the solution of most problems involving three-component systems.

## GAS–METAL EQUILIBRIUM

Since the production, melting, and heat treatment of metals are carried out in the presence of gases (oxygen, nitrogen, and hydrogen, for example), it is evident that

---

* When three phases are in equilibrium, as in this instance, their relative amounts are such that the three-phase triangle would balance in a horizontal plane if it were supported at the point representing the system composition (20% $A$ and 60% $B$) and if the weights of the corresponding phases were applied at the vertices of the triangle. This visualization is the analog of Fig. 7.7 for two-phase equilibrium. Construction of the line $mn$ (through the *beta*-phase and system composition points) reduces the problem of determining the *beta* phase to that of applying the ordinary lever law.

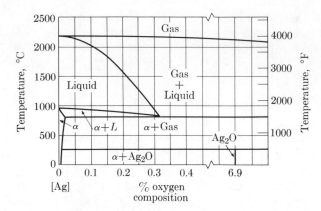

**Fig. 7.33** Silver-oxygen equilibrium diagram showing the silver end of the diagram. This diagram is for atmospheric pressure.

gas–metal reactions are possible. Many aspects of these reactions can be understood most easily by means of the appropriate gas–metal equilibrium diagram. Basically such a diagram is the same as that for an ordinary binary metal system, but it is to be expected that the gas phase and therefore the pressure variable are of greater importance.

As an example of a gas–metal diagram, the equilibrium relations between silver and oxygen are shown in Fig. 7.33 for a pressure of one atmosphere. This system is a good illustration of the importance of a gas phase, since "spitting" can occur during the freezing of silver. The diagram shows that pure silver will melt if it is heated above 961°C, and that the liquid phase will continue at this temperature even if it dissolves oxygen gas from the surrounding atmosphere. Given sufficient time at the melting point, the composition of the liquid will change from 0% oxygen to about 0.3% oxygen. If the liquid is then cooled, the liquid phase reaches eutectic composition during solidification and then decomposes into the alpha solid solution (almost pure silver) and the gas phase. The gas bubbles rise to the surface of the solidifying mass and produce the spitting of liquid as they escape. Since this gas evolution results in a casting of poor quality, the melting of silver is preferably carried out in an atmosphere from which oxygen is excluded.

While pressure has relatively little effect on solid and liquid equilibria, it is an important factor in reactions involving gases. For example, the maximum solubility of oxygen in the alpha silver phase at a given temperature is greatly increased at higher pressures. In fact the whole appearance of Fig. 7.33 is quite different at much higher or much lower pressures; the $Ag_2O$ phase, for instance, is absent at low pressures. Although data are usually not available for the complete effect of pressure on gas–metal systems, the variation of *maximum* gas solubility $S$ with the pressure $p$ is given by *Sievert's law* for diatomic gases:

$$S = K\sqrt{p}, \tag{7.10}$$

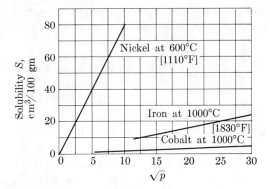

**Fig. 7.34** The influence of pressure on the maximum solubility of hydrogen in solid metals. (Adapted from C. J. Smithells.)

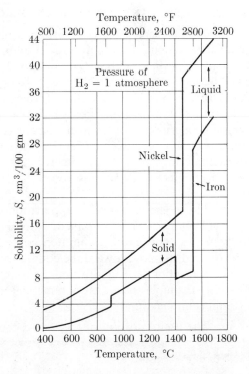

**Fig. 7.35** The influence of temperature on the maximum solubility of hydrogen in solid and liquid nickel and iron.

where $K$ is a constant that depends on temperature. Similarly, the effect of temperature on solubility at constant pressure is given by

$$S = Ae^{-Q/RT}, \qquad (7.11)$$

where $A$ and $Q$ are constants, $R$ is the universal gas constant, and $T$ is absolute temperature in Kelvin. Data for applying these two equations to many gas–metal problems are available in the literature. Figures 7.34 and 7.35 show some results on hydrogen solubility.

Figure 7.35 indicates that the solubility of iron for hydrogen increases suddenly at 910°C, when the iron changes from body-centered cubic to face-centered cubic. A corresponding decrease occurs at 1400°C, when the iron again reverts to the body-centered-cubic structure. The large difference between the solubility of liquid and solid metals is the basis of a method for decreasing the gas content of molten metals prior to casting. The temperature of the molten alloy is allowed to fall to just below the freezing range, and as a result much of the dissolved gas is "precipitated out." If the alloy is then quickly heated to the casting temperature and cast, there is insufficient time for the gas content to increase again to the maximum value. As a result, the gas content of the final casting is significantly reduced.

## TYPICAL INDUSTRIALLY IMPORTANT EQUILIBRIUM DIAGRAMS

The diagrams used to illustrate the type of equilibrium usually found in metal systems were chosen for their simplicity. The average equilibrium diagram, however, is made up of a number of these simple reactions (solid-solution, peritectic, eutectic, and so on) and frequently *looks* complicated. Actually, such a diagram is no more difficult to analyze than one of the reactions of which it is composed, since any composition-temperature point lies only in a single reaction region. Figure 7.36 illustrates this simplifying principle for the iron-carbon system, in which solid-solution formation and peritectic, eutectic, and eutectoid reactions occur. The analysis of alloy 1, for example, involves only a peritectic reaction during solidification; there is an independent eutectoid decomposition of the gamma solid solution at lower temperatures.

A more serious problem in the commercial application of equilibrium diagrams is the fact that equilibrium is often not attained. Although each alloy system requires separate study in this connection, a common governing principle gives assurance that every alloy tends to move toward the equilibrium condition. Thus, even in the absence of ideal reactions the equilibrium diagram is an indispensable part of alloy calculations. Examples based on commercial alloys will illustrate the practical use of equilibrium diagrams.

**The iron-carbon system.** The useful portion of the iron-carbon equilibrium diagram is shown in Fig. 7.37. Iron-carbon alloys are usually divided into three categories on the basis of composition:

1. *Irons*, in which the carbon content is very low and has negligible effect on properties

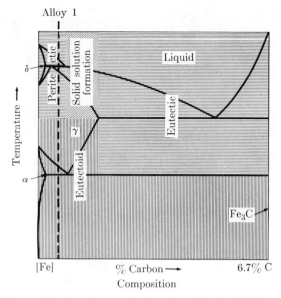

**Fig. 7.36** Iron-carbon equilibrium diagram broken down into its simple component reactions.

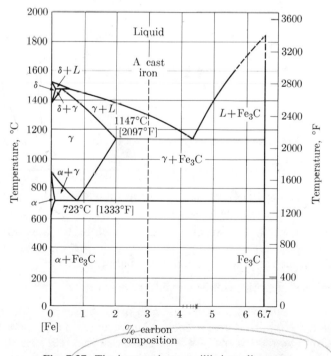

**Fig. 7.37** The iron-carbon equilibrium diagram.

*reproduce & label*

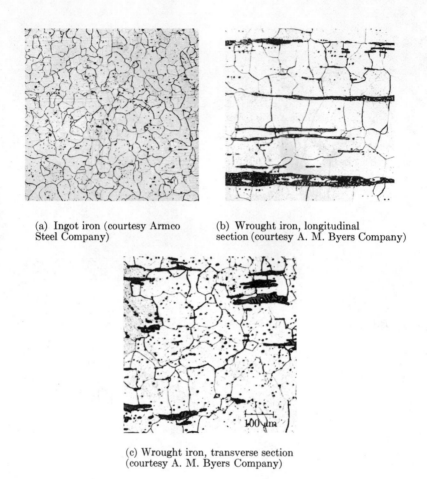

(a) Ingot iron (courtesy Armco          (b) Wrought iron, longitudinal
Steel Company)                          section (courtesy A. M. Byers Company)

(c) Wrought iron, transverse section
(courtesy A. M. Byers Company)

**Fig. 7.38**   Photomicrographs showing the structures of two types of iron, ingot iron and wrought iron.

2. *Steels*, in which the carbon content is important and usually in the range of 0.1 to 1.5% with 2.0% as a maximum value

3. *Cast irons*, in which the carbon content is such as to cause some liquid of eutectic composition to solidify. The minimum carbon content is therefore about 2%, while the practical maximum is about 4.5%.

Each of these categories will be considered separately.

*Irons.*   For special purposes, such as research problems, *electrolytic iron* and *carbonyl iron* of 99.99% purity are available. However, commercial *ingot iron*, Fig. 7.38(a), contains about 0.1% of various impurities (about 0.01% carbon). Iron of even this purity is relatively expensive to produce, and its use is restricted to appli-

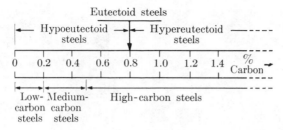

**Fig. 7.39** Important steel classifications based on carbon content.

cations where its superior ductility, corrosion resistance, electrical conductivity, or magnetic permeability with respect to the lower-cost steels is needed. Ingot iron in the form of galvanized or enameled sheets is used for such purposes as roofing and siding.

*Wrought iron* is not strictly an *iron* but consists of about 3% slag distributed in an iron matrix. During the hot-working operations involved in making the final wrought iron shapes, the slag particles become elongated, as shown in Fig. 7.38(b), and cause pronounced directional properties. These slag fibers account for the good fatigue-resistant properties of this material and perhaps contribute to its corrosion resistance. A principal application of wrought iron is in piping that handles mildly corrosive liquids like salt water.

The phase analysis of a material like ingot iron is especially easy. The metal is essentially a single phase at all temperatures, with the transition from one phase to another occurring at about the equilibrium temperature. At temperatures between 910 and 1400°C the iron is in the face-centered-cubic gamma-iron condition. For example, it is hot-rolled as gamma iron. On cooling, it transforms into body-centered-cubic alpha iron; alpha iron grains appear in the photomicrograph of Fig. 7.38(a). The properties of ingot iron are relatively unaffected by the rate of cooling from high temperatures, and a water-quenched specimen is about as soft as one that has been slowly cooled.

*Steels.* The term *plain carbon steel* describes steels in which alloying elements play only a small part in determining the properties. These steels can be classified in a number of ways, two of which are shown in Fig. 7.39. It will be recalled that a steel of eutectoid carbon content, 0.8%, forms a completely pearlitic structure on slow cooling. A *hypo*eutectoid steel contains *proeutectoid ferrite* and pearlite, while a *hyper*eutectoid steel contains *proeutectoid cementite* and pearlite. These divisions are too large for many purposes, and the low, medium, and high carbon classes of steel are of more practical interest. The significance of these groupings will be illustrated in discussing cast steels and wrought steels.

*Cast steels.* The mechanical properties of a casting tend to be poorer than those of the same alloy in the wrought form, but factors such as an intricate shape or a small number of units to be manufactured may favor a steel casting. Typical uses

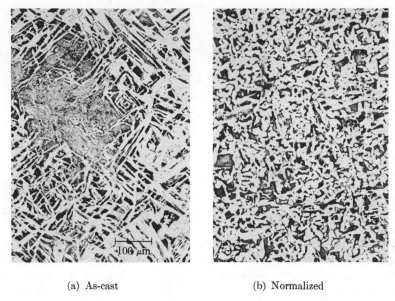

(a) As-cast                              (b) Normalized

**Fig. 7.40** Microstructure of a 0.3 % carbon steel cast in a one-inch-thick section. (Courtesy M. F. Hawkes)

of steel castings are in agricultural and excavating equipment. A high carbon content increases the strength of a casting but decreases its ductility; therefore most steel castings are in the medium carbon range. During slow cooling of the molten steel after it has been cast in the sand mold, the first solid phase that begins to form is delta. During further cooling, however, this is replaced by coarse austenite, which later decomposes into a Widmanstätten ferrite pattern, Fig. 7.40(a). The (white) ferrite appears at the austenite grain boundaries and as Widmanstätten plates with the (dark) pearlite. Relatively poor mechanical properties are associated with this structure. If the casting is heated into the austenite region (about 870°C, in practice), the entire structure changes to relatively fine austenite grains. On air cooling, this austenite decomposes into a correspondingly fine mixture of ferrite (white) and pearlite (dark), Fig. 7.40(b), with better properties. The heat treatment, consisting of heating a steel into the austenite region of the equilibrium diagram and then air cooling, is called *normalizing*.

*Wrought steels.* Since the mechanical properties of cast alloys are improved by plastic deformation, most steel is used in the form of *wrought* alloys produced by hot-working the cast ingot. Hot rolling is the lowest-cost and most common hot-working process. The temperature of hot rolling is in the austenite region; therefore a hot-rolled steel is essentially in the normalized condition after it air cools from the temperature of hot-working. Wrought steels are classified as low-, medium-, and high-carbon alloys.

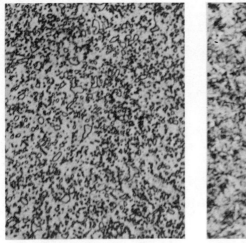 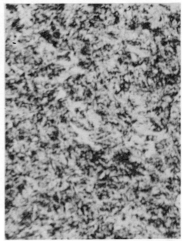

(a) Spheroidized structure; spheres of carbide in a ferrite matrix

(b) Structure produced by quenching from 815°C[1500°F], fine martensite needles and a few undissolved carbides

**Fig. 7.41** Microstructures of a 1% carbon tool steel. (Courtesy A. J. Scheid, Columbia Tool Steel Company)

Low-carbon steels are not hardened by heat treatment. A common example, ordinary mild steel, contains about 0.2% carbon and is used as a general purpose, low-cost construction material. A decrease in carbon content improves the ductility. For this reason killed and rimming steels containing about 0.1% carbon are used in large quantities in the automotive industry for press forming into automobile fenders and bodies.

Medium-carbon steels are stronger than low-carbon steels and can be further strengthened by heat treatment. For example, a forging made of 0.45% carbon steel can be given about twice the strength of a low-carbon steel and still retain adequate ductility. The heat treatment frequently produces a fine pearlite harder than the coarse pearlite that results from equilibrium cooling. The heat treatment of steel is considered in detail in Chapter 10.

The carbon content of high-carbon steels seriously reduces the ductility, and this type of steel is used only when strength or hardness is more important than ductility. Consequently, high-carbon steels are always given a hardening heat treatment. In a metal cutting tool, such as a drill, high hardness is needed to keep the cutting edge sharp. The 1.0% carbon steel that can be used for making a drill is too difficult to machine in the normalized condition. Even in the annealed (furnace-cooled) condition, high-speed machining is not commercially feasible. Therefore the iron carbide plates in the pearlite are caused to "ball up" by prolonged heating just below the eutectoid temperature. The *spheroidized* structure,

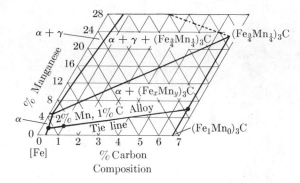

**Fig. 7.42** The position of a 2% manganese, 1% carbon steel in a modified room-temperature section through the iron-manganese-carbon equilibrium diagram.

Fig. 7.41(a), produced by this spheroidizing heat treatment is easily machined. After the drill has been made by a machining operation, it is heated to 815°C (1500°F) to change the steel to the austenite phase. Water quenching from this temperature does not prevent the decomposition of the austenite but causes it to occur at a low temperature, about 250°C (500°F), and results in the formation of hard *martensite*, Fig. 7.41(b), rather than soft pearlite. The nature of martensite is discussed in Chapter 9; essentially it is body-centered-cubic iron that has formed almost instantaneously from the austenite. The carbon atoms, which were dissolved in the austenite, are trapped in the body-centered-cubic structure and harden it by distorting the lattice. To relieve the stresses produced by the quenching operation, the drill is *tempered* at 200°C (400°F) for an hour. The tempering heat treatment also allows the carbon atoms to *begin* the process of forming $Fe_3C$, but the tempered martensite now present in the drill is still hard and has little ductility.

Wrought carbon steel is also used in the cold-worked condition. Of the many purposes for which cold-working may be used some of the more important are (1) to improve the machinability of low-carbon steel by decreasing its ductility, (2) to produce steel with a good surface finish and close dimensional tolerance, and (3) to increase the strength of steel in such forms as cold-drawn wire for suspension bridge cables.

*Alloy steels.* When alloying elements are added to steel, both solid-solution and compound formation usually occur, as in most metal systems. However, the presence of both iron and carbon in the steel leads to the peculiar circumstance that the solid solution is formed with the iron, while the compound is usually a carbide. If only a single alloying element is being considered, information on the phases present in a given alloy can be obtained from the corresponding ternary equilibrium diagram. Figure 7.42 shows a phase analysis of a 2% manganese, 1% carbon steel, using an isothermal section of the iron-manganese-carbon diagram.

$$\begin{aligned}
&\text{Point—}2.0\%\ \text{Mn},\ 1.0\%\ \text{C alloy at room temperature}\\
&\text{Phases—}\alpha \qquad\qquad \text{and} \qquad\qquad (Fe_x Mn_y)_3 C\\
&\text{Compositions—}1.5\%\ \text{Mn},\ 0.3\%\ \text{C} \qquad 6.0\%\ \text{Mn},\ 6.7\%\ \text{C}\\
&\text{Amounts—}89\% \qquad\qquad\qquad \frac{1.0-0.3}{6.7-0.3} \times 100 = 11\%
\end{aligned}$$

If similar phase analyses are made of steels of various manganese contents (see Problem 21), it is found that the manganese content of the alloy influences not only the manganese content of the solid solution but also that of the carbide.

Most commercial alloy steels must be treated as at least four-component systems, and few of the corresponding equilibrium diagrams are available. However, qualitative information on the tendency of the individual alloying elements in such steels to form a solid solution or a carbide phase can be given in a table like Table 7.1. A given alloying element (manganese, for example) will typically form both a solid solution and a carbide. The position of the X in Table 7.1 with respect to the lines representing (complete) solid-solution or carbide formation roughly denotes the proportion of the alloying element that enters each of these phases. The X for manganese, for example, indicates that about three-quarters of the manganese in a typical steel is in solid solution in the iron, while the remaining one-quarter is in the carbide phase. A number of elements, such as silicon, dissolve entirely in the solid solution. Others, like titanium, are almost completely combined in carbide phases.

Each alloying element in steel produces specific effects. However, a number of general property changes are associated with the formation of any carbide phase or of any solid solution. Since the carbide particles are hard, they increase wear

**TABLE 7.1**

Distribution of Alloying Elements in Steels

| Alloying element | Tends to form Solid solution | | | | Carbide |
|---|---|---|---|---|---|
| Phosphorus | X | | | | |
| Silicon | X | | | | |
| Aluminum | X | | | | |
| Nickel | X | | | | |
| Cobalt | | X | | | |
| Manganese | | | X | | |
| Chromium | | | | X | |
| Molybdenum | | | | | X |
| Tungsten | | | | | X |
| Vanadium | | | | | X |
| Niobium | | | | | X |
| Titanium | | | | | X |

TABLE 7.2

Effect of Alloying Elements in Strengthening Ferrite

| Alloying element | Approximate increase in yield strength of ferrite per 1% alloying element, $10^7$ N/m$^2$ |
|---|---|
| Phosphorus | 70 |
| Beryllium | 60 |
| Silicon | 11 |
| Manganese | 10 |
| Nickel | 8 |
| Titanium | 7 |
| Aluminum | 4 |
| Molybdenum | 3 |
| Tungsten | 2 |
| Vanadium | 2 |
| Cobalt | 2 |
| Chromium | 1 |

To convert from N/m$^2$ to lb/in$^2$, multiply by $1.450 \times 10^{-4}$.

resistance, decrease machinability, and slightly increase the hardness of the alloy. These particles may also prevent grain coarsening when the steels are heated into the austenite region. A few carbides are capable of markedly hardening the steel by a precipitation-type reaction (see Chapter 10), but usually these particles act like inert granules in the steel. On the other hand, when an alloying element is present in solid solution, it very actively influences most of the properties of the steel. Not only does it appreciably increase hardness and strength, as shown in Table 7.2, but it also affects corrosion resistance, electrical properties, and that most important quality, *hardenability*, which is considered in detail in Chapter 10.

*Cast Irons.* Perhaps even more strikingly than the steels, the cast irons show the influence of factors other than those shown in the equilibrium diagram in determining the final structure and properties of an alloy. Fundamental to this fact is the instability of iron carbide. At all temperatures the following reaction tends to occur:

$$Fe_3C \rightarrow 3Fe + C \text{ (graphite).} \qquad (7.12)$$

At low temperatures, however, the reaction occurs so slowly that iron carbide is metastable and can exist for thousands of years. At higher temperatures, especially if silicon is present as an alloying element in the iron, rapid graphitization of the iron carbide can take place. If graphitization is complete, the iron-graphite equilibrium diagram (Fig. 7.44) adequately represents the actual structure. For many commercial alloys in which only partial decomposition of the iron carbide occurs, both the iron-carbon diagram, Fig. 7.37, and the iron-graphite diagram must be used.

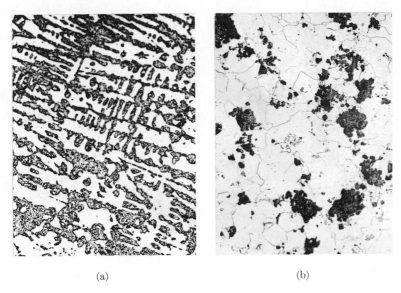

(a)                                           (b)

**Fig. 7.43**  Photomicrographs of (a) a white cast iron and (b) a ferritic malleable cast iron produced by heat treatment of white cast iron.  (Courtesy H. A. Schwartz, National Malleable and Steel Castings Company)

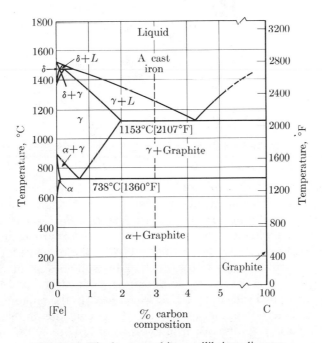

**Fig. 7.44**  The iron-graphite equilibrium diagram.

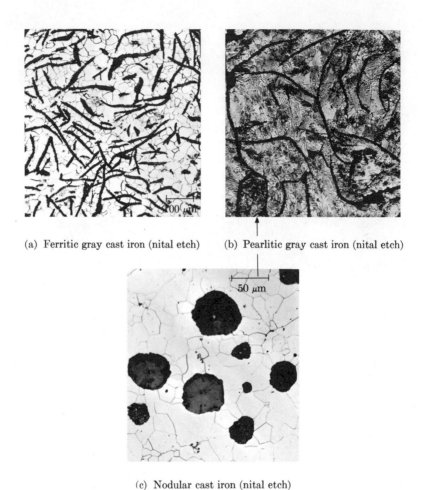

(a) Ferritic gray cast iron (nital etch)    (b) Pearlitic gray cast iron (nital etch)

(c) Nodular cast iron (nital etch)

**Fig. 7.45**    The microstructures of three types of cast iron.  (Courtesy International Nickel Company)

*White cast iron.*  If a cast iron containing 3.0% carbon is rapidly cooled from the molten condition, no graphitization occurs.  Because of the white appearance of its fracture surface, the alloy is called *white cast iron*, Fig. 7.43(a).  The manner of solidification of a white cast iron can be determined approximately from the iron-carbon diagram, Fig. 7.37.  In cooling to 1147°C the austenite forms from the molten alloy as dendrites; about 50% of the alloy solidifies in this manner.  At 1147°C the liquid reaches the eutectic composition, 4.3% carbon, and solidifies as a eutectic of austenite and cementite, called *ledeburite*.  In the photomicrograph of Fig. 7.43(a), this eutectic appears mostly as white cementite and surrounds the fernlike dendrite arms.  As the cast iron cools from 1147 to 723°C, the austenite changes in carbon content from 2.0 to 0.8% by precipitating iron carbide.  This additional iron carbide builds up largely on cementite particles already present. At 723°C the austenite changes to the eutectoid of ferrite and cementite, pearlite. In Fig. 7.43(a) the dark areas are pearlite (too fine to be resolved at the magnification employed) and the white areas are cementite. The analysis used here is only approximate because (1) the alloy is cooled too rapidly for equilibrium to be established, and (2) the ternary Fe-C-Si diagram should have been used,* since white cast irons contain about 1% silicon. In spite of these approximations the analysis is adequate for many purposes.

Rapid cooling prevents graphitization of the cementite in white cast iron, but if the casting is reheated to about 870°C graphite is slowly produced in a distinctive form known as *temper carbon.* The resulting alloy, Fig. 7.43(b), is called *malleable cast iron.* The matrix of the alloy may be ferrite, or it may be pearlite if the alloy is more rapidly cooled from a temperature above 723°C at the end of the malleableizing treatment. The hard, brittle cementite makes white cast iron undesirable except at the surface of some gray iron castings, where it is useful for wear resistance. Nevertheless, large tonnages of white iron castings are made for conversion into malleable iron castings, which are widely employed in industrial and farm machinery.

*Gray cast iron.* The most common type of cast iron is *gray cast iron,* in which flakes of graphite form during the solidification of the casting. An approximate analysis of the solidification of a 3.0% carbon alloy can be made with the aid of the iron-graphite equilibrium diagram, Fig. 7.44. As the molten alloy cools to 1153°C, dendrites of the austenite phase form within the liquid. At 1153°C the liquid reaches eutectic composition and solidifies as a eutectic of austenite and graphite. A phase analysis just below 1153°C yields

Point—3.0% carbon alloy at 1150°C
Phases—*austenite* and *graphite*
Compositions—2.0% C 100% C
Amounts—99% $\dfrac{3-2}{100-2} \times 100 = 1.0\%$

As the solidified alloy continues to cool, additional graphite forms from the austenite, while the austenite approaches the eutectoid composition, 0.7% carbon. Just above the eutectoid temperature the phase analysis is

Point—3.0% carbon alloy at 740°C
Phases—*austenite* and *graphite*
Compositions—0.7% C 100% C
Amounts—97.7% $\dfrac{3-0.7}{100-0.7} \times 100 = 2.3\%$

---

* Silicon plays such a significant part in determining the structure of cast irons (see Fig. 7.46, for example) that it is more exact to consider that cast irons are iron-carbon-silicon alloys, not simply iron-carbon alloys. The binary iron-carbon diagram is used in this introductory treatment as a convenient simplification.

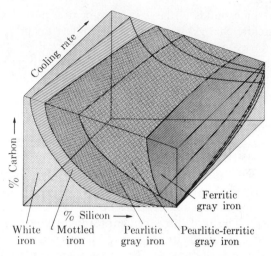

**Fig. 7.46** The effect of cooling rate and composition on the structure of cast iron. Mottled cast iron is a mixture of the white and the gray cast iron structures.

The eutectoid austenite may decompose into a mixture of ferrite and graphite, but this additional graphite merely adds to existing graphite flakes, rather than appearing as a separate eutectoid microconstituent. Figure 7.45(a) shows the microstructure of such a *ferritic* gray cast iron. It is noteworthy that etching is not necessary to reveal the graphite in a cast iron. The microstructure of Fig. 7.45(a) is not greatly changed by etching. It is possible to cause the eutectoid austenite to decompose into pearlite (or into a mixture of ferrite and pearlite) and thus produce a *pearlitic* gray cast iron like that shown in Fig. 7.45(b).

Two factors determine what structure (white, ferritic gray, pearlitic gray, etc.) a cast iron will have on solidification. These factors are *composition* and *cooling rate*. Figure 7.46 is a schematic structural diagram of the effects of carbon and silicon contents and of cooling rate on the microstructure of cast irons. It shows that an increase in cooling rate decreases the tendency to form graphite. An increase in carbon or silicon content, on the other hand, promotes graphitization. Some alloying elements, nickel for example, have essentially the same effect as silicon; that is, they increase the graphitization tendency. Other alloying elements (chromium and molybdenum) stabilize the iron carbide and thus decrease the graphitization tendency. Additional benefits of alloy additions to cast iron are (1) refining the graphite flake size, (2) strengthening the matrix, and (3) improving machinability by preventing the formation of white or mottled iron at edges or thin sections of the casting.

Since a cast iron containing *coarse* graphite flakes has low strength, various means are used to modify the graphite distribution. Finer graphite flakes tend to form when the alloy is *superheated* just before casting. A finer flake size can also

**TABLE 7.3**

Typical Mechanical Properties of Various Iron-Carbon Alloys

| Alloy | Tensile strength, $10^7$ N/m² | Yield strength, $10^7$ N/m² | Elongation in 50 mm, % | Modulus of rupture, $10^7$ N/m² |
|---|---|---|---|---|
| Ferritic gray cast iron | 17 | — | 0.5 | 34 |
| Pearlitic gray cast iron | 31 | — | 0.5 | 45 |
| Nodular cast iron | 55 | 34 | 5.0 | — |
| Malleable cast iron | 38 | 24 | 18.0 | — |
| Cast steel (as cast) | 51 | 23 | 19.0 | |
| Cast steel (normalized) | 52 | 26 | 24.0 | |
| Machine steel, 0.2% C | 34 | 19 | 35.0 | |
| Hardened and tempered 0.45% C steel | 69 | 45 | 28.0 | |
| Wrought iron | 34 | 21 | 30.0 | |
| Ingot iron | 31 | 21 | 40.0 | |
| Cold drawn 0.80% C steel | 138 | 103 | 5.0 | |
| Spheroidized 1.0% C steel | 55 | 31 | 30.0 | |
| Hardened and tempered 1.0% C steel | 172 | 138 | 1.0 | |

→ Increasing cost (rough estimate for finished part)

To convert from N/m² to lb/in², multiply by $1.450 \times 10^{-4}$.

be produced by *inoculating* the molten alloy. Typical inoculants are ferrosilicon and calcium-silicon, very small amounts of which are effective in reducing flake size, presumably by causing rapid nucleation. Graphite nodules rather than graphite flakes will form if the molten alloy is treated with magnesium or cerium. The resulting *nodular cast iron*, Fig. 7.45(c), has superior strength and ductility.

The mechanical properties of cast irons are determined by the standard tests, discussed in Chapter 4, but in addition the *transverse bend test* is often employed. In a typical test of this type a bar 30 mm in diameter is supported on two knife edges 0.45 m apart and is broken by a load applied midway between the two supports. Significant data obtained from this test are the maximum deflection at the center of the bar before fracture and the load necessary to cause fracture. The breaking load is often reported directly, but it may be converted to a *modulus of rupture*, which is closely related to the breaking stress in the outer fiber:

$$\text{Modulus of rupture} = Mc/I, \tag{7.13}$$

where $M$ is the maximum bending moment, $c$ is the distance from the center of the bar to the outer fiber, and $I$ is the section moment of inertia. The modulus of rupture is useful in designs involving lengths of cast iron unsupported in the center, such as lengths of pipe subjected to heavy loading.

*Properties of iron-carbon alloys.* Since iron-carbon alloys are in a price range significantly below that of other structural materials (brass, aluminum alloys, etc.), naturally their use is widespread. However, within this iron-carbon group there are also cost differences which, together with properties, determine the choice of material for a given application. The final cost of a completed metal part is determined by so many factors that the classification here can only be considered suggestive. In Table 7.3 various iron-carbon alloys are listed in approximate order of cost together with typical mechanical properties. Alloy steels are discussed in Chapter 10.

**Copper alloys.** The low strength of copper restricts its use as an engineering material. Fortunately, there are a number of ways in which copper can be strengthened. These include cold-working, order hardening, precipitation hardening, and solid-solution formation. Solid-solution formation is the most versatile, the cheapest, and the most widely used of these hardening methods.

*Brasses.* The equilibrium diagram of Fig. 7.47 shows that almost 40% zinc can dissolve in solid copper to form the alpha solid solution. The corresponding changes in tensile strength and percent elongation are plotted in Fig. 7.48. Since zinc is less expensive than copper, the alloys are actually cheaper than pure copper, but the corrosion resistance of the brasses is generally inferior to that of copper. *Dezincification*, the loss of zinc in reducing atmospheres at high temperatures or in certain corrosive media, and *season cracking*, grain-boundary corrosion in cold-worked alloys, are two special corrosion phenomena present in brasses that contain more than 15% zinc. (See Chapter 11.)

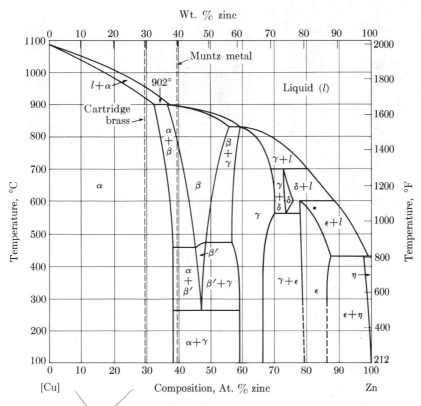

**Fig. 7.47** The copper-zinc equilibrium diagram. The $\beta'$-phase is the ordered $\beta$-phase, in which the copper and zinc atoms occupy particular positions in the crystal lattice.

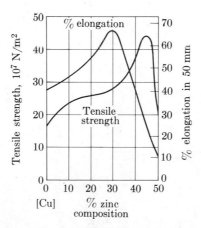

**Fig. 7.48** The tensile strength and percent elongation of annealed copper-zinc alloys.

**TABLE 7.4    Properties of Several Brasses**

| Alloy | Composition, % Cu | Zn | Sn | Pb | Mn | Tensile strength, $10^7$ N/m² | Yield strength, $10^7$ N/m² | Elongation in 50 mm, % | Typical applications |
|---|---|---|---|---|---|---|---|---|---|
| | | | | | | Wrought alloys (annealed) | | | |
| Gilding metal | 95 | 5 | | | | 21 | 7 | 40 | Coins, jewelry base for gold plate |
| Commercial bronze | 90 | 10 | | | | 23 | 7 | 45 | Grillwork, costume jewelry |
| Red brass | 85 | 15 | | | | 25 | 7 | 48 | Weatherstrip, plumbing lines |
| Low brass | 80 | 20 | | | | 26 | 8 | 52 | Musical instruments, pump lines |
| Cartridge brass | 70 | 30 | | | | 28 | 7 | 65 | Radiator cores, lamp fixtures |
| Yellow brass | 65 | 35 | | | | 30 | 11 | 55 | Reflectors, fasteners, springs |
| Muntz metal | 60 | 40 | | | | 37 | 14 | 40 | Large nuts and bolts, architectural panels |
| Low-leaded brass | 64.5 | 35 | | 0.5 | | 30 | 11 | 55 | Ammunition primers, plumbing |
| Medium-leaded brass | 64 | 35 | | 1.0 | | 30 | 11 | 52 | Hardware, gears, screws |
| Free-cutting brass | 62 | 35 | | 3.0 | | 30 | 12 | 51 | Automatic screw machine stock |
| Admiralty metal | 71 | 28 | 1 | | | 37 | 15 | 65 | Condenser tubes, heat exchangers |
| Naval brass | 60 | 39 | 1 | | | 40 | 19 | 45 | Marine hardware, valve stems |
| Manganese bronze | 58.5 | 39.2 | 1 | 1 Fe | 0.3 | 45 | 21 | 33 | Pump rods, shafting rods |
| | | | | | | Cast alloys | | | |
| Cast red brass | 85 | 5 | 5 | 5 | | 23 | 12 | 25 | Pipe fittings, small gears, bearings |
| Cast yellow brass | 60 | 38 | 1 | 1 | | 28 | 10 | 25 | Hardware fittings, ornamental castings |
| Cast manganese bronze | 58 | 39.7 | 1 Al | 1 Fe | 0.3 | 48 | 19 | 30 | Propeller hubs and blades |

When maximum corrosion resistance is not essential, *cartridge brass* (70% Cu, 30% Zn) gives a good combination of strength and ductility. For example, its good cold-working characteristics are used in the deep drawing of a flashlight case. The machinability of such alpha brasses can be improved by the addition of lead, but only with some sacrifice of cold-working properties. *Admiralty metal* is essentially cartridge brass with 1% tin replacing part of the zinc. Its improved corrosion resistance makes it suitable for use in condenser tubes. *Red brass* (85% Cu, 15% Zn) is used for applications such as plumbing pipe, where its good corrosion resistance is needed. Still better corrosion resistance is given by commercially pure copper in such applications as chemical process equipment. Since copper has high electrical conductivity (exceeded only by silver), large quantities of copper are used also by the electrical industries.

With increase in zinc content above the alpha solid-solution range, the structure of brasses changes to a mixture of the alpha and beta phases. This two-phase structure has relatively poor ductility, but high strength because the beta-phase matrix is strong at room temperature. The *Muntz metal* composition (60% Cu, 40% Zn) combines the strength of the alpha and beta mixture at room temperature with the increased ductility of the beta-phase structure at high temperatures. Figure 7.47 shows that Muntz metal consists of only the beta phase at temperatures above 760°C. The beta phase is soft and plastic in this range of temperatures, and therefore Muntz metal can be hot rolled, extruded, or forged. During cooling to room temperature the two-phase structure is produced by precipitation of alpha. Figure 7.49 shows the microstructures of Muntz metal after various treatments. It is seen that precipitation of the alpha phase can be partially prevented by rapidly cooling the beta phase, Fig. 7.49(b). Such control of phase changes is typical of solid-state reactions, and in this case it can be used to obtain smaller alpha particles by using a later precipitation heat treatment, Fig. 7.49(c). *Naval brass*, an alloy with improved corrosion resistance for such uses as marine hardware, is obtained by replacing about 1% of the zinc in Muntz metal by tin.

The brasses considered so far have been *wrought* brasses, that is, brasses that are used after being rolled or forged, etc. Brasses that are to be used in the *cast* condition must have, in addition to adequate strength and ductility, good machinability and good casting qualities. These properties are improved by the addition of elements such as lead and tin; therefore simple binary copper-zinc alloys are seldom used for castings. *Cast red brass* (85% Cu, 5% Zn, 5% Sn, 5% Pb) is an example of an alloy that has good corrosion resistance; it also has good antifriction (bearing) properties. *Cast yellow brass* (60% Cu, 38% Zn, 1% Sn, 1% Pb) is a lower-cost material for general use, for example, in plumbing supplies. The properties of these and other brasses are given in Table 7.4.

*Bronzes.* The word *bronze* has come to mean so many different things (witness *commercial bronze* and *manganese bronze* in the table of brasses) that it is useless to hold to its original definition as an alloy of copper and tin. In addition to tin-bronzes, there are aluminum-bronzes, silicon-bronzes, beryllium-bronzes, and

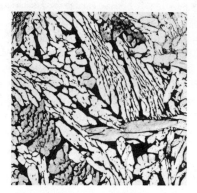

(a) Extruded alloy slowly cooled.
Alpha phase is light, beta is dark

(b) Extruded alloy reheated to the
beta region and quenched.
Alpha phase is dark, beta is light

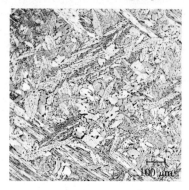

(c) Extruded and quenched alloy
reheated to precipitate additional
alpha phase

**Fig. 7.49** Photomicrographs of Muntz metal (60% Cu, 40% Zn).   (Courtesy Anaconda American Brass Company)

**TABLE 7.5**

Solid Solution Formation in Bronzes

| System | Maximum solid solubility | Maximum tensile strength of solid solution, $10^7$ N/m$^2$ | Elongation in 50 mm, % |
|---|---|---|---|
| Copper-tin | 15% Sn | 48 | 70 |
| Copper-aluminum | 9% Al | 52 | 65 |
| Copper-silicon | 5% Si | 48 | 25 |
| Copper-nickel | 100% | 52 | 45 |
| Copper-beryllium | 2% Be | 50 | 50 |

**TABLE 7.6    Properties of Several Bronzes**

| Alloy | Condition | Composition, % | | | | | Tensile strength, $10^7$ N/m² | Yield strength, $10^7$ N/m² | Elongation in 50 mm, % | Typical applications |
|---|---|---|---|---|---|---|---|---|---|---|
| | | Cu | Sn | Zn | Pb | | | | | |
| *Tin-Bronzes* | | | | | | | | | | |
| 5% phosphor-bronze | Wrought, cold-worked | 94.8 | 5 | | | 0.2 P | 63 | 61 | 5 | Diaphragms, springs, switch parts |
| 10% phosphor-bronze | Wrought, cold-worked | 89.8 | 10 | | | 0.2 P | 74 | 68 | 10 | Bridge bearing plates, special springs |
| Leaded tin-bronze | Sand cast | 88 | 6 | 4.5 | 1.5 | | 26 | 11 | 35 | Valves, gears, bearings |
| Gun metal | Sand cast | 88 | 10 | 2 | | | 28 | 11 | 30 | Fittings, bolts, pump parts |
| *Aluminum-Bronzes (eutectoid hardening)* | | | | | | | | | | |
| 5% aluminum-bronze | Wrought, annealed | 95 | | | | 5 Al | 41 | 17 | 66 | Corrosion-resistant tubing |
| 10% aluminum-bronze | Sand cast | 86 | | | | 3.5 Fe 10.5 Al | 52 | 24 | 20 | |
| Same | Sand cast and hardened | | | | | | 69 | 31 | 12 | Gears, bearings, bushings |
| *Silicon-Bronze* | | | | | | | | | | |
| Silicon-bronze, type A | Wrought, annealed | 95 | | | | 1 Mn 3 Si | 39 | 14 | 63 | Chemical equipment, hot water tanks |
| *Nickel-Bronzes* | | | | | | | | | | |
| 30% cupro-nickel | Wrought, annealed | 70 | | | | 30 Ni | 41 | 17 | 45 | Condenser, distiller tubes |
| 18% nickel-silver | Wrought, annealed | 65 | | 17 | | 18 Ni | 39 | 17 | 42 | Table flatware, zippers |
| Nickel-silver | Sand cast | 64 | 4 | 8 | 4 | 20 Ni | 28 | 17 | 15 | Marine castings, valves |
| *Beryllium-Bronzes (precipitation hardening)* | | | | | | | | | | |
| Beryllium-copper | Wrought, annealed | 98 | | | | 0.3 Co 1.7 Be | 48 | 21 | 50 | Springs, nonsparking tools |
| Same | Wrought, hardened | | | | | | 121 | 90 | 5 | |

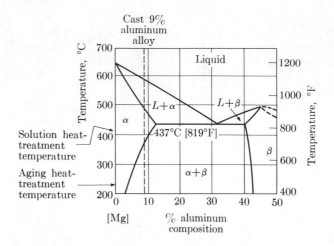

**Fig. 7.50**  A portion of the magnesium-aluminum equilibrium diagram.

many more.  A few complications exist even in this system of naming, since the terms usually employed are *beryllium-copper* and *nickel-silver* rather than beryllium-bronze and nickel-bronze.

The bronzes cost more than brass, but such superior properties as corrosion resistance and strength justify their use in many applications.  Almost all the improvement in corrosion resistance and much of the increase in strength is the result of solid-solution formation, the extent of which is shown in Table 7.5.  Some of the bronzes can be further strengthened by heat treatments involving precipitation hardening or eutectoid decomposition, which are discussed in later chapters. Information on several bronzes is given in Table 7.6.  An important property of nickel-silvers is their silvery color, which promotes their use as tableware and in restaurant and dairy equipment.  The high-strength alloys, aluminum-bronze and beryllium-copper, cost appreciably more than the standard tin-bronzes.

**Magnesium alloys.**  Commercial magnesium alloys illustrate many aspects of practical alloying.  Magnesium metal produced by the electrolysis of molten salts has a tensile strength of only $7 \times 10^7$ N/m² in the cast condition.  The strength can be increased severalfold by the use of various alloying elements, but the ductility and corrosion resistance tend to be lowered by some of these added elements. Although very many specific magnesium alloys are used for a variety of applications, most of them contain aluminum as the principal strengthening alloying element, with zinc and manganese added for improved corrosion resistance.  Since the zinc and manganese usually dissolve in the alpha and beta solid solutions, it is possible to interpret many magnesium alloy structures satisfactorily on the basis of the magnesium-aluminum equilibrium diagram, Fig. 7.50.

**TABLE 7.7**

Compositions and Properties of Typical Magnesium Alloys

| Use | Am. Soc. of Testing Materials designation | Composition, % | | | | Condition | Tensile strength, $10^7$ N/m$^2$ | Yield strength, $10^7$ N/m$^2$ | Elongation in 50 mm, % |
| --- | --- | --- | --- | --- | --- | --- | --- | --- | --- |
| | | Al | Zn | Mn | Zr | | | | |
| Sand and permanent mold casting | AZ92A | 9.0 | 2.0 | 0.1 | | As-cast | 17 | 10 | 2 |
| | | | | | | Solution treated | 27 | 10 | 10 |
| | | | | | | Aged | 27 | 14 | 3 |
| Die casting | AZ91C | 9.0 | 0.7 | 0.2 | | As-cast | 23 | 14 | 3 |
| Sheet | AZ31B | 3.0 | 1.0 | 0.45 | | Annealed | 26 | 15 | 15 |
| | | | | | | Hard | 29 | 21 | 8 |
| | | | | | | Extruded | 26 | 20 | 15 |
| Structural shapes | AZ80A | 8.5 | 0.5 | 0.2 | | Extruded | 34 | 25 | 12 |
| | | | | | | Extruded and aged | 38 | 28 | 8 |
| | ZK60A | | 5.5 | | 0.5 | Extruded | 34 | 26 | 12 |
| | | | | | | Extruded and aged | 37 | 30 | 10 |

To convert from N/m$^2$ to lb/in$^2$, multiply by $1.450 \times 10^{-4}$.

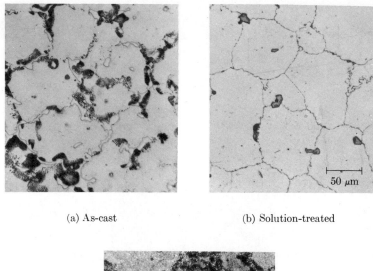

(a) As-cast                    (b) Solution-treated

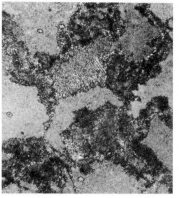

(c) Solution-treated and aged

**Fig. 7.51** Photomicrographs of the cast magnesium alloy AZ92A. (Courtesy P. F. George, Dow Chemical Company)

*Cast Magnesium Alloys.* A typical alloy for sand or permanent mold casting is AZ92A, whose composition* and properties are given in Table 7.7. Although no eutectic structure should form during the equilibrium cooling of a 9% aluminum composition, according to the diagram of Fig. 7.50, the structure of the cast alloy, Fig. 7.51(a), shows clear evidence of the solidification of eutectic liquid around the alpha solid solution. Massive particles of the beta solid solution form during solidification of the eutectic liquid. This type of eutectic structure (in which the two

---

* Note that the principal alloying elements and their approximate amounts are indicated by the ASTM designation. Thus AZ92A contains 9% aluminum and 2% zinc.

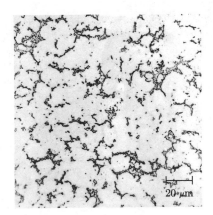

**Fig. 7.52** Microstructure of die-cast magnesium alloy AZ91C. (Courtesy P. F. George, Dow Chemical Company)

solid phases tend to form large, separate pieces) is called a *divorced eutectic*. During further cooling of the solidified casting, a lamellar precipitate forms at the grain boundaries and around the beta-phase particles. It will be recalled (Fig. 7.18) that nonequilibrium cooling is responsible for the presence of liquid of eutectic composition during the solidification of this alloy.

Since beta is a brittle intermediate phase, the ductility and strength of the alloy can be improved by dissolving this phase by means of a *solution heat treatment*. The alloy is well within the alpha-phase region of the diagram at 410°C, and after 20 hours at this temperature the alloy remains almost completely single phase on air cooling to room temperature, Fig. 7.51(b). The strength of the casting can be further improved, but with a loss in ductility, by causing the beta phase to precipitate in a finely divided form in the alpha grains. The equilibrium diagram shows that a mixture of alpha and beta phases is stable at room temperature; therefore it is only necessary to heat the supersaturated alpha solid solution to a temperature at which the atoms have sufficient mobility to cause some precipitation of beta phase to take place. This *precipitation* or *aging heat treatment* is carried out by heating the solution-treated alloy to 220°C for 5 hours. Figure 7.51(c) shows the microstructure that results from this treatment.

A preferred magnesium alloy for die casting is AZ91C (Table 7.7). Its microstructure, Fig. 7.52, shows eutectic structure surrounding the alpha solid-solution dendrites. The finer structure, compared with the sand-cast magnesium alloy, is a result of the faster cooling rate of a die casting.

*Wrought Magnesium Alloys.* A typical magnesium alloy available in sheet form is AZ31B (Table 7.7). When the sheet is to be formed, by deep drawing for example, it is supplied in the *annealed temper*. The corresponding microstructure is shown in Fig. 7.53(a). The strength of the alloy is increased if it is subjected to cold rolling,

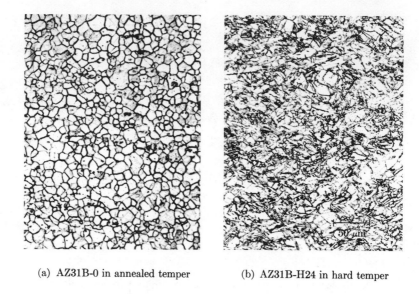

(a)  AZ31B-0 in annealed temper        (b)  AZ31B-H24 in hard temper

**Fig. 7.53**  Microstructures of wrought magnesium alloy AZ31B.

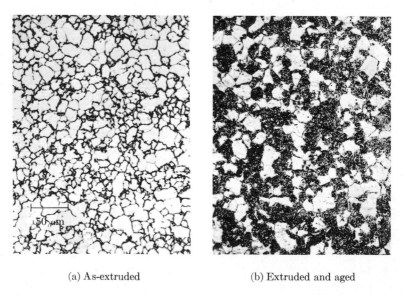

(a) As-extruded                        (b) Extruded and aged

**Fig. 7.54**  Microstructure of wrought magnesium alloy AZ80A.  (Courtesy P. F. George, Dow Chemical Company)

and the *hard temper* material produced in this manner is superior for applications not involving a forming operation or shock loading. The microstructure of the cold-rolled alloy, Fig. 7.53(b), shows the distorted grains produced by the severe plastic deformation. The small amount of beta phase present in this alloy is usually found at the grain boundaries.

Structural shapes of magnesium alloys are usually formed by extrusion or forging rather than by hot rolling. A typical alloy for these applications is AZ80A (Table 7.7). Since the extruded alloy is air cooled from the extruding temperature, 385°C, its structure at room temperature consists of alpha solid solution in which beta phase has precipitated at the grain boundaries, Fig. 7.54(a). Heat treatment of this alloy consists simply in aging the extruded alloy for about 20 hours at 175°C. Figure 7.54(b) presents the microstructure of the aged alloy showing a fine precipitate of the beta phase. A high-strength extrusion alloy, ZK60A, contains about 5.5% zinc and 0.5% zirconium. The fine grain size produced and maintained by the zirconium is responsible for the excellent mechanical properties of this alloy, Table 7.7. Magnesium alloys for use at elevated temperatures exploit rare-earth metals or thorium as alloying elements.

Magnesium alloys are used principally where light weight is important, and so find extensive application in aircraft. Many uses of sheet magnesium alloys depend on their good resistance to atmospheric corrosion. The higher initial cost in comparison with steel has been the principal factor limiting the extensive use of magnesium alloys.

**Titanium alloys.** The attractive properties of titanium alloys are their high strength-to-weight ratio (density 4.5 $g/cm^3$) and their excellent corrosion resistance in many media. The alloying and heat treatment of titanium are dominated by the allotropic transformation at 885°C (1625°F) from the low-temperature hexagonal alpha phase to the body-centered-cubic beta phase. Some alloying elements (Al, Sn, C, O, N) stabilize the alpha phase and can be used to make alloys based on this phase. The most important commercial alloy of the alpha type is Ti-5Al-2.5Sn. Oxygen, carbon, and nitrogen are usually considered as impurities in titanium, but small quantities of oxygen are used to promote improved strength in unalloyed titanium and in most alloys. Most alloying elements stabilize the beta phase and cause it to exist as the stable phase at temperatures considerably below 885°C. This behavior is the basis of the commercial alpha-beta alloys whose heat treatment will now be considered.

The most important beta-stabilizing elements are molybdenum, vanadium, chromium, manganese, and iron. Since each of these elements has a similar effect on the solid-phase reactions in titanium, it is convenient to discuss the behavior of titanium alloys using the generalized phase diagram shown in Fig. 7.55. Strictly speaking, this diagram is correct only for Ti-Mo and Ti-V alloys, since a eutectoid reaction exists for the other three metals at the position shown. However, the eutectoid reaction is sluggish, and for purposes of general discussion, its effects can

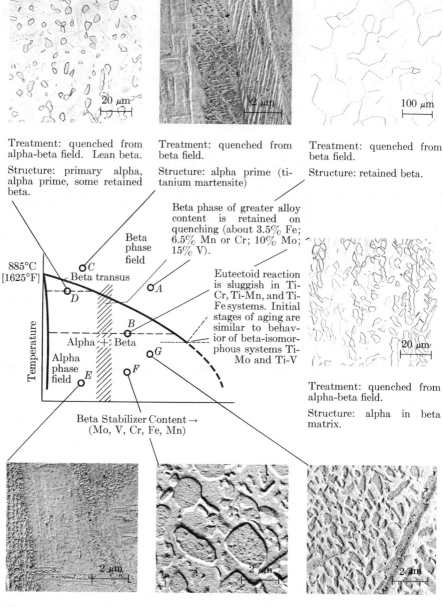

Treatment: quenched from alpha-beta field. Lean beta.

Structure: primary alpha, alpha prime, some retained beta.

Treatment: quenched from beta field.

Structure: alpha prime (titanium martensite)

Treatment: quenched from beta field.

Structure: retained beta.

Beta phase of greater alloy content is retained on quenching (about 3.5% Fe; 6.5% Mn or Cr; 10% Mo; 15% V).

Eutectoid reaction is sluggish in Ti-Cr, Ti-Mn, and Ti-Fe systems. Initial stages of aging are similar to behavior of beta-isomorphous systems Ti-Mo and Ti-V.

885°C [1625°F]

Beta phase field

Beta transus

C

D

A

B

G

Alpha + Beta

Alpha phase field

E

F

Temperature

Beta Stabilizer Content →
(Mo, V, Cr, Fe, Mn)

Treatment: quenched from alpha-beta field.

Structure: alpha in beta matrix.

Treatment: quenched from beta field, aged at temperature shown.

Structure: tempered alpha prime.

Treatment: quenched from alpha-beta field, aged at temperature shown.

Structure: primary alpha plus finer alpha precipitate in beta matrix.

Treatment: quenched from beta field, aged at temperature shown.

Structure: alpha precipitate in beta matrix.

be neglected. The shaded zone in the diagram marks the dividing line between "lean" alloys, in which the beta phase is not retained on quenching, and alloys "rich" in solute, which do retain the beta phase on quenching (from the solution-heat-treatment temperature). The importance of this distinction will become evident in the following consideration of possible structures in titanium alloys.

Point $A$ in Fig. 7.55 represents a rich alloy that has been heated into the beta-phase field. The coarse beta grains are retained on quenching, as shown in the corresponding photomicrograph. A fine grain size, which gives better properties, can be obtained if a rich alloy is fabricated (plastically deformed) and heated within the alpha-plus-beta region and then quenched, structure $B$. Fabrication of alpha-beta alloys within the beta region, which is becoming quite a popular practice for some alloys, results in acicular alpha microstructure quite unlike any of those shown in Fig. 7.55. The beta phase transforms on cooling by a nucleation and growth process to result in either fine or coarse needles or plates of alpha, depending on alloy content and on cooling rate from above the beta transus to room temperature. If a lean beta phase is quenched, it transforms martensitically to give a structure called alpha prime ($\alpha'$) or titanium martensite, structure $C$. Unlike martensite in steel, titanium martensite is not hard. It is a constituent phase after certain heat treatments for some alpha-beta alloys such as Ti-6Al-4V, Table 7.8. The beta phase in lean alloys quenched from the alpha-plus-beta region transforms partially to the alpha prime phase, structure $D$. This treatment has the advantage of ensuring a fine grain size and optimum combinations of strength and ductility.

High strength in combination with adequate ductility is obtained by aging solution-heat-treated and quenched alloys in the 480 to 590°C (900 to 1100°F) range. The aging treatment causes a transformation of the nonequilibrium alpha prime and beta phases and results in a fine dispersion of the alpha and beta phases. The structure produced by this treatment in a lean alloy is shown by $E$, while $F$ and $G$ show the structures of rich alloys.

An important aspect of the heat treatment of titanium alloys is the occurrence of the omega ($\omega$) phase in beta at temperatures below about 510°C. Alloys that contain much omega are likely to be brittle, and therefore it is important that the treatment be adequate for minimizing the occurrence of this phase. Solution heat treatments should be terminated with cooling rates sufficiently rapid to pass quickly through the temperature range of the beta-to-omega transformation; also, the times and temperatures of aging should be adjusted to minimize the amount of omega phase. As the time of aging is increased, the omega phase forms and then gradually disappears and is replaced by the alpha phase, Fig. 7.56. It is necessary to age beyond the maximum hardness to minimize the effects of the omega phase and therefore to maximize ductility. Aging temperatures of not less than 480°C are

**Fig. 7.55**  Generalized phase diagram for alpha-beta titanium alloys and examples of structures produced by various heat treatments after fabrication in the alpha-beta field. (Courtesy P. D. Frost, Battelle Memorial Institute)

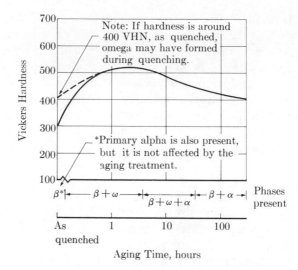

**Fig. 7.56** Correlation of hardness and phase composition for an alpha-beta titanium alloy quenched from the alpha-beta region and aged at 425°C (800°F). (After P. D. Frost)

usually employed, and aging times are adjusted for the combinations of strength and ductility desired. The mechanical properties of commercial titanium and several alloys are listed in Table 7.8. One of the disadvantages of titanium is its high price relative to other structural metals, but with the increased use of titanium, this disadvantage is diminishing.

## IMPURITIES IN METALS

In most day-to-day applications of metals and alloys, only the base metal and the principal alloying elements are of direct interest. However, when a production-line casting begins to develop serious gas pockets, the gaseous impurities hydrogen, nitrogen, and oxygen suddenly become very important. Solid impurities are even more common, and even a fraction of one percent of an unwanted element can destroy the useful properties of commercial alloys. For example, 0.01% arsenic causes difficulties in the production of lead-antimony alloy cable sheath. Sulfur is a nonmetallic impurity that causes steels to develop cracks when rolled at high temperatures. As discussed previously, the cause of this *hot shortness* is the melting of an iron sulfide compound in the steel.

All commercial alloys contain impurities. However, it is only when the amounts of certain critical impurity elements exceed a specific value that objectionable effects are produced. Although each base metal represents a separate, specialized problem, the impurities usually get into the final metal article in the following ways: (1) in the original smelting operation from impurities in the ore, flux, and

TABLE 7.8    Mechanical Properties of Titanium and its Alloys

| Composition, weight percent | Alloy type | Heat treatment* | Yield strength, $10^7$ N/m$^2$ | Tensile strength, $10^7$ N/m$^2$ | Elongation in 50 mm, % | Reduction in area, % |
|---|---|---|---|---|---|---|
| Unalloyed titanium (very low oxygen) | Alpha | Annealed 2 hours 875°C (1600°F) air cooled | 14 | 26 | 62 | 87 |
| Commercial titanium (about 0.23 oxygen) | Alpha | Annealed 2 hours 790°C (1450°F) air cooled | 59 | 66 | 36 | 53 |
| Ti-5Al-2.5Sn | Alpha | Annealed 2 hours 850°C (1560°F) air cooled | 79 | 90 | 17 | 43 |
| Ti-6Al-4V | Alpha-beta | Annealed 2 hours 870°C (1600°F) FC to 590°C (1100°F) air cooled | 85 | 92 | 15 | 35 |
| Ti-6Al-4V | Alpha-beta | Solution treated plus aged 1 hour 930°C (1700°F) WQ + 8 hours 480°C (900°F) air cooled | 105 | 120 | 13 | 50 |
| Ti-6Al-6V-2Sn | Alpha-beta | Solution treated plus aged 1.5 hours 890°C (1630°F) WQ + 4 hours 565°C (1050°F) air cooled | 130 | 134 | 11 | 25 |
| Ti-4.5Sn-6Zr-11.5Mo | Beta | Solution treated 1 hour 720°C (1325°F) WQ | 79 | 99 | 24 | 65 |
| Ti-4.5Sn-6Zr-11.5Mo | Beta | Solution treated plus aged 1 hour 720°C (1325°F) WQ + 8 hours 430°C (900°F) air cooled | 127 | 137 | 15 | 36 |

* Cooling rates: FC = furnace cool, WQ = water quench.    To convert from N/m$^2$ to lb/in$^2$, multiply by $1.450 \times 10^{-4}$.

fuel; (2) during the later melting operations by contamination from the furnace or fuel, or from scrap-metal additions; (3) in the course of heating for hot-working, solution treatment, etc., from furnace or fuel contaminants.

Successful commercial practice consists *not* in the expensive task of completely eliminating undesirable elements, but in adequately controlling them. Usually this is done by maintaining the impurity elements below a safe minimum, but frequently it is more economical to add a second element to convert the impurity to a relatively harmless form. Thus, sulfur in steel can be caused to form harmless manganese sulfide (which actually improves machinability), if a little manganese is added to the steel during manufacture.

## PROBLEMS

**1.** a) Assuming that the solution is ideal ($\Delta H = 0$), calculate $\Delta G$ at $N_2 = 0, 0.2, 0.5, 0.8,$ and 1.0, and show that the curve of $\Delta G$ versus $N_2$ is similar to those in Fig. 7.1(a).

b) Explain how a slight non-ideality of the solution could account for the "tilting" of the curves in Fig. 7.1(a).

**2.** a) Sketch a *P-X* diagram for a temperature above the triple points of both metals in Fig. 7.2.

b) Sketch a *P-X* diagram for a temperature between the two triple points. Use the phase rule to show that there are no additional degrees of freedom (once the temperature has been fixed) for a binary system containing three phases, and therefore the three phases exist in equilibrium only at a given pressure.

**3.** Sketch a two-dimensional picture of the relative positions of (about twelve) atoms of a 50% copper–50% nickel alloy (a) at 1400°C, (b) at 1200°C, and (c) at 20°C. (See Fig. 6.1.)

**4.** Derive Eq. (7.5). (Note that the principle of the derivation is sketched in the text.)

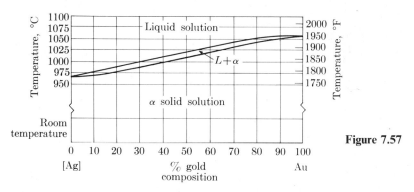

Figure 7.57

**5.** Consider an alloy consisting of 30% gold and 70% silver, Fig. 7.57.

a) At 1100°C: (1) What phase(s) are present? (2) What is the chemical composition of each phase? (3) What percentage of each phase is present?

b) Make a similar phase analysis at 1000°C.

c) Make a similar phase analysis at 990°C.

d) Make a similar phase analysis at 900°C.

e) Make a similar phase analysis at room temperature.

f) Sketch a possible microstructure for each of these five temperatures.

**6.** Prove that the tangent line in Fig. 7.10 gives the free energy of mixtures of the $\alpha$- and $\beta$-phases for alloys in the composition range between $X_\alpha$ and $X_\beta$. Use the lever law to determine the amounts of the two phases that exist at a given composition $X$, and then write the expression for the free energy as the sum of contributions from the two phases, using $\Delta G_\alpha$ and $\Delta G_\beta$ to represent the free energy per unit amount of each phase. Show that the resulting expression is the equation of the tangent line in question.

**7.** Consider the equilibrium diagram of Fig. 7.12.

a) Using alloy 1, determine the percentage of alpha solid solution in the eutectic microconstituent.

b) Make phase analyses of alloy 2, (1) at a temperature just above the liquidus curve, (2) at temperature $T_1$, (3) at temperature $T_2$, (4) at a temperature just below the eutectic reaction horizontal.

c) Make phase analyses of alloy 3, (1) at temperature $T_1$, (2) at a temperature slightly below the liquidus curve, (3) at a temperature slightly above the solidus curve, (4) at temperature $T_2$, and sketch a possible microstructure for each of these four temperatures.

d) How does the solidification of alloy 3 compare with the solidification of an alloy in a complete solid solubility diagram, such as Fig. 7.6?

**8.** Use the diagram of Fig. 7.17 to make the following phase analyses of a nickel alloy containing 3% carbon at $24 \times 10^8 \, \text{N/m}^2$ pressure: (1) 1650 K, (2) 1500 K, and (3) 600 K.

**9.** Using the equilibrium diagram of Fig. 7.19:

a) Make phase analyses of alloy 1 at (1) 1650°C, (2) 1200°C, and (3) 1150°C, and sketch a possible microstructure for each of these three temperatures.

b) Make phase analyses of alloy 2 at (1) 1200°C, (2) 1150°C, and (3) 980°C, and sketch a possible microstructure for each of these three temperatures.

c) Make phase analyses of the 20% platinum alloy at (1) 1200°C, (2) 1090°C, and (3) 980°C, and sketch a possible microstructure for each of these three temperatures.

**10.** An incongruently melting intermetallic compound often produces a peritectic reaction in its equilibrium diagram. With the aid of Fig. 7.58, sketch a possible equilibrium diagram in this

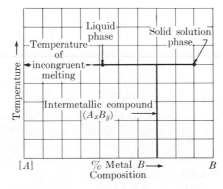

Figure 7.58

case. [*Note*: A eutectic reaction would almost always occur on the left-hand side of such an equilibrium diagram.]

11. a) Sketch a diagram similar to Fig. 7.22, except with the liquidus and solidus lines sloping upward, and show that $k_0$ is then greater than one.

b) Explain why the curve for $k = 5$ decreases with increase in g in Fig. 7.23.

12. For slow rates of solidification the effective distribution coefficient $k$ is essentially equal to $k_0$. Use a sketch of the variation of solute concentration in the liquid to explain why the value of $k$ is between $k_0$ and unity for faster rates of solidification.

13. Using the equilibrium diagram of Fig. 7.26, make phase analyses of a 25% magnesium alloy at (a) 500°C, (b) 400°C, and (c) 200°C, and sketch a possible microstructure for each of these three temperatures.

14. Using the equilibrium diagram of Fig. 7.27, make phase analyses of a 0.2% C alloy at

a) 900°C,        b) 850°C,        c) 724°C,        d) 722°C.

e) What percent of the alloy is pearlite at 722°C?

Sketch a possible microstructure for each of these four temperatures.

15. a) Is it possible to cause the eutectic microconstituent to form in alloy 1, Fig. 7.28?

b) What is the minimum silver content of an alloy in which the eutectic structure appears on equilibrium cooling?

16. In a face-centered-cubic space lattice there are three times as many face-centered positions as there are cube-corner positions. From this fact would you conclude that *ordering* of a face-centered-cubic solid solution of metal $B$ in metal $A$ would be more likely at the composition $A_3B$ or at $A_2B$? Why?

17. In the triangular diagram of Fig. 7.31 show the location of these alloys:

a) 100% $A$,        0% $B$,        0% $C$
b)  20% $A$,        30% $B$,        50% $C$
c)  33% $A$,        33% $B$,        34% $C$

18. Using Fig. 7.32, make phase analyses of the alloy 25% $A$, 25% $C$, 50% $B$ at

a) $T_1$,        b) $T_2$,        c) $T_3$,        d) $T_4$.

e) Estimate the amount of pseudoeutectic structure at $T_4$.

19. Make a phase analysis of ingot iron at (a) 1600°C, (b) 1500°C, (c) 1000°C, and (d) 700°C, using the equilibrium diagram of Fig. 7.37.

20. If a hypereutectoid steel is slowly cooled from the austenite region, the primary cementite crystals form a continuous layer in the austenite grain boundaries. This brittle network seriously reduces the ductility of the steel at room temperature. What heat treatment would be most effective in removing this cementite network?

21. Use Fig. 7.42 to make a phase analysis of (a) a 1% C, 6% Mn alloy, and (b) a 2% C, 2% Mn alloy. Draw tie lines according to the pattern illustrated in Fig. 7.32. Compare these results with the result of the analysis of the 1% C, 2% Mn alloy given in the text. What conclusions can you draw concerning (c) the effect of carbon content and (d) the effect of manganese content on the carbide-forming tendency of manganese?

**22.** Assume that the yield strength of commercially pure iron is $14 \times 10^7 \, \text{N/m}^2$. What would be the yield strength of a normalized high-strength low-alloy steel containing 0.10% C, 0.90% Mn, 0.3% Si, 0.10% P, and 0.02% S? The strengthening effect of carbon in normalized steels is about the same as that of phosphorus.

**23.** Show by means of a sketch based on the equilibrium diagram of Fig. 7.50 why a eutectic structure appears in the photomicrograph of the magnesium-9% aluminum alloy, Fig. 7.51(a). (See Fig. 7.18.)

## REFERENCES

Gordon, P. *Principles of Phase Diagrams in Materials Science*, New York: McGraw-Hill, 1968. An introductory treatment emphasizing the thermodynamic principles.

Hansen, M. and K. Anderko, *Constitution of Binary Alloys*, New York: McGraw-Hill, 1958. First Supplement (by R. P. Elliott), 1965. Second Supplement (by F. A. Shunk), 1969. Useful compilation of phase diagrams and of related experimental data on binary metallic alloys.

Prince, A., *Alloy Phase Equilibria*, New York: American Elsevier, 1966. A more advanced treatment of many aspects of phase equilibria in metals, emphasizing three- and higher-component systems.

Rhines, F. N., *Phase Diagrams in Metallurgy*, New York: McGraw-Hill, 1956. An introduction to the construction and use of binary and ternary phase diagrams.

CHAPTER 8

# DIFFUSION IN METALS

# INTRODUCTION

A massive, rigid block of metal has some unexpected characteristics. In the chapter on electronic structure it will be shown that this dense body is mostly empty space. In the present chapter an equally surprising fact will be discussed, that there is continuous, long-range motion of the atoms that form such a rigid body. This motion or *diffusion* of atoms occurs under all conditions, but it is of greatest interest when it causes pronounced changes in alloys. Examples of these changes are the surface hardening of steel through the addition of carbon atoms by diffusion at about 925°C (1700°F), and the pronounced strengthening of certain aluminum alloys resulting from the diffusion of copper atoms at 200°C (400°F) to form particles of precipitate. We will now consider the essential features of diffusion in metals as preparation for extensive application of this phenomenon in the remainder of the book.

# THEORY

The cause of diffusion is easily understood; it is simply that the atoms in a solid are continually jumping from one position in the structure to a neighboring position. The mechanism by which jumping occurs will be explained later, but it will be helpful first to consider how atomic jumping is related to large-scale diffusion phenomena. In a pure metal, since all the atoms are alike, it is difficult to detect any evidence of diffusion. However, if certain atoms are distinguishable by virtue of being radioactive isotopes of the metal in question, then it is possible to measure the rate of *self-diffusion* at which the atoms diffuse among themselves.

Figure 8.1 is a schematic illustration of self-diffusion between a central region containing a uniform concentration $c_1$ of radioactive atoms and two adjoining regions initially containing only normal atoms. It is known that the diffusion behavior of normal and of radioactive atoms is essentially the same. By one of the mechanisms discussed below, each of the atoms will tend to jump from its position, as shown in Fig. 8.1(a), to one of the neighboring positions. The atomic distribution shown in Fig. 8.1(b) might exist after an average of one jump per atom. In the case of the radioactive atoms, for example, there were initially four such atoms in each column, and each atom had the possibility of jumping up, down, left, or right (in the two-dimensional case illustrated). It is easy to verify that random jumping would leave the number of radioactive atoms unchanged in columns 4 and 5. However, a radioactive atom would leave column 3 and enter column 2, and similarly an atom would leave column 6 and enter column 7. The concentration curve in Fig. 8.1(b) is a convenient method for recording this spreading of radioactive atoms through the specimen. With further jumping the atoms continue to spread (see Problem 1) until a uniform distribution of atoms is finally produced, Fig. 8.1(c). Although the atoms then continue to jump about as before, the concentration curve remains unchanged and it is no longer possible to observe diffusion as a change in the concentration curve.

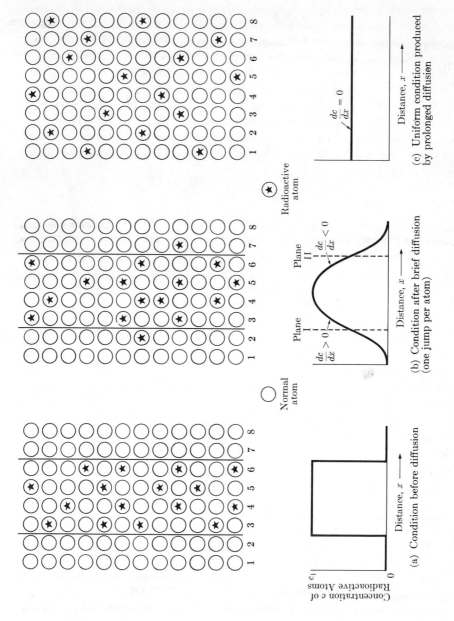

**Fig. 8.1** The process of self-diffusion in metals: Atomic distributions (above) and the corresponding concentration curves (below) for three stages in the diffusion process.

If the starred atoms in Fig. 8.1 are a dissolved alloying element, the same picture represents the essential features of diffusion in a substitutional solid solution. Even if the solute atoms occupy the interstices in the matrix lattice (carbon atoms in iron, for example) diffusion occurs by the same type of virtually random jumping, see Problem 2. For this reason two fundamental equations, Fick's first and second laws, give the basis for almost all diffusion phenomena of interest in metallurgy.

**Fick's laws.** The concentration gradient $dc/dx$ determines whether or not (observable) diffusion occurs. Thus, where $dc/dx = 0$, as in Fig. 8.1(c), diffusion does not occur. However, in the portion of Fig. 8.1(b) where $dc/dx$ is positive ($dc/dx > 0$), a net flow of solute atoms occurs across plane I in the negative $x$-direction, while where $dc/dx$ is negative the net flow is in the positive $x$-direction. The corresponding quantitative relation can be obtained in the following way. Let $c_2$ and $c_3$ be the concentrations of solute atoms at positions 2 and 3, Fig. 8.1(b), on either side of a reference plane at an arbitrary time $t$ during the diffusion process. If the concentration $c$ is the number of atoms per unit volume (atoms/m$^3$), then the number of atoms per unit area in each atomic plane (represented by a column of atoms in Fig. 8.1) is $n = c\alpha$, where $\alpha$ is the jump distance. Each solute atom jumps $\Gamma$ times per second on the average and in one of six directions in three dimensions. Therefore, $\Gamma c_2 \alpha/6$ solute atoms jump per second from plane 2 to plane 3 across unit area of the reference plane. The oppositely directed atomic flow is $\Gamma c_3 \alpha/6$. The difference of these two quantities is the net flux, $J$, of solute atoms,

$$J = \tfrac{1}{6}\Gamma\alpha(c_2 - c_3) \qquad \text{atoms/(m}^2 \cdot \text{s)}. \qquad (8.1)$$

The relation of the two concentrations can usually be approximated as $c_3 = c_2 + (dc/dx)dx$. Since $dx = \alpha$ in the present case, Eq. (8.1) becomes

$$J = -\frac{1}{6}\Gamma\alpha^2 \frac{dc}{dx}. \qquad (8.2)$$

This equation, *Fick's first law*, is usually written in the form

$$J = -D\frac{dc}{dx}, \qquad (8.3)$$

where $D$ is the diffusion coefficient,

$$D = \tfrac{1}{6}\Gamma\alpha^2 \qquad \text{m}^2/\text{s.} \qquad (8.4)$$

Fick's first law simply states that the rate of diffusion is directly proportional to the concentration gradient, $dc/dx$. Although Eq. (8.4) was derived here for a simple-cubic structure, it applies also for diffusion in BCC and FCC solid solutions. Because the jump distance $\alpha$ is $\sqrt{3}a_0/2$ in BCC metals and $\sqrt{2}a_0/2$ in FCC metals,

Eq. (8.4) can also be written in terms of the lattice parameter $a_0$:

$$D = \tfrac{1}{8}\Gamma a_0^2 \quad \text{(BCC)}, \tag{8.5}$$

$$D = \tfrac{1}{12}\Gamma a_0^2 \quad \text{(FCC)}. \tag{8.6}$$

The jump frequency $\Gamma$ depends on the local chemical environment and therefore usually varies with change in the concentration of solute. Consequently $D$ shows a similar variation.

A useful property of a randomly jumping atom is given by the theory of random walk.* During the time $t$, in which the atom completes $\Gamma t = n$ jumps each of length $\alpha$, the total distance $R$ that the average atom travels from its original location is

$$R = \alpha\sqrt{n}. \tag{8.7}$$

Substitution of this result in Eq. (8.4) gives the interesting relation,

$$R = (6Dt)^{1/2}, \tag{8.8}$$

describing the average distance, $R$, of diffusion of atoms during diffusion for time $t$.

Since Fick's first law, Eq. (8.3), does not include time as a variable, a second form is often needed to treat technical problems. Consider the accumulation of solute in the volume $dx \times 1 \text{ m}^2$ between two reference planes separated by the distance $dx$. The flux across the first plane is given by Eq. (8.3), while that across the second plane can be written

$$J' = J + \frac{dJ}{dx}\,dx. \tag{8.9}$$

The accumulation per second, $dA$, is the excess of the flux into the volume over the flux out of the volume,

$$dA = J - J' = -\frac{dJ}{dx}\,dx = \frac{d}{dx}\left(D\frac{dc}{dx}\right)dx. \tag{8.10}$$

The accumulation of atoms at the rate $dA$ per second in the infinitesimal volume $dx \times 1 \text{ m}^2$ is equivalent to the rate of change in concentration $dc/dt$, since the concentration is referred to unit volume. Because the concentration is now a function of both $x$ and $t$, we use partial derivatives and write

$$dA/dx = \partial c/\partial t. \tag{8.11}$$

Substitution of Eq. (8.11) in Eq. (8.10) give *Fick's second law* for diffusion in the

---

* A proof is given by P. G. Shewmon, *Diffusion in Solids*, pp. 47–52, New York: McGraw-Hill, 1963.

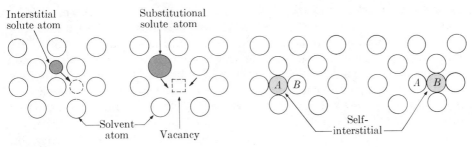

Interstitial
solute atom

Substitutional
solute atom

Solvent atom    Vacancy

Self-interstitial

(a) Interstitial diffusion of an interstitial solute in a metallic matrix

(b) Vacancy diffusion of the solute in a substitutional solid solution

(c) Self-interstitial (or interstitialcy) diffusion in an alkali metal. At first the self-interstitial is atom $A$; next it is atom $B$, and so on.

**Fig. 8.2** Schematic illustrations of the three most common mechanisms of diffusion in metals and solid solutions.

$x$-direction,

$$\frac{\partial c}{\partial t} = \frac{\partial}{\partial x}\left(D\,\frac{\partial c}{\partial x}\right). \tag{8.12}$$

If the variation of the diffusion coefficient with concentration is neglected, Eq. (8.12) takes the more convenient form,

$$\partial c/\partial t = D\,\partial^2 c/\partial x^2 \tag{8.13}$$

For solid solutions based on a cubic crystal structure,* Eq. (8.13) becomes in three dimensions,

$$\partial c/\partial t = D(\partial^2 c/\partial x^2 + \partial^2 c/\partial y^2 + \partial^2 c/\partial z^2) = D\nabla^2 c. \tag{8.14}$$

**Mechanisms of diffusion.** Diffusion in an interstitial solid solution is especially simple for two reasons. First, the solvent atoms jump only about $10^{-4}$ as often as the small solute atoms, so diffusion of the solvent can often be neglected. Second, the solute atoms can be pictured as merely jumping from one interstice in the matrix lattice to a neighboring interstice, Fig. 8.2(a). Only for special purposes must the character of the interstitial spaces (described in Figs. 1.19 and 1.20) be taken into consideration. Important interstitial solutes in metallic alloys are carbon, boron, hydrogen, nitrogen, and oxygen.

In a pure metal or a substitutional solid solution a special mechanism must operate to permit diffusion to occur. In particular, two adjacent atoms are *not* able to exchange positions directly. The most important mechanism involves vacancies, discussed in Chapter 3. Compared to the interstitial mechanism the

---

* Diffusion in noncubic crystal structures is considered by P. G. Shewmon, *Diffusion in Solids*, pp. 32–37, New York: McGraw-Hill, 1963.

*vacancy mechanism* reveals the following new features. (1) The rate of diffusion depends directly on the fraction, $N_V$, of vacant lattice sites. (2) When a vacancy is adjacent to a solute atom, Fig. 8.2(b), either the solute atom or a solvent atom may jump into the vacant lattice site (leaving its own site vacant in the process). Usually the probability of a solute–vacancy exchange is different from the probability of a solvent–vacancy exchange, and therefore the corresponding partial diffusion coefficients, $D_{solute}$ and $D_{solvent}$, have different values. (3) Since $N_V$ is typically only about $10^{-4}$ at the temperature of diffusion, a given atom is a neighbor of a vacancy only infrequently. However, once an atom exchanges with a vacancy, this atom is again in position for a second possible exchange (which would cancel the effect on the diffusion process of the original exchange). This phenomenon, known as *correlation*, produces nonrandomness in the jumping of atoms, even in self-diffusion. In terms of the (random) jump frequency, $\Gamma_V$, of the vacancies, the self-diffusion coefficient of a pure metal is

$$D = \tfrac{1}{6} f \alpha^2 \Gamma_V N_V, \qquad (8.15)$$

where $f$ is the *correlation factor* that takes account of the phenomenon of correlation. The value of $f$ is 0.78 for FCC metals and 0.73 for BCC metals.

If the defect responsible for diffusion is an *excess* atom (rather than a missing atom, the vacancy) the mechanism is termed *self-interstitial* or *interstitialcy*. This mechanism is closely analogous to the vacancy mechanism, but is rarely the dominant mode of diffusion because of the large amount of energy required to produce self-interstitials in most metals. The exceptions are the alkali metals, which are BCC and therefore not closely packed, and which also have small inner-electron cores relative to their atomic diameters.

**Thermal activation.** Two factors account mainly for the large increase in rate of diffusion with increase in temperature: the increase in the jump frequency $\Gamma$ of the atoms (or atom-vacancy pairs); and the increase in the equilibrium concentration $N_V$ of vacancies (or other defects), thus increasing the number of atom-vacancy pairs available for jumping. We begin by considering the first of these factors and will use interstitial diffusion as an example since $N_V$ is not involved in this mechanism. The same kind of reasoning applies, however, to atomic jumping via the vacancy mechanism.

*Interstitial Diffusion.* An interstitial atom must possess an exceptional amount of energy before it can jump out of its interstice. The atom at interstice 1 in Fig. 8.3(a), for example, is usually confined near the bottom of its potential-energy valley (see Fig. 1.6). The *average* thermal energy $kT$ of the atom is far too small to permit it to jump to a neighboring interstice. Because of statistical fluctuations, however, of the total of $N$ atoms, a certain number $n_i$ have at any instant a specified energy $\varepsilon_i$; for example, the energy corresponding to the height of the barrier in Fig. 8.3. The ratio $n_i/N$, which represents the probability $p_i$ that a given atom has

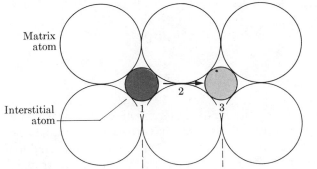

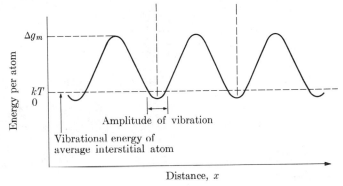

(a) An interstitial atom initially at interstice 1 must pass through the high-energy position 2 to complete the jump to interstice 3

(b) The variation in energy with position of the interstitial atom

**Fig. 8.3** Schematic illustration of the energy required for the motion (jumping) of an interstitial atom.

energy $\varepsilon_i$ is given by the classical distribution of energy,

$$p_i = \frac{n_i}{N} = \frac{1}{Q} \exp\left(-\varepsilon_i/kT\right). \tag{8.16}$$

To understand the meaning of $Q$ (known as the partition function), which is defined by Eq. (8.18), imagine that a given atom possesses energy $\varepsilon_i$ because of thermal vibrations. Since energy is quantized, the value of $\varepsilon_i$ can be expressed as

$$\varepsilon_i = (i + \tfrac{1}{2})hv \tag{8.17}$$

where $i$ is the vibrational quantum number and $hv/2$ is the "zero-point energy"; $h$ is Planck's constant and $v$ is the frequency of vibration. An expression for the partition function is

$$Q = \sum_{i=0}^{\infty} \exp\left(-(i + \tfrac{1}{2})h\nu/kT\right)$$

$$= \exp(-h\nu/2kT) \sum_{i=0}^{\infty} \exp(-ih\nu/kT), \tag{8.18}$$

If the energy barrier in question has an energy, $\varepsilon_r$, corresponding to a vibrational quantum number $r$, those atoms having an energy equal to or greater than $\varepsilon_r$ can surmount the barrier; that is, they can make a diffusional jump. This total number of atoms is conveniently expressed as the total probability, $p$, that a given atom has at least energy $\varepsilon_r$. From Eq. (8.16),

$$p = \sum_{i=r}^{\infty} \frac{n_i}{N} = \sum_{i=r}^{\infty} \frac{1}{Q} \exp(-\varepsilon_i/kT). \tag{8.19}$$

In view of the definition of $\varepsilon_i$, Eq. (8.17),

$$p = \frac{1}{Q}\{\exp[-(r + \tfrac{1}{2})h\nu/kT] + \exp[-(r + \tfrac{3}{2})h\nu/kT + \cdots]\}$$

$$= \frac{1}{Q}\{\exp(-h\nu/2kT)\exp(-rh\nu/kT)[1 + \exp(-h\nu/kT) + \exp(-2h\nu/kT) + \cdots]\}. \tag{8.20}$$

Substitution of the expression for $Q$ from Eq. (8.18) gives the final expression for the desired probability,

$$p = \exp(-u/kT) \tag{8.21}$$

where $u = rh\nu$ is the internal energy per atom. The corresponding thermodynamic quantity for a mole of atoms is $U = Nu$, where $N$ is Avogadro's number. The energy $u$ differs from $\varepsilon_r$ only by the factor of the zero-point energy, $h\nu/2$. This difference is of no practical significance since thermodynamic quantities are usually referred to some reference state (and are then denoted by $\Delta U, \Delta H, \Delta G$, etc.)

For diffusion under the usual conditions of constant temperature and pressure, the appropriate thermodynamic quantity is the Gibbs free energy, $\Delta G$, rather than simply the internal energy, $\Delta U$. By analogy with Eq. (8.21) one obtains the useful expression,

$$p = \exp(-\Delta G_m/RT). \tag{8.22}$$

Correspondingly, a given atom must acquire the energy $\Delta g_m = \Delta G_m/N$ if it is to move (jump). The jump frequency, $\Gamma$, of an interstitial atom can now be obtained as the product

$$\Gamma = (\nu z)p = \nu z \exp(-\Delta G_m/RT). \tag{8.23}$$

Here $\nu$ is the frequency of vibration of an atom in a metal ($\sim 10^{13}$ hertz) and $z$

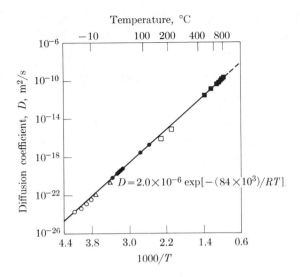

**Fig. 8.4** The large variation of diffusion coefficient with temperature is shown by these data, obtained by various experimental techniques, on the diffusion of carbon in BCC iron. (After C. A. Wert)

is the number of neighboring interstitial sites into which a given atom can jump ($z$ is four for carbon diffusing in BCC iron). Recall from Chapter 6 that the change in free energy $\Delta G$ at constant temperature is defined in terms of the enthalpy $\Delta H$ and the entropy $\Delta S$,

$$\Delta G = \Delta H - T\,\Delta S. \tag{8.24}$$

Therefore, when Eq. (8.23) is substituted in Eq. (8.4), the final expression for the diffusion coefficient can be written,

$$D = \tfrac{1}{6}\alpha^2 vz \exp\left(\Delta S_m/R\right) \exp\left(-\Delta H_m/RT\right) \tag{8.25}$$

All of the quantities except the final exponential term are usually combined into a single constant, $D_0$, and $\Delta H_m$ is replaced by the activation energy, $Q$, in joules per mole, giving the form usually employed,

$$D = D_0 \exp\left(-Q/RT\right) \tag{8.26}$$

The data for the diffusion of carbon in BCC iron, Fig. 8.4, furnish a good example of the variation of $D$ with temperature described by this equation.

*Vacancy Mechanism of Diffusion.* An additional quantity, $N_V^0$, the equilibrium concentration of vacancies, is needed for a description of vacancy diffusion. The appropriate expression, Eq. (6.56), is

$$N_V^0 = \exp\left(-\Delta G_f/RT\right) = \exp\left(\Delta S_f^{xs}/R\right) \exp\left(-\Delta H_f/RT\right) \tag{8.27}$$

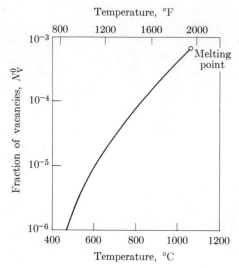

**Fig. 8.5** The variation of the fraction of vacancies in gold as a function of temperature. (Data of R. O. Simmons and R. W. Balluffi)

Typical data on the variation of the fraction of vacancies $N_V^0$ with temperature are shown in Fig. 8.5. Near the melting point of gold $N_V^0$ is nearly $10^{-3}$, but it decreases rapidly with decreasing temperature and is only $10^{-6}$ at 450°C (840°F).

As described above in connection with Fig. 8.2(b), an adequate theory of substitutional diffusion should provide separate diffusion coefficients for the solute and solvent atoms. We will limit the present treatment, however, to a description of the general type of equation involved, which also applies to the single interdiffusion coefficient $\tilde{D}$ commonly used in treating technical substitutional-diffusion problems. When Eq. (8.15) is developed in a manner similar to Eq. (8.25), the result is

$$D = \tfrac{1}{6} f\alpha^2 vz \exp\left(\frac{\Delta S_m + \Delta S_f^{xs}}{R}\right) \exp\left[-\frac{(\Delta H_m + \Delta H_f)}{RT}\right]. \qquad (8.28)$$

If $\Delta H_m + \Delta H_f$ is replaced by the activation energy, $Q$, in joules per mole, and the remaining quantities combined into the single constant $D_0$, an expression identical with Eq. (8.26) is obtained. Thus diffusion constants of all types can be described by two constants, $D_0$ and $Q$, which are determined experimentally. Several examples are given in Table 8.1.

## SOLUTIONS OF THE DIFFUSION EQUATIONS

Fick's two laws of diffusion, Eqs. (8.3) and (8.13), are differential equations and must be solved to obtain an algebraic equation describing the variation of concentration as a function of distance and (usually) of time. The ease of solution depends on

**TABLE 8.1**

Approximate Values of $D_0$ and $Q$ for Several Diffusion Systems

| Diffusing metal | Matrix metal | $D_0$, $10^{-5}$ m$^2$/s | $Q$, $10^3$ J/mol |
|---|---|---|---|
| Carbon | $\gamma$-iron | 2.0 | 140 |
| Carbon | $\alpha$-iron | 0.20 | 84 |
| Iron | $\alpha$-iron | 19 | 239 |
| Iron | $\gamma$-iron | 1.8 | 270 |
| Nickel | $\gamma$-iron | 4.4 | 283 |
| Manganese | $\gamma$-iron | 5.7 | 277 |
| Copper | Aluminum | 0.84 | 136 |
| Zinc | Copper | 2.1 | 171 |
| Silver | Silver (volume diffusion) | 7.2 | 190 |
| Silver | Silver (grain-boundary diffusion) | 1.4 | 90 |

To convert joules to calories, multiply by 0.239.

such factors as the geometry of the specimen in question, the initial conditions of concentration in the specimen, and the "boundary conditions" [for example, whether diffusion penetrates the entire (finite) specimen or affects only a portion of an (infinite) specimen]. The solutions can be listed in three categories: (1) Those that can be obtained easily by elementary manipulations of the diffusion equations. Problem 7 is an especially simple example. The treatment of oxidation in the following section illustrates a typical procedure. A majority of the diffusion problems encountered by an engineer can be simplified to permit at least an order-of-magnitude solution in this manner. (2) A variety of technically important diffusion problems have been solved by more advanced mathematical techniques, and the solutions conveniently collected in reference volumes.* Solutions of Fick's second law for cylinders and spheres are examples. (3) Diffusion problems of many types, such as those involving concentration-dependent diffusion coefficients, are solved most conveniently using a digital computer (or perhaps an analog computer). Examples can be found in technical journals, but usually each special problem must be solved individually.

**Fick's first law.** Since this equation, $J = -D \, dc/dx$, does not explicitly consider the change of concentration with time, its use is limited to steady-state diffusion

---

* Two of the most useful are: J. Crank, *The Mathematics of Diffusion*, London: Oxford University Press, 1964; and H. S. Carslaw and J. C. Jaeger, *Conduction of Heat in Solids*, 2nd ed., London: Oxford University Press, 1959.

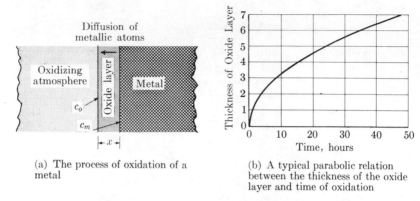

(a) The process of oxidation of a metal

(b) A typical parabolic relation between the thickness of the oxide layer and time of oxidation

Fig 8.6  The oxidation of metals, a diffusion process.

(in which the concentration remains constant at each point, see Problem 7), or to problems in which the significant parameters do not change with time. A typical example is the following analysis of one type of oxidation of metals.

Experimental data on the high-temperature oxidation of metals frequently show a parabolic relation between the time of oxidation and the amount of oxide formed. With the aid of Fig. 8.6(a), this behavior can be explained in terms of a diffusion process. It is assumed that diffusion of only the metal atoms need be considered. Furthermore, the difference between the metal concentration $c_o$ at the oxide surface and the concentration $c_m$ at the metal surface has a constant value $\Delta c$. Therefore, at any thickness $x$ of the oxide layer the rate of transfer of metal atoms per unit area is

$$\frac{dm}{dt} = -D\frac{\Delta c}{x}. \tag{8.29}$$

However, $dm$ and the increase in thickness, $dx$, are proportional; that is, $dx = K\,dm$. Therefore Eq. (8.29) can be written

$$dx = -D'\frac{\Delta c}{x}\,dt, \tag{8.30}$$

where $D'$ is $KD$ and is a constant when $D$ is assumed to be constant. This equation can be solved as follows:

$$\int_{x=0}^{x=x} x\,dx = -D'\Delta c\int_{t=0}^{t=t} dt, \tag{8.31}$$

$$x^2 = -2D'\Delta ct,$$

$$x^2 = K't. \tag{8.32}$$

$K'$ is positive, since the minus sign cancels the minus sign of the concentration difference $\Delta c$. The thickness of the oxide layer increases as the square root of the time of oxidation. An illustration of this relationship is given in Fig. 8.6(b).

**Fick's second law.** We will consider only the simpler form, Eq. (8.13), for a constant* value of $D$. A solution, $c = f(x, t)$, of this equation must satisfy both the partial differential equation of Eq. (8.13) and also the initial and boundary conditions in question. We begin with a relatively simple solution to illustrate these features.

*Thin-Film Solution.* A common experimental procedure, especially when radioactive tracers are used, is to deposit a thin layer of metal $A$ at the end of a long bar of metal $B$. Often two such specimens are joined to produce a thin "sandwich" of $A$ between two bars of $B$. Mathematically, the initial concentration of $A$ atoms is represented as $M$ atoms per square meter, located on the plane at $x = 0$. In the case of a sandwich-type specimen, the following solution gives the concentration as a function of distance $x$ into metal $B$ after a given time $t$ of diffusion,

$$c = \frac{K}{\sqrt{t}} \exp(-x^2/4Dt), \tag{8.33}$$

$K$ is a constant. By differentiation, one can show that Eq. (8.33) is a solution of the differential equation, Eq. (8.13). Since the total initial number of atoms, $M$, merely becomes redistributed during diffusion,

$$M = \int_{-\infty}^{\infty} c\, dx, \tag{8.34}$$

and therefore,

$$M = 2K\sqrt{\pi D}. \tag{8.35}$$

Substitution of Eq. (8.35) in Eq. (8.33) gives the final form of the solution,

$$c = \frac{M}{2\sqrt{\pi Dt}} \exp(-x^2/4Dt) \tag{8.36}$$

which satisfies the boundary conditions existing in the specimen.

If, instead of being a sandwich, the specimen consists of the same amount $M$ at one end of a bar, then the concentration is greater by a factor of two,

$$c = \frac{M}{\sqrt{\pi Dt}} \exp(-x^2/4Dt), \tag{8.37}$$

since the atoms diffuse only in the *positive* x-direction.

---

* When the diffusion coefficient changes appreciably with the concentration of the diffusing metal, the curve of concentration versus distance of diffusion can be analyzed by a graphical procedure, the *Boltzmann-Matano Method*. See P. G. Shewmon, *Diffusion in Solids*, p. 28, New York: McGraw-Hill, 1963.

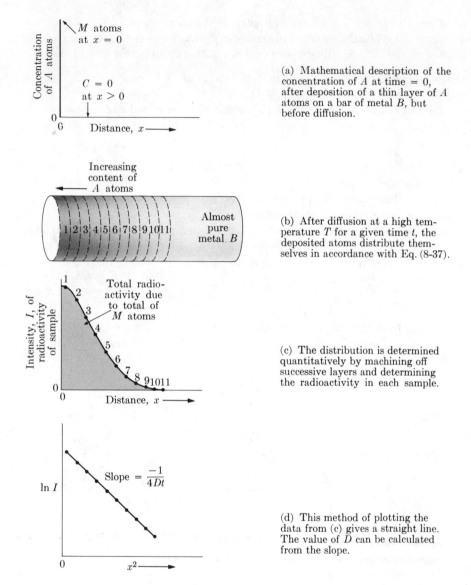

(a) Mathematical description of the concentration of $A$ at time $= 0$, after deposition of a thin layer of $A$ atoms on a bar of metal $B$, but before diffusion.

(b) After diffusion at a high temperature $T$ for a given time $t$, the deposited atoms distribute themselves in accordance with Eq. (8-37).

(c) The distribution is determined quantitatively by machining off successive layers and determining the radioactivity in each sample.

(d) This method of plotting the data from (c) gives a straight line. The value of $D$ can be calculated from the slope.

**Fig. 8.7** A typical method for using the thin-film solution to determine a tracer diffusion coefficient.

A typical use of Eq. (8.37) is in the determination of the tracer diffusion coefficient of metal $A$ in metal $B$, illustrated in Fig. 8.7. The thin layer of radioactive $A$ atoms, Fig. 8.7(a), is deposited by a method such as electroplating. Diffusion at a temperature about nine-tenths of the melting temperature of $B$ for several hours produces a convenient depth of diffusion ($\sim 1$ mm) into metal $B$, Fig. 8.7(b). The quantitative diffusion profile, Fig. 8.7(c), is determined by sectioning the bar and measuring the intensity, $I$, of the radioactivity in each layer. Since the radio-activity is directly proportional to the number of $A$ atoms, Eq. (8.37) can be written

$$I = k \exp(-x^2/4Dt), \tag{8.38}$$

where $k$ replaces the previous constants (including $D$ and $t$, which are constant for a given experiment). After taking the natural logarithm of both sides of Eq. (8.38) we can write the equation of a straight line,

$$y = mx \qquad\qquad + b$$

$$\ln I = (-1/4Dt)x^2 + \ln k. \tag{8.39}$$

Therefore, the diffusion coefficient $D$ can be calculated from the slope of the line obtained by plotting $\ln I$ versus the corresponding values of $x^2$, Fig. 8.7(d).

*Infinite Solid.* Many technically important instances of diffusion occur in a limited region near the junction (at $x = 0$) of two solid bodies. The two bodies behave for mathematical purposes as though they extend to $-\infty$ and $+\infty$, respectively, and therefore this type of problem is called the infinite solid. An example is the diffusion couple, Fig. 8.8(a), formed by an alloy of composition $c_0$ and a pure metal. If the alloy is pictured as consisting of a continuous series of thin layers, Fig. 8.8(b), then the solution for this diffusion problem is easily obtained from Eq. (8.36), the solution for a thin film. Consider the typical layer shown in the figure. After diffusion for time $t$, this layer makes the following contribution to the concentration at position $x$,

$$\frac{c_0 d\xi}{2\sqrt{\pi Dt}} \exp(-\xi^2/4Dt). \tag{8.40}$$

The effect of all the layers (which start at $\xi = X$ and extend to $\xi = \infty$) is given by the summation (integration),

$$c = \int_x^\infty \frac{c_0}{2\sqrt{\pi Dt}} \exp(-\xi^2/4Dt)\, d\xi = \frac{c_0}{\sqrt{\pi}} \int_{x/2\sqrt{Dt}}^\infty \exp(-\eta^2)\, d\eta, \tag{8.41}$$

where

$$\eta = \xi/2\sqrt{Dt}. \tag{8.42}$$

The integral in Eq. (8.41) is essentially the error function,

$$\left( \operatorname{erf}(z) = \frac{2}{\sqrt{\pi}} \int_0^z \exp(-\eta^2)\, d\eta. \right) \tag{8.43}$$

Its values are commonly given in mathematical tables and are shown on the right-hand ordinate of Fig. 8.8(c) as a function of its argument $z$ ($z \equiv x/2\sqrt{Dt}$ in the present instance).

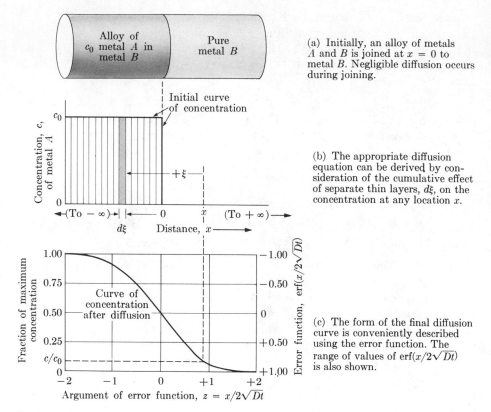

(a) Initially, an alloy of metals $A$ and $B$ is joined at $x = 0$ to metal $B$. Negligible diffusion occurs during joining.

(b) The appropriate diffusion equation can be derived by consideration of the cumulative effect of separate thin layers, $d\xi$, on the concentration at any location $x$.

(c) The form of the final diffusion curve is conveniently described using the error function. The range of values of $\mathrm{erf}(x/2\sqrt{Dt})$ is also shown.

**Fig. 8.8** The diffusion couple, an important example of diffusion in an infinite solid, and its relation to diffusion from a thin layer. (Adapted from J. Crank)

The integral of Eq. (8.41) can be expressed as the difference,

$$\int_{z}^{\infty} \exp(-\eta^2)\, d\eta = \int_{0}^{\infty} \exp(-\eta^2)\, d\eta - \int_{0}^{z} \exp(-\eta^2)\, d\eta$$

$$= \mathrm{erf}(\infty) - \mathrm{erf}(z) = 1 - \mathrm{erf}(z). \tag{8.44}$$

Therefore the final solution for the concentration in a diffusion couple is

$$c = \frac{c_0}{2}[1 - \mathrm{erf}(x/2\sqrt{Dt})], \tag{8.45}$$

or,

$$\frac{c}{c_0} = \tfrac{1}{2}[1 - \mathrm{erf}(x/2\sqrt{Dt})]. \tag{8.46}$$

The values from Eq. (8.46) are shown as the left-hand ordinate in Fig. 8.8(c).

*Semi-Infinite Solid.* Diffusion problems involving only a single uniform solid (that is, either half of the specimen of Fig. 8.8(a)) possess a semi-infinite geometry; the specimen is pictured as extending from $x = 0$ toward $x = +\infty$. The solutions are analogous to those for the infinite solid; the two types are: (1) essentially the right-hand half of the curve of Fig. 8.8(c),

$$c/c_0 = 1 - \text{erf}(x/2\sqrt{Dt}), \tag{8.47}$$

to describe the *addition* of $A$ atoms (from the gas phase) to the surface of metal $B$ at $x = 0$; and (2) essentially the left-hand half of the curve of Fig. 8.8(c) [except that $x/2\sqrt{Dt}$ is increasing, rather than decreasing],

$$c/c_0 = \text{erf}(x/2\sqrt{Dt}) \tag{8.48}$$

to describe the *removal* of $A$ atoms (via the gas phase) from the surface of an alloy of $A$ and $B$. Problem 9 treats decarburization of steel by the use of Eq. (8.48).

An example of the application of Eq. (8.47) is in the carburizing of iron, in which a constant carbon content is maintained at the surface of the iron, Fig. 8.9(a), by a suitable atmosphere such as natural gas. The initial carbon content of the iron plate is zero at all distances from the surface. As the carbon atoms go into solution at the surface of the iron, they are free to begin diffusing farther into the plate. The maximum concentration, $c_0$, of carbon atoms in solution in iron at 927°C (1700°F) can be read from the iron-carbon diagram (see Chapter 7) and is about 1.3% carbon. The carburizing atmosphere can build up this carbon concentration at the *surface* almost immediately, so that the carbon penetration curve shortly after the beginning of carburizing is of the form shown in Fig. 8.9(b).

To calculate the carbon penetration curve resulting from diffusion for 10 hours at 927°C we need the values of $D$ and $t$. From Table 8.1 the diffusion coefficient for carbon in $\gamma$-iron is found to be $1.5 \times 10^{-11}$ m$^2$/s. The time $t$ is $10 \times 60 \times 60 = 3.6 \times 10^4$ seconds. For these values the quantity $z = x/2\sqrt{Dt}$ can be written

$$z = \frac{x}{2\sqrt{1.5 \times 3.6 \times 10^{-7}}} = (6.8 \times 10^2)x. \tag{8.49}$$

The steps used in calculating the value of $c$ at 0.4 mm intervals below the surface of the steel plate are summarized in Table 8.2. A plot of these results gives the curve of carbon concentration versus distance below the surface, which is shown in Fig. 8.9(c). Because the value of $D$ actually increases with carbon concentration, this plot does not coincide exactly with experimental data, but it gives an approximation that is useful for many purposes.

In practical carburizing operations it is usually required that a certain minimum carbon content be produced at a given depth below the surface. From the form of the variable $z = x/2\sqrt{Dt}$ it can be concluded that the *time* of carburizing necessary to produce the given carbon content at the given depth increases with the square of the depth and is inversely proportional to the diffusion coefficient.

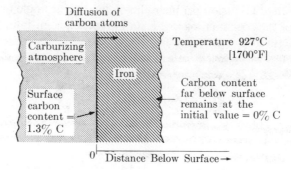

(a) Schematic representation of the procedure
used in carburizing an iron plate

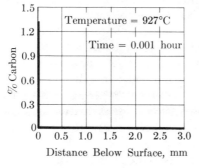

(b) Carbon penetration curve near
the beginning of the carburizing
process

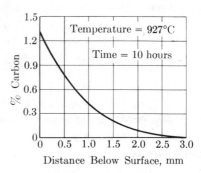

(c) Carbon penetration curve
after carburization for 10 hours

**Fig. 8.9**  The diffusion of carbon in iron during carburizing.

**Diffusion in alloys: the Kirkendall effect.** When the tracer diffusion coefficients,
$D^*$, are determined for both components in an alloy (using the method of Fig.
8.7, for example) the $D^*$ values are unequal, and both vary with composition.
Data for the gold-nickel system are given in Fig. 8.10; the gold atoms diffuse more
rapidly than the nickel atoms at all concentrations. The presence of a concentration
gradient modifies the diffusion behavior through a thermodynamic factor, and
the appropriate diffusion coefficient is then given by the equation

$$D_{Ni} = D_{Ni}^* \left( 1 + \frac{d \ln \gamma_{Ni}}{d \ln N_{Ni}} \right), \tag{8.50}$$

where $D_{Ni}$ is called the *intrinsic* diffusion coefficient for nickel and $\gamma_{Ni}$ is the thermodynamic activity coefficient of nickel in the gold-nickel alloy containing an atomic
fraction $N_{Ni}$ of nickel. A similar equation can be written for $D_{Au}$, the intrinsic

**TABLE 8.2**

Summary of Calculations for Plotting the Curve of Figure 8.9(c)

| Distance $x$ below the surface, $10^{-3}$ m | $z = x/2\sqrt{Dt}$ $= (6.8 \times 10^2)x$ | erf$(z)$ | $1 - $ erf$(z)$ | Carbon concentration, $c = 1.3\,[1 - $ erf$(z)]$ |
|---|---|---|---|---|
| 0.0 | 0.000 | 0.000 | 1.000 | 1.30 |
| 0.4 | 0.272 | 0.300 | 0.700 | 0.91 |
| 0.8 | 0.544 | 0.558 | 0.442 | 0.57 |
| 1.2 | 0.816 | 0.752 | 0.248 | 0.32 |
| 1.6 | 1.088 | 0.876 | 0.124 | 0.16 |
| 2.0 | 1.360 | 0.946 | 0.054 | 0.07 |
| 2.4 | 1.632 | 0.979 | 0.021 | 0.03 |
| 2.8 | 1.904 | 0.993 | 0.007 | 0.01 |
| 3.2 | 2.176 | 0.998 | 0.002 | 0.00 |

diffusion coefficient for gold. The factor in parentheses in Eq. (8.50) ordinarily differs from unity by a factor of less than 10.

Like the self-diffusion coefficients, the intrinsic diffusion coefficients for the two components in an alloy have different values. Consequently, when two metals interdiffuse there is a net transport of material across the plane that initially separated the two specimens. This phenomenon, known as the *Kirkendall effect*, can be revealed by the type of experiment shown schematically in Fig. 8.11. A specimen is made by joining a gold and a nickel bar so that diffusion will occur across the marked interface. The inert markers may be pieces of very fine tungsten wire lying in the plane of joining. During a diffusion anneal of many hours at a sufficiently high temperature, such as 900°C (1650°F), interdiffusion of the gold and nickel will occur and will change the concentration distribution as shown. However, since $D_{Au}$ is greater than $D_{Ni}$, more gold atoms than nickel atoms will have diffused past the inert markers. Consequently, the inert markers will no longer be at their original position but will have moved toward the gold end of the specimen. This movement is known as the *Kirkendall shift*.

The fact that the flux of gold is greater than that of nickel means that a net flux of vacancies occurs across the Kirkendall interface. In the absence of reactions involving the vacancies, this flux of vacancies would lead to a large excess of vacancies in one portion of the specimen and a deficiency in another portion. In fact, the vacancies react in various ways to maintain their concentration at essentially the equilibrium value everywhere in the specimen. The simplest reaction, which occurs when the concentration of vacancies drops slightly below the equilibrium value, is the formation of a vacancy at a dislocation (or other discontinuity in the crystal lattice, such as a grain boundary). The reverse of this reaction, the destruction of a vacancy, can occur in regions where the vacancy

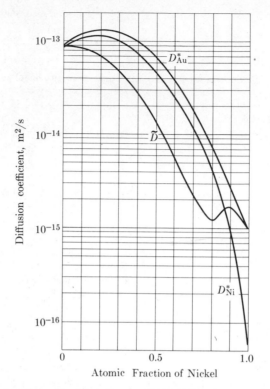

**Fig. 8.10** Diffusion data for gold-nickel alloys at 900°C (1650°F). The relation between $\tilde{D}$ and the two tracer coefficients is given by Eq. (8.52). (Data of J. E. Reynolds, B. L. Averbach, and M. Cohen.)

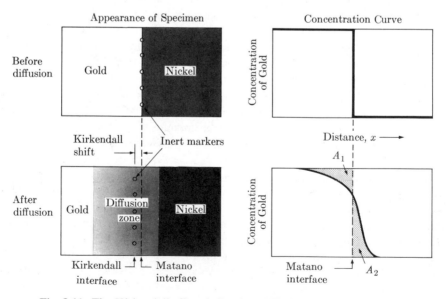

**Fig. 8.11** The Kirkendall effect during interdiffusion of gold and nickel.

concentration is above the equilibrium value (supersaturation of vacancies). If supersaturation is relatively high, then additional types of reactions become important for the destruction of vacancies. One of these reactions is the "precipitation" of vacancies to form small voids within the crystal lattice. A region of porosity resulting from this reaction is commonly found on that side of the Kirkendall interface where the vacancies are supersaturated.

The plane at the position of the original joining of the gold and nickel is called the *Matano interface* and has the important property that the two areas $A_1$ and $A_2$ are equal.* $A_1$ and $A_2$ represent the amounts of nickel and of gold, respectively, that have crossed this interface. Therefore if the Matano interface is used in describing diffusion behavior, only a single diffusion coefficient $D$ need be used. It can be shown that the relation between $\tilde{D}$ and $D_{Au}$ and $D_{Ni}$ is

$$\tilde{D} = N_{Ni}D_{Au} + N_{Au}D_{Ni}, \tag{8.51}$$

and between $\tilde{D}$ and $D_{Au}^*$ and $D_{Ni}^*$ is

$$\tilde{D} = (N_{Ni}D_{Au}^* + N_{Au}D_{Ni}^*)\left(1 + \frac{d \ln \gamma_{Ni}}{d \ln N_{Ni}}\right). \tag{8.52}$$

The values of $\tilde{D}$ for various compositions in the gold-nickel system are shown in Fig. 8.10. To distinguish $\tilde{D}$ from the other two types of diffusion coefficient, it may be called the *interdiffusion* or *mutual diffusion* coefficient. However, since it is the quantity usually employed in diffusion calculations, it is most often referred to simply as the diffusion coefficient.

## FACTORS THAT INFLUENCE DIFFUSION

The diffusion coefficient $\tilde{D}$ is usually not a constant. Ordinarily $\tilde{D}$ is a function of many variables such as temperature, concentration, and crystal structure. In a given instance all these variables are assumed to have specified values, and the diffusion coefficient is then a definite number. However, in arriving at this number it is frequently necessary to consider the individual effects of one or more of these variables. Furthermore, in addition to lattice (or volume) diffusion, which we have been considering, significant amounts of diffusion may occur by "short circuits" along grain boundaries, surfaces, or dislocations. Diffusion may result, not only from the presence of a concentration gradient, but also because of an electric field, a temperature gradient, or the local state of stress.

**Concentration.** Since it is common practice to assume for mathematical convenience that the diffusion coefficient is independent of concentration, it is useful to know the error involved in this assumption. The data of Fig. 8.10 for the gold-

---

* This treatment neglects such effects as the formation of porosity in the diffusion zone or possible changes in length because of alloying.

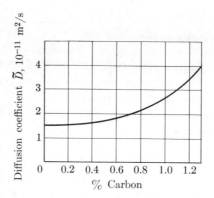

**Fig. 8.12** The variation with concentration of the diffusion coefficient for carbon diffusing in iron at 927°C (1700°F).

nickel system show an extreme variation of $\tilde{D}$ with concentration, while Fig. 8.12 shows that there is only a relatively small change in the diffusion coefficient of carbon in austenite up to 1.3% C, the limit of solubility at the temperature in question. Even in systems for which $\tilde{D}$ is strongly concentration-dependent, little error is involved in assuming that $\tilde{D}$ is constant, provided diffusion occurs in a dilute solution or over a small range of concentration. For example, diffusion in a solution of a few atomic percent gold in nickel at 900°C can be calculated adequately using $\tilde{D} = 1 \times 10^{-15}$, while the value $\tilde{D} = 9 \times 10^{-14}$ would be a good approximation for a calculation of diffusion in a dilute solution of nickel in gold.

**Crystal structure.** Because of the importance of the allotropic transformation of iron from body-centered cubic to face-centered cubic at high temperatures, the effect of this change in crystal structure on the rate at which solute atoms diffuse in iron has been studied. At a given temperature these diffusion processes and self-diffusion of the iron occur about one hundred times more rapidly in ferrite (body-centered cubic) than in austenite.

Another effect of crystal structure is the variation of the diffusion coefficient with crystal direction in a single crystal of the solvent metal. Such anisotropy is nearly or completely absent in cubic metals, but bismuth (rhombohedral space lattice) shows a ratio of about one thousand in its self-diffusion constants measured parallel and perpendicular to the c-axis. Moreover, if a crystal structure is distorted either by elastic strains or by extensive plastic deformation, the rate of diffusion is usually increased.

**Impurities.** The presence of small amounts of additional metals usually has a relatively small effect on the diffusion of solute atoms in a solvent metal. This fact is often useful in treating problems in which diffusion is only one of several possible variables. For example, it can be concluded that the strong influences of

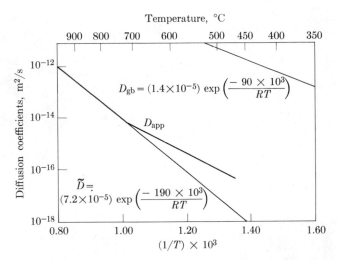

Temperature, °C

**Fig. 8.13** The effect of grain-boundary diffusion on the apparent diffusion coefficient, $D_{app}$, in polycrystalline silver. $\tilde{D}$ was measured on single crystals. (Adapted from data of R. E. Hoffman and D. Turnbull)

alloying elements on the hardenability of steel (Chapter 10) must be the result of factors other than large changes in the rate of carbon diffusion.

**Grain size.** Since grain-boundary diffusion is faster than that within the grains as discussed below, it is to be expected that the over-all diffusion rate would be higher in a fine-grained metal. However, in the usual range of grain sizes it is not necessary to take grain size into account in making diffusion calculations.

**Short-circuit diffusion.** Because the regularity of the crystal lattice of a metal is disturbed at a grain boundary, near a free surface, and adjacent to a dislocation line, diffusion by the vacancy mechanism is greatly enhanced in these regions. Both the number of vacancies and their mobility may be larger as a result of the local disruption of crystalline regularity. For this reason the activation energy, $Q$, is less for grain-boundary diffusion than for volume diffusion. The data for diffusion in silver, Table 8.1, are representative. $Q$ is still lower for surface diffusion and for "pipe" diffusion along grain boundaries. This difference in $Q$ values explains why short-circuit diffusion plays a significant role only at diffusion temperatures below about three-quarters of the absolute melting point, $T_f$K. Only a small fraction of the cross section of a typical metal is employed by a given short-circuit mechanism. For grain-boundary diffusion the fraction is about $10^{-5}$ (see Problem 10). Therefore, the ratio $D_{gb}/\tilde{D}$ must approach $10^5$ before the amount of material transported along the grain boundaries is comparable to that diffusing in the volume of the grains. At high temperatures the ratio is less than this value, but with decrease in temperature $D_{gb}$ decreases less than $\tilde{D}$ because of the difference in their $Q$ values. As shown by the data for silver in Fig. 8.13, the

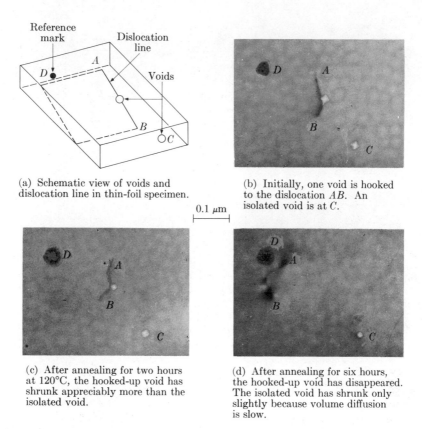

(a) Schematic view of voids and dislocation line in thin-foil specimen.

(b) Initially, one void is hooked to the dislocation $AB$. An isolated void is at $C$.

0.1 $\mu$m

(c) After annealing for two hours at 120°C, the hooked-up void has shrunk appreciably more than the isolated void.

(d) After annealing for six hours, the hooked-up void has disappeared. The isolated void has shrunk only slightly because volume diffusion is slow.

**Fig. 8.14** Electron micrographs showing the effect of diffusion along a dislocation line in accelerating the elimination of a void in an aluminum foil. (Courtesy T. E. Volin and R. W. Balluffi, Cornell University)

apparent diffusion coefficient, $D_{app}$, begins to deviate from $\tilde{D}$ at a sufficiently low temperature because of the increasing contribution of grain-boundary diffusion.

At still lower temperatures, where both volume diffusion and grain-boundary diffusion may be ineffective, significant atomic transport can occur along the "pipe" of disturbed lattice surrounding a dislocation line. Figure 8.14(b) shows voids about 300 Å in diameter that were formed in an aluminum foil by the precipitation of vacancies during quenching from a high temperature. If a void is connected to the surface of the specimen by a dislocation line, Fig. 8.14(a), the counterflow of vacancies and atoms is able to occur more easily. Such a void disappears more quickly on heating, Fig. 8.14(c), than does a void surrounded by perfect lattice, Fig. 8.14(d).

**Driving forces for diffusion.** Not only a concentration gradient, but also an electric field or a temperature gradient can cause net diffusive fluxes in metals and alloys.

All three of these influences can be encompassed in a unified treatment in the following manner. We consider an interstitial solute for simplicity and begin with the effect of an electric field. If a solute atom has a net charge of $Z^*$ electron units, $e$, then an applied electric field, $E$, exerts a force,

$$F = EZ^*e \tag{8.53}$$

on each solute atom. Let the mobility, $B$, be defined as the average velocity of a solute atom when acted on by unit force. The velocity, $v$, is then

$$v = FB = BEZ^*e. \tag{8.54}$$

If the concentration of solute, $c$, has units of atoms/m$^3$, then the diffusive flux, $J$, is simply

$$J = cv = cBEZ^*e. \tag{8.55}$$

$\tilde{D}$ is related to $B$ by a relation similar to Eq. (8.50),

$$\tilde{D} = kTB\left(1 + \frac{d \ln \gamma}{d \ln N}\right) \tag{8.56}$$

where $k$ is Boltzmann's constant. Therefore Eq. (8.55) gives a complete description of diffusion in an electric field provided the experimental coefficient $Z^*$ (known as the *effective charge*) and $\tilde{D}$ are known for the interstitial alloy in question.

If the proper expression for the force, $F$, due to a concentration gradient or due to a temperature gradient is used in the above development, the same reasoning gives the equation for the corresponding flux. The two forces are known to be

$$F_{\text{concn}} = -\frac{kT}{c}\frac{dc}{dx}\left(1 + \frac{d \ln \gamma}{d \ln N}\right) \tag{8.57}$$

$$F_{\text{temp}} = -\frac{Q^*}{T}\frac{dT}{dx}. \tag{8.58}$$

$Q^*$ is an experimental coefficient called the *heat of transport*. Therefore, the total flux produced in the presence of driving forces of all three kinds is

$$J = -B\left[kT\frac{dc}{dx}\left(1 + \frac{d \ln \gamma}{d \ln N}\right) + \frac{cQ^*}{T}\frac{dT}{dx} - cEZ^*e\right]. \tag{8.59}$$

Typical values of $Z^*$ and $Q^*$ are listed in Table 8.3. Usually a steep temperature gradient (about $10^5$ K/m) or a large electric current (about $10^7$ A/m$^2$) is required to produce appreciable diffusive effects.

## APPLICATIONS

**Metal bonding.** The bonding of metals, which includes such processes as galvanizing, welding, and metal-cladding, is a good example of an industrial application of diffusion principles. Although a continuous metallic bond may be formed

**TABLE 8.3†**

Typical Values of $Z^*$ and $Q^*$

| Matrix metal | Solute element | Temperature °C | Temperature °F | $Z^*$ | $Q^*$ $10^3$ J/mol |
|---|---|---|---|---|---|
| | | *Pure Metals* | | | |
| Silver | (Vacancies) | 800 | 1472 | −24 | ∼0 |
| Gold | (Vacancies) | 900 | 1652 | −33 | −36 |
| Copper | (Vacancies) | 900 | 1652 | −14 | +21 |
| Platinum | (Vacancies) | 1600 | 2912 | −0.3 | +66 |
| | | *Alloys* | | | |
| Iron | Carbon | ∼1000 | ∼1830 | +4 | −8 |
| Copper | Cobalt | ∼1000 | ∼1830 | −32 | +17 |
| Zinc | Silver | ∼ 400 | ∼ 750 | −3 | +4 |

† Data adapted from Y. Adda and J. Philibert, *La Diffusion dans les Solides*, Paris: Presses Universitaires, 1966.

under many different conditions between two different metals, in every case it is necessary that some diffusion occur. Since diffusion can take place only in a solid solution, only two metals that have appreciable solid solubility can be bonded. Thus, because lead is practically insoluble in iron, it is possible to heat steel in molten lead baths without danger of forming an adherent lead layer on its surface. The opposite problem arises when it is desirable to protect sheet iron with a thin coating of lead. To obtain adequate bonding it is necessary to add to the molten lead a small amount of a second metal that alloys with iron. The most popular addition is tin, and the resulting alloy is widely used in making *terne* plated sheet steel for roofing.

A complicating factor in some bonding processes is the presence of brittle intermetallic compounds between the two metals being joined. When diffusion occurs between two metals, the primary solid solutions are formed; and in addition, those intermediate phases that exist at the diffusion temperature also form. This behavior is shown in Fig. 8.15 for the coating of steel with zinc to produce a *galvanized* sheet. The galvanized coating is produced in about fifteen seconds, while the carefully prepared sheet steel is passed through a molten zinc bath at 450°C. During this time the three intermediate solid phases indicated by the

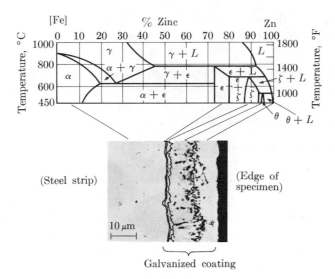

**Fig. 8.15** Correlation of the phases present in a galvanized coating produced at 450°C and the phases present in the corresponding temperature section of the iron-zinc diagram. (Courtesy D. H. Rowland, Carnegie-Illinois Steel Corporation.)

iron-zinc diagram are developed as so-called diffusion bands. Two-phase regions do not form in the course of the growth of diffusion bands; they merely constitute the interface between the one-phase regions. The resistance to flaking of a galvanized sheet when it is bent is improved by decreasing the amount of the brittle intermediate phases present. This decrease can be accomplished by decreasing the total coating thickness or by adding a suitable alloying element, such as aluminum, to the molten zinc bath.

Many different methods are used to attach a layer of one metal to a different base metal. In making *Alclad* or *Pureclad* sheets, a protective layer of pure aluminum is bonded to a strong aluminum alloy by hot rolling a "sandwich" made of slabs of the two materials. Care must be taken to prevent excessive diffusion in this case, since the good corrosion resistance of the pure aluminum is lost if it is contaminated by the additional elements present in the strong alloy. No problem of brittle compound formation is met in producing Alclad, but when a layer of copper is to be attached to aluminum, special precautions must be taken to ensure a ductile product. One successful procedure is hot rolling the bimetal combination to disintegrate the brittle layer as it begins to form.

The term *welding* refers to a method for joining two separate pieces of metal to make a single, rigid assembly. The following four categories encompass most of the numerous types of welding processes.

1. *Fusion welding* produces local melting of the areas to be joined. Its variants include gas-, arc-, electroslag-, electron-beam-, and laser-beam welding. The

solidified metal in the weld "bead" has a cast structure, and the heat-affected zone adjacent to the weld may also have a structure and properties different from the original metals (which are usually in the wrought condition).

2. *Pressure welding* joins the two surfaces through the action of plastic deformation (plus diffusion, if at a high temperature). Ultrasonic bonding, diffusion bonding, isostatic bonding, electrical-resistance or spot welding, and hammer welding are examples of this method.

3. *Explosive welding* employs an explosive charge to propel one metal against the other at very high velocity. The resulting extreme deformation creates a jet of metal that abrades the surfaces of the two metals just before they impinge and bond.

4. *Flow welding* uses a low-melting alloy (plus a suitable fluxing agent) to join the two pieces of metal without appreciably altering them. Two common examples are *soldering* and *brazing*, in which a molten alloy wets (and may partially diffuse into) the solid metals and produces bonding. Solders have melting ranges below 425°C (800°F), while brazes melt above this temperature. Typical compositions and applications are listed in Table 8.4. The low-melting solders are convenient to use and produce little distortion. However, their shear strength is only about $3 \times 10^7 \text{ N/m}^2$ (5000 lb/in²). The brazes are about ten times stronger and are widely used for such purposes as fastening sintered carbide inserts in machine tools. In most cases the presence of intermediate alloy layers is not objectionable in these joints, as high ductility is not necessary.

**Homogenization.** A chemically inhomogeneous solid solution tends to become homogeneous through diffusion. This process has a vanishingly small rate at low temperatures, and in practice a special heat treatment is used to produce homogenization. In many commercial alloys homogenization is achieved rather easily. Examples of this treatment were given in Chapter 7 for cast aluminum-copper and magnesium-aluminum alloys. In some instances the diffusion coefficient of the alloying element plays a decisive role. Thus, at a moderate temperature the rapid diffusion of zinc in copper produces effective homogenization of cast brasses, while at the same temperature nickel diffuses so slowly in copper that it is difficult to eliminate segregation in cast cupronickels. Only by plastically deforming cupronickels, and thereby decreasing the diffusion distance, can rapid homogenization be obtained at this temperature.

As an example of the application of diffusion data to the problem of eliminating chemical inhomogeneity, the subject of banding in steel will be considered. The use of the term "banding" to describe chemical heterogeneity in rolled steels arises from the presence of closely spaced light and dark bands in the microstructure of some of these steels. These bands represent areas of segregation of alloying elements during freezing of the ingot. During rolling the segregated areas are elongated and compressed into narrow bands. This segregation can be

**TABLE 8.4**

Characteristics of Typical Solders and Brazing Filler Metals

| Material | | Composition | Melting range | | Typical uses |
|---|---|---|---|---|---|
| | | | °C | °F | |
| Solders: | 35-65 solder | 35Sn 65Pb | 183–245 | 361–473 | Wiping solder |
| | 45-55 solder | 45Sn 55Pb | 183–226 | 361–439 | For automobile radiator cores; general-purpose solder |
| | 60-40 solder | 60Sn 40Pb | 183–187 | 361–368 | For use where temperature requirements are critical |
| | Cd-Ag solder | 95Cd 5Ag | 337–390 | 639–734 | High-temperature solder |
| Brazing filler metals: | RBCuZn-A | 57Cu 42Zn 1Sn | 890–905 | 1630–1650 | General-purpose alloy for steel, copper alloys, and nickel alloys |
| | BCu-1a | 99+ Cu | 1083 | 1981 | Furnace brazing alloy |
| | BCuP-2 | 93Cu 7P | 710–795 | 1310–1460 | For self-fluxing brazing of copper alloys |
| | BAg-1 | 45Ag 24Cd 16Zn 15Cu | 610–620 | 1125–1145 | Low-melting, free-flowing braze for general-purpose work |
| | BAg-8 | 72Ag 28Cu | 780 | 1435 | For applications, such as vacuum tubes, where volatile elements are harmful |
| | BAlSi-2 | 91Al 8Si 1Fe | 580–615 | 1070–1135 | General-purpose alloy for brazing aluminum |
| | BMg-1 | 89Mg 9Al 2Zn | 445–600 | 830–1110 | General-purpose alloy for brazing magnesium |
| | BNi-1 | 73Ni 14Cr 3B 4Si bal. mostly Fe | 975–1040 | 1790–1900 | Heat-resistant alloy used for fabricating stainless steels or high-nickel alloys for jet engines |

eliminated only if the alloying elements diffuse from regions in which their concentration is high to regions in which their concentration is low.

The mathematical treatment of diffusion between adjacent regions of high and low alloy content is simplified if it is assumed that the alloy concentration varies sinusoidally with distance about the average value, so that

$$c = c_m \sin \frac{\pi x}{l}, \tag{8.60}$$

where $c$ is the variation from the average concentration at the point $x$, $c_m$ is the initial maximum variation from the average concentration, $x$ is the distance in meters, and $l$ is the distance between a region of maximum concentration and an adjacent region of minimum concentration. Using this expression for concentration, it can be shown that a solution of the second Fick law (Eq. 8.13) is

$$c = c_m \sin \frac{\pi x}{l} \exp\left(-\pi^2 \tilde{D} t / l^2\right). \tag{8.61}$$

In this equation, $c_m$ is a constant. Sin $\pi x/l$ represents the sinusoidal variation of concentration that was assumed in using Eq. (8.60). For the present purpose only a maximum value of this function need be considered; that is, sin $\pi x/l$ can be set equal to unity. Evidently, then, the factor that controls the decrease in degree of inhomogeneity with increasing time $t$ is $\exp\left(-\pi^2 \tilde{D} t / l^2\right)$, which decreases from unity toward zero with increasing time. Thus the time required to produce a given degree of homogenization increases with the square of the diffusion distance $l$ and is inversely proportional to the diffusion constant $\tilde{D}$ of the segregated alloying element. If banding due to manganese segregation occurs, it is extremely difficult to produce effective homogenization in large ingots because of the large diffusion distance involved. After the ingot has been rolled, the value of $l$ in Eq. (8.61) is reduced to about 0.03 mm, and diffusion is greatly accelerated. Even in this case, however, many hours of heating at an excessively high temperature, 1200°C (2200°F), are required to remove banding.

**Anelasticity and damping.** Metals and alloys generally transmit vibrational energy efficiently compared with other structural materials such as concrete, wood, or plastics. Under some circumstances, however, vibrational energy may be effectively absorbed by a metallic structure by the action of various mechanisms. In any event, if a metallic rod is set in vibration by hitting it with a hammer, the amplitude of this vibration decreases with time. This characteristic is called the *damping capacity.* One of the many proposed measures of this property is the *damping increment* Δ, defined as

$$\Delta = \frac{\Delta W}{W}, \tag{8.62}$$

where $W$ is the total vibrational energy per cycle, and $\Delta W$ is the amount of vibrational energy lost in the form of heat during one cycle.

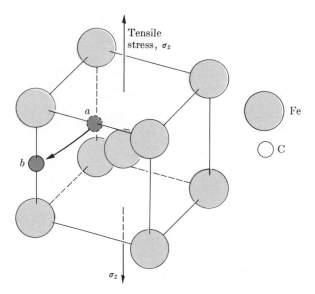

**Fig. 8.16** The atomic basis of internal friction for an alloy of carbon in BCC iron. At the stage shown in the vibration cycle, the carbon atom jumps preferentially to position *b* (in a direction of expansion).

Provided that the frequency of vibration is in a suitable range, a principal mechanism of damping may involve the jumping of individual atoms. This phenomenon, part of the larger topic of *internal friction*, may be used to measure diffusion coefficients under special conditions; for example, at unusually low temperatures. In the case of interstitial carbon atoms in BCC iron, Fig. 8.16, the interstitial positions cease to be equivalent in the presence of the vibrational stresses. If the *z*-axis is momentarily the direction of tension, then the interstitial sites along the *z*-axis are low-energy positions for the carbon atoms, which then tend to jump into these sites. A quarter-cycle later in the vibration cycle the *y*-axis may be in tension, and the carbon atoms then have a preference for sites along this axis. For a given applied stress, the excess of carbon atoms in preferred sites is limited to a certain maximum quantity, $\Delta n_{max}$. After any time *t*, the actual excess $\Delta n(t)$ approaches this limit exponentially,

$$\Delta n(t) = \Delta n_{max}[1 - \exp(-t/\tau)], \tag{8.63}$$

where $\tau$ is the experimentally determined relaxation time. The diffusion coefficient is related to $\tau$ through Eq. (8.4),

$$D = \frac{\Gamma \alpha^2}{6} = \frac{a_0^2}{18\tau} \tag{8.64}$$

since $\alpha = a_0/\sqrt{2}$ and $3\Gamma/2 = 1/\tau$. Values of *D* determined by internal-friction measurements are included in Fig. 8.4.

**TABLE 8.5**

Damping Capacities of Several Metals and Alloys at Low Stresses

| Material | Damping decrement $\Delta$ (room temp.) |
|---|---|
| Lead | $144 \times 10^{-4}$ |
| Tin | $108 \times 10^{-4}$ |
| Nickel | $93 \times 10^{-4}$ |
| Copper | $70 \times 10^{-4}$ |
| 95% Cu, 5% Sn | $14 \times 10^{-4}$ |
| 70% Cu, 30% Zn | $4 \times 10^{-4}$ |
| Iron | $40 \times 10^{-4}$ |
| 0.2–1.5% C steel, annealed | $10 \times 10^{-4}$ |
| 0.8% C steel, hardened | $40 \times 10^{-4}$ |
| 0.8% C steel, hardened and tempered | $20 \times 10^{-4}$ |
| Gray cast iron | $60 \times 10^{-4}$ |
| Zinc | $15 \times 10^{-4}$ |
| Molybdenum | $10 \times 10^{-4}$ |
| Magnesium | $4 \times 10^{-4}$ |
| Aluminum | $1 \times 10^{-4}$ |

Many mechanisms besides atomic diffusion are responsible for the absorption of vibrational energy. A major contribution is associated with the motion of dislocations under various conditions; for example, the bowing of pinned dislocations, the motion of dislocations at grain boundaries, and the special behavior at interfaces. In ferromagnetic materials (discussed in Chapter 12) energy is dissipated by the realignment of magnetic domains. Since each mechanism involves a different relaxation time and is often strongly dependent on stress, the damping capacity of even a particular alloy varies greatly with such factors as frequency, stress, and temperature.

With the exception of the bronzes used in bells, a high damping capacity is almost always desirable in industrial metals. Turbine blades, for example, are subjected to thousands of impulses per minute and will build up significantly greater stresses at critical resonant speeds if their damping capacity is not sufficient to dissipate energy quickly. Because of the large number of factors that influence damping capacity, it is difficult to list quantitative values that can be used directly in design calculations. A more feasible procedure is to list relative damping capacities that have been determined under similar conditions for a number of materials. Such a tabulation is given in Table 8.5 for small amplitudes of vibration, that is, for low stresses. Under these conditions nickel is one hundred times better than aluminum in damping out vibrations. The damping decrement increases rapidly with increasing stress; for example the $\Delta$ value for gray cast iron increases from 0.006 to 0.3 as the stress increases from a value near zero to about $7 \times 10^7$ $N/m^2$ ($10 \times 10^3$ lb/in$^2$).

## PROBLEMS

**1.** Predict a possible concentration curve that might exist after a second round of jumps in Fig. 8.1. Start with Fig. 8.1(b) and use the same argument that was employed in obtaining this figure from Fig. 8.1(a). It is convenient to use fractional parts of atoms to express the probabilities in this case.

**2.** a) Sketch a square array of matrix atoms, modeled after Fig. 8.1. Add a uniform concentration of small interstitial atoms (in the interstices or openings in the matrix lattice) in the central section, comparable to Fig. 8.1(a).

b) Explain why the interstitial atoms are able to diffuse.

c) Sketch a possible distribution of the atoms after each interstitial atom has jumped four times.

**3.** a) To make a rough check of Eq. (8.7), obtain "experimental" data on the jumping of an atom in three dimensions by rolling a die to determine the random direction of each jump ($\bullet$ = + x-direction,  $\vdots$ = − x-direction, ... $\vdots\vdots$ = − z-direction). Calculate the value of $R$ after $n$ jumps (that is, $n$ rolls of the die) for $n = 9$, 25, and 49. Plot your data and compare it with the theoretical equation.

b) How many rolls of the die would be necessary to reduce the statistical uncertainty to one percent?

**4.** The Laplacian of $c$, $\nabla^2 c$, is given in Eq. (8.14) for Cartesian coordinates. Using the relations $x = r \cos \theta, y = r \sin \theta, r = (x^2 + y^2)^{1/2}, z = z$, show that the Laplacian in cylindrical coordinates is

$$\nabla^2 c = \frac{1}{r}\left[ \frac{\partial}{\partial r}\left( r \frac{\partial c}{\partial r} \right) + \frac{\partial}{\partial \theta}\left( \frac{1}{r}\frac{\partial c}{\partial \theta} \right) + \frac{\partial}{\partial z}\left( r \frac{\partial c}{\partial z} \right) \right]. \tag{8.65}$$

**5.** a) If a concentration gradient of vacancies $dc_V/dx$ exists in a pure metal, show that a flux of vacancies $J_V$ must occur according to Eq. (8.3).

b) Show that the diffusion coefficient $D_V$ of the vacancies is larger than the self-diffusion coefficient, $D$, of the metal atoms, Eq. (8.15), by the factor $(fN_V)^{-1}$.

**6.** Use the value of $D_0$ given in Fig. 8.4 to determine the value of $\Delta S_m/k$ for the diffusion of carbon atoms in BCC iron. Show that the jump distance $\alpha$ is $a_0/2$ in this case, where $a_0$ is the lattice constant of BCC iron, Table 1.2.

**7.** Hydrogen diffuses through metals so rapidly that it is often difficult to keep this gas under pressure at high temperatures.

a) Set up the first Fick law for the general problem of hydrogen storage in a metallic container when hydrogen is diffusing through the wall of the container at steady-state conditions. [*Hint*: Assume that the container has a surface area $A$ and a wall thickness $b$, and let $D_H$

represent the diffusion coefficient of hydrogen in the metallic wall.  Use Eq. (7.10) for the variation with pressure of the solubility of a gas in a metal.]

b) Explain what each term in the equation means.

c) Suggest conditions for minimizing the loss of hydrogen by diffusion.

8. There is an advantage in carburizing steel parts at 870°C (1600°F) rather than at 927°C (1700°F), since a finer grain size can be obtained in the finished parts.

a) Calculate the diffusion coefficient of carbon in $\gamma$-iron at this temperature.  (Use the data of Table 8.1.)

b) What carburizing time gives the same results at 870°C that 10 hours gives at 927°C? (Neglect the change in carbon solubility.)

c) Using 0.3% carbon as a measure of the depth of carburizing, what fraction of the depth at 927°C is produced by 10 hours' carburizing at 870°C?  Use 1.2% carbon as the maximum solubility at 870°C, and neglect the effect of the lower limit of solubility of carbon in austenite.

9. Equation (8.48) can be used to make approximate calculations of the degree of *decarburization* that a steel will experience.  Use the data of Table 8.2 to plot the curve of carbon concentration versus distance that would exist at the surface of a 1.3% carbon steel that had been decarburized for 10 hours.  Assume that the carbon concentration at the surface is zero.

10. It is usually assumed that diffusion occurs by the grain-boundary mechanism in a region only about 5 Å wide.

a) Show that the ratio of the grams of metal transported across a given plane by grain-boundary diffusion to the grams of metal transported by volume diffusion is approximately $(10^{-9}/d)(D_{gb}/D_v)$, where $d$ is the average grain diameter in meters and $D_{gb}$ and $D_v$ are the coefficients of grain-boundary and volume diffusion, respectively.

b) Using the data of Table 8.1, calculate the ratio $D_{gb}/D_v$ for silver at 927°C and at 727°C.

c) For a grain size of $10^{-4}$ m, could the effect of grain-boundary diffusion be detected at either 927°C or at 727°C?  Assume that experimental error is $\pm 5\%$.

11. As an economy measure it might be suggested that lead be substituted for the usual lead-tin solders used in joining iron.  Would such substitution be practical?  Why?

12. Consider the problem of increasing the chemical homogeneity of a solid solution of zinc in copper.

a) Can inhomogeneity be completely removed by diffusion in a practical time?

b) Why?

c) Use the data of Table 8.1 to calculate the diffusion coefficient for zinc diffusing in copper at 815°C (1500°F).

d) If the maximum variation from the average zinc content is 5% zinc, and if the distance between a region of maximum zinc content and a region of minimum zinc content is 0.1 mm, use Eq. (8.61) to estimate the time necessary to decrease the maximum variation to 1% zinc.

e) Would cold-working a cast alloy before a high-temperature treatment increase or decrease the rate of homogenization?

f) Why?

# REFERENCES

Cottrell, A. H., *An Introduction to Metallurgy*, London: Edward Arnold, 1967.  Chapters 6, 12, and 19 give supplementary material on diffusion and energy of activation.

Lancaster, J. F., *Metallurgy of Welding, Brazing and Soldering*, London: Allen and Unwin, 1965.  Gives a detailed treatment of these subjects including heat flow, gas-metal reactions, and practical aspects of bonding of metals.

Shewmon, P. G., *Diffusion in Solids*, New York: McGraw-Hill, 1963.  Presents a thorough, mathematical treatment of diffusion in metals, alloys, and ionic crystals.

Wert, C. A. and R. M. Thomson, *Physics of Solids*, 2nd ed., New York: McGraw-Hill, 1970. Chapters 3 and 4 discuss imperfections in solids and their relation to various diffusion phenomena.

# CHAPTER 9

# PHASE TRANSFORMATIONS

During almost every step in the production and treatment of metallic alloys, phase transformations are involved. Liquid → solid transformations determine the character of cast alloys. The transformation, strained alloy → annealed alloy, occurs during hot-working and as a part of special recrystallization treatments. The hardening of steel and of aluminum alloys (although each is caused by a separate mechanism) are both connected with the change of an alloy from a metastable state to a relatively stable state. As a preparation for studying some of the basic features of this important but wide-ranging subject of phase transformations, it will be helpful to look briefly at the possible causes of metastability and at the various ways in which the metal or alloy can react in moving toward a stable condition.

## OVERALL VIEW OF PHASE TRANSFORMATIONS

For simplicity the following treatment is restricted to a single, initially uniform phase, but the essential results are applicable to such technically important alloys as highly alloyed steels in which second-phase particles are also present in the transforming phase. The initial phase may be a pure metal, but many of the reactions to be discussed can only occur in suitable liquid solutions or solid solutions. The distinction will usually be clear from the context. The result of a phase transformation is a phase (or a mixture of phases) that differs significantly from the initial phase. The product phase usually has a different crystal structure; this structure can be designated as the $\beta$ phase to distinguish it from the initial $\alpha$ phase. In many cases, however, the product "phase" differs from the initial phase only in chemical composition (as in spinodal decomposition), in amount of surface energy (sintering of metal powders), in amount of interfacial energy (grain growth) or in amount of strain energy (recrystallization of cold-worked alloys). This more general interpretation of a phase transformation permits a unified treatment of the kinetics of various processes in solid alloys.

**Sources of instability of a phase.** For processes occurring at constant temperature and pressure, the thermodynamic criterion of equilibrium is $\Delta G = G_\beta - G_\alpha = 0$. However, if the Gibbs free energy $G_\alpha$ of the initial phase is greater than $G_\beta$ (here $\beta$ represents the product phase(s) even if $\alpha$ and $\beta$ are identical crystallographically), then $\Delta G$ is negative and the $\alpha$ phase tends to react and form the $\beta$ phase. The rate of reaction depends on kinetic factors, of course, and may be so slow as to be unobservable.

Many influences may cause $G_\alpha$ to be greater than $G_\beta$, but the principal ones of interest in metallurgy are the following.

1. A change in temperature that moves the alloy (in the phase diagram) from the region where $\alpha$ is the stable phase to the region of stability of the $\beta$ phase. An example is the cooling of a liquid metal to a temperature below its melting point. (A change in pressure, or possibly in magnetic or electrical field, can have the same effect, but these variables are of less technical interest.)

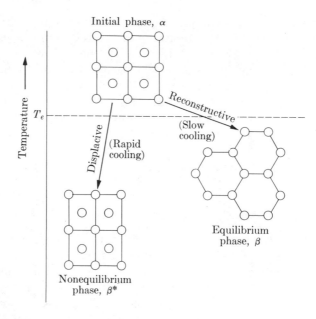

**Fig. 9.1** Schematic comparison of the changes in crystal structure accompanying a reconstructive transformation and a displacive transformation. $T_e$ is the temperature of the $\alpha \rightarrow \beta$ transformation under equilibrium conditions.

2. The presence of strain energy in the $\alpha$ phase; for example, as a result of cold-working. The metal or alloy reacts under this driving force by undergoing the processes of recovery and recrystallization described in a later section.

3. The presence of interfacial energy in the $\alpha$ phase; for example, as the result of a relatively small grain size. In this case the response is a process of grain growth.

4. The presence of extra surface energy because of the existence of the $\alpha$ phase in the form of many small particles; an example is a metal powder employed in powder-metallurgy. The response of the system is the process of sintering, described later, in which the particles agglomerate and eliminate most of the free surface.

As the quoted examples suggest, the ability of an alloy to move from a condition of instability to one of greater stability depends on its environment, especially on the temperature.

**Categories of phase transformations.** A given alloy, such as a steel, can move from a condition of instability by any one of several distinctly different phase transformations. This richness of the spectrum of phase transformations contributes substantially to the variety of properties that can be conferred on a given alloy by appropriate choice of thermal and mechanical treatments. A major distinction is  between *reconstructive and displacive* transformations, Fig. 9.1.

*Reconstructive Transformations.* In a reconstructive transformation a "rebuilding" (crystallographic and/or chemical) is required to convert the $\alpha$ phase into the $\beta$ phase. Such reconstruction can be accomplished only through the motion of *individual* atoms. This atomic motion may be only diffusion (as in spinodal decomposition) or may be primarily an interfacial reaction in which an atom moves across the $\alpha/\beta$ interface (the interface between the $\alpha$ and $\beta$ phases). In general, both diffusion and interfacial reaction must occur, but often one of the processes requires almost all of the free energy available for the transformation and therefore controls the overall rate of transformation. The atoms composing the $\alpha$ phase can be rearranged, not only in the form of the equilibrium $\beta$ phase, but also in the form of various *transition phases* ($\beta'$, $\beta''$, etc.) if the kinetics of such alternative transformations are more favorable. A subsequent reconstructive transformation can then transform $\beta'$, for example, into the equilibrium $\beta$ phase.

*Displacive (Martensitic) Transformations.* In contrast to the wide spectrum of reconstructive transformations of the $\alpha$ phase, the possibility of even *one* displacive transformation may not exist in a given alloy system. The conditions are extremely restrictive, since the crystal structure of the $\alpha$ phase must transform (without motion of individual atoms) into a crystal structure $\beta^*$ which has a lower energy. In the case of steel, for example, $\beta^*$ is almost identical crystallographically to the equilibrium $\beta$ phase but has a significantly different chemical composition. As discussed more fully in a later section, a displacive transformation occurs by the cooperative movements of a large number of neighboring atoms. A high driving force is required for this purpose. The usual source of driving force (that is, of free energy $\Delta G$ of transformation) in such cases is a lowering of the temperature of the alloy. Ordinarily, however, an alternative transformation will occur and prevent the driving force from becoming great enough to cause a displacive transformation. A typical exception occurs when an alloy is cooled very rapidly (for example, by quenching in water) so that the time available for individual atomic movements is greatly reduced. The $\alpha$ phase may then be retained almost unchanged to a low temperature; consequently its free energy may become sufficiently greater than that of the $\beta^*$ structure that the appropriate displacive transformation occurs. The $\beta^*$ (martensite) structure in steel has great hardness and can be further treated (tempered) to give a desirable combination of strength and toughness.

*Metastability.* Three degrees of stability are usefully distinguished. If $G_\alpha < G_\beta$ the $\alpha$ phase is *stable* relative to the $\beta$ phase; an example is the liquid ($\alpha$) phase relative to the solid ($\beta$) phase at temperatures above the melting point. If $G_\alpha > G_\beta$ but an energy barrier $\Delta G_+$ exists between the system in state I ($\alpha$ phase) and state II ($\beta$ phase), Fig. 9.2, then the $\alpha$ phase is *metastable* relative to the $\beta$ phase. An especially simple example of such an energy barrier was given in Fig. 8.3 for the diffusive jumping of an interstitial atom. If $G_\alpha > G_\beta$ but no energy barrier exists, the $\alpha$ phase is said to be *unstable*. In this hypothetical case, which is useful for theoretical arguments, the $\alpha$ phase would transform instantly to the $\beta$ phase. It follows that

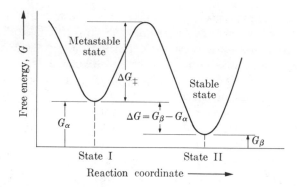

**Fig. 9.2**  Schematic illustration of the energy barrier $\Delta G_{\ddagger}$ that separates a metastable phase, $\alpha$, from any of its possible transformation products, $\beta$.

any phase available for technical use is either stable or metastable; the word instability refers to a metastable phase since unstable phases do not exist.

*Homogeneous Transformations.*  A homogeneous reaction is one that occurs within a single homogeneous (but not necessarily uniform) phase. The standard example is a chemical reaction in the gas phase. This type of transformation is rare in solid alloys, but one example is spinodal decomposition, described later in connection with Fig. 9.12. Since the "$\beta$" phase in this case is crystallographically identical with the metastable $\alpha$ phase, the $\alpha \rightarrow \beta$ transformation satisfies the definition of a homogeneous reaction. An essential feature is the *absence* of a sharp interface between the $\alpha$ and $\beta$ phases. The distinction between the phases is merely a difference in chemical composition, and this difference occurs as a gradual transition.

*Heterogeneous Transformations.*  The crucial feature in heterogeneous transformations is the presence of an interface. If the phases involved in a transformation have different crystal structures, a major portion of the transformation occurs at the interface between the phases. Even if the two phases have the same structure and differ only in amount of strain energy (as in the case of recrystallization), the major portion of the transformation occurs at the interface between the $\alpha$ and "$\beta$" phases. This interface represents a heterogeneity in the system from the viewpoint of kinetics; therefore, such a transformation should not be treated as being homogeneous.

Since at the start of a heterogeneous phase transformation only the (nominally homogeneous) $\alpha$ phase exists, a crucial aspect is the manner in which the $\beta$ phase first appears within the $\alpha$ phase and thus creates the interface at which transformation can continue. Extensive study of this *nucleation* phenomenon in heterogeneous phase transformations has shown that it is almost always catalyzed in technical processes. The word *catalyst* is used here in the broad sense of a heterogeneity within the $\alpha$ phase. In most cases tiny particles of impurity are the source of catalytic action, but lattice imperfections or concentrations of strain are believed to be effective for some types of transformations in solid phases.

The two stages in an overall heterogeneous phase transformation can be considered from two significantly different points of view. *Nucleation and growth* emphasizes the genesis of nuclei of the $\beta$ phase within the metastable $\alpha$ phase and the subsequent growth of the particles of $\beta$ phase. *Boundary migration*, on the other hand, focuses attention on the interface (boundary) between the phases and on factors, such as relative orientation of two solid phases, that can influence the rate of migration. The significance of the nucleation step in this view is the appearance of a mobile boundary. Each of these two viewpoints has advantages for certain purposes, as will be clear in later treatments.

## VAPOR → LIQUID AND LIQUID → SOLID TRANSFORMATIONS

Because kinetic theory gives a simple, satisfactory model of the vapor phase, the vapor → liquid transformation is well understood. This transformation is interesting in itself for phenomena such as vaporization of metals in vacuum, but it gains additional importance because it is the basis for the treatment of liquid → solid transformations. These transformations are in turn the key to understanding the *cast* structures produced by the solidification of liquid alloys.

**Vapor → liquid transformations.** The condition for thermodynamic stability of the vapor of a pure metal at any temperature $T$ is easily specified in terms of the pressure $P_0$ at which it is in equilibrium with the liquid metal. If the actual pressure $P$ is less than $P_0$, the vapor is stable and has no tendency to transform to the liquid phase. On the other hand, if $P$ is greater than $P_0$, the formation of unit volume of liquid is accompanied by a negative change in free energy,† $\Delta G_v$; therefore, transformation tends to occur. Even if a vapor has a sufficiently high pressure that the vapor → liquid reaction is possible thermodynamically, the initial formation (nucleation) of a drop of liquid is inhibited by a barrier, represented by the energy of its interface with the vapor. Let $\gamma$ be the energy needed to create one square meter of interface. Then the overall change in free energy, $\Delta G$, accompanying the formation of a spherical drop of liquid of radius $r$ is

$$\Delta G = \underbrace{4\pi r^2 \gamma}_{\substack{\text{(energy needed to} \\ \text{create interface)}}} + \underbrace{\tfrac{4}{3}\pi r^3 \, \Delta G_v.}_{\substack{\text{(energy released by volume} \\ \text{of condensing phase)}}} \tag{9.1}$$

The first term in this expression is always positive but the second term will be negative when $\Delta G_v$ is negative for the phase transformation being considered. Since the terms contain $r^2$ and $r^3$, respectively, they vary with increasing radius of

---

† The subscript $v$ indicates that $\Delta G_v$ is the change in chemical free energy accompanying the formation of *unit volume* of the product phase (the liquid phase in the present case). Tabulations of the free energy of a given substance generally give the value (relative to a reference state) for one gram molecular weight of substance.

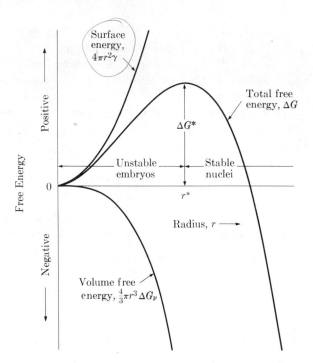

**Fig. 9.3**  Variation of the free energy of a drop of liquid as a function of its radius.

the drop in the manner shown in Fig. 9.3; the $r^3$ term is smaller initially but becomes greater for larger values of $r$.  Thus the sum $\Delta G$ goes through a maximum at some critical radius $r^*$, whose value (see Problem 5) is

$$r^* = -\frac{2\gamma}{\Delta G_v}. \tag{9.2}$$

The corresponding value of $\Delta G$ is

$$\Delta G^* = \frac{16\pi\gamma^3}{3(\Delta G_v)^2}. \tag{9.3}$$

The drops of liquid described by the curve of $\Delta G$ versus $r$ arise because of random collision of two atoms to form a pair, the addition of a third atom to the group in another collision, and so on through successive collisions.  Thus at any instant there will tend to be drops of the whole range of sizes, and it can be shown that in each cubic meter of vapor the number of drops of radius $r^*$ is

$$n^* = n \exp(-\Delta G^*/kT), \tag{9.4}$$

where $n$ is the number of atoms per cubic meter in the vapor phase and $k$ is Boltzmann's constant.  Drops of radius $r^*$ or greater differ from smaller drops in that the addition of one more atom to these larger drops is accompanied by a decrease in the free energy of the drop below the maximum value $\Delta G^*$. Thus a drop larger than $r^*$ will grow spontaneously, and for this reason it is called a *nucleus*. Drops smaller than $r^*$ will tend to decrease in size and can grow only if they are subjected to the proper combination of chance collisions. These drops are called *embryos*.

Since the number of critical nuclei $n^*$ is given by Eq. (9.4), the rate of formation of stable nuclei can be determined from the rate at which these nuclei are struck by an additional vapor atom. If $z$ is the number of vapor atoms striking one $m^2$ per second, then the rate of nucleation per cubic meter per second, $I_v$, is

$$I_v = z \times n^* 4\pi (r^*)^2, \tag{9.5}$$

since the second factor is the area of critical-nucleus surface available for collision. Using the value of $z$ obtained from kinetic theory (see Problem 6) and the value of $n^*$ from Eq. (9.4), we find that the rate of nucleation is

$$I_v = \frac{nP4\pi(r^*)^2}{\sqrt{2\pi MRT}} \exp\left(-\Delta G^*/kT\right), \tag{9.6}$$

where $P$ is the pressure of the vapor phase and $M$ is the kilogram molecular weight.

When appropriate values are used in Eq. (9.6), it is found that extremely large amounts of supercooling would be required to obtain a practical rate of nucleation of drops in the volume of the vapor. This is the process of *homogeneous nucleation*. In practice, before homogeneous nucleation has a chance to occur, drops of liquid form on foreign material that serves as a nucleation catalyst. Suitable materials for this purpose are the walls of the container or dust particles in the vapor. The mechanism of the catalytic action is similar to that described (in the following section) for catalyzed nucleation in the freezing of a liquid.

The formation of a solid phase directly from the vapor is similar to the formation of a liquid, but here there is a problem not only in promoting nucleation but also in causing growth to continue. Assuming that a stable crystallite has been produced by catalyzed nucleation, further growth occurs by the addition of atoms from the vapor to incomplete crystal planes. If the crystal were perfect this process would soon lead to the development of atomically smooth crystal surfaces on which an additional plane of atoms could form only with great difficulty. The presence of a dislocation intersecting the surface of the growing crystallite avoids this problem, since a "step" is then a permanent feature of the crystal face (Fig. 9.4a) and offers a site for the addition of atoms from the vapor phase (Fig. 9.4b). Figure 9.4 shows the special case of a pure screw dislocation. Evidence for the existence of dislocations in growing crystals is afforded by the spiral growth pattern (Fig. 9.4c) found in many instances.

Other transformations involving the vapor phase are the evaporation of liquid metals and the sublimation of solid metals. These processes are the basis for the commercial refining of many metals, including zinc, magnesium, arsenic, and antimony. Excessive volatilization of alloying elements can sometimes cause

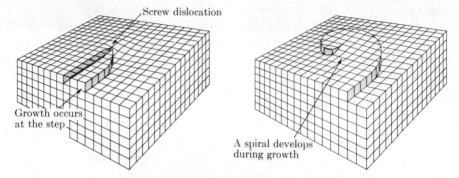

(a) and (b) A screw dislocation produces a "step" in the crystal surface, at which additional atoms can attach themselves. The resultant growth causes the step to assume a spiral form.

(c) Growth spiral in a single crystal of paraffin. The step is one molecule high.

**Fig. 9.4** Dislocations in the growth of crystals. (Electron micrograph courtesy H. F. Kay, University of Bristol.)

difficulty during melting in vacuum or even during vacuum annealing of solid metals. The loss of zinc from brass and the loss of manganese from sheet steel are examples of the latter problem. It is possible to predict the rate of vaporization of a metal into a vacuum by means of the following equation, derived from kinetic theory (see Problem 7):

$$G = \frac{P}{17.14\sqrt{T/M}},$$

(9.7)

where $G$ is the kilograms of metal vaporized per square meter per second, $P$ is the vapor pressure in $N/m^2$, $T$ is the temperature in K, and $M$ is the kilogram atomic weight of the metal. Since the vapor pressure is the principal factor determining the rate of vaporization, a useful reference point is provided by the fact that usually a vapor pressure of about $3 \times 10^2 \, N/m^2$ must be attained to make distillation commercially feasible.

There are two common circumstances that require modification of Eq. (9.7). If the metal being vaporized is in an inert atmosphere rather than in a vacuum, the rate of evaporation is reduced in direct proportion to the pressure of the inert gas. The magnitude of this effect is illustrated by the fact that the rate of evaporation of magnesium at its melting point is reduced by a factor of 100 by the presence of $10^2 \, N/m^2$ pressure of argon. Equation (9.7) must also be modified if it is to be applied to a component of an alloy rather than to a pure metal. As a rough approximation it can be assumed that Raoult's law is valid; namely, that the vapor pressure of a metal in solution is given by its mole fraction multiplied by its vapor pressure in the pure state, $P^0$ (Table 6.1). By definition, an ideal solution obeys Raoult's law. However, if data on the thermodynamic activities $a$ of a nonideal solution are available, the correct vapor pressure can be calculated as

$$P = aP^0,$$

(9.8)

provided that the activities are referred to the pure metal as the standard state. For example, the activity of zinc in brass containing 0.3 mole fraction zinc is only 0.07 at 750°C (1380°F); hence in this case the true vapor pressure is significantly lower than the value given by the Raoult's law approximation.

**Liquid → solid transformations.** As an introduction to this important class of phase transformations, we consider a quantitative example of the thermodynamic driving force in question, using 1 $m^3$ of liquid zinc as the system to be analyzed. At the melting point of zinc, $T_f$, the liquid phase and the solid phase are in equilibrium and there is no tendency for the transformation to occur; that is,*

$$\Delta G_v = \Delta H_v - T_f \, \Delta S_v = 0,$$

$$\Delta S_v = \frac{\Delta H_v}{T_f}.$$

(9.9)

---

* A subscript $v$ is used on the quantities $\Delta G_v$, $\Delta H_v$, etc., in the following treatment as a reminder that they refer to 1 $m^3$ (unit volume) of material.

If it is assumed that $\Delta H_v$ and $\Delta S_v$ do not vary with temperature, then the free-energy change accompanying the solidification of liquid zinc at any temperature $T$ can be determined as

$$\Delta G_v = \Delta H_v - T\,\Delta S_v$$
$$= \Delta H_v \left( \frac{T_f - T}{T_f} \right), \tag{9.10}$$

where the value of $\Delta S_v$ given by Eq. (9.9) has been used. Since $7 \times 10^8$ joules of heat are given off during the solidification of 1 m$^3$ of zinc, $\Delta H_v = -7 \times 10^8$ and $\Delta G_v$ is negative (that is, the solidification reaction is possible) provided the temperature $T$ is less than the melting point $T_f$.

It is well known that a reaction such as solidification does not occur at temperatures only infinitesimally below the equilibrium temperature; rather, an appreciable undercooling is required before a practical rate of reaction is produced. This behavior is a result of the mechanism by which phase transformations occur, and the *amount* of undercooling distinguishes between homogeneous nucleation and catalyzed nucleation.

Homogeneous nucleation in a liquid can be analyzed in a manner analogous to that employed above for the vapor phase. However, in this case the equation corresponding to Eq. (9.5) is

$$I_l = (n^* m^*)v, \tag{9.11}$$

where $m^*$ is the number of atoms in the liquid phase that are neighbors of a given critical nucleus; therefore the quantity in parentheses is the total number of atoms that have an opportunity to combine with the number $n^*$ of critical nuclei per cubic meter. The frequency $v$ with which a typical liquid atom will cross the interface and become part of the solid nucleus can be determined by transition-state theory. The derivation is given here as a simple example of the use of this important tool for kinetic analyses.

For use with the sketch of Fig. 9.2, the passage of an atom $A$ from the liquid phase into the critical (solid) nucleus is written as the chemical reaction,

$$A_l \rightarrow A_{\ddagger} \rightarrow A_s. \tag{9.12}$$

A basic assumption of the theory is that the atoms $A_{\ddagger}$ in the transition state are in equilibrium with the reactants (the atoms in the liquid, $A_l$). Consequently, the equilibrium constant $K$ can be written in the usual manner,

$$K = \frac{a_{A_{\ddagger}}}{a_{A_l}} = \exp\left( -\Delta G_{\ddagger}/kT \right). \tag{9.13}$$

Here $a_A$ is the thermodynamic activity of chemical species $A$; $\Delta G_{\ddagger}$ is the free energy of activation and is evaluated for the condition that $A_l$ and $A_{\ddagger}$ are in their standard chemical states. For a pure substance $a = 1$, so for the liquid phase of a pure metal $a_{A_l} = 1$. In view of the definition of activity, Eq. (6.32),

$$a_{A_{\ddagger}} = \gamma_{\ddagger} N_{A_{\ddagger}}. \tag{9.14}$$

With these substitutions, Eq. (9.13) becomes

$$N_{A_{\ddagger}} = 1/\gamma_{\ddagger}\exp\left(-\Delta G_{\ddagger}/kT\right), \tag{9.15}$$

thus giving an expression for the concentration of atoms in the transition state. The theory now assumes that these "activated" atoms undergo reaction to the final product, $A_s$, at the universal rate $kT/h$, where $k$ is Boltzmann's constant and $h$ is Planck's constant. Therefore, the rate of reaction in units of atoms/(m³·s), is given by

$$\text{rate} = \frac{kT}{h}\frac{N}{\gamma_{\ddagger}V}\exp\left(-\Delta G_{\ddagger}/kT\right). \tag{9.16}$$

Here $N$ is Avogadro's number and $V$ is the molar volume [m³/(kg mol)] of the liquid. Division of the rate of reaction by the total number of atoms per cubic meter, $N/V$, then gives the rate (or frequency $v$) of reaction of a typical atom,

$$v = \frac{kT}{h\gamma_{\ddagger}}\exp\left(-\Delta G_{\ddagger}/kT\right). \tag{9.17}$$

Substitution of this expression in Eq. (9.11) gives the final equation for the rate of homogeneous nucleation,

$$I_l = \frac{nm^*kT}{h\gamma_{\ddagger}}\exp\left[-(\Delta G^* + \Delta G_{\ddagger})/kT\right]. \tag{9.18}$$

Here $n$ is the number of atoms per cubic meter in the liquid phase and $\Delta G^*$ enters through Eq. (9.4).

When calculations are made using Eq. (9.18), a number of simplifying assumptions are usually adopted (see Problem 9). These include: grouping of the unknown factor $\gamma_{\ddagger}$ with $m^*$ and arbitrarily assigning the value $m^*/\gamma_{\ddagger} = 1$: neglecting $\Delta G_{\ddagger}$ with respect to $\Delta G^*$; and assigning an arbitrary value, such as $I_l = 1$ nucleus/(m³·s), as the rate for which solidification can be observed experimentally. Calculations of this type demonstrate that a metal must be undercooled approximately 100° below its melting point before homogeneous nucleation will occur. Although this type of nucleation can be produced in the laboratory, it is clear that the solidification of most metals does not occur by this mechanism, since undercooling is ordinarily only several degrees. As in the case of the vapor phase, catalyzed nucleation in a liquid occurs in practice either on a nucleating substance floating in the liquid or on the walls of the container. The following discussion shows that differences in interfacial energies account for the relative ease of catalyzed nucleation compared with homogeneous nucleation.

If the embryos of the solid phase are assumed to form as spherical caps on a planar nucleating surface in a manner analogous to that shown in Fig. 6.20, then the equilibrium among the interfacial energies is given by the analog of Eq. (6.62), namely

$$\gamma_{SL} = \gamma_{\alpha S} + \gamma \cos\theta. \tag{9.19}$$

Here $S$ refers to the nucleating surface and $\alpha$ refers to the solid phase being formed;

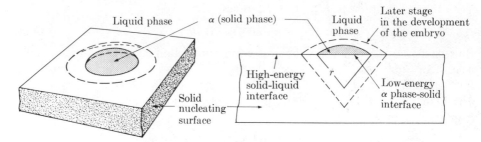

**Fig. 9.5** Two stages in the development of an embryo of the $\alpha$ (solid phase) from the liquid phase. (a) Sketch of a spherical cap of the $\alpha$ phase and a later stage in its development. (b) A cross section through (a), showing that high-energy interface is replaced by low-energy interface during development of the embryo.

$\gamma$ is equivalent to $\gamma_{\alpha L}$. There is now another free-energy term that enters into Eq. (9.1), besides $\gamma$ and $\Delta G_v$. As the embryo develops and covers a larger area on the nucleating surface, Fig. 9.5, the high-energy $S$-$L$ interface is replaced by the $\alpha$-$S$ interface. Per square meter of growth of the $\alpha$-$S$ interface, there is a change in free energy equal to the difference in the two interfacial energies: $(\gamma_{\alpha S} - \gamma_{SL})$. From Eq. (9.19) this free-energy change can be evaluated as

$$\gamma_{\alpha S} - \gamma_{SL} = -\gamma \cos \theta. \tag{9.20}$$

When this additional energy term is taken into account Eq. (9.1) has the form

$$\Delta G_s = \left(\pi r^2 \gamma + \frac{\pi r^3 \Delta G_v}{3}\right)(2 - 3\cos\theta + \cos^3\theta) \tag{9.21}$$

and the critical values of $r$ and $\Delta G_s$ are

$$r_s^* = -\frac{2\gamma}{\Delta G_v}\sin\theta = r^* \sin\theta, \tag{9.22}$$

$$\Delta G_s^* = \Delta G^* \frac{(2 + \cos\theta)(1 - \cos\theta)^2}{4}, \tag{9.23}$$

where $r^*$ and $\Delta G^*$ are the values for homogeneous nucleation given by Eqs. (9.2) and (9.3). The equation for the rate of catalyzed nucleation is obtained by an argument analogous to that used in connection with Eq. (9.18) for homogeneous nucleation:

$$I_s = n_s \frac{kT}{h}\exp\left[-(\Delta G_s^* + \Delta G_{\ddagger})/kT\right]. \tag{9.24}$$

Here $n_s$ is the number of atoms of the liquid adsorbed on 1 m$^2$ of the solid nucleating surface. It can usually be assumed that a monolayer of liquid atoms are adsorbed on this surface. Problem 10 illustrates how heterogeneous nucleation can explain the small amounts of undercooling that occur in the actual solidification of metals.

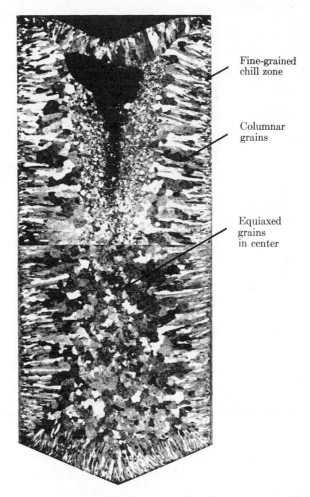

Fine-grained
chill zone

Columnar
grains

Equiaxed
grains
in center

**Fig. 9.6** A typical ingot structure; a macroetched cross section through an iron-silicon alloy ingot. (Courtesy *Metallography* by Cecil H. Desch, Longmans, Green and Co., Ltd.)

**Growth of crystals into a liquid phase.** Once the nuclei of the solid phase have formed, they grow by consuming the liquid phase and so produce the final solidified alloy. The resulting structure may have a variety of forms, depending on the conditions that exist during growth of the crystals. For example, an ingot cast in a metallic mold usually has the structure shown in Fig. 9.6, consisting of a fine-grained chill zone at the surface, a region of columnar grains, and a central zone of equiaxed grains. The explanation of this diverse behavior is to be found in three factors: (1) crystallographic features of growth, (2) temperature distribution during freezing, and (3) solute redistribution between the liquid and solid phases during the freezing

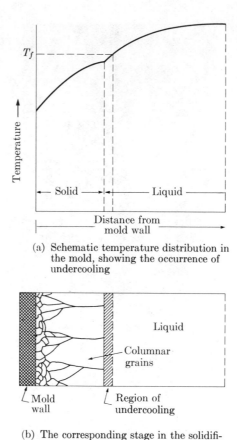

(a) Schematic temperature distribution in the mold, showing the occurrence of undercooling

(b) The corresponding stage in the solidification of the liquid metal

**Fig. 9.7** Columnar grain growth during the freezing of a pure metal whose melting point is $T_f$.

of an alloy. The solidification of a pure metal will be discussed first, and then the additional effect of alloying will be considered.

When a liquid metal is placed in a cold metallic mold, the liquid near the surface of the mold is quickly cooled below the equilibrium melting temperature $T_f$ and stable nuclei are formed both on the mold wall and in the body of the liquid near the wall. These nuclei grow very rapidly into grains of roughly spherical shape, liberating the large latent heat of fusion. This heat plus the superheat of the liquid metal must be dissipated through the mold wall and the layer of solidified metal. Therefore, the rate of growth of the crystals soon becomes limited by the rate of heat removal rather than by the inherent growth velocity (the latter is extremely large). The temperature distribution in the solidifying metal at this stage is shown in Fig. 9.7(a). The small amount of undercooling in advance of the growing grains does not permit nucleation of new grains, and many of the grains in the chill zone give

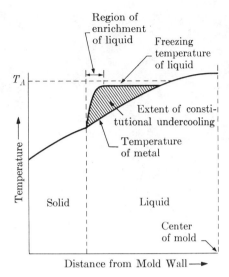

Fig. 9.8  Constitutional undercooling in a liquid alloy having an original freezing temperature $T_A$.

way to more favorably oriented grains. These two factors give rise to the columnar structure shown in Fig. 9.7(b). The columnar grains tend to have a characteristic crystal direction as their axis, and they advance uniformly into the liquid. In pure metals the type of solidification shown in Fig. 9.7 continues to the center of the mold.

*Constitutional Undercooling.* In the freezing of many alloys an entirely different pattern of undercooling is produced, even though the temperature distribution is similar to that for pure metals. The additional cause of undercooling in this case is the rejection of solute by the freezing solid and the consequent enrichment of the liquid near the growing interface. In Chapter 7 it was shown that the enriched liquid then has a lower freezing temperature (for the usual case that $k < 1$). Therefore, if the growth of the solid is to continue, the temperature of the liquid in contact with it must be low enough to maintain the necessary amount of undercooling, Fig. 9.8. Since the freezing temperature of the liquid rises with increasing distance from the surface of the solid, it is possible for the amount of undercooling to increase for a distance into the liquid before decreasing to zero. This behavior is called *constitutional undercooling*, since it involves a change in the constitution (composition) of the liquid.

The initial stage of solidification of an alloy is similar to that for a pure metal (Fig. 9.7b). During the period of columnar growth, however, the solid-liquid interface is not planar but tends to develop protuberances, Fig. 9.9. This behavior results because the constitutionally undercooled liquid allows portions of each grain to grow rapidly. In the course of this growth a new solute distribution is produced that

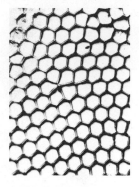
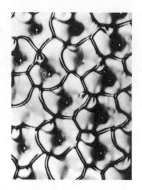
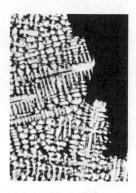

(a) Cellular structure in a
lead-0.05% tin alloy
(×100)

(b) Dendritic structure in
a lead-0.75% tin alloy
(×50)

(c) Dendrites that grew in
casting of monel metal
(×25)

**Fig. 9.9**  Examples of nonuniform growth fronts in the solidification of metals. (Photographs (a) and (b) courtesy W. A. Tiller, Stanford University; (c) courtesy C. W. Mason, Cornell University.)

discourages similar rapid growth by neighboring portions of the grain.  Because of the symmetry of the diffusion process which controls the new solute distribution, the protuberances have a uniform spacing.  If the amount of undercooling is small, the protuberances are only mounds on the surface of the growing crystal and are revealed as a "cell" structure (Fig. 9.9a) if the liquid is removed.  If the thickness of undercooled liquid is large, then the protuberances can grow well in advance of the completely solidified portions, and a *dendritic* pattern of growth is obtained, Fig. 9.9(b).  Under certain conditions of growth a dendrite consists not only of the main axis of growth but also of secondary and higher-order axes that branch off the existing axes.  A well-developed dendrite is shown in Fig. 9.9(c).

Constitutional undercooling also explains the occurrence of the central, equiaxed grains in Fig. 9.6.  The reasoning on which Fig. 9.8 is based leads to the prediction that eventually all the liquid remaining in the central portion of the mold will be undercooled.  The degree of undercooling is often great enough to nucleate new grains within the liquid, grains that then grow into the undercooled liquid by the dendritic mode of solidification.

*Structure of Cast Alloys.*  Depending on the requirements placed on a commercial casting (including the factor of competitive cost) it may have various structural features, including columnar grains, dendrites, and segregation of alloying elements.  On the other hand, the best combination of properties (including strength, toughness, stability, and directional uniformity) is usually obtained when the microstructure consists of fine, uniform, equiaxed grains of the various phases in an alloy.  Several means can be employed to promote the formation of a desirable cast structure, but the essential goal is the following: *to cause uniform nucleation throughout the solidifying alloy,* rather than to permit long-range growth in the form either of columnar grains or of dendrites.

Of the numerous methods employed to promote uniform nucleation, the following are both effective and relatively economical. A *nucleating agent* can be added to the liquid alloy just before pouring the casting and can serve as a catalyst for nucleation. An example is the addition of metallic titanium to molten alloys of aluminum. Within the molten bath the titanium reacts with dissolved carbon to form tiny particles of titanium carbide; these particles are effective nucleation catalysts and help produce a much finer grain size. A second method is the use of an alloying element that encourages the development of constitutional undercooling; the presence of copper in aluminum alloys has this effect. As explained above, constitutional undercooling encourages the nucleation of new grains in front of a planar solid/liquid interface, but it also permits dendritic growth. The more important of these two aspects may be dendritic growth, with the additional requirement that the long, dendritic filaments be broken apart within the body of the liquid. The resulting fragments then serve as nuclei for the growth of new grains, thus preventing full development of the initial dendrite. Breaking of the dendrites tends to occur because of convective currents in the liquid metal, but the process can be accelerated by such means as mechanical agitation of the mold which holds the solidifying casting, ultrasonic vibrations induced in the liquid, or magnetic stirring of the liquid.

Other factors that must be adequately controlled during a casting operation are the evolution of gases, the occurrence of segregation, and the amount of shrinkage. As Fig. 7.35 shows for hydrogen in iron or nickel, many gases have a high solubility in liquid alloys but the solubility decreases to a relatively low value in solid alloys. Consequently, during solidification gases tend to come out of solution and to form pockets of gas called blowholes or pinholes. This condition in a casting is termed *porosity.* Even if such gross defects are avoided, in the case of hydrogen the gas may come out of solution at a later stage of processing of the solid alloy and cause thin cracks (hairline cracks) or indirectly lead to premature fracture. An effective method for the removal of dissolved gases, including hydrogen, oxygen, and nitrogen, is to expose the liquid alloy to a vacuum in one of the commercial methods of *vacuum degassing.*

Our study of phase diagrams has shown that segregation of alloying elements (usually enriching the liquid phase) is an inevitable part of the process of solidification. Additional segregation caused by the action of gravity can occur if the solid phase has a density substantially different from the liquid. For example, solid grains of antimony forming in a lead-rich alloy tend to float to the surface of the heavy liquid. Casting techniques that result in fine grain size are usually effective in restricting segregation to microscopic dimensions. Such segregation is less harmful; also, it can more easily be diminished by a homogenization heat treatment as described in Chapter 8. *Shrinkage* occurs for two reasons during the production of a casting. First, the volume occupied by the solid alloy is about 5%

less than the liquid, even at almost the same temperature. Second, during cooling the solid alloy continues to contract at a rate of about 1% per 1000°C. These large amounts of shrinkage can create cavities within a casting and can cause fracture (hot tearing) or distortion. Shrinkage cavities are commonly avoided by arranging for a supply of liquid metal (via feeding heads or *risers*) until the main portion of the casting has solidified.

## TRANSFORMATIONS IN THE SOLID STATE

Undoubtedly the most important advantages of alloys depend on phase transformations that occur within the solid alloy. Some examples are: Steel in a *soft* condition is machined to the form of a complex die, and the die is then *hardened* by martensitic transformation; an intricate gear is mass produced directly from iron powder by simply forming the powder in a die and then sintering; alloy sheet for a transformer is prepared with the grains precisely oriented as a result of recrystallization of the cold-rolled sheet. The theory developed for vapors and liquids provides the basis for understanding such solid-state transformations, but several new features must also be considered.

**Special features of solid-state transformations.** The crystalline structure of solid phases influences their phase transformations in several ways. (1) The atoms are more firmly bound than in the liquid phase and therefore diffuse more slowly by a factor of about $10^{-5}$ even at temperatures near the melting point. (2) Often alternative crystalline structures (transition lattices) can form, permitting transformation via one or more metastable phases. (3) The behavior of an initial phase relative to a product phase may be largely determined by crystallographic matching at their mutual interface; for example, a transition lattice may form preferentially because of the possibility of an energetically favorable orientation relationship relative to the initial phase. (4) Nucleation catalysts can include, in addition to foreign particles, various structural catalysts such as point, linear, or planar defects or regions of high local strain. (5) The total energy now includes strain energy, which results either from overall change in volume (if the product phase occupies a volume greater or less than that of the initial phase) or from local incompatibility of atomic bonding across an interface. (6) Growth of nuclei may be limited either by a process at the growing interface (as in liquid $\rightarrow$ solid transformations) or by a process of diffusion within the body of the phases.

In view of the many features that influence technically important processes of phase transformation in typical, complex commercial alloys, an empirical method employing an Avrami equation is used to describe overall transformations. Examples of these various features will be illustrated numerous times in the following treatment of solid-state transformations. It will be helpful, however, to begin with brief explanations of the role of strain energy and the character of an Avrami equation.

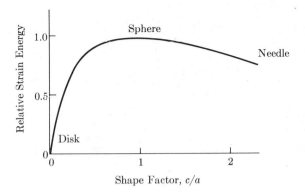

**Fig. 9.10**   Dependence of relative strain energy on the shape of ellipsoidal particles of second phase that have an incoherent interface with the surrounding matrix; $a$ is the equatorial diameter and $c$ is the polar axis of the ellipsoid.   (After F. R. N. Nabarro)

*Strain Energy*   The change in free energy $\Delta G$ accompanying homogeneous nucleation of spherical particles, Eq. (9.1), can be modified as follows to take account of two factors related to strain energy:

$$\Delta G = V(\Delta G_v + \Delta G_\varepsilon) + fV^{2/3}\gamma. \tag{9.25}$$

$V$ is the volume of the particle and $\Delta G_\varepsilon$ is the total strain energy (per unit volume of particle) in both the matrix phase and in the particle.   $\Delta G_\varepsilon$ is positive, whereas $\Delta G_v$ is negative.   If the matrix/particle interface is considered to be an ordinary, incoherent boundary, then the value of $\Delta G_\varepsilon$ varies with the shape of the particle, Fig. 9.10, where the shape is defined by the ratio $c/a$ of the semiaxes $(a, a, c)$ of an ellipsoid.   A low value of $\Delta G_\varepsilon$ accompanies a disc-shaped particle, but the amount of interfacial area is then larger than that for the same volume of a spherical shape. The relative amount of boundary is determined by the shape factor, $f$, which is large for discs or needles.   For an incoherent interface the interfacial energy is large (on the order of 1 J/m$^2$) and is relatively insensitive to changes in orientation or in composition.   Therefore, if the second-phase particles have an incoherent interface, the advantage of low strain energy for a disc shape is offset by the high total surface energy.

Often the term *coherent interface* includes both fully coherent interfaces, Fig. 9.11(a), as well as semi-coherent interfaces, Fig. 9.11(b).   In contrast, an incoherent interface has virtually no atomic matching between the phases, Fig. 9.11(c).   A description of atomic matching must include both the *planes* (hkl) across which matching exists and the *directions* [hkl] in the matching planes along which the atoms are in registry.   A simple example of a fully coherent interface would be the close-packed plane, (111) in the FCC lattice matched with the (0001)

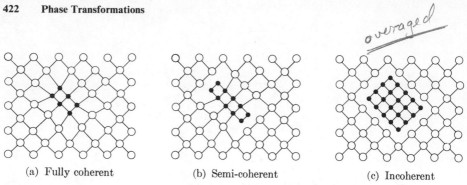

*overaged*

(a) Fully coherent          (b) Semi-coherent          (c) Incoherent

**Fig. 9.11** Schematic comparison of the nature of a coherent (or semi-coherent) interface and of an incoherent interface.

in the HCP lattice (see Fig. 1.11), if the atomic spacings were identical in the two phases:

$$(0001)_{HCP}\|(111)_{FCC} \qquad [2\bar{1}\bar{1}0]_{HCP}\|[\bar{1}10]_{FCC}. \qquad (9.26)$$

A similar orientation relationship applies to a semi-coherent boundary except that additional information is needed on the density (number per square meter of interface) of interfacial dislocations. An estimate of this density can be obtained from the mismatch, $\delta$, in interatomic spacing,

$$\delta = \frac{a_\alpha - a_\beta}{a_\alpha}, \qquad (9.27)$$

where $a_\alpha$ and $a_\beta$ are the interatomic distances in the initial ($\alpha$) and product ($\beta$) phases in a particular direction of interest. Because much elastic energy is required to overcome the effects of mismatch over large distances, an interface tends to have a structure in which periodic interfacial dislocations compensate for the mismatch $\delta$. For example, if $\delta = 0.05$, then a dislocation tends to occur at intervals of approximately 20 atom positions, thus minimizing the elastic energy. The energy of a fully coherent interface is only about 0.1 J/m². That of a semi-coherent interface gradually increases above this value (with an accompanying increase in the density of interfacial dislocations) but seldom exceeds about 0.5 J/m², half the energy of an incoherent interface.

When Eq. (9.25) is interpreted for coherent particles, the following picture is obtained. The critical size (volume, $V$) of embryo can be obtained at the cost of little strain energy, $\Delta G_\varepsilon$, by the formation of thin discs. Even though the shape factor, $f$, is unfavorable, the total interfacial energy is low because of the low value of $\gamma$ for coherent phases. Depending on the size of the critical embryo, the initial nuclei may be fully coherent or may contain interfacial dislocations and thus be semi-coherent. In typical hardenable alloys, a product phase (transition lattice) exists which possesses a favorable orientation relation with the initial phase but has a higher free energy than the *equilibrium* phase. Consequently, the alloy has a tendency to continue the process of phase transformation until the equilibrium

**TABLE 9.1**

Values of $n$ for Use in an Avrami Equation*

| Type of Transformation | | Value of $n$ |
|---|---|---|
| Cellular | Constant rate of nucleation | 4 |
| | Zero rate of nucleation | 3 |
| | Nucleation at grain edges | 2 |
| | Nucleation at grain boundaries | 1 |
| Precipitation | Particles growing from small dimensions: | |
| | 1. Constant rate of nucleation | 2.5 |
| | 2. Zero rate of nucleation | 1.5 |
| | Thickening of needles | 1 |
| | Thickening of plates | 0.5 |

* Adapted from J. W. Christian, *The Theory of Transformations in Metals and Alloys*, New York: Pergamon Press, 1965.

condition is reached. If, after suitable heat treatment, the alloy is used at room temperature, the rate of such additional reaction is negligibly slow. In the case of alloys employed in service at high temperatures, however, the effect of reactions toward the equilibrium condition may be significant.

*Empirical Equations.* In view of the many interacting factors that determine the manner in which a typical phase transformation proceeds, a complete analysis of the kinetics of the reaction is virtually impossible. Fortunately, a single equation is able to describe for technical purposes the fraction, $y$, of completion of an overall transformation at constant temperature. This *Avrami* equation is conveniently written,

$$y = 1 - \exp(-kt)^n. \tag{9.28}$$

The rate coefficient $k$ depends on the temperature (and of course on the pertinent properties of the initial phase; composition, grain size, etc.). The coefficient $n$ has the values shown in Table 9.1 for various types and conditions of transformation.

The nature of the two types of transformation listed in Table 9.1 can be described, using the diagrams of Fig. 9.12. The essential characteristic of a *cellular transformation* is the crystallographic reconstruction of the *entire* initial phase. An especially clear example of such a transformation is a eutectoid decomposition, Fig. 9.12(a), in which the initial $\gamma$ phase completely disappears and is replaced by a mixture of the $\alpha$ and $\beta$ phases. In the course of this transformation, a closely spaced array of $\alpha$ and $\beta$ plates* grows in the form of a cell or nodule into the $\gamma$

* The plate configuration is typical, but many other configurations of the two phases occur. An important one consists of rods of one phase embedded in the second phase, Fig. 10.9(b).

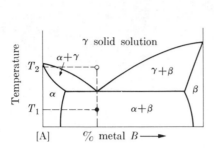

(a) The γ phase in a eutectoid diagram must transform by a cellular reaction.

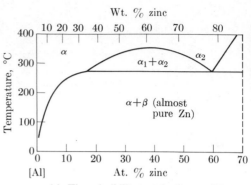

(c) The miscibility gap in the α solid solution indicates the presence of a spinodal curve as explained in (d).

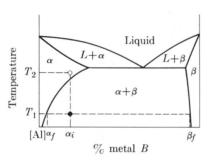

(b) The formation of particles of β phase within the α matrix phase is an example of a precipitation reaction.

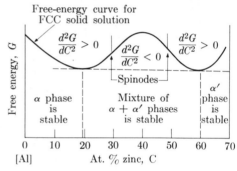

(d) Schematic curve of free energy versus composition at 300°C in the Al-Zn system. In the range of composition between the two spinodes, spinodal decomposition is possible.

**Fig. 9.12** Diagrams useful for describing the character of three principal types of reconstructive transformations.

phase, described later in connection with Fig. 9.16. The majority of phase transformations are cellular, the reconstruction of the initial phase occurring at the interface with the product phase or phases. Usually the rate of transformation is determined by the rate of the interfacial reaction, but in some cases diffusional processes are rate controlling. The other major type of transformation is a *precipitation reaction*, Fig. 9.12(b), characterized by the fact that a portion of the original phase (α) remains at the end of the reaction,

$$\alpha_i(\text{initial}) \rightarrow \beta + \alpha_f(\text{final}). \tag{9.29}$$

The transformation of $\alpha_i$ into $\alpha_f$ involves reconstruction compositionally but *not* crystallographically. For example, in Fig. 9.12(b) the composition of the α phase changes according to Eq. (9.29) as the equilibrium amount of $\beta$ phase precipitates. This change in composition occurs by diffusion over the relatively long distance separating an average region of $\alpha$ phase from the most accessible $\alpha/\beta$ interface (at which crystallographic reconstruction of some of the $\alpha$ phase into the $\beta$ phase occurs). Consequently, the rate of a precipitation reaction is almost always controlled by the rate of diffusion, either volume diffusion or some type of short-circuit diffusion. The overall reaction described by Fig. 9.12(b) and Eq. (9.29) can also occur by a cellular transformation in the manner described later in connection with Fig. 9.16. For precipitation hardening of technical alloys (see Chapter 10), special means must sometimes be employed to minimize the cellular mode of transformation to obtain full advantage of the precipitation reaction.

Two additional types of reconstructive transformations are of less technical importance at present. In a *massive transformation* the $\alpha$ phase is reconstructed crystallographically to the $\beta$ phase without change in composition. A simple example is the change of pure $\gamma$ iron (FCC) to $\alpha$ iron (BCC) when the temperature is decreased below the transition temperature, 910°C. The atoms need only move individually across the phase interface, and therefore the rate of transformation is high. In some alloy systems one can find a pair of points $\alpha(T_2, C)$, $\beta(T_1, C)$, such that the two phases have the same composition $C$ at two different temperatures. If the $\alpha$ phase can be cooled from $T_2$ to $T_1$ sufficiently rapidly that no alternative transformation has opportunity to occur, then the conditions for a massive transformation from $\alpha$ to $\beta$ are present. *Spinodal decomposition* is interesting scientifically because it is an example of a homogeneous reaction; that is, it involves only one phase (from the crystallographic point of view). The essential characteristic of this transformation involves the special shape of the free-energy curve, Fig. 9.12(d), that describes the relation between two phases having the same crystal structure but different compositions. An example of two such phases are $\alpha$ and $\alpha'$ in the Al-Zn system near 300°C, Fig. 9.12(c). By analogy with Fig. 7.10, a tangent line must exist since the $\alpha$ and $\alpha'$ phases are in equilibrium within the miscibility gap. Therefore, the free-energy curve must have the central maximum shown, and consequently $d^2G/dC^2$ must be negative within a certain range of composition. This range is determined by the *spinodes* (where $d^2G/dC^2 = 0$) and is known as the spinodal region.

**Spinodal decomposition.** In this and the subsequent three sections we consider some of the principal features of homogeneous (spinodal) transformations and of reconstructive heterogeneous transformations (cellular and precipitational). Displacive transformations are then discussed separately in the following section.

Ordinarily a volume $V$ of $\alpha$ solid solution of composition $C$ would not be expected to undergo the reaction,

$$V\alpha_C \rightarrow \frac{V}{2}\alpha_{(C-\Delta C)} + \frac{V}{2}\alpha_{(C+\Delta C)}. \tag{9.30}$$

And in fact, for a free-energy curve of normal shape, Fig. 7.10, this reaction is

0.1 μm

**Fig. 9.13** Electron micrograph showing the nature of the compositional regions corresponding to fluctuations in composition resulting from spinodal decomposition of an alloy of 6 at. % titanium in copper. (Courtesy of R. W. Newman, Florida Institute of Technology)

thermodynamically impossible. The corresponding change in free energy $\Delta G_v$ is

$$\Delta G_v = G \,(\text{products}) - G \,(\text{reactants})$$

$$= \frac{V}{2}\left[ G_\alpha + \frac{dG_\alpha}{dC}(-\Delta C) + \frac{d^2 G_\alpha}{dC^2}\left(\frac{-\Delta C}{2}\right)^2 \right.$$

$$\left. + G_\alpha + \frac{dG_\alpha}{dC}(\Delta C) + \frac{d^2 G_\alpha}{dC^2}\left(\frac{\Delta C}{2}\right)^2 \right] - V G_\alpha$$

$$= V \frac{d^2 G}{dC^2}(\Delta C)^2. \tag{9.31}$$

Because the first term in the expansion of $G_\alpha$ cancels, the second term must also be included as shown here. Since the usual free-energy curve for a solid solution is concave upward, $d^2 G/dC^2$ is positive; therefore $\Delta G_v$ is positive and the reaction of Eq. (9.30) cannot occur. Within a spinodal region, Fig. 9.12(d), $d^2 G/dC^2$ is negative, and in this special case the decomposition of a solid solution is thermodynamically possible as far as the chemical free energy $\Delta G_v$ is concerned. This means that any random fluctuation in composition produces a relatively stable mixture of two different compositions. In fact, diffusion now tends to occur *up* the gradient of concentration ("uphill" diffusion*) and to increase the magnitude of the difference in concentration. The final result is a three-dimensional array of tiny volumes (only about 50 Å on an edge) of α phase with two different average

---

* This type of diffusion obeys the necessary thermodynamic requirement; it is *down* the gradient of chemical potential.

compositions, one greater than the composition of the equilibrium $\alpha$ and the other less than that of the equilibrium $\alpha'$, Fig. 9.13.

Three factors determine whether a spinodal decomposition can occur in a given phase. First, the initial composition must lie in the region between two spinodes, Fig. 9.12.(d). This condition is rarely satisfied. Second, the additional *gradient energy* (associated with the concentration gradients between the compositional regions) must be supplied. Finally, appreciable strain energy must also be supplied by the available chemical energy, $\Delta G_r$, for systems in which the lattice constant changes appreciably with composition. Thus, spinodal decomposition is a rare phenomenon in alloys, but future research may demonstrate its special areas of usefulness.

**Nucleation.** We begin our treatment of heterogeneous transformations with a consideration of the phenomenon of nucleation of the product phase ($\beta$) within the initial phase ($\alpha$). As in the analogous case of liquid $\rightarrow$ solid transformation, research has shown that *homogeneous* nucleation rarely occurs in practice. Instead, catalyzed nucleation occurs at such *nucleating sites* as impurity particles, grain boundaries, and dislocations. As a preparation for a useful, qualitative analysis of competitive nucleation among alternative product phases, we generalize the concept of phase equilibria in two respects (especially the significance of solvus curve, first introduced in Fig. 7.12). First, an equilibrium can involve a metastable phase as well as a stable one. In fact, the equilibrium involving $Fe_3C$ in the Fe-C system (recall Fig. 7.27) is an equilibrium of a metastable phase. Figure 9.14(a) shows the relation of the equilibrium of the stable graphite phase to that of the metastable $Fe_3C$ phase. The $\gamma/Fe_3C$ solvus is different from the $\gamma$/graphite solvus; the eutectoid reaction is also affected. Except under unusual conditions the graphite phase does not form in steels, so the $Fe_3C$ phase is the one of practical interest. The second extension of the concept of equilibrium is the introduction of surface energy. Large particles of the $Fe_3C$ phase are present during the usual experimental determinations of phase equilibria. But as shown in Problem 11, the total energy of a phase is appreciably influenced by surface energy only if the particles of the phase are less than about 10 μm in diameter; therefore, effects of surface energy do not appear in the usual phase diagram. A qualitative estimate of the influence of particle size can be easily made, however, from free-energy curves (see Problem 12). Such an estimate of the equilibrium behavior of 1 μm particles of $Fe_3C$ is shown by the dotted lines in Fig. 9.14(a).

The application of these two aspects of phase equilibria to nucleation of a second phase is illustrated in Fig. 9.14(b) for precipitation in the Al-Cu system. The equilibrium of the stable phase, $\theta$, was previously discussed in connection with Fig. 7.18. Since the size of GP (Guinier–Preston) zones is about 0.01 μm and of the particles of the $\theta''$ transition lattice is about 0.1 μm, their equilibria depend sensitively on particle size as well as on the free energy ($\Delta G_v + \Delta G_\varepsilon$), Eq. (9.25), associated with the volume of the particle. Consequently, the solvus

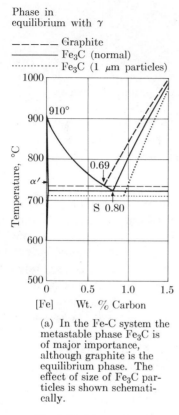

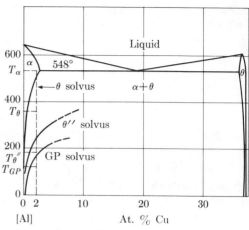

(a) In the Fe-C system the metastable phase Fe₃C is of major importance, although graphite is the equilibrium phase. The effect of size of Fe₃C particles is shown schematically.

(b) Typical equilibria of two of the metastable phases in the Al-Cu system are shown here on the phase diagram for the stable phase, $\theta$.

**Fig. 9.14**  Examples of metastable equilibria in technically important alloy systems.

lines for the GP zones and for the $\theta''$ phase apply only for particles of a given shape and of a given size. In spite of these limitations, such solvus lines are useful in treating the complex question of competitive nucleation in heterogeneous phase transformations.

In terms of solvus lines, the thermodynamic condition for nucleation [that $\Delta G$, Eq. (9.25), be negative] is simply that the initial alloy be taken below the solvus line of the phase in question. For example, if a 2 a/o Cu Alloy, Fig. 9.14(b), is cooled from $T_\alpha$ where the $\alpha$ phase is stable to $T_\theta$ where $\theta$ (but not $\theta''$ or GP zones) is in equilibrium with the $\alpha$ phase, then only the $\theta$ phase can nucleate and subsequently grow. At the temperature shown, nucleation occurs very slowly; also, the resulting structure is of little technical importance. Alternatively, the alloy can be cooled moderately quickly to $T_{\theta''}$; although both $\theta$ and $\theta''$ are thermodynamically possible products, the disc-like shape of $\theta''$ gives this phase more favorable kinetics, so the major product after a short time of reaction is $\theta''$. Given a sufficiently long time, the stable $\theta$ phase would form and cause the $\theta''$ to dissolve.

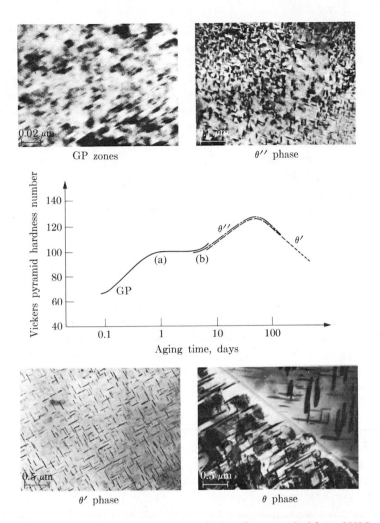

**Fig. 9.15** The change in hardness of an Al–2 at. % Cu alloy quenched from 550°C and aged at 130°C. The corresponding stages in the four competitive precipitation reactions are indicated, and typical microstructures are illustrated. (Data of J. M. Silcock, T. Heal, and H. Hardy. Electron micrographs courtesy J. Nutting, University of Leeds.)

The same type of argument can be repeated for $T_{GP}$. The competition now involves three reaction products, of which the GP zones have the fastest initial rate of nucleation.

**Precipitation reactions.** The processes of nucleation and subsequent growth have been especially thoroughly studied for precipitation reactions: the behavior of an Al-2 a/o Cu alloy, Fig. 9.15, can serve as a typical example. The first step in producing a precipitation reaction is *solution heat treatment* (in a range of temperature between about 525 and 575°C (975 and 1065°F) for Al-2 a/o Cu). In addition to putting the copper atoms in solution in the $\alpha$ phase, this treatment produces a high fraction of vacancies (about $2 \times 10^{-4}$ at 550°C (1020°F)). During rapid cooling to the aging temperature, 130°C (270°F), virtually all these vacancies come out of solution by one of various processes. The process of most interest is the formation of tiny clusters of vacancies; for example, four vacancies in the configuration of a tetrahedron. The number of these vacancy clusters is so enormous (see Problem 13) that their mean spacing $\lambda$ is only $\sim 100$ Å. If each tetrahedron serves as the nucleus of a GP zone, then the average copper atom need diffuse only one-fourth of this minute distance to complete the process of forming an initial equilibrium distribution of GP zones within the partially depleted $\alpha$ solid solution. A time of about one day (Problem 14) is required to complete this stage of precipitation, point (a) on the curve of Fig. 9.15. During this initial stage, if the temperature is increased so as to be somewhat above the GP solvus, Fig. 9.14(b), the GP zones gradually disappear, a phenomenon known as *reversion*. This behavior demonstrates that GP zones are an equilibrium phase.

For the same reason that smaller particles of $Fe_3C$ have higher energy, Fig. 9.14(a), the smaller particles of any phase tend to disappear, thus permitting the larger (more stable) particles to grow. This process of "coarsening" of particles of a second phase is known as *Ostwald ripening*. To a first approximation the total amount of second phase doesn't change during coarsening and therefore the number of particles decreases. A more accurate analysis, however, must consider the change in position of the GP solvus with increase in particle size and the consequent increase in the total amount of GP zones. Ostwald ripening terminates at point (b) on the curve of Fig. 9.15 because of the appearance of a more stable, competitive phase, $\theta''$. The GP zones gradually disappear as the $\theta''$ phase (and later the $\theta'$ phase) continues to form.

Basically, the $\theta''$ and $\theta'$ transition phases can grow at the expense of a GP zone because they have lower energy. Also, since their solvus curves lie at successively lower solute contents, these two phases permit additional solute atoms (copper) to leave the solid solution. Nucleation of $\theta''$ and $\theta'$ may require a dislocation; in any event, the number of nucleation sites is lower than for GP zones by an enormous factor. Eventually, however, the nucleation and growth of these two phases dominate the overall process of precipitation, as shown by the hardness curve and micrographs of Fig. 9.15. The reason for the maximum in hardness (and strength) is discussed in the next chapter. The stable phase, $\theta$, cannot be observed even after a year of aging at 130°C; at higher temperatures, however, its rate of formation can also be studied.

**Cellular transformations.**  As described earlier, Fig. 9.12, a cellular transformation leads to the complete disappearance of the initial phase. This is true whether the product is a single phase (as in the polymorphic transformation of FCC $\gamma$-iron to BCC $\alpha$-iron) or a mixture of two or more phases. The technically important

(a) Strongly preferential nucleation of $\beta$ phase; eventually the $\alpha$ phase nucleates in regions of high solute content. The bainite reaction is an example; see page 477.

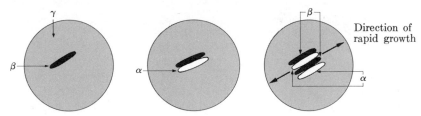

(b) Initial nucleation of the $\beta$ phase at a nucleation catalyst in the body of the initial phase; $\alpha$ phase nucleates on the particle of $\beta$ phase. An example is the formation of pearlite within a grain of austenite.

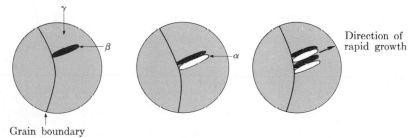

Grain boundary

(c) Initial nucleation of the $\beta$ phase at a grain boundary; $\alpha$ phase nucleates on the particle of $\beta$ phase; cell (nodule) grows toward center of grain. Pearlite usually forms in this manner; see Fig. 10.17.

**Fig. 9.16** Schematic illustrations of three modes of cellular transformation of an initial phase ($\gamma$) into two product phases ($\alpha$ and $\beta$).

reaction, $\gamma \rightarrow \alpha + \beta$, is considered here as an example. This reaction describes a eutectoid decomposition (for example, austenite $\rightarrow$ pearlite), but it also applies to several other types of transformations including the grain-boundary transformation that often occurs in competition with a precipitation reaction (see Fig. 6.23).

Cellular transformations, even in the same alloy, can occur by various modes, Fig. 9.16, depending on the conditions of reaction. Imagine that the transformation is analogous to a precipitation reaction. One of the two product phases nucleates first, Fig. 9.16(a), the $\beta$ phase for example. The surrounding region of

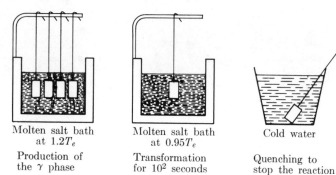

Molten salt bath
at $1.2T_e$

Production of
the $\gamma$ phase

Molten salt bath
at $0.95T_e$

Transformation
for $10^2$ seconds

Cold water

Quenching to
stop the reaction

(a) Schematic illustration of an experimental procedure for
determining the amount of transformation at temperature $0.95T_e$
for $10^2$ seconds.

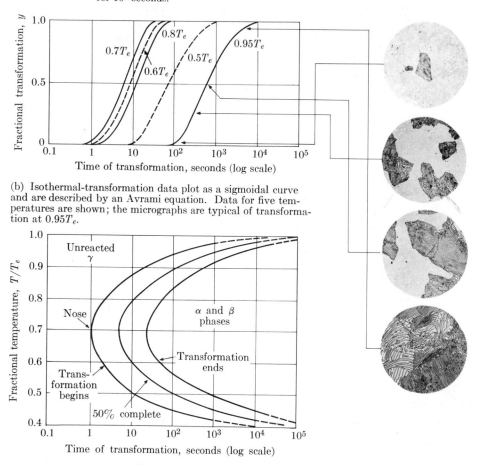

(b) Isothermal-transformation data plot as a sigmoidal curve
and are described by an Avrami equation. Data for five temperatures are shown; the micrographs are typical of transformation at $0.95T_e$.

(c) The data from the curves in (b) can be summarized in a TTT
diagram of this type.

$\gamma$ phase becomes depleted in solute, Fig. 9.16(a), so additional nucleation of $\beta$ phase particles tends to occur only at an appreciable distance from the initial particle. If the $\gamma \rightarrow \alpha + \beta$ reaction is to be completed, evidently the $\alpha$ phase must form, even though its nucleation may be considerably more difficult than that of the $\beta$ phase. A chemical composition favorable for the $\alpha$ phase exists in the depleted region surrounded by $\beta$ phase particles. Eventually $\alpha$ nucleates, probably using the surface of the $\beta$ phase as a nucleation site. In a more common mode of transformation, Fig. 9.16(b), the $\alpha$ phase nucleates easily on the $\beta$ phase shortly after a particle of the $\beta$ phase has formed on a nucleation catalyst. Suitable growth of one phase past the other in three dimensions (plus additional nucleation in some cases) results in a pattern of alternate plates or rods that constitute one cell (or nodule) of transformation product. The cell grows more rapidly into the initial $\gamma$ phase in the direction normal to the mutual interface of both the $\alpha$ and $\beta$ phases with the $\gamma$ phase. At this interface the solute can be redistributed from the average composition of the $\gamma$ phase to the low value of the $\alpha$ phase and the high value of the $\beta$ phase by three possible processes of diffusion: diffusion along the interface; diffusion in the $\gamma$ phase; diffusion in the $\alpha$ and/or $\beta$ phase. The most common mode of cellular transformation, Fig. 9.16(c), employs a grain boundary in the $\gamma$ phase as a nucleating site. The reaction then proceeds as in (b).

TTT (time-temperature-transformation) diagrams, Fig. 9.17, are usefully employed to describe the overall process of heterogeneous transformation of all types. Problem 15 is an application to precipitation reactions. A *TTT diagram* is simply a convenient graphical representation of data on a given transformation in a given alloy obtained over the range of temperatures of interest for the transformation in question. The upper limiting temperature is the temperature $T_e$ at which the initial phase $\gamma$ can first be in equilibrium with the product phase or phases ($\alpha$ and $\beta$, at the eutectoid temperature in the example considered here). Only at some temperature, such as $0.95\,T_e$, appreciably below the equilibrium temperature does the transformation proceed at a rate fast enough to be of practical interest. The course of reaction can be determined experimentally, Fig. 9.17(a), and plotted as the curve labeled $0.95\,T_e$ in Fig. 9.17(b). In view of the two adjustable parameters in an Avrami equation, Eq. (9.28), virtually any experimental transformation curve can be fit by an equation of this type. The curve is always of the sigmoidal type shown. Reaction at a lower temperature, $0.8\,T_e$, occurs more rapidly since the thermodynamic driving force is then greater. Usually this reaction curve can be fit with the same $n$ parameter in the Avrami equation but with a larger value of the rate coefficient $k$. At some temperature, such as $0.7\,T_e$,

**Fig. 9.17** Illustration of the procedure for obtaining and plotting data on isothermal transformation in the form of a time-temperature-transformation diagram. The behavior shown here is essentially the transformation of the austenite phase ($\gamma$) in an iron-carbon alloy of eutectoid composition.

the rate of reaction reaches a maximum value. At lower temperatures, $0.6 \, T_e$ and $0.5 \, T_e$, the individual atomic motions required for a reconstructive transformation become progressively slower and eventually cause the rate of reaction to become negligibly small.

The desired overall picture of the kinetics of the transformation can be obtained by plotting, as a function of temperature, the time for initiation ($\sim 2\%$ completion), half-reaction, and completion (actually, $\sim 98\%$ completion). When the data of Fig. 9.17(b) are plotted in this manner, the result is the corresponding TTT diagram, Fig. 9.17(c). The diagram for a given reaction, such as the formation of pearlite from austenite, is usually plotted in a single figure together with the diagrams for other reactions (bainite and martensite, in the case of steels) to give an overall transformation diagram for the alloy in question, Fig. 10.16. The diagram for a given transformation has two principal uses. First, the position of the *nose* of the diagram indicates the rate at which the alloy must be cooled from $T_e$ to avoid reaction at the higher temperatures. The alloy is then able to react at a lower temperature to the desired, alternative transformation product. Second, the curves for initiation and completion of transformation can be used to design heat treatments appropriate for obtaining useful properties in the product of transformation. This aspect is treated in Chapter 10.

**Displacive transformations.**  Just as with reconstructive transformations, displacive transformations can occur in various ways depending on the alloy and the conditions of heat treatment. In most cases, however, the following general features are observed. As indicated in Fig. 9.1, the initial phase ($\alpha$) must be greatly undercooled (usually by quenching) past the region of temperature in which an alternative reconstructive transformation can occur. As a result of this undercooling, a large chemical driving force is available to produce a transition phase, $\beta^*$, by a special process; namely, the coordinated movement of many atoms in the initial phase, each atom shifting only a fraction of an atom spacing with respect to its neighbors. This process, characteristic of displacive transformations, has several significant features. First, ordinary diffusion need not be involved; hence the term *diffusionless* transformation. Second, thermal activation is usually unnecessary, a fact expressed by the term *athermal.* Third, the joint movement of groups of atoms leads to local irregularities that appear as an uneven surface on a transformed specimen.

Because of the importance of martensite in steels, the term *martensitic* is usually employed by metallurgists to refer to displacive transformations. The essential nature of the transformation in this case is shown in Fig. 9.18. The body-centered tetragonal structure of martensite is able to form from the FCC austenite with only a slight relative movement of the atoms because of the correspondence between these two lattices, Fig. 9.18(a), first pointed out by E. C. Bain. The $\beta_c$ lattice that corresponds to the austenite, however, differs appreciably in lattice constants from the martensite phase, $\beta^*$, that actually forms. A homo-

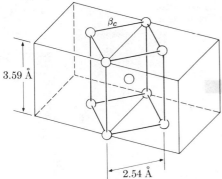
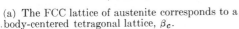
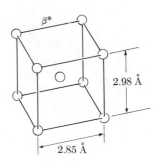

(a) The FCC lattice of austenite corresponds to a body-centered tetragonal lattice, $\beta_c$.

(b) A homogeneous strain is required to convert the lattice parameters of $\beta_c$ to those of martensite, $\beta^*$.

**Fig. 9.18** Two aspects of the formation of martensite in a 1% carbon steel. (a) The Bain lattice correspondence, and (b) the homogeneous strain to the actual lattice parameters of the martensite phase.

geneous strain, Fig. 9.18(b), must occur to effect the necessary change in shape.

A region of martensite forms as a relatively thin plate that propagates at almost the speed of sound through the grain of austenite. The plane in the austenite (habit plane) on which the plate of martensite forms is virtually undistorted and unrotated. This behavior can be understood only if an additional shape-change strain accompanies the homogeneous strain. Various mechanisms can account for the necessary change in shape, but slip or twinning on a fine scale is known to occur in certain steels and in other alloys that undergo displacive transformations, Fig. 9.19. The procedure for strengthening steels by the martensitic reaction is discussed in Chapter 10. Displacive reactions in other systems generally do not produce technically important increases in strength.

## RECOVERY, RECRYSTALLIZATION, AND GRAIN GROWTH

When a metal or alloy has been severely deformed (or damaged by irradiation from a nuclear reactor), it is in a metastable, high-energy state. The rate and manner in which such a metal attains a more stable configuration depends on the time and temperature of the treatment (called *annealing*, in general) to which it is subjected. The various possible transformations are reconstructive and thermally activated, but they are conveniently divided into three overlapping categories: recovery, recrystallization, and grain growth. As the name suggests, recrystallization is the process of forming a set of new, relatively perfect grains from the initial deformed metal. Grain growth results in an increase in the size of the average grain. For simplicity, the present treatment considers single-phase alloys primarily.

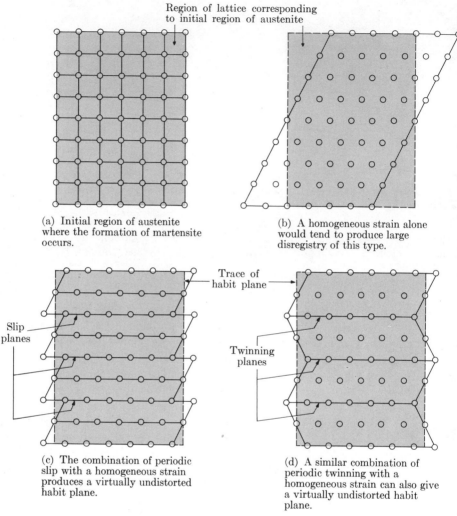

Region of lattice corresponding
to initial region of austenite

(a) Initial region of austenite where the formation of martensite occurs.

(b) A homogeneous strain alone would tend to produce large disregistry of this type.

Trace of habit plane

Slip planes

Twinning planes

(c) The combination of periodic slip with a homogeneous strain produces a virtually undistorted habit plane.

(d) A similar combination of periodic twinning with a homogeneous strain can also give a virtually undistorted habit plane.

**Fig. 9.19** Schematic illustration of the manner in which slip or twinning on a fine scale can compensate for the homogeneous strain in the formation of a region of martensite.

**Recovery.** The excess energy in a severely deformed metal is associated with structural features of many kinds, including point defects and dislocations. If such a metal is slowly heated, the excess "stored" energy is gradually released, Fig. 9.20, while various structural changes occur in the range of *recovery*, which includes all the changes prior to recrystallization. Although the structural changes associated with recovery cannot be seen with the optical microscope, the electron microscope permits the identification of two characteristic changes in the arrangement of dislocations. If deformation is performed under special conditions such

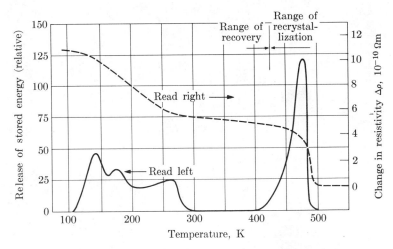

**Fig. 9.20** Measurements of the release of stored energy by cold-worked copper reveal the occurrence of several processes in the temperature range of recovery. The copper was deformed 54% at −195°C prior to being gradually heated to higher temperatures. (Data of L. Clareborough *et al.*)

that slip occurs only on a single slip plane, Fig. 9.21(b), then the dislocations are able to rearrange themselves into simple low-angle boundaries by glide and climb, a process known as *polygonization*. Ordinary polycrystalline, cold-worked metals, however, must recover by more complex dislocation reactions. The result is the formation of a *subgrain structure*, Fig. 9.22, in which the center of each subgrain is relatively perfect. The subgrain boundaries consist of groups of dislocations arranged in low-energy configurations such as prismatic networks.

Several factors must be considered in technical applications of recovery processes. At a given temperature the rate of recovery is fast initially and drops off at longer times. Thus the amount of recovery that occurs in a practical time increases with increasing temperature. In a given cold-worked metal the individual properties recover at different rates and attain various degrees of completion. The behavior of tungsten, shown in Fig. 9.23, is typical. In this case, for example, annealing for one hour at 400°C (750°F) produces a 70% relief of residual stresses, whereas the hardness is essentially unchanged by this heat treatment.

The principal application of recovery heat treatments is in stress-relieving cold-worked alloys to prevent stress-corrosion cracking or to minimize the distortion produced by residual stresses. Usually, some stress relief can be achieved without greatly affecting the mechanical properties, as the curves of Fig. 9.23 show in the case of tungsten. However, complete removal of residual stresses usually requires temperatures high in the recovery range. Such high-temperature treatment is commonly used for cast or welded steel parts. Recovery may affect the course of subsequent recrystallization of cold-worked metal, since the tendency toward recrystallization is lowered when appreciable recovery has occurred.

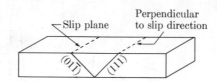

(a) Orientation of the single crystal before being bent

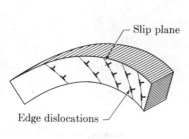

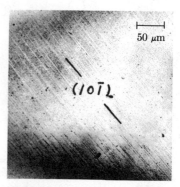

(b) Slip steps on the surface produced by the motion of dislocations on slip planes during bending of the crystal

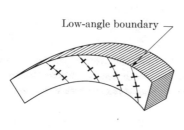

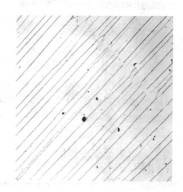

(c) Alignment of the dislocations into polygonized boundaries after annealing for 24 hours at 950°C

**Fig. 9.21** The process of polygonization, shown here for a crystal of iron-3% silicon alloy. (Courtesy C. G. Dunn, Research and Development Center, General Electric Company)

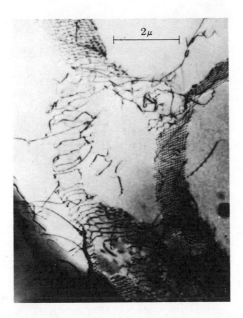

**Fig. 9.22** Electron micrograph of a cell structure in aluminum, produced by recovery for 300 hours at 160°C after an initial cold-working of 10%. (Courtesy T. J. Headley, University of Florida)

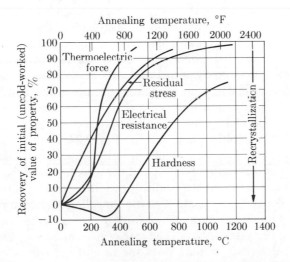

**Fig. 9.23** Approximate property variations produced by annealing cold-worked tungsten for one hour at temperatures below that necessary for complete recrystallization.

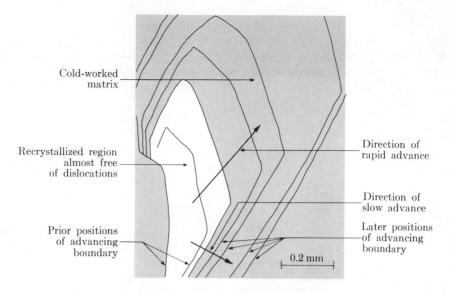

**Fig. 9.24**  Schematic illustration of the migration of a boundary into the cold-worked matrix in pure aluminum.  An intermediate position of the boundary is shown.  (Adapted from an electron micrograph of B. B. Rath and H. Hu)

**Recrystallization.**  Although a portion of the strain energy of a severely cold-worked metal is released during recovery, Fig. 9.20, the major part remains and serves as the driving force for the process of recrystallization.  The overall process appears to be the nucleation of strain-free grains within the cold-worked matrix and the subsequent growth of these grains, Fig. 9.26.  This point of view leads to the nucleation-and-growth analysis of recrystallization that is considered in the following section.  Electron microscopic studies, however, suggest that the fundamental process in recrystallization is the migration of boundaries separating the cold-worked matrix from a region that is essentially free of dislocations.  The origin of a mobile boundary (corresponding to nucleation) is still uncertain, but such a boundary may originate in a location where a region relatively free of dislocations is adjacent to a high-angle boundary.  A mobile boundary then migrates through the cold-worked matrix, Fig. 9.24, creating behind it the recrystallized structure.  The velocity of migration depends on such factors as the type and density of dislocation pattern and the orientation of the boundary.

The rate of growth during recrystallization is independent of time but increases with the degree of deformation and with the annealing temperature.  The presence of both soluble and insoluble impurities usually decreases the rate of growth, and hence the rate of recrystallization.  The rate of nucleation increases more rapidly with increasing deformation than does the rate of growth.  Consequently, the final grain size after recrystallization decreases with increase in the amount of initial deformation.  The recrystallized grains usually are not randomly oriented but have their crystal axes lying near certain favored directions with respect to the cold-worked grains. For example, when severely cold-rolled copper is recrystallized the new grains have a *preferred orientation* such that the [100] direction is parallel to the direction in which rolling was done, and the (001) plane lies in the rolling plane. In this case the preferred orientation is different from that developed by cold-working. Since pronounced alignment of crystal directions can produce anisotropic behavior in polycrystalline metals, preferred orientations are often objectionable; texture strengthening (see Chapter 10) and grain-oriented magnetic sheet (Chapter 12) are notable exceptions, however. Complete randomness of crystalline orientation can rarely be achieved, but the percentage of oriented grains can be reduced to a low value by various means. Small amounts of alloying elements may be effective for this purpose; about one percent tin essentially eliminates preferred orientation in copper. Also, a small amount of cold work and a low recrystallization temperature result in a minimum of preferred orientation.

In general the structure that forms during recrystallization is essentially the same as the structure that existed before cold-working. The commercial importance of recrystallization arises from the fact that the properties of an alloy after recrystallization are about the same as those it had before cold-working. For example, when an alloy is subjected to a cold-working operation such as deep-drawing, it becomes harder and less ductile (see Fig. 10.2), and it is therefore difficult to continue the forming operation. If the partially formed article is given a recrystallization anneal, the alloy is returned to its original condition of good ductility and easy deformation, and the article can then be given additional deep-drawing.

*Isothermal Recrystallization.* An important aspect of recrystallization in structural materials is the loss of strength that accompanies appreciable disappearance of the cold-worked grains. While it is evident that a work-hardened alloy should not be subjected to conditions that will produce recrystallization in service, it is often difficult to establish the exact range of permissible temperatures and times. For example, although the term *recrystallization temperature* is useful for many purposes, it does not refer to a definite temperature below which recrystallization is impossible. Rather, *the recrystallization temperature of a given alloy is the temperature at which the highly cold-worked alloy completely recrystallizes in about one hour.* Such recrystallization temperatures are listed in Table 9.2 for several metals and alloys. Very pure metals are seen to have low recrystallization temperatures compared with impure metals or alloys.

The wide use of the term *recrystallization temperature* reflects the fact that the recrystallization process is more sensitive to changes in temperature than to variations in time at constant temperature. The maximum *temperature* is usually fixed in engineering structures, but the *time* may reach extremely large values (scores of years). Therefore it is desirable to have a representation of recrystallization that takes time, as well as temperature, into account. The isothermal recrystallization curve shown in Fig. 9.25 meets this need.

**TABLE 9.2**

Approximate Recrystallization Temperatures for Several Metals and Alloys

| Material | Recrystallization temperature | |
|---|---|---|
| | °C | °F |
| Copper (99.999%) | 120 | 250 |
| Copper (OFHC) | 200 | 400 |
| Copper-5% zinc | 320 | 600 |
| Copper-5% aluminum | 290 | 550 |
| Copper-2% beryllium | 370 | 700 |
| Aluminum (99.999%) | 80 | 175 |
| Aluminum (99.0%+) | 290 | 550 |
| Aluminum alloys | 320 | 600 |
| Nickel (99.99%) | 370 | 700 |
| Nickel (99.4%) | 600 | 1100 |
| Monel metal (Nickel + 30% copper) | 600 | 1100 |
| Iron (electrolytic) | 400 | 750 |
| Low-carbon steel | 540 | 1000 |
| Magnesium (99.99%) | 65 | 150 |
| Magnesium alloys | 230 | 450 |
| Zinc | 10 | 50 |
| Tin | −3 | 25 |
| Lead | −3 | 25 |

The experimental points for constructing such a curve are obtained by placing ten or twenty specimens of the cold-worked alloy in a constant-temperature bath or furnace. Specimens are removed periodically, and their percentages of recrystallization are determined by microscopic examination. Figure 9.26 shows the gradual structural changes that occur. The initial cold-worked alloy had a relatively fine grain size. This material forms the background in Fig. 9.26(a). By the time a specimen of the cold-worked alloy had been heated for 50 hours at 310°C, several of the recrystallized grains had grown large enough to be seen, Fig. 9.26(a). On further heating, the existing recrystallized grains continued to grow and additional recrystallized grains also appeared, Figs. 9.26(b) and 9.26(c). After about 100 hours at this temperature recrystallization was complete, Fig. 9.26(d). A plot of percent recrystallization versus the time of annealing forms the isothermal recrystallization curve. A characteristic feature of such a curve is the incubation period that precedes the first visible recrystallization.

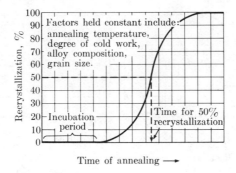

Fig. 9.25  A typical isothermal recrystallization curve.

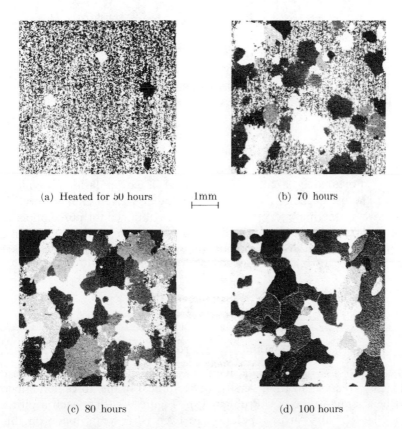

(a) Heated for 50 hours       1mm       (b)  70 hours

(c)  80 hours                             (d)  100 hours

Fig. 9.26  The course of isothermal recrystallization at 310°C in 99.5% aluminum that has been cold-worked 5%. (Courtesy W. A. Anderson, Alcoa Research Laboratory)

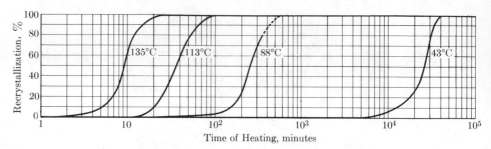

(a) Isothermal recrystallization of pure copper cold-rolled 98%

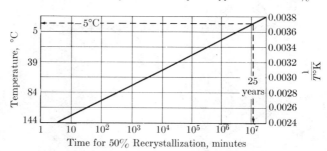

(b) Plot for extrapolation from the data given by the four curves of (a)

**Fig. 9.27** Isothermal recrystallization of 99.999% pure copper. (After Decker and Harker.)

It is evidently impractical to determine experimentally whether a given cold-worked alloy will recrystallize in a long period of engineering interest, e.g., twenty-five years. Fortunately, a convenient method of extrapolation is available for estimating the useful life of a cold-worked alloy at a given temperature. Figure 9.27(a) shows isothermal recrystallization curves obtained for pure copper at four different temperatures. An Arrhenius equation (such as Eq. (9.16)) gives the following type of relation between the rate of a reaction and the temperature in Kelvin

$$\text{Rate} = Ae^{-B/T}, \tag{9.32}$$

where $A$ and $B$ are constants. Using

$$\text{Rate} = \frac{1}{\text{time for 50\% recrystallization}} \tag{9.33}$$

as a measure of the reaction rate, it follows that a plot of the logarithm of the time for 50% recrystallization versus the reciprocal of the absolute temperature should be a straight line (see Problem 18). Figure 9.27(b) is such a plot of the data of Fig. 9.27(a). An extrapolation to twenty-five years made on this plot gives the surprising information that the cold-worked copper would be 50% recrystallized in this time even if it were refrigerated at $-5°C$.

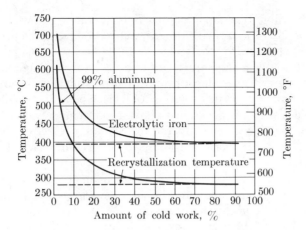

**Fig. 9.28**  Effect of the degree of cold-working on the temperature necessary to cause recrystallization during heating for one hour.  The position of the usual "recrystallization temperature" is indicated.

If the constants $A$ and $B$ are known,* Eqs. (9.32) and (9.33) can be used to calculate the time for recrystallization at any temperature.  For example, the time (in minutes) for 50% recrystallization of cold-worked OFHC (oxygen-free high-conductivity) copper wire at 100°C (373 K) can be estimated with aid of the values $A = 10^{12}$ min$^{-1}$ and $B = 1.5 \times 10^4$ for this material.  Equation (9.32) can then be written

$$\text{Rate} = 10^{12}e^{-(1.5 \times 10^4)/373}, \tag{9.34}$$

$$\text{Rate} = 10^{12}e^{-40.2} = 10^{12}10^{-17.45} = 10^{-5.45},$$

$$\text{Rate} = 0.35 \times 10^{-5} \text{ min}^{-1}. \tag{9.35}$$

When Eq. (9.33) is used to express the rate, Eq. (9.35) can be put in the form

$$\text{Time for 50\% recrystallization} = 2.9 \times 10^5 \text{ min}. \tag{9.36}$$

Comparison of this result with the recrystallization temperature for OFHC copper (Table 9.2) shows that recrystallization is more than 1000-fold slower at 100°C than at 200°C.

*Factors Influencing Recrystallization.*  A number of factors besides time and temperature influence the recrystallization process.  The most important of these are (1) alloying elements, (2) grain size, and (3) degree of cold work.  Figure 9.28 shows typical effects of the degree of cold work on the temperature necessary to

---

* These two constants can be calculated for a given cold-worked metal if the times for 50% recrystallization at two temperatures are determined.  Equation 9.32 can then be written for each of the two temperatures, and these two equations can be solved for $A$ and $B$.

cause recrystallization. The rapid increase in temperature below about 20 percent cold work is notable. The capacity for recrystallization produced by a given degree of cold work is influenced by the grain size of the metal; a fine grain size causes recrystallization to occur in a shorter time or at a lower temperature. Various working processes, such as rolling, drawing, and pressing, also produce somewhat different effective amounts of deformation for a given percentage reduction in cross section. If cold-working is performed above room temperature, its effectiveness is reduced in proportion to the temperature. Similarly, a recovery treatment may reduce the tendency for recrystallization to occur during a subsequent high-temperature anneal.

The common means of increasing the useful temperature range of a cold-worked metal is by suitable alloying. Usually the addition of a small percentage of a soluble alloying element sharply raises the recrystallization temperature of the base metal. With further increase in alloy content, however, the recrystallization temperature usually attains a maximum value and may then decrease. Thus the optimum addition of magnesium to aluminum is one percent, and of zinc to copper is $5\%$. Although the various alloying elements tend to follow this general pattern in their influence on recrystallization, there is a wide variation in the effectiveness of specific addition metals. For example, beryllium and zirconium are outstanding in increasing the recrystallization temperature of aluminum, whereas molybdenum and tungsten are especially useful in steel.

The property changes that occur as a result of heating a cold-worked metal or alloy may be the combined effects of recovery, recrystallization, and grain growth. Thus in Fig. 9.29 ductility (expressed as reduction in area) increases sharply during recrystallization and finally decreases after pronounced grain growth occurs. On the other hand, the electrical resistance diminishes markedly during the recovery stage and somewhat further on subsequent recrystallization. The pattern of property change is dependent not only on the property being considered but also on the nature of the alloy and its history of mechanical and thermal treatment. In the case of hardness and strength the trends of these important design properties may be significantly different from the corresponding change in the percent recrystallization that has occurred. Thus there is sometimes good reason for determining the hardness or strength values directly during an anneal, rather than depending on deductions from the observed amount of recrystallization.

**Secondary recrystallization.** Under some conditions an alloy that has formed a fine-grained structure by (primary) recrystallization will experience pronounced grain growth if it is heated at a higher temperature. The mechanism involves the rapid migration of the boundaries of a few of the primary recrystallized grains, with the result that the majority of the primary grains are consumed and very large secondary grains are created. A stage in this process is shown for an iron-silicon alloy in Fig. 9.30. Since the specks in the photomicrograph are etch pits indicating the presence of dislocations, it is evident that in this alloy the secondary

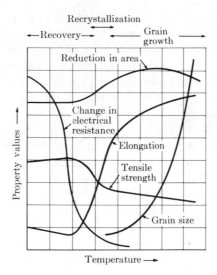

**Fig. 9.29** Typical property changes produced by heating a cold-worked metal for one hour at elevated temperatures.

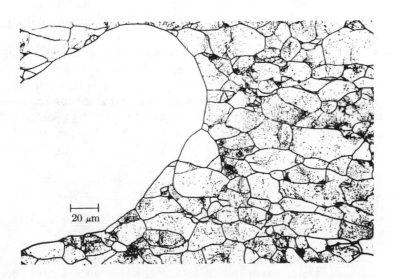

**Fig. 9.30** A large secondary grain growing into the primary-recrystallization grain structure in an iron-3% silicon alloy. (Courtesy C. G. Dunn, General Electric.)

grains are more perfect than the initial recrystallized grains. Other names for this phenomena are discontinuous grain growth and abnormal grain growth.

Secondary recrystallization can occur by at least two different mechanisms. The first requires the existence of a wide range of grain sizes in the initial structure. If a single large grain is surrounded by an array of smaller grains (diameters less than half that of the large grain), the corresponding configuration of grain boundaries is unstable. The large grain grows into its neighbors, obtaining the driving force from their grain-boundary energy. The second mechanism can operate in an array of grains of uniform size and depends on impediments to the growth of most of the grains. An example of such an impediment is the energy of the metal/gas surface in thin sheets. This energy is anisotropic and therefore favors the growth of those grains having orientations associated with a low surface energy, {100} planes parallel to the surface in the case of the iron-3% silicon alloy of Fig. 9.30. A suitable combination of deformation and annealing can lead to secondary recrystallization and produce a sheet containing large grains, each having a {100} plane in the surface. Since this orientation has desirable magnetic properties, "grain-oriented" sheet of this type is useful for such applications as transformer cores.

**Grain growth.** When an ordinary, strain-free metal or alloy is heated at a sufficiently high temperature, the grain boundaries slowly migrate and produce a uniform increase in grain size. This process is known as *normal grain growth*, and it occurs far more slowly than the types of boundary migration considered above in connection with the annealing of cold-worked metals (namely, motion of subgrain boundaries, growth of primary recrystallized grains, and abnormal grain growth during secondary recrystallization).

The driving force for normal grain growth is the energy associated with grain boundaries. When the grain size increases, the total grain-boundary area decreases, and consequently the energy of the metal is lowered. An equation describing the kinetics of grain growth can be obtained by assuming that the instantaneous rate of growth $dD/dt$ is proportional to the grain-boundary energy per unit volume of the metal, which in turn is proportional to $1/D$, where $D$ is the grain diameter. This reasoning leads to the expression

$$\frac{dD}{dt} = K\frac{1}{D},\qquad(9.37)$$

where $K$ is a constant. It is shown in Problem 22 that the integrated form of this equation is

$$D^2 - D_0^2 = Ct.\qquad(9.38)$$

Although Eq. (9.38) applies in only a few ideal cases, an analogous equation approximates the grain-growth behavior of single-phase alloys when the grain size is small compared with the dimensions of the specimen. This empirical

**TABLE 9.3**

Approximate Values of the Isothermal Grain-Growth Parameters $n$ and $C$

| Material | Temperature, °C | °F | $n$ | $C$ |
|----------|-----------------|-----|-----|-----|
| Aluminum (99.99%) | 400 | 750 | 0.1 | $8 \times 10^{-9}$ |
| | 500 | 930 | 0.2 | $3 \times 10^{-3}$ |
| | 600 | 1110 | 0.3 | $6 \times 10^{-1}$ |
| Cartridge brass (70% Cu, 30% Zn) | 500 | 930 | 0.2 | $13 \times 10^{-11}$ |
| | 600 | 1110 | 0.2 | $9 \times 10^{-9}$ |
| | 700 | 1290 | 0.2 | $6 \times 10^{-7}$ |
| Iron | 500 | 930 | 0.1 | $2 \times 10^{-15}$ |
| | 600 | 1110 | 0.2 | $6 \times 10^{-11}$ |
| | 800 | 1470 | 0.5 | $4 \times 10^{-4}$ |
| Carbon steel (austenite) 0.8% C | 760 | 1400 | 0.1 | $6 \times 10^{-16}$ |
| | 870 | 1600 | 0.2 | $2 \times 10^{-8}$ |
| | 980 | 1800 | 0.23 | $2 \times 10^{-6}$ |

equation is

$$D^{1/n} - D_0^{1/n} = Ct, \tag{9.39}$$

where $D$ is the grain diameter in millimeters that exists after heating for $t$ minutes at constant temperature, $D_0$ is the initial grain diameter, and $n$ and $C$ are independent of time but vary with factors such as composition and temperature. $C$ increases steadily with temperature, but the value of $n$ appears to approach 0.5 as an upper limit with increasing temperature. Approximate values of $n$ and $C$ are given in Table 9.3 for several metals and single-phase alloys. The grain growth occurring with increasing *time* at constant temperature is small compared with that occurring with increasing *temperature* at constant time (usually one hour at temperature). Figure 9.31 shows the grain-growth behavior of two steels as the austenitizing temperature is increased. The grains of the coarse-grained steel grow in the manner characteristic of a single-phase alloy.

The restricted grain growth at low heating temperatures shown by fine-grained steel is typical of the behavior of alloys that contain a finely dispersed second phase. In steels, nonmetallic inclusions formed during the deoxidation of the liquid steel tend to restrain grain growth. Usually the larger the number of second-phase particles, the smaller the grain size in metals containing a dispersed phase. Above a certain high temperature, however, *grain coarsening* occurs, in which a few large grains partially or completely replace the fine grains that exist over a wide range of lower temperatures. A second possible cause of restricted grain growth is the presence of strong preferred orientation of the initial set of grains. This restrictive influence, too, is overcome at high temperatures, and grain coarsening occurs. The final grain size of a cast alloy is determined during the solidification process,

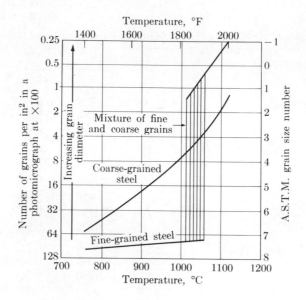

**Fig. 9.31** Typical grain-growth behavior of fine-grained and coarse-grained steels on being heated for one hour at various temperatures.

and usually it cannot be changed by heat treatment, since grain growth is essentially lacking in such alloys.

## POWDER METALLURGY AND SINTERING

A mass of small metallic particles, each about 1 μm in size, tends to undergo a transformation into a single piece of metal, employing as driving force virtually all the surface energy associated with the original small particles. Technical processes of powder metallurgy, although based on this simple mechanism, are quite varied and invariably more complex. Essentially, this method of forming metal parts consists in compacting a powder in a die to form the desired shape, Fig. 9.32, and then heating (*sintering*) the resulting *compact* to cause agglomeration of the particles. During the complex mechanical deformation accompanying compaction, the particles may be bonded by simple adhesion, van der Waals forces, or by the use of binders added for this purpose. The process of sintering may be affected, in addition to surface energy, by chemical interactions among alloy powders, by the action of foreign constituents such as oxide layers on the particles, by gases present in the sintering furnace, and by the presence of a film of liquid that wets the particle (in the special technique of liquid-phase sintering).

The mechanisms of typical processes of sintering take account of driving forces from such additional factors as grain boundaries, dislocations, vacancies,

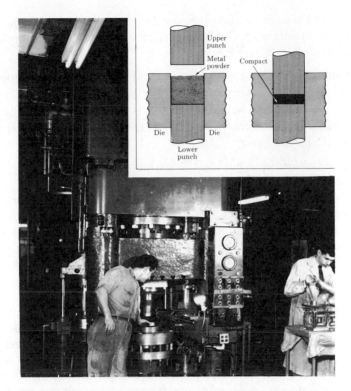

**Fig. 9.32** A 500-ton self-contained oil-hydraulic forming and containing press for powder metallurgy.   The method of forming a powdered-metal compact is shown in the inset. (Designed and built by Hydropress, Inc., New York.)

and trapped gases.  The atoms and vacancies are transported by several mechanisms during the various stages of the complete sintering operation.   Diffusion along the surface is known to be important in the earliest stage (the formation of a neck between two adjacent particles), but lattice diffusion of the vacancies also occurs.  Vaporization and condensation makes a contribution in some systems. At a later stage, when a network of grain boundaries has formed, diffusion along these boundaries is an additional mechanism.  Plastic deformation must accompany the changes in configuration of the particles during sintering and is especially significant if an external stress is applied.  Both grain-boundary glide and also creep by a vacancy method contribute to the deformation.

Powder-metallurgical techniques find many applications in mass production. Because metal powders are costly, these applications tend to be limited to such purposes as the following.

1. To achieve properties that cannot be realized by other processes, for example in porous bearings and cemented carbides

2. To eliminate machining and other finishing operations by holding very close tolerances and maintaining excellent surface conditions, for example in small machine parts such as bushings and gears

3. To produce high melting point (refractory) metals such as tungsten, molybdenum, and tantalum. An example of this type of product is *heavy metal*, a tungsten alloy containing 6% nickel and 4% copper. It has a density of 17 and is used for radium containers, balancing weights, and other applications where high density is needed.

A number of factors influence the design of powder metallurgy products. Since pressures of the order of $30 \times 10^7$ N/m² ($40 \times 10^3$ lb/in²) are used in the pressing operation, it is evident that commercially practical cross-sectional areas are fairly small. Also, the powder does not behave like a fluid under the compacting force, and excessive thicknesses must not be specified if low-density, low-strength areas are to be avoided. For the same reason, undercut areas and holes perpendicular to the compression direction cannot be molded conveniently. However, the cross section in the direction of pressing may be quite complex; thus gears and cams are important powder metallurgy products.

The mechanical properties of pressed and sintered powders approach those of castings but are significantly poorer than those of wrought alloys. Consequently this process is not suitable for highly stressed parts or for those subject to impact loads. Properties approximating those of the wrought alloy can be achieved by additional pressing of the sintered compact, but this step adds to the manufacturing cost and is not common. Tolerances of 0.02 mm are maintained on the diameter of self-lubricating porous bearings, an important powder metallurgy product.

## PROBLEMS

**1.** Thermodynamics takes a practical approach in describing a given alloy in terms of *phases*.

a) Consider a determination of the specific heat of a specimen, performed by observing the increase in temperature of a water bath initially at $T_1$ when the specimen (initially at $T_2$) is dropped into it. Explain why, in the absence of any other information, the specimen is considered to be simply a single substance (one phase) even though it may actually be an alloy of the $\alpha$ and $\beta$ phases.

b) Explain why a metal powder, even if composed of only a single metal of the highest purity, may be considered to consist of two phases for the purpose of describing the process of sintering. In this process, about $10^{15}$ particles per cubic meter agglomerate to form a single piece of metal. [*Hint*: A given particle is considered to consist of two phases, one phase being the main central portion.]

**2.** In a manner analogous to a change in temperature, a change in pressure can move an alloy from a region where the $\alpha$ phase is stable to a region where the $\beta$ phase is stable. The behavior of iron is analogous to that of carbon, Fig. 6.2. At pressures above $1.3 \times 10^{10}$ N/m² iron exists as an unusual phase, HCP. Devise a process for "pressure treating" (analog of heat treating) an alloy of iron that would make use of the transformation Fe(HCP) $\rightarrow$ Fe(FCC).

**3.** As shown in Fig. 9.18, the martensitic transformation in iron is possible because an (almost) BCC lattice can be formed from the lattice points of the FCC lattice. Make a similar analysis of the HCP lattice to see if there is a possibility for a displacive transformation in alloys that have the HCP lattice at high temperatures.

**4.** Alternative definitions can be used for a phase:

a) that it is a *uniform* region; for example, with respect to composition;

b) that it is merely *homogeneous* and may contain gradients in composition. Which definition is more appropriate for analyses of phase transformations? Explain.

**5.** a) Determine the value of $r$ at which $\Delta G$ in Eq. (9.1) attains its maximum value. Recall the use of the derivative of a function in this connection.

b) Determine the maximum value of $\Delta G$.

**6.** Obtain an expression for $z$ in Eq. (9.5), using the results of simplified kinetic theory, Eqs. (6.40) and (6.42). [Equation (6.40) will refer to the number of atoms in 1 m³ if $l$ is set equal to unity.] Note that the number of atoms striking an area of 1 m² can be considered to be the number per cubic meter that are moving in the direction of a given coordinate, multiplied by the distance moved per second (the velocity $u$). Compare your result with that obtained by more refined kinetic theory,

$$z = P/\sqrt{2\pi MRT}.$$

**7.** Use the method of the previous problem to determine the rate of vaporization of a metal in a vacuum. It can be assumed that the number of kilograms vaporized is equal to the number of atoms striking unit area multiplied by the weight of each atom. Compare your result with that obtained by more refined kinetic theory,

$$G = P/\sqrt{(2\pi RT)/M}.$$

**8.** a) Use Eq. (9.10) to demonstrate that the sign of $\Delta G_v$ correctly predicts that the reaction Zn(liquid) $\rightarrow$ Zn(solid) is impossible above the melting point of zinc.

b) Considering the value of $\Delta H$ in this case, criticize the naive concept that the heat liberated in a reaction is the driving force for the reaction.

**9.** Assuming that $\gamma$ (liquid-solid interfacial energy) for zinc is 0·06 J/m², and that $\Delta G_+$ can be neglected in comparison with $\Delta G^*$, calculate the degree of supercooling, $\Delta T = T_f - T$, at which the rate of homogeneous nucleation, $I_l$, first reaches an appreciable value, say 1 nucleus per second per m³. $I_l$ varies extremely rapidly with change in $\Delta T$, and suitable assumptions can be used to keep the mathematics quite simple.

**10.** a) Under actual conditions liquid zinc does not nucleate homogeneously, as assumed in the previous problem, but heterogeneously. Assuming that the actual undercooling is observed to be 3°C, solve the equation for the rate of heterogeneous nucleation to obtain the corresponding value of $\theta$. See the previous problem for comments on the mathematical procedure.

b) Make a sketch showing how nucleation occurs in this case.

11. Taking $10^8$ J/m$^3$ as a typical value of the chemical free energy of a second phase particle and 1 J/m$^2$ as the surface energy, calculate the radius of a spherical particle for which the surface energy is 1% of the chemical energy.

12. Using a sketch of free energy curves for the $\gamma$ and $Fe_3C$ phases analogous to Fig. 7.10, show that a 10% increase in the free energy of the $Fe_3C$ phase due to surface energy shifts the solvus line to higher carbon contents. $Fe_3C$ has a higher free energy than the $\gamma$ phase.

13. Assuming that the fraction of vacancies in solution at 130°C is negligible compared to the value $2 \times 10^{-4}$ at 550°C:

a) show that the fraction of tetrahedral clusters of vacancies (if this is the only type of cluster) is $0.5 \times 10^{-4}$;

b) calculate the average spacing of the tetrahedra in terms of atom spacings, $d$, and also in meters (for aluminum, $d = 3$ Å).

14. a) Using the data for diffusion of copper in aluminum, Table 8.1, calculate the diffusion coefficient at 130°C ($\sim 400$ K).

b) The equation,

$$x^2 = 4Dt \qquad (9.40)$$

gives a rough estimate of the time $t$ required for diffusion to occur over a distance $x$. Calculate the distance copper can diffuse in one day ($\sim 10^5$ seconds) at 130°C.

c) Compare this result with the diffusion distance for an average copper atom (one-quarter of the spacing of vacancy tetrahedra).

15. a) Using the data of Figs. 9.14(b) and 9.15, devise three TTT curves for the overall transformation to the three reaction products, GP, $\theta''$, and $\theta'$. Plot all three curves in a single figure.

b) On your figure show three temperatures of isothermal reaction at which GP, $\theta''$, and $\theta'$, respectively, would be the initial product of transformation.

16. Criticize this statement: "Recovery of cold-worked tungsten occurs at 500°C."

17. The presence of an alloying element in solid solution usually decreases the rate of nucleation in the recrystallization process. Recalling that both nucleation and growth occur in recrystallization, explain why, in spite of this effect on rate of nucleation, the grain size after recrystallization may be *decreased* by the addition of such an alloying element.

18. Show how Eqs. (9.32) and (9.33) determine what quantities should be plotted to obtain the straight-line extrapolation used in Fig. 9.27(b).

19. OFHC copper that has had its yield strength increased by a factor of four through cold-working is proposed for use in an electrical device operating at 150°C. A safety factor of two is to be employed in determining the maximum allowable stress.

a) Using values of $A = 10^{12}$ min$^{-1}$ and $B = 1.5 \times 10^4$ in Eq. (9.32), calculate the approximate useful life of the electrical device.

b) Why should recovery effects be considered?

c) Why is it unnecessary to consider the effects of grain growth?

20. List a number of means that might be used to raise the temperature at which a cold-worked copper article, which must have good electrical conductivity, will recrystallize in five years.

**21.** Give the advantages of plotting recrystallization data
 a) at constant temperature, and
 b) at constant time.

**22.** It can be shown that Eq. (9.38) is obtained when Eq. (9.37) is applied to the problem of the growth of a grain whose diameter is $D_0$ at time $t = 0$.

 a) Algebraically rearrange Eq. (9.37) so that $D$ and $dD$ are on one side of the equation and $dt$ is on the other side.

 b) Integrate both sides of this equation over the interval between zero time and some arbitrary time $t$. Note that the corresponding limits on the grain diameter are $D_0$ and $D$.

 c) Combine the constants into a single constant $C$.

**23.** As a useful approximation it is often assumed that no grain growth occurs in a steel heated at 760°C, but that grain growth does occur at temperatures near 870°C. Taking an initial grain size of 0.05 mm, calculate the grain growth that would occur on heating a 0.8% carbon steel one hour at each of these temperatures.

## REFERENCES

Burke, J., *The Kinetics of Phase Transformations in Metals*, New York: Pergamon Press, 1965. Gives a brief, overall treatment at an intermediate level.

Christian, J. W., *The Theory of Transformations in Metals and Alloys*, New York: Pergamon Press, 1965. The authoritative, advanced treatment of this topic.

Fine, M. E., *Phase Transformations in Condensed Systems*, New York: Macmillan, 1964. Gives a detailed description of the nucleation and growth of precipitates in solids.

*Mechanisms of Phase Transformations in Crystalline Solids*, London: Metal and Metallurgy Trust, 1969. Various authors discuss recent advances in this area at an advanced level.

Shewmon, P. G., *Transformations in Metals*, New York: McGraw-Hill, 1969. A thorough coverage of phase transformations at an intermediate level.

Tiller, W. A., "Solidification," in *Physical Metallurgy*, 2nd ed., R. W. Cahn (Ed.), p. 403, Amsterdam: North Holland Publishing Company, 1970. A detailed treatment of the theory of the solidification of metals and alloys.

# CHAPTER 10

# STRENGTHENING MECHANISMS AND PROCESSES

## INTRODUCTION

We can usefully begin our study of methods for strengthening metals and alloys with a brief consideration of the need for strength in various structures. As shown clearly by modern architecture, good design reflects the properties of available materials, especially the crucial factor of cost. Ceramic materials, such as concrete and brick, are the least expensive means for supporting a compressive load; their extensive use in construction is a reflection of this fact. Similarly, the use of plastics for many lightly stressed articles (telephones, containers, toys, etc.) is due to the low initial and manufacturing costs of this type of material. With metallic materials also, by far the largest tonnage is employed in the form of inexpensive cast iron, ordinary hot-rolled steel, and easily produced alloys of the nonferrous metals such as aluminum and copper. For some designs, however, the superior properties of a more expensive material are amply justified, even from the standpoint of economics. Corrosion resistance, discussed in Chapter 11, is often a major factor in such a decision, but improved strength under various conditions of service is generally of principal importance in engineering design.

Strengthening of metallic materials can be viewed in two different ways: in terms of a practical process; and from the standpoint of theory and mechanism. Although present-day theory gives many useful insights into the nature of the strengthening mechanisms, only rarely is it able to make quantitative predictions of the behavior of complex, technical alloys. For example, strengthening by cold-working is such a simple process that the ancient cavemen used it to strengthen lumps of brass in the act of beating the alloy into a useful shape. We will see that modern theory explains much about the nature of cold-working and is a valuable guide in the development of new processes; nevertheless, most aspects of the technology of cold-working have no direct link to the theory. Strengthening processes that depend on heat treatments (such as the hardening of steel or of aluminum alloys) have a closer connection to theory because of their dependence on phase diagrams and because much is known about the atomic mechanisms that lead to strengthening. Even here, we must recognize that technological practice is the major factor determining the successful development and use of complex alloys.

##  STRENGTHENING BY DEFORMATION

**Process of cold-working.** A process of cold-working differs basically from hot-working (discussed in Chapter 4) in producing persisting deformation of the crystalline structure. This important difference is illustrated in Fig. 10.1 for the rolling of brass (which had previously been converted from the cast to the wrought structure). Recrystallization occurring simultaneously with the plastic deformation accounts for the fact that the grain structure remains essentially unaffected during hot-working. An important consequence is the constancy of the flow stress $S_0$, with the result that the work required for deformation, Eq. (4.35), remains at a

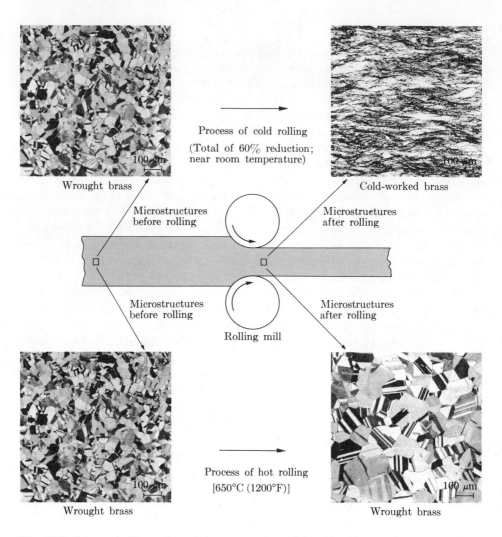

**Fig. 10.1** Schematic illustration of the manner in which cold-rolling produces a deformed crystalline structure, whereas hot-rolling (with accompanying recrystallization) leaves the structure essentially unaffected, except for a possible change in grain size. (Photomicrographs courtesy Anaconda American Brass Company)

relatively low value. In contrast, the stress needed to continue a process of cold-working increases greatly with increase in degree of deformation. This behavior is shown in terms of true stress $\sigma$ in the tension-test curve of Fig. 4.14(a); more commonly it is measured by the increase in hardness accompanying increasing amounts of cold work, Fig. 10.2. Quantitative measures of work hardening are given in Table 4.2 in the form of the strain-hardening exponent. The technological

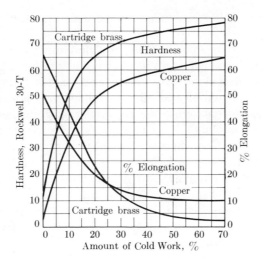

**Fig. 10.2** The changes in hardness and ductility produced by cold-working copper and cartridge brass (70% Cu, 30% Zn).

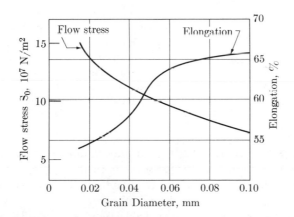

**Fig. 10.3** Effect of grain size on the flow stress and ductility (% elongation) of cartridge brass tested in tension. To convert from N/m² to lb/in² multiply by $1.450 \times 10^{-4}$.

importance of this type of strengthening is indicated by the fact that the tensile strength of cartridge brass is doubled by 60% cold-working. However, cold deformation has the disadvantage of causing a pronounced decrease in ductility, shown in Fig. 10.2 by reduction in the percent elongation during a subsequent tension test. Because of this embrittling effect, most metals begin to crack if cold-working exceeds a certain amount. An intermediate anneal avoids this difficulty since the resulting recrystallization returns the structure to the initial, unstrained condition.

Grain size is often an important factor determining optimum performance of an alloy in a cold-forming operation. A specified grain size can usually be obtained by suitable control of the recrystallization processes during prior fabrication of the alloy sheet to be cold formed. Brass intended for cold-working operations is an example of commercial grain-size control. Figure 10.3 shows that the ease of working of cartridge brass (70% Cu, 30% Zn) increases with increasing grain size. Not only is the ductility improved but the flow stress is lowered as well. On the other hand, problems with surface appearance may be encountered as the grain size increases, since the nonuniformity of deformation from grain to grain becomes plainly visible. Figure 10.4 shows the "orange peel" surface that occurs on coarse-grained metals that are subjected to severe deformation. The choice of grain size in this case is therefore a compromise, and Table 10.1 shows the values that are recommended for various applications.

**TABLE 10.1**

Recommended Grain Sizes in Brass for Cold-Forming Operations

| Grain diameter, mm | Type of cold-forming operation |
| --- | --- |
| 0.015 | Light forming |
| 0.025 | Shallow drawing |
| 0.035 | Average drawing |
| 0.050 | Deep-drawing |
| 0.100 | Severe drawing on thick sheet |

Other aspects of strengthening by deformation must be taken into account in a wide variety of commercial processes—extrusion, deep-drawing, and wire drawing, for example. Cold-drawn wires have the advantage, not only of greatly increased strength, but also of only slightly impaired electrical conductivity. Corrosion resistance is sometimes drastically reduced by cold-working, and the resulting stress corrosion (see Chapter 11) is of industrial importance. *Preferred orientation* of the metal grains is an inevitable result of the slip and twinning processes that occur during cold-working. During the cold drawing of aluminum wire, for example, grains that were initially oriented at random are deformed in such a way that they tend to have a [111] crystal direction along the wire axis. Steps are often taken to control the degree of preferred orientation produced in a given alloy, sometimes to increase fts magnitude, as in magnetic materials, but more often to prevent its leading to inhomogeneous deformation in later working operations.

Original brass sheet

Cupped specimen

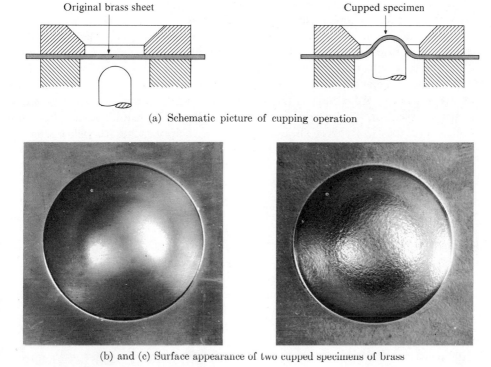

(a) Schematic picture of cupping operation

(b) and (c) Surface appearance of two cupped specimens of brass

100 μm

(d) and (e) Corresponding microstructures of the original brass sheet

**Fig. 10.4** The effect of grain size on the surface appearance of cold-drawn brass sheet. The microstructures reveal that a coarse grain size causes the development of "orange peel" surface. (Courtesy H. L. Burghoff, Chase Brass and Copper Company)

A useful aspect of preferred orientation is its relation to *texture strengthening*, particularly in highly anisotropic hexagonal metals such as titanium, zirconium, beryllium, and magnesium. When sheets of these metals are cold rolled, the resulting preferred orientation (texture) has the (0001) planes preferentially in the plane of the sheet. The slip directions for the easily activated slip systems then also lie in the plane of the sheet; consequently, deformation perpendicular to the sheet is greatly restricted. Advantage is taken of this fact in the design of hemispherical pressure vessels, in which the major stresses are biaxial tensions in the plane of an element of the metal sheet. The limiting stress then approaches the yield strength in compression; for beryllium sheet this quantity has the high value of $40 \times 10^7$ N/m$^2$ ($60 \times 10^3$ lb/in$^2$). The same mechanism also makes such sheets resistant to denting. Texture strengthening is less effective in cubic metals because of the many slip systems available, but even here it can be used to increase the drawability of sheet steel, for example.

**Mechanisms of strain hardening.** It will be recalled that deformation of metals occurs principally by the motion of existing dislocations and by the creation of many additional dislocations (for example, through the action of Frank-Read sources). Although strain hardening is not yet fully understood, its essential feature is the interference with the motion of a given dislocation by "obstacles" posed by neighboring dislocations. The nature of the obstacles and their relative importance varies greatly with grain structure, crystallography, and conditions of deformation. For example, in stage I deformation of single crystals (see Chapter 4), the stress remains relatively constant with increasing deformation because the active dislocations are moving on parallel sets of slip planes. The principal mode of interaction between dislocations is through their stress fields, and a characteristic behavior is the grouping of active slip planes into closely spaced groups (slip bands), while adjacent slip bands are separated by relatively large distances, on the order of 1 μm. When several intersecting slip systems operate during the deformation of a single crystal (stage II), interference among neighboring dislocations is much greater. Consequently, the slope of a stress-strain curve increases appreciably at the transition between stage I and stage II because of the greater strengthening that accompanies a given amount of deformation in stage II. The new strengthening mechanisms that accompany the intersection of dislocations are the formation of kinks, jogs, and dipoles on initially straight dislocations, and the creation of sessile dislocations by the reaction of previously glissile dislocations. Interference among dislocations is especially pronounced in polycrystalline metals because of the geometric requirement that a minimum of five slip systems must operate within each grain to accomplish an arbitrary change in shape and yet maintain continuity across each grain boundary. Furthermore, the grain boundaries serve as an additional type of impediment at which pile-up of dislocations can occur. The dislocation reactions in question have been considered in Chapter 3.

The structures of cold-worked metals and alloys, even polycrystalline ones that

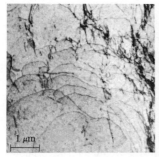

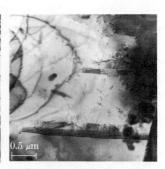

(a) Distribution of dislocations is relatively uniform at small strains.

(b) Cell walls of high dislocation density, about $10^{16}/m^2$, form at large strains.

(c) Stainless steel, because of its low stacking-fault energy, shows a more uniform distribution of dislocations at large strains.

**Fig. 10.5** Electron micrographs showing the structures developed in a single crystal of copper [(a) and (b)] and in stainless steel (c) by cold-working. (Micrographs (a) and (b) courtesy H. Mughrabi and U. Essmann, respectively, University of Stuttgart)

have been severely deformed, show an unexpected feature. Relatively large volumes of the microstructure are essentially free of dislocations, the dislocations being concentrated in the surrounding regions, thus forming a cell structure. This structure develops with increasing deformation in the manner shown in Fig. 10.5. For small amounts of strain, the dislocation density is relatively uniform, Fig. 10.5(a). As the amount of strain increases, tangled dislocations tend to segregate into regions of high dislocation density. At sufficiently large strains, Fig. 10.5(b), the dislocations are mainly in high-density cell walls, leaving the center of the cell almost completely free of dislocations. The size of the cells and the strain needed to form them varies with the alloy and with such factors as temperature and rate of straining. Cell structures are characteristic of alloys with high stacking-fault energies, because the dislocations are undissociated and can cross-slip easily. In contrast, alloys with low stacking-fault energies have a more uniform distribution of dislocations after large amounts of cold work, Fig. 10.5(c). Furthermore, the dislocations tend to remain straight and to create pile-ups (that is, planar arrays) at such obstacles as grain boundaries, dispersed phases, or sessile dislocations.

Although the rate of strain hardening cannot yet be accurately predicted for commercial alloys, some general guidelines can be given. The effect of crystal structure can be seen especially easily in the behavior of hexagonal metals. When an HCP single crystal is deformed, Fig. 1.5(c), only the primary slip system is activated. The active dislocations are confined to a single set of parallel planes, and they eventually run out of the crystal at a free surface. As a result, both the density of dislocations and the extent of mutual interference are small; therefore, strain hardening is minimal. A similar pattern underlies the plastic deformation of polycrystalline HCP metals, but their behavior is complicated by the occurrence of twinning and the operation of additional slip systems. Nevertheless, HCP alloys generally have a lower rate of strain hardening than do FCC or BCC alloys. The latter crystal structures permit the operation of many slip systems both in single crystals and in polycrystalline specimens. The interacting dislocations form obstacles to the motion of other dislocations, which in turn pile up and increase the shear stress, $\tau$, required to continue the process of deformation. An estimate of the value of $\tau$ can be obtained from the following relation,

$$\tau = \tau_0 + \alpha\mu b\rho^{1/2} \tag{10.1}$$

where $\tau_0$ is the shear stress in the absence of work hardening, $\alpha$ is a factor of about one-half, $\mu$ is the shear modulus, $b$ is the Burgers vector, and $\rho$ is the average density of dislocations. In severely deformed metals, the value of $\rho$ in the tangled cell walls can be as high as $10^{16}/m^2$. The rate of strain hardening (determined experimentally as the coefficient $n$, Table 4.2, or more generally as the local slope $d\sigma/d\varepsilon$ of a stress-strain curve) usually decreases as the temperature of testing is increased. The addition of a soluble alloying element may either increase or decrease the rate of strain hardening, but the final hardness of a cold-worked alloy is almost always greater than that of the similarly cold-worked pure metal because of the higher yield strength of the alloy.

## EFFECT OF GRAIN SIZE ON STRENGTH

Like cold-working, decreasing the grain size is a method of strengthening that can be applied to pure metals as well as to alloys. In contrast, the remaining strengthening methods to be discussed in this chapter require the presence of alloying elements. Typically, processes used to strengthen commercial alloys depend on more than one strengthening mechanism, and in practice the individual effects may be difficult to separate. In particular, the action of grain boundaries is similar to that of cell boundaries and to the effect of certain kinds of second-phase particles: All reduce the size of the individual element of volume within which a given dislocation can move freely. Since the strengthening effect of grain boundaries can be distinguished reasonably well from other factors in the case of pure metals, there is an advantage in obtaining an understanding of this aspect of the overall strengthening of alloys.

Data on a wide variety of metals and alloys show a common pattern of increase in strength with decrease in grain size. The functional relation involves the square root of the grain size, as shown by the data of Fig. 10.6. These data were obtained by performing tension tests on a series of samples of the same steel, each sample having been previously given its individual grain size by suitable preliminary treatments. The cells of dislocation structure discussed in the previous section show a similar linear dependence of yield stress on cell size, although the slope of the line is slightly lower.

The appropriate equation can be derived in terms of the piling-up of dislocations at grain boundaries. An external force can activate a dislocation source

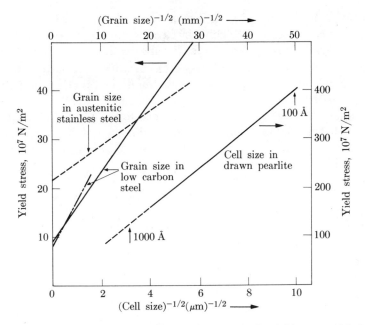

**Fig. 10.6** Typical experimental data showing the increase in yield stress with decrease in grain size or cell size. (Data of W. B. Morrison, of R. A. Grange, and of J. D. Embury *et al.*)

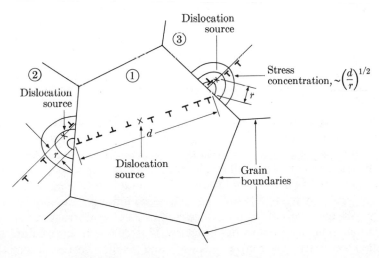

**Fig. 10.7** Schematic illustration of the interaction between dislocations in adjacent grains, which leads to the Hall-Petch relation.

(say, in grain 1, Fig. 10.7) and send a series of dislocations toward the adjacent grain boundaries. Since the neighboring grains are less favorably oriented for slip with respect to this applied force, the dislocations start to pile up at the grain boundaries. Eventually the stress concentration associated with a pile-up activates slip at a source some distance $r$ within grain 2 (or 3). Elasticity theory gives $(c/r)^{1/2}$ as the stress concentration at a distance $r$ from a crack of length $c$; in this case the length of "crack" is proportional to the number of piled-up dislocations, Fig. 4.19(a), and therefore to the grain diameter, $d$. Corresponding to an applied stress $\sigma$, an additional stress $\sigma'$ acts on a source within a neighboring grain:

$$\sigma' = K(\sigma - \sigma_i)(d/r)^{1/2}. \tag{10.2}$$

Here $K$ is a proportionality factor and $\sigma_i$ is the friction stress of the lattice, which opposes the usual gliding of a dislocation. $\sigma'$ reaches the value necessary to activate the source when the applied stress $\sigma$ equals the yield stress, $\sigma_y$. Equation (10.2) can therefore be put in the form,

$$\sigma_y = \sigma_i + kd^{-1/2}, \tag{10.3}$$

where the constant $k$ includes the average value of $r^{1/2}$. This relationship between yield strength and grain diameter is known as the *Hall-Petch equation*.

An illustration of the technical application of strengthening by grain refinement is given by ultra-fine-grained steels. Grain refinement is achieved by either: (1) severe hot deformation in the austenitic condition followed by controlled cooling to obtain further grain refinement from the $\gamma \to \alpha$ phase transformation with a minimum of grain growth; or (2) a series of rapid heating and cooling cycles to produce a sequence of $\alpha \to \gamma$ and $\gamma \to \alpha$ phase transformations. The resulting grain size is in the range near 1 $\mu$m and, as shown by the data of Fig. 10.6, substantially increases the yield strength. This mechanism of strengthening plus precipitation hardening is the basis of high-strength structural steels that permit substantial reduction of weight in many types of construction.

## STRENGTHENING BY ALLOYING

When one or more alloying elements are added to a given base metal, a wide variety of strengthening possibilities become available. A convenient classification of this important subject is based on the distinction between *solid-solution strengthening* and *second-phase strengthening*. If an alloying element simply cooperates with the matrix atoms in forming a solid solution (either random or ordered), the strengthening processes are relatively simple. More commonly, an alloying element combines with the atoms of the base metal to form a crystal structure significantly different from that of the pure base metal. This new crystal structure may be dispersed in the matrix phase in various ways and result in corresponding changes in properties. Special alloys and particular heat treatments are employed to take advantage of these possibilities for strengthening almost all of the common metals.

In addition, plastic deformation may be joined with a heat treatment in the form of a *thermomechanical treatment* to produce special combinations of strength and ductility.

**Solid-solution strengthening.** Solute atoms, either substitutional or interstitial, act as atomic-sized obstacles to the motion of dislocations. The resulting strengthening depends on the interaction between a solute atom and a dislocation, and on the number of solute atoms per unit area of the slip plane. The latter quantity is proportional to the square root of the solute concentration. The principal factors determining the degree of solute–dislocation interaction are relative atomic size, relative valence, and certain chemical and physical differences between the solvent and solute atoms. The action of these factors can be described in terms of elastic, electronic, and chemical effects on the motion of a dislocation through the alloy.

The effectiveness of a solute atom in impeding a dislocation through elastic interactions depends on three factors. First, the larger the difference in size between solute and matrix atoms, the larger is the elastic distortion. Second, if the solute produces a *tetragonal* distortion (as in the case of the interstitial solution of carbon in BCC iron), the elastic interaction is especially effective. The stress fields are then able to interact with both screw and edge dislocations (recall that *spherically symmetrical* stress fields interact only with edge dislocations). Finally, the solute atom may produce a distinctly different local value of the shear modulus, $\mu$. Because of the self-energy of a dislocation line, Eq. (3.14), a dislocation is attracted by the solute (if $\mu$ is lowered) or repelled by the solute (if $\mu$ is raised).

The electronic effects come into play because of the tendency of the valence electrons in an alloy to move from regions of compression to regions of tension in the stress field of a dislocation. The resulting local electric dipole is greater the higher the valence of the matrix atoms. The extent of its interaction with the solute depends principally on the difference in valence between the solute and the matrix. Chemical effects, arising from various sources, cause the solute atoms to interact with the region of stacking fault that is part of virtually all dislocations (as a result of their dissociation).

*Solute Atmospheres.* We have seen that solute atoms can interact with dislocations in various ways; one of these leads to accumulation of solute atoms (an atmosphere) around a given dislocation line. Such an atmosphere acts to immobilize or *pin* the dislocation line, with the result that a higher than normal stress is required to initiate plastic deformation. If the dislocation can become unpinned (free of the atmosphere), then deformation can continue at the usual, lower stress. The ability to form an effective atmosphere varies widely among solute elements and often is of negligible importance. An example of a strongly bound atmosphere is that of carbon (or nitrogen) in BCC iron. Because the interstitial solute produces a tetragonal pattern of lattice distortion, it interacts strongly with a dislocation line. As a result, the solute atom is bound to the dislocation line with an energy of about 0.5 eV per atom. Very small amounts of carbon or nitrogen are able to pin all of the dislocations (see Problem 2). Furthermore, even if unpinning is accomplished (by slight cold rolling, called skin rolling), the high rates of diffusion of carbon and nitrogen cause the atmospheres to gradually reform even at room temperature.

Solute atmospheres explain the occurrence of yield point phenomena, previously described in Fig. 4.15 for steels but also observed to some extent in several other alloys. The presence of a solute atmosphere requires the stress to rise to the upper yield point (well above the usual yield stress) before deformation begins. Those dislocations in regions of complex stress, near the gripped portion of the specimen, first become unpinned. They can then move at about the usual yield stress, so the stress on the specimen falls to a value known as the lower yield stress. Additional dislocations are generated in the local region in question, and the resulting local deformation (Lüders band) can be seen on the surface of the specimen. The main unyielded portion of the specimen resists deformation at the lower yield stress, but a gradual unpinning action occurs because of stress concentration at the boundaries with the yielded regions. Consequently, the entire specimen eventually experiences a deformation of several percent by this process of nonuniform yielding. This phenomenon causes surface roughness (stretcher strains) when susceptible sheet steel is given a typical cold-forming operation. In those portions of the steel that are subjected to deformation of only a few percent, differences of surface level are apparent and are not hidden by subsequent painting of the finished product. Skin rolling is a practical means for preventing stretcher strains.

*Order Hardening.* As we have seen in Chapters 6 and 7 atomic ordering may occur to varying degrees in alloys, but significant effects of ordering usually occur only when the concentration of solute atoms is about 20% or more. An essential concept here is that of antiphase boundary (APB). When two kinds of atoms form an ordered arrangement, Fig. 10.8(b), domains are formed within which the ordering is "in phase." In this case the word phase refers to a relative lattice position occupied by one of the atoms. An adjoining domain is characterized by a different phase, and a boundary (the APB) can be traced between the two domains, Fig. 10.8(c). Because the ordered state represents a low-energy condition for the solid solution in question, the energy increases with amount of APB. We will now consider two different ways in which the domain structure increases the strength of an alloy.

A dislocation of a special kind, the superdislocation, is needed for deformation of an ordered alloy. The unit Burgers vector $\mathbf{b_o}$ for a total dislocation in an ordered structure is typically twice as large as the Burgers vector $\mathbf{b_d}$ of the disordered structure, Fig. 10.8(a). Since the energy of a dislocation is proportional to $b^2$, total dislocations $\mathbf{b_o}$ are improbable. Deformation within an ordered domain, therefore, occurs as a two-step process. First a dislocation equal to $\mathbf{b_d}$ passes

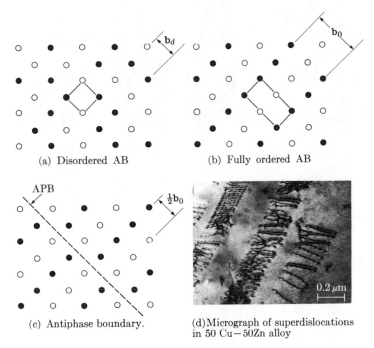

(a) Disordered  AB

(b) Fully  ordered  AB

(c) Antiphase boundary.

(d) Micrograph of superdislocations in 50 Cu—50Zn alloy

**Fig. 10.8** Schematic illustration of one type of ordered structure at the 50A—50B composition, showing the nature of an antiphase boundary (APB). Typical electron-microscope images of superdislocations are illustrated.

through the domain, Fig. 10.8(c), but the result is the creation of APB. The passage of a second dislocation of the same kind eliminates the region of high-energy APB and restores the ordered structure. This behavior of a superdislocation, Fig. 10.8(d), is analogous to that of the usual partial dislocation with a ribbon of stacking fault. Examples of strengthened ordered phases are found in superalloys, where second-phase particles become more effective obstacles to the passage of a dislocation because of ordering.

Ordered domains can produce strengthening by a second mechanism, basically the same as dispersion hardening, discussed in the following section. In addition to changes in identity of nearest neighbors on ordering, several alloys experience appreciable changes in atomic spacing; for example, cubic crystals become tetragonal. If such an alloy is initially disordered (as a result of quenching from a suitably high temperature), the process of ordering will occur when the alloy is held at a moderate temperature within the region of stability of the ordered phase, given by the phase diagram. Ordering begins at many centers. As these grow and eventually impinge on one another they create not only an array of domains (typically of four different "phases") but also a pattern of lattice distortions due to tetragonality. The resulting strengthening effect can be large. An interesting

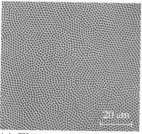

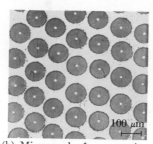

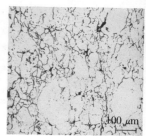

(a) Electron micrograph showing the cross-section of long rods of Al₃Ni phase in a matrix of aluminum solid solution, produced by directional solidification of the Al-Al₃Ni eutectic.

(b) Micrograph of a composite showing the cross-section of long boron fibers embedded in an aluminum alloy. The bright central area represents the thin tungsten wire on which the boron fiber was formed by vapor deposition.

(c) Micrograph of a SAP alloy, consisting of about 7 wt.% Al₂O₃ powder dispersed in an aluminum matrix. (*Courtesy R. G. Connell, University of Florida.*)

**Fig. 10.9** Examples of distributions of phases produced by three special procedures. In each instance the dispersed phase(s) strengthens the matrix. [Parts (a) and (b) courtesy United Aircraft Research Laboratories.]

example is the CoPt composition in which the strengthening results in magnetic "hardness." This alloy has such exceptionally high coercive force that it finds use in critical applications in spite of the high cost of platinum.

**Strengthening by second-phase particles.** Often the best practical combinations of properties, including high strength, are obtained from the presence of one or more phases in the metallic matrix, which is usually a solid solution. Various manufacturing processes are employed in producing multiphase metallic materials. Most are produced by "ordinary alloying," involving melting, addition of alloying elements, casting, and hot forming. Important special arrangements of the second phase, however, can be obtained only by departures from this ordinary procedure. Conventional casting, for example, must be replaced by controlled *directional solidification* to produce aligned plates or rods of a second phase, Fig. 10.9(a). A similar aligned configuration of a second phase can be produced in a *composite*, Fig. 10.9(b), which is manufactured by infiltrating a molten alloy into a bundle of high-strength fibers, typically boron or carbon. "Alloying" in the ordinary sense need not occur in a composite. Powder-metallurgical techniques permit the fine dispersion of spherical oxide particles throughout a metallic matrix. Al₂O₃ in aluminum, known as SAP (sintered aluminum powder), Fig. 10.9(c), and ThO₂ in nickel are examples.

Since the character of the second phase and of the matrix mainly determine the properties of a material, strengthening mechanisms can be treated without regard to the process by which the microstructure in question was produced. From this point of view the principal factor is the nature of the dispersed particles. Among the various types of particles, three merit special attention. First is the

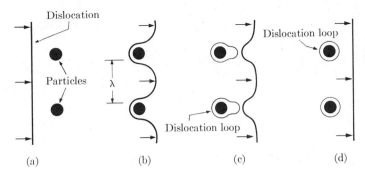

**Fig. 10.10** A dislocation line being forced through second-phase particles lying in its slip plane. This mechanism results in Orowan strengthening.

transition configuration between the individual atom (in solid solution) and a recognizable second phase; the concepts "cluster" and "Guinier-Preston (GP) zone" (see Chapter 9) describe stages in this transition. The other two types of particle (usually a second phase) are distinguished by their behavior under the deformation that accompanies the motion of dislocations. *Impenetrable particles* are strong enough to resist penetration by a moving dislocation; *deformable particles* can be penetrated by a dislocation and therefore share in the overall deformation.

*Impenetrable Particles.* When a moving dislocation encounters "strong" particles (Fig. 10.10), the dislocation line first bends around the particles. If the particles are able to resist penetration, then the externally applied stress forces the dislocation to pass between adjacent particles. Segments of the dislocation line are constrained to form a dislocation loop around each particle, thus freeing the main dislocation line to continue its passage through the alloy. This mechanism is known as *Orowan strengthening* after the man who first described it. The applied shear stress $\tau$ necessary to force dislocations between particles is (see Problem 4)

$$\tau = \frac{2T}{b\lambda}, \tag{10.4}$$

where $\lambda$ is the mean-free spacing of the particles and $T$ is the force needed to overcome the line tension of the dislocation. Substitution of the expression for $T$ given by Eq. (3.18) leads to the desired relation,

$$\tau = \frac{\mu b}{2\pi\lambda} B \ln\left(\frac{\lambda}{2r_0}\right). \tag{10.5}$$

$B$ is unity for screw dislocations, $(1 - v)^{-1}$ for edge dislocations, and $(1 - v)^{-1/2}$ as an average for mixed dislocations. The success of Eq. (10.5) in describing the strengthening of copper by a dispersion of BeO particles is shown in Fig. 10.11.

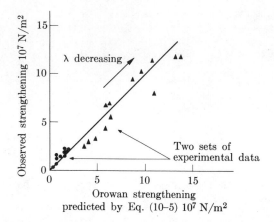

**Fig. 10.11** Comparison of observed and calculated strengthening of copper by BeO particles characterized by various values of mean-free spacing, $\lambda$. (After M. F. Ashby)

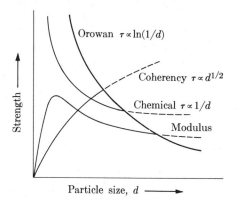

**Fig. 10.12** Qualitative curves describing strengthening as a function of particle size for three principal mechanisms involving deformable particles. The Orowan mechanism (involving impenetrable particles) sets the upper limit of strengthening for a given particle size. (After P. M. Kelly)

*Deformable Particles.* The apparent "strength" of a particle depends greatly on the inter-particle spacing, and thus on the average size of the particles. During precipitation hardening, for example, the total volume of precipitated particles is almost constant. In the first stages of precipitation, therefore, the tiny particles are very closely spaced, so Eq. (10.5) predicts that a high shear stress would be required to by-pass these obstacles. Actually, the yield stress of such an alloy is limited by deformation of the particles. Figure 10.12 illustrates qualitatively the contributions of three mechanisms of strengthening in this case as a function of particle size. If the precipitation-hardening heat treatment is carried so far that very large particles are produced, these particles are easily by-passed and yielding occurs at a low stress.

The complex behavior of deformable particles can often be explained in terms of the three mechanisms introduced in Fig. 10.12. *Coherency* between the crystal structure of the matrix and of a cluster, GP zone, or second-phase particle causes straining of the lattice due to differences in lattice spacing. The difficulty of moving a dislocation through the region of strain increases with increase in mismatch $\delta$ [see Eq. (9.27)] between the two lattices. *Chemical hardening* refers to two effects connected with the difference in chemical composition between the particle and the matrix. Since the matrix/particle surface has an associated interfacial energy, a corresponding force must act when a particle is sheared and additional interface is produced. The second effect is related to the change in distribution of solute atoms caused by a process of slip. For example, if the atoms were initially ordered, then unit slip produces an antiphase domain boundary and an increase in energy. The elastic shear *modulus* is singled out as the third effect because of its strong influence on such basic properties of dislocations as the line tension and self-energy. For example, as a dislocation attempts to pass through a region of second phase that has a higher modulus than the matrix, the dislocation experiences a resistance, since its energy must increase.

## HEAT TREATMENT OF STEEL

The hardening of steel is the best known example of strengthening of metallic materials. In view of its commercial importance, this topic is presented here in sufficient detail to serve as a background for industrial practice. We begin with a consideration of treatments closely related to *equilibrium* conditions of a steel, building on the foundation of the iron-carbon phase diagram (Chapter 7). *Nonequilibrium* decomposition of austenite is the next topic, and it leads naturally to a description of the quantitative aspects of the hardening and tempering of typical alloys.

**Equilibrium conditions.** Because of the importance of the iron-carbon phase diagram, Fig. 10.15 (or the corresponding ternary or higher diagram involving one or more alloying elements), it is convenient to begin with heat treatments that are closely related to the equilibrium diagram. The microstructures in question are rarely strictly "in equilibrium"; for example, graphite is the true equilibrium phase, but it appears in steels (as the objectionable phenomenon of *graphitization*) only after prolonged service at high temperatures. Nevertheless, a useful qualitative distinction can be made between the hard, strong products of nonequilibrium reactions (such as martensite, Fig. 10.13a) and the relatively soft "equilibrium" products such as pearlite and spheroidite. The equilibrium heat treatments include austenitizing, annealing, normalizing, and spheroidizing. Diagrams other than the equilibrium diagram must be used in considering hardening heat treatments that require rapid cooling to obtain a nonequilibrium structure.

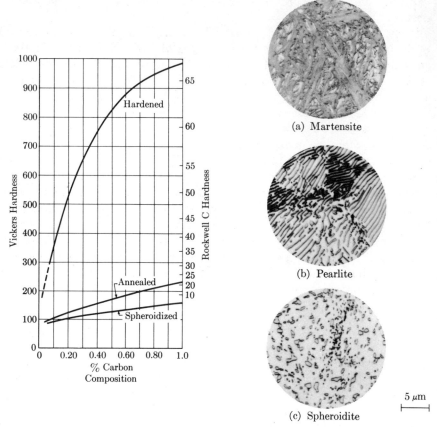

(a) Martensite

(b) Pearlite

(c) Spheroidite

5 μm

**Fig. 10.13** The approximate hardness of carbon steels after three different heat treatments, and the corresponding microstructure in a 0.8% C steel. (After E. C. Bain. Courtesy United States Steel Research Laboratory)

*Austenitizing.* The face-centered-cubic form of iron, austenite, that exists at high temperatures is capable of dissolving as much as two percent carbon. When the austenite decomposes on cooling, only a small fraction of its carbon content can dissolve in the body-centered-cubic iron (ferrite) that exists at low temperatures, and the bulk of the carbon appears in another form, $Fe_3C$ for example. The exact nature of austenite decomposition varies with the type of steel heat treatment, but almost all heat treatments require that austenite be produced as a first step in heat-treating operations. The equilibrium diagram of Fig. 10.15 shows the lowest temperatures at which 100% austenite can be obtained. This temperature varies with carbon content; with increasing carbon content it decreases along the $A_3$ line to a minimum value $A_1$ at the eutectoid composition (0.8% carbon) and then increases along the $A_{Cm}$ line. However, these lowest temperatures are not used in practical heat treatments because of the slow reaction rates they produce.

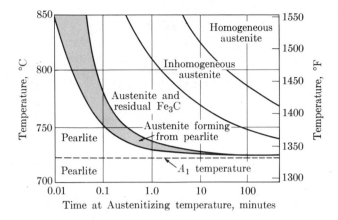

**Fig. 10.14** Approximate times necessary for the isothermal formation of austenite in a normalized eutectoid steel at various austenitizing temperatures. (After G. A. Roberts and R. F. Mehl)

Figure 10.14 shows several important aspects of the austenitizing of eutectoid steels as a function of time and temperature. The temperature must be somewhat above the equilibrium temperature $A_1$ to produce austenitization in a reasonable time. The first stage in the production of homogeneous austenite is the nucleation and growth of austenite from pearlite. However, after complete disappearance of the pearlite some carbide particles remain in the austenite, and even after the carbides have dissolved, their trace remains for some time in the form of inhomogeneities of carbon concentration in the austenite.

Every stage in the formation of homogeneous austenite is accelerated by (1) increasing the temperature, and (2) increasing the fineness of the initial carbide particles. Both these factors are taken into account in commercial heat-treating operations, but temperature is the more important. The temperature used for hypoeutectoid and eutectoid steels is about 60°C (100°F) above the minimum temperature for 100% austenite, and the corresponding time is one hour per 25 mm of cross section of the bar being heated. Much of this time is used in raising the bar to the temperature range in which austenitization occurs. Among the reasons for keeping the austenitizing temperature as low as possible is the increased tendency toward (1) cracking and distortion, (2) oxidation and decarburization, and (3) grain growth, as the temperature is raised. Also, higher temperatures and longer heat-treating add to the cost of the process.

At the high carbon contents of hypereutectoid steels the temperature needed for obtaining 100% austenite is frequently quite high. Fortunately, austenite suitable for hardening in such cases is obtained at about 775°C (1425°F). The small amount of equilibrium carbide dispersed in the austenite has little effect on the final mechanical properties, and it may actually improve wear resistance. Of course if the carbide forms an embrittling network at the grain boundaries,

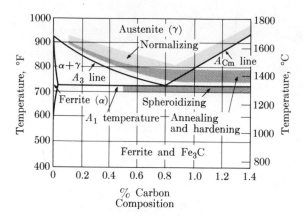

**Fig. 10.15** A portion of the iron-carbon equilibrium diagram showing the temperature ranges usually employed in various heat treatments.

it is necessary to eliminate this undesirable structure by a preliminary heating above the $A_{Cm}$ line followed by moderately fast cooling. Very low-carbon steels, too, must be heated to high temperatures to austenitize them completely, but steels containing less than about 0.2% carbon respond poorly to hardening heat treatment and are seldom heated for this purpose.

*Annealing.* The word *annealing* has the general meaning of *holding at a high temperature* and is therefore used to describe scores of heat treatments. However, in steel heat-treating it usually refers to furnace cooling from the austenitizing range shown in Fig. 10.15. This treatment is used primarily to soften certain steels to the point of optimum machinability, but it may also be used for grain refinement or chemical homogenization. It might be well to review at this point the treatment of phase analyses of the annealing of steels of various carbon contents, discussed in Chapter 7 (especially Fig. 7.27). (See Problem 5 of the present chapter.) It will be recalled that proeutectoid reactions produce primary ferrite or cementite during equilibrium cooling of hypoeutectoid or hypereutectoid steels. These reactions tend to occur also during faster cooling, although in decreased amount. For simplicity this chapter is limited to eutectoid steels in which the proeutectoid reactions are absent.

*Normalizing.* The name of this heat treatment was perhaps suggested by the fact that steels are *air cooled* from the usual manufacturing operations, like hot rolling and forging, carried out on steels in the austenite range. Therefore to put a steel in the normal condition it is air cooled from the range of austenitizing temperatures shown in Fig. 10.15. The properties of such normalized steels depend on the chemical composition and on the cooling rate; the latter in turn is determined by the size of the bar being considered. Although these two factors can produce considerable variation in the hardness of normalized steels, usually the micro-structure contains relatively fine pearlite, and a phase analysis based on equilibrium cooling is not greatly in error for low- and medium-carbon steels. The principal uses of normalizing are to refine the structure of steel castings and of overheated steels and to eliminate the carbide network at the grain boundaries of hypereutec-toid steels. These structural changes improve the engineering properties of the alloys.

*Spheroidizing.* The spheroidized structure shown in Fig. 10.13 produces a steel with minimum hardness and maximum ductility. At high carbon or alloy contents this structure is needed for optimum machinability and for the prevention of cracking during cold-forming operations. A number of heat treatments can be used to produce a spheroidized structure, but all of them are relatively long and costly. Since a fine initial carbide size accelerates spheroidization, normalized steels are often chosen as the material to be processed. Spheroidization of the fine pearlite takes place when the steel is heated to just below the $A_1$ temperature and held there for about ten hours. Other methods involve holding the steel alternately just above and just below the $A_1$ temperature, or employ a high-temperature isothermal transformation of the austenite.

**Nonequilibrium decomposition of austenite.** Many heat treatments of steel involve conditions of reaction so far removed from equilibrium that equilibrium diagrams are of only limited use. Their usefulness is restricted to fixing the austenitizing temperature and predicting the phases that are *eventually* obtained at a given composition and temperature. The principal source of information on the actual process of austenite decomposition under nonequilibrium conditions is a TTT diagram, previously described in Fig. 9.17. Other names for this diagram are S-curve or C-curve.

*TTT diagram.* A given TTT diagram describes the decomposition of austenite in a given steel. Therefore, for each steel composition there is a different diagram. Furthermore, such factors as the grain size of the austenite and the presence of inclusions or other inhomogeneities can change the diagram for a given steel composition, although these complications are frequently neglected. The diagram for eutectoid carbon steel, Fig. 10.16, shows that the austenite decomposes in two radically different ways. At high temperatures the reaction products form *with increasing time* at constant temperature. Martensite, on the other hand, forms only *with decreasing temperature* and has almost no tendency to continue forming at constant temperature. These two modes of austenite decomposition will be considered next.

*Pearlite and Bainite.* Under the conditions used in determining a TTT diagram, both pearlite and bainite form *isothermally* from austenite; that is, they form with increasing time at constant temperature. As Fig. 10.16 shows, pearlite is the decomposition product of austenite at temperatures between $A_1$ and about the

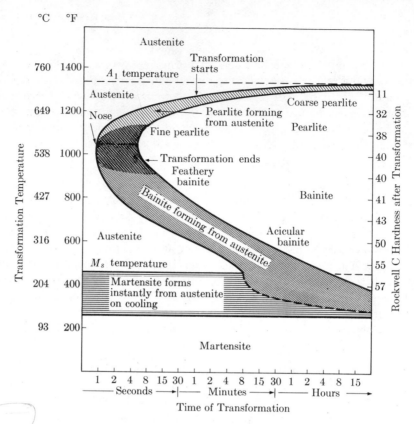

**Fig. 10.16** The TTT diagram for the decomposition of austenite in a eutectoid carbon steel.

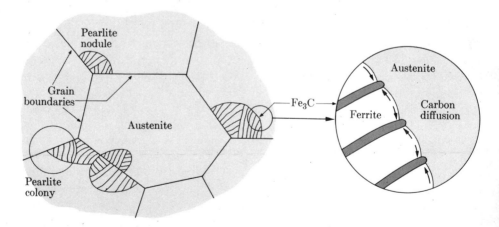

**Fig. 10.17** Schematic illustration of the transformation of austenite to pearlite. After preferential nucleation at a grain boundary, a colony of lamellar plates of cementite ($Fe_3C$) and ferrite grow by the diffusion mechanism shown. Several colonies comprise a rounded nodule of pearlite.

(a) Feathery bainite formed at
495°C[925°F].  Fine pearlite is also
present in the martensite matrix.

(b) Acicular bainite formed at
290°C[550°F].  The matrix is martensite.

**Fig. 10.18**  Bainite formed isothermally at two different temperatures in a eutectoid steel. (Courtesy United States Steel Research Laboratory)

"nose" (or "knee") of the TTT diagram.  When austenite is cooled to a temperature within this range, there is an initial period in which no evidence of pearlite formation can be detected.  This behavior is similar to the incubation period in the recrystallization of cold-worked metals.  Eventually, decomposition of the austenite begins with the formation of a region of cementite that acts as the nucleus for a colony of pearlite.  This nucleation occurs preferentially at the grain boundaries of the austenite or at an inhomogeneity such as an inclusion.  The joint growth of neighboring colonies leads to a nodule of pearlite, Fig. 10.17.  The interface between the growing pearlite and the austenite is incoherent.  It advances at a constant velocity (provided that the temperature is constant) and tends to produce a hemispherical nodule.  The mechanism of growth of pearlite involves the rejection of carbon by the growing ferrite plates and its incorporation into the cementite. This process requires diffusion of the carbon by one or more of three paths; through the austenite, through the ferrite, or along the interface. The lower the temperature of transformation, the larger is the amount of chemical energy driving the reaction. Consequently a finer spacing of the pearlite is possible, since the necessary interfacial energy can be supplied.  Because the distance of diffusion is less, the overall growth process can occur more rapidly.  This reasoning explains why the rate of formation of pearlite increases and the interlamellar spacing decreases as the temperature of transformation is lowered from $A_1$ to the knee of the TTT curve.

In a temperature range near the nose of the TTT diagram both pearlite and bainite tend to form, as seen in Fig. 10.18(a).  However, at temperatures below this range and above the $M_s$ point (where martensite starts to form) bainite is the only product of isothermal decomposition of austenite. The aspect of bainite changes with the temperature at which it forms. Figure 10.18 shows the *feathery* bainite obtained in the upper part of the temperature range and the *acicular* (needlelike) bainite produced by lower reaction temperatures. The process of formation of bainite is similar to that of pearlite in that it is characterized by a sigmoidal curve, Fig. 9.17(b), and the product of reaction is a ferrite-carbide aggregate. The initial nuclei, however, are ferrite particles in the grain boundaries of the austenite. In feathery bainite the carbide particles form between plates of ferrite; in acicular bainite the carbide precipitates within the ferrite. The TTT diagrams of some alloy steels show two noses, one of which is associated with pearlite formation and the other with bainite formation. In these cases it may be possible for bainite to form during a quenching operation, but in carbon steels isothermal reaction is necessary to produce the characteristic bainitic structure.

*Martensite.* Although the austenite in a eutectoid steel is unstable at all temperatures below $A_1$, it has been seen that the face-centered-cubic austenite does not change *immediately* to body-centered-cubic ferrite (plus cementite) on being cooled to temperatures only moderately below $A_1$. At these temperatures the decomposition occurs by nucleation and growth of the equilibrium phases, and the diffusion process involved takes a relatively long time. At a sufficiently low temperature, however, the driving force that tends to cause the transformation from face-centered-cubic to body-centered-cubic iron becomes so strong that the change occurs by a displacive transformation (see Chapter 9) without diffusion of carbon. Since the amount of carbon present is much more than can be held in solid solution, the body-centered-cubic lattice is distorted into a tetragonal *martensite*. The interstitially dissolved carbon atoms cause solid-solution hardening, as described previously, but the principal sources of the strength and hardness of martensite are the elastic distortions produced during the formation of the platelets of martensite. Because the tetragonality of martensite increases with carbon content, the hardness also depends strongly on carbon content (Fig. 10.13).

The upper temperature at which martensite begins to form is called the $M_s$ point. At this temperature only about one percent martensite has formed during cooling of the austenite. As austenite is cooled to lower and lower temperatures the amount of martensite increases, until at the $M_f$ point the steel is essentially all martensite (99%) after rapid cooling. If austenite is rapidly cooled to a temperature between $M_s$ and $M_f$, the austenite that has not changed to martensite *during cooling* transforms *isothermally* to bainite during prolonged holding at the reaction temperature. Carbon and most other alloying elements decrease both the $M_s$ and $M_f$ temperatures, so that in certain steels even the $M_s$ point is below room temperature. Certain properties of alloy steels with an $M_f$ point below room temperature can be improved by the use of a refrigeration treatment that causes additional martensite to form. Although it has been shown that the austenite-to-martensite

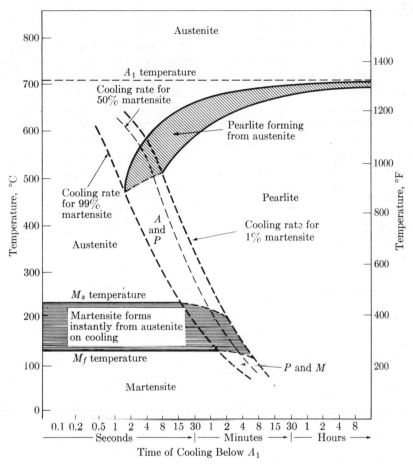

**Fig. 10.19** An approximate CCT (continuous cooling transformation) diagram for eutectoid carbon steel.

reaction requires a finite time, martensite formation in commercial steels is practically instantaneous.

*Continuous Cooling Diagram.* Commercial heat-treating operations on large pieces of steel involve cooling from the austenitizing temperature at rates that are low compared with those used in obtaining TTT curves. We should not assume, therefore, that a given piece of steel can be cooled instantly to any temperature (to produce martensite, for example). It would be useful to have a diagram showing the changes that occur in a given steel during cooling at the various rates produced by commercial quenching operations. The complications in the actual cooling of metals make the construction of such a diagram extremely difficult, but an approximation to it is the diagram showing the transformation of austenite at the rates of cooling in various positions in a Jominy bar (see Fig. 10.23). By analogy with the more common TTT diagram, this diagram may be termed a CCT (continuous cooling transformation) diagram.

Although few accurate CCT diagrams are available, Fig. 10.19 is an approximation to the one for a eutectoid carbon steel. Figure 10.16 is the corresponding TTT diagram. The most important cooling rates on the CCT diagram are those that result in 99% and in 50% martensite. The first of these, the critical cooling rate, determines the position of the nose of this diagram, the time for which is about twice the time for the nose of the TTT diagram. The cooling rate that produces 50% martensite is very significant in practical heat-treating, since full hardening is usually defined in terms of 50% martensite at the center of the bar. No bainite forms in a eutectoid carbon steel during continuous cooling, and even in alloy steels the bainite reaction may be slowed down one thousandfold as a result of changes that take place in the earlier stages of continuous cooling.

**Hardening heat treatments.** The basic requirements for hardening a steel by means of the eutectoid-decomposition reaction can be summarized as follows:

1. *Adequate carbon content.* In view of the nature of martensite hardness, it is clear that some carbon must be present to produce hardening of a steel. Figure 10.13 shows that the maximum hardness increases with increasing carbon content, but since the ductility decreases rapidly, the carbon content is held near 0.45% in many engineering steels. For wear resistance the carbon content may be increased to over 1.0%, for example in tool and die steels.

2. *Nonequilibrium austenite decomposition.* Inasmuch as austenite decomposition under equilibrium conditions may produce coarse pearlite that is relatively soft, various degrees of hardening are represented by fine pearlite, bainite, and martensite. Although the term *steel hardening* usually refers to the production of martensite, the other two reaction products are frequently desired. The rate at which austenite must be cooled to produce a given nonequilibrium reaction product is widely variable, but quenching is always used.

Tempering is discussed on page 490, but it can be noted here that tempering is *not* used to harden a steel. The purpose of tempering is to restore a portion of the ductility that is lost by hardening, and often an appreciable softening is produced by the tempering heat treatment. It is also notable that alloying elements are not necessary for steel hardening and do not significantly increase the *hardness** of martensite obtainable in a steel of given carbon content. On the other hand, carbon is an essential component if an alloy of iron, possibly with one or more alloying elements, is to be capable of hardening by eutectoid decomposition.

---

* The *hardness* of martensite, which is unaffected by alloying elements, must be distinguished from the *hardenability* of a steel, a property strongly increased by the principal alloy additions to steel. Hardenability refers to the *depth* of martensite that can be produced by given heat-treating conditions, and is not concerned with the *hardness* of the martensite so obtained.

*Quenching.* The rate at which a given portion of a steel bar cools from the austenitizing temperature depends on two factors: (1) the temperature to which the surface of the bar is cooled by the quenching medium, and (2) the rate of heat flow in the steel bar itself. This rate is relatively low, so that the center of a large bar of steel would cool slowly even if the surface of the bar could be cooled instantly to room temperature. Although heat flow is evidently important in steel hardening, it is subject to little variation and will not be considered further.

If the only object in commercial quenching of steel were to lower the surface temperature of the bar as rapidly as possible, a water solution would be a suitable cooling medium. It will be seen later that cooling in water often approximates the *ideal quench*, instantly lowering the surface temperature of the bar to room temperature; but an additional requirement in quenching is that cracking and excessive distortion must be prevented. Since the large temperature gradients that cause distortion are an inevitable consequence of using a drastic quenching medium, it is often necessary to use a milder cooling medium such as oil or an air blast.

**TABLE 10.2**
Approximate $H$ Values for
Several Quenching Mediums

| Type of quench | $H$ value |
| --- | --- |
| Ideal | $\infty$ |
| Agitated brine | 5 |
| Still water | 1 |
| Still oil | 0.3 |
| Cold gas | 0.1 |
| Still air | 0.02 |

*Hardenability.* The aim of a quenching operation is to produce the desired non-equilibrium decomposition product of austenite, usually martensite. *Hardenability* plays a decisive role in such successful hardening. In the absence of adequate hardenability the most drastic quench is incapable of producing martensite in a steel bar of a given size. A convenient method of separating the effects of quenching and of hardenability on martensite formation involves the use of $H$ values* for various types of quenching. The ideal quench, which would instantly cool the surface of a hot bar to room temperature, has an $H$ value of infinity ($\infty$). Actual quenching mediums have $H$ values from 10 to 0.02, and their effectiveness compared with the ideal quench can be determined. The $H$ values of several quenching mediums are given in Table 10.2.

*Ideal Critical Diameter.* The definition of hardenability is conveniently given in terms of a specific measure of hardenability, that is, the *ideal critical diameter* $D_I$.

---

* $H$ represents the cooling power of the quenching medium; it does not stand for hardenability.

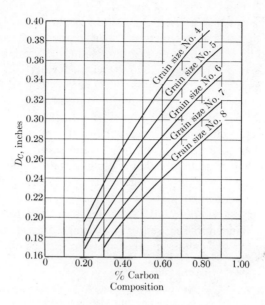

**Fig. 10.20** A plot of the base diameter $D_C$ used in calculating the ideal diameter of a steel. (Courtesy M. A. Grossman)

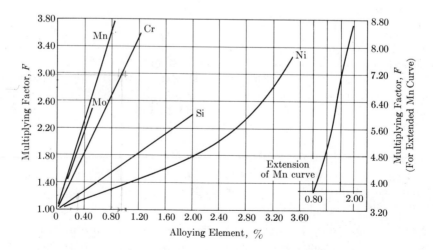

**Fig. 10.21** Multiplying factors $F$ for several of the common alloying elements. (Courtesy Boyd and Field)

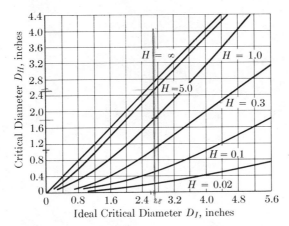

**Fig. 10.22** The relation between ideal critical diameter and critical diameter $D_H$ that can be fully hardened using a quenching medium with a given cooling power $H$.

This hardenability value for a given steel is *the diameter in inches\* of a cylindrical bar that will form 50% martensite at the center during an ideal quench.* (The 50% martensite value is arbitrarily chosen for convenience in measuring.) The significance of this hardenability value is that a bar of this steel with a diameter *larger* than the ideal critical diameter cannot be hardened all the way through its cross section even by an infinitely rapid quench. Thus hardenability is an index of the depth to which martensite can be formed in a given steel as the result of a given hardening treatment. Hardenability is not an indication of the *hardness* of a steel. The hardness of martensite is determined by carbon content, and a high-carbon steel may have a relatively low hardenability.

The ideal critical diameter can be determined experimentally, but it can also be calculated from the chemical composition of the steel. For the latter purpose two types of factor are required: a base diameter $D_C$ that depends on carbon content and grain size,† Fig. 10.20, and multiplying factors $F$ for each of the alloying elements, Fig. 10.21. The ideal critical diameter is the product of $D_C$ and all the multiplying factors for the alloying elements in the given steel. Thus the ideal critical diameter of steel with a No. 8 grain size and containing 0.5% C, 0.6% Mn, 1% Cr, and 2% Ni is

$$D_I = D_C \times F_{Mn} \times F_{Cr} \times F_{Ni},$$

$$D_I = 0.22 \times 3.00 \times 3.17 \times 1.77 \tag{10.6}$$

$$= 3.70 \text{ inches.}$$

---

\* Historically, the treatment of hardenability was developed in terms of inches (rather than millimeters). To convert from inches to millimeters, multiply by 25.4.

† The meaning of the ASTM grain size numbers is shown by the ordinates of Fig. 9.31.

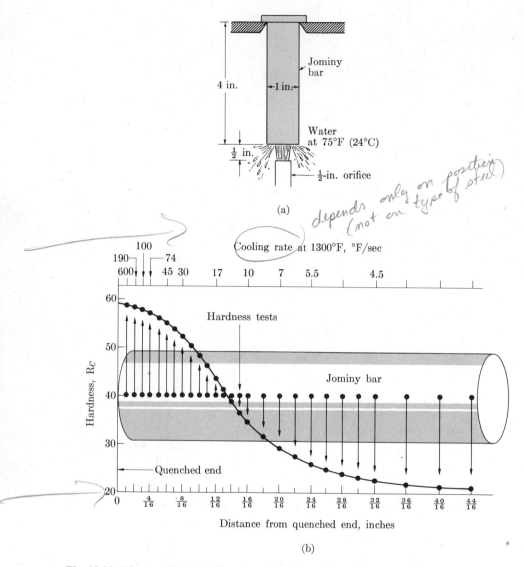

*depends only on position (not on type of steel)*

(a)

(b)

**Fig. 10.23** The steps in obtaining a Jominy curve. (a) Procedure for end-quenching a Jominy bar after transfer to the quenching jig from the austenitizing furnace. (b) Jominy curve obtained by plotting the hardness values measured along the length of the quenched bar.

Since the ideal critical diameter represents the size bar that could be fully hardened (50% martensite at the center) by an *ideal* quench, it is evident that only smaller bars of this steel can be fully hardened by such *actual* quenching mediums as oil or water. Figure 10.22 shows the relation between *ideal* critical diameter

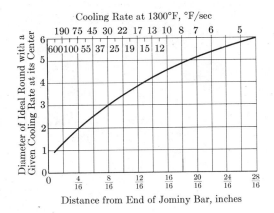

Fig. 10.24 The correspondence between cooling rates at the center of ideal rounds and at various distances from the quenched end of a Jominy bar.  To convert from °F/s to °C/s, multiply by 0.55.

and the critical diameter $D_H$ that can be fully hardened by means of a quenching medium with a given $H$ value.  From this chart it can be determined that for a steel with $D_I - 3.70$ inches, $D_{1.0} = 2.80$ inches and $D_{0.3} = 1.75$ inches.  That is, a 2.80-inch bar of this steel can be fully hardened by water quenching, and a 1.75-inch bar by oil quenching.

*Jominy Test.*  Hardenability is a property of steel, and a given heat* of steel has a definite hardenability, which may be expressed in terms of ideal critical diameter.  However, other measures of hardenability, such as that given by a *Jominy* test, may also be used.  This test is standardized and can be performed easily by both users and manufacturers of steel.  Figure 10.23(a) shows the procedure for end-quenching the standard Jominy bar after it has been given a suitable austenitizing treatment.  This type of quenching produces in the bar a wide range of cooling rates that decrease from a maximum at the quenched end.  Figure 10.24 gives the cooling rate at various distances from the quenched end and also shows the corresponding size of ideal round that has the same cooling rate *at its center*.  Hardness tests along the length of the end-quenched bar are plotted to obtain a Jominy curve, Fig. 10.23(b).

Typical Jominy curves for a number of AISI (American Iron and Steel Institute) steels, Table 10.3, are shown in Fig. 10.25(a).  A Jominy curve is strictly valid only for a given heat of the specified steel, since the permitted range in chemical composition allows an appreciable range in hardenability.  Figure 10.25(b) shows the *hardenability band* within which the hardenability curves for all heats of AISI 4145*H* steel will lie.  The wide possible variation in hardenability among bars of

---

* A *heat* is the "batch" ($\sim$ 100 tons) of steel produced by a steel-making process.

**TABLE 10.3**

Brief Descriptions of Some AISI Steels

| AISI number | Average alloy content* | Average hardenability of 0.45% carbon steel; ideal critical diameter in inches | Approximate cost relative to carbon steel |
|---|---|---|---|
| 10xx | None | 1.1 | 1 |
| 13xx | 1.8–2.0% Mn · | 3.3 | 1.2 |
| 41xx | 0.5–1.0% Cr, 0.2–0.3% Mo | 6.3 | 1.3 |
| 43xx | 0.5–0.8% Cr, 1.8% Ni, 0.3% Mo | 7.5 (0.40% C) | 1.6 |
| 51xx | 0.8–1.1% Cr | 3.3 | 1.2 |
| 61xx | 0.8–1.0% Cr, 0.1–0.2% V | 3.7 | 1.4 |
| 86xx | 0.6% Ni, 0.5–0.7% Cr, 0.2% Mo | 4.0 | 1.4 |
| 87xx | 0.6% Ni, 0.5% Cr, 0.3% Mo | 4.4 | 1.4 |
| 92xx | 1.4–2.0% Si, 0.6–0.9% Mn, 0–0.7% Cr | 2.7 (0.55% C) | 1.3 |

* All the steels contain less than 1.0% Mn, 0.05% P, 0.05% S, 0.35% Si, and residual amounts of other alloying elements unless otherwise noted.

steel of the same AISI grade must be considered in choosing a steel to meet a hardenability requirement. In mass production heat treatments it is generally uneconomical to choose a steel of such high hardenability that even bars with the minimum hardenability would be satisfactory, since this would require a more expensive steel. However, to avoid excessive rejections because of inadequate hardening, a suitable steel must possess a reasonably high average hardenability.

The decomposition of austenite in a portion of a quenched bar is governed by the cooling rate at that place in the bar. Correlation of the cooling rates at various depths below the surface of an actual steel part with the corresponding cooling rates in a Jominy test makes it possible to predict the hardness distribution that would be produced by a given quenching medium. Often the hardness (and strength) must be above a specified value throughout a given steel part. Since the center hardness is the minimum value in a steel bar, attention can often be restricted to this point in the bar. The Jominy method allows consideration of center hardnesses different from the hardness corresponding to 50% martensite used in defining ideal diameter. However, in practice the "half-hard" criterion is widely used in Jominy calculations, and the *Jominy distance* is defined as the distance from the quenched end of the Jominy bar, at which 50% martensite forms. In alloy steels containing 0.45% carbon the hardness of the 50% martensite structure is about 45 Rockwell C and is a few points lower if no alloying elements are present.

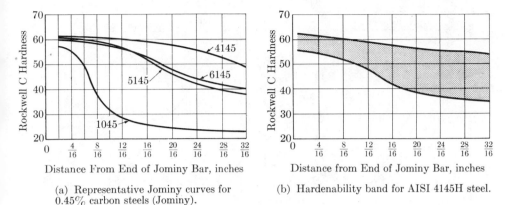

(a) Representative Jominy curves for 0.45% carbon steels (Jominy).

(b) Hardenability band for AISI 4145H steel.

**Fig. 10.25** Two common methods of presenting data on Jominy hardenability.

*Interrupted Quenching.* The rapid cooling to room temperature ordinarily used to cause martensite formation has the disadvantage of setting up severe stresses in the steel. These stresses are the combined effect of contraction during cooling and expansion caused by the martensite reaction, and they can easily cause cracking or distortion of heat-treated parts. The very slow cooling rate produced by cooling steel in still air from the austenitizing temperature decreases quenching stresses to a minimum, but steels with sufficient hardenability to form martensite on air cooling are expensive. Such *air-hardening steels* are chosen for complex dies for which preliminary machining costs are high; in these cases it is economical to protect the investment in machining operations by the use of a steel that can be hardened with a minimum danger of distortion or cracking. Almost equal freedom from quenching stresses can be obtained in steels of moderate hardenability by an interrupted quenching procedure. If the steel can be rapidly cooled past the nose of the CCT diagram, Fig. 10.19, it can then be cooled slowly through the martensite reaction or be allowed to form lower bainite.

The most generally useful type of interrupted quenching is called *martempering.** It is a *hardening* operation that produces martensite; it is not *tempering.* The steel is quenched in a molten salt bath held at a temperature near the $M_s$ point, Fig. 10.26(a). When the entire piece has reached the bath temperature and before the bainite reaction starts, the steel is removed from the bath and air cooled. The martensite that forms on air cooling is just as hard as that formed by water quenching, but serious stresses are not set up. Although subsequent tempering is not required to remove quenching stresses, it is customarily used to increase ductility. A variation of martempering is *timed quenching,* in which the steel is quenched in oil or water for a time sufficient to decrease its average temperature to about the

---

* Also called *marquenching.*

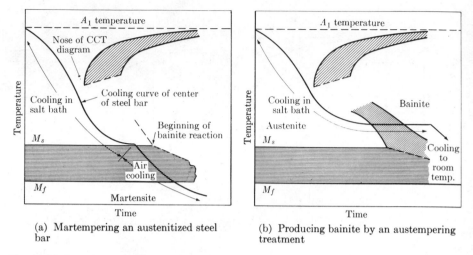

(a) Martempering an austenitized steel bar

(b) Producing bainite by an austempering treatment

**Fig. 10.26** Descriptions of two types of interrupted quenching by use of the CCT diagram.

$M_s$ temperature. It is then allowed to air cool as in martempering. The hardenability required for martempering is slightly greater than that needed for oil quenching, while timed quenching requires no increase in hardenability over that for oil or water quenching.

The process of *austempering* involves the formation of bainite rather than martensite. Figure 10.26(b) shows that the quenching procedure is much like that used for martempering, except that the molten lead or salt bath is usually at a higher temperature and the steel is held in the bath until transformation to bainite is complete. Since the quenching bath of higher temperature has lower cooling power, a higher hardenability is needed for austempering than for martempering. Tempering is rarely needed after austempering. Besides decreasing quenching stresses, this heat treatment produces relatively high impact resistance in the high-hardness range, where hardened and tempered steels may be somewhat brittle.

**Tempering.** After completion of the hardening heat treatment, it is usually desirable to increase the ductility of the steel. This is accomplished by tempering, which consists of heating the hardened steel to some temperature below $A_1$ for about one hour to produce tempered martensite. Essentially, the tempering reaction can be pictured as the change from *carbon atoms* dispersed in the martensite to precipitated *carbide particles* of increasing size. Figure 10.27 indicates a number of stages by which this change is believed to occur. At low tempering temperatures a hexagonal-close-packed carbide (called epsilon carbide) begins to form, and with this rejection of carbon the crystal structure of martensite changes ultimately from tetragonal to the body-centered-cubic characteristic of ferrite. Concurrently

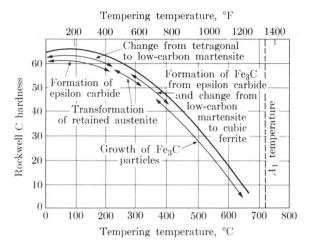

**Fig. 10.27** Changes in structure and hardness that accompany the tempering of eutectoid carbon steel.

with the formation of carbide particles, the structure of the matrix changes significantly. The initial martensitic structure, which has a high density of dislocations and interfaces, tends to undergo a process of reorganization analogous to the recovery and recrystallization of a cold-worked metal.

At high temperatures the epsilon carbide is converted to $Fe_3C$, which is orthorhombic in crystal structure. Prolonged tempering just below the $A_1$ temperature results in large $Fe_3C$ particles in a ferrite matrix, a *spheroidite* structure. The tempering reaction is a function of both temperature and time, but temperature is far the more important variable, and the time of tempering usually adopted is about one hour. Tempering also serves to eliminate the residual stresses in the hardened steel. These are of two types: (1) the macroscopic stresses produced during quenching, and (2) the microscopic stresses inherent in the martensitic structure. The origin of residual stresses and their relief by heating are discussed in Chapter 2.

Full hardening followed by tempering results in a combination of properties generally superior to that achieved by other treatments. However, in practice some pearlite or bainite may be mixed with the martensite if the local cooling rate is less than the critical rate necessary to avoid the nose of the CCT diagram. Also, retained austenite may be present if the $M_f$ temperature is below room temperature. In either case, poorer properties are obtained after tempering. In certain high-alloy steels, the retained austenite is converted to brittle martensite on cooling from the tempering temperature, and a second tempering treatment is helpful in such cases.

Although raising the tempering temperature usually produces an improvement in ductility, it also lowers the strength and hardness. Hence the useful upper tempering temperature is limited. For example, plain carbon and low-alloy tool steels that require high hardness for wear resistance are usually tempered below 250°C (500°F), and only a slight decrease in hardness results from this treatment, Fig. 10.27. On the other hand, the superior ductility needed in medium carbon AISI steels for use in automobiles, airplanes, and other machines requires high tempering temperatures that significantly decrease the hardness. Alloying elements have little effect on tempering below 250°C, but their influence in retarding softening at higher temperatures is sometimes very great, as in the case of 18-4-1 high-speed tool steel. This steel contains 18% tungsten, 4% chromium, and 1% vanadium in addition to 0.70% carbon. After quenching (usually in oil) and tempering at about 550°C (1000°F), this steel has a hardness of $R_C$ 65. It can retain this high hardness during prolonged use at temperatures near 500°C; for example, as a tool bit heated as a result of high-speed machining.

## PRECIPITATION HARDENING

The necessary condition for precipitation from solid solution is merely the existence of a sloping solvus line, discussed previously in connection with Fig. 7.28. Therefore precipitation occurs to some degree in almost all alloy systems and to a marked extent in hundreds of known cases. There is no doubt that virtually any metal can be made to precipitation harden by the addition of a properly chosen alloying element, and still further hardening should be possible in ternary or higher component alloys. Examples of this hardening process have already been given in discussing copper-beryllium and magnesium alloys, and additional commercial applications are listed in a later section.

**Heat treatment.** The entire process of producing a precipitation-hardened alloy may be divided into three parts: (1) choice of the composition, (2) solution heat-treating, and (3) precipitation heat-treating. The development of commercial precipitation-hardening alloy compositions is a long, difficult task, but it is possible to state some of the principles underlying such development. The equilibrium diagram of Fig. 10.28 is typical of a system that may show hardening as a result of precipitation of the beta phase from the supersaturated alpha solid solution. While the maximum hardening effect is probably produced here at six percent metal B, the limit of solubility of metal B in metal A, some hardening can occur in the entire range of compositions over which the alpha-plus-beta phase field extends at low temperatures. In practice, compositions other than those capable of maximum hardening are used. The casting properties of cast alloys are often improved by the presence of appreciable eutectic liquid during solidification. Therefore a composition such as nine percent metal B may be used. The composition of wrought alloys may be held to about four percent metal B to allow the alpha phase to be obtained for hot-working. In many cases the maximum working temperature is that of the binary eutectic or of a still lower-melting ternary eutectic in complex alloys.

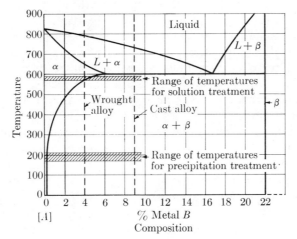

**Fig. 10.28** Equilibrium diagram on which are indicated compositions of a wrought and a cast alloy suitable for precipitation hardening.

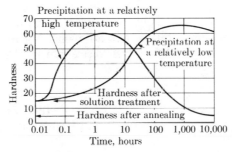

**Fig. 10.29** The course of hardening during precipitation heat treatment at two different temperatures.

A similar heat treatment is used for both cast and wrought alloys, except that longer times or higher temperatures must be used for the more slowly reacting cast materials. The purpose of the first step, *solution treatment*, is to dissolve a maximum amount of the second phase in the alpha solid solution and then to retain this solution down to room temperature. This is accomplished by (1) heating the alloy to a high temperature, but below the temperature at which excessive grain growth or melting of one of the constituents would occur, (2) holding for a fraction of an hour to almost a day to allow solution to take place, and (3) quenching in water to obtain the supersaturated solid solution at room temperature. The hardness after solution treatment is relatively low, Fig. 10.29, but it is higher than that of the slowly cooled, annealed material.

The full hardness of these alloys is developed during the *precipitation treatment*, in which the supersaturated solution undergoes changes that lead eventually to the formation of the second phase (see Fig. 9.15). In some cases the precipitation may occur in a reasonable time at room temperature, and the alloy is then said to be *naturally* aging. Usually it is necessary to age the alloy *artificially* by holding it in a temperature range such as that shown in Fig. 10.28. The exact temperature used for the precipitation heat treatment is determined by two factors: (1) the time for appreciable reaction and (2) the property of principal interest. The time factor must be of reasonable length for industrial heat treatment. (The higher the temperature, the shorter the time.) In regard to the second factor it should be understood that different properties change at different rates during precipitation. For example, the strength properties tend to reach higher maximum values at lower precipitation temperatures. The course of aging at two different temperatures is shown in Fig. 10.29. Properties such as hardness attain a maximum value during precipitation at a given temperature and then gradually decrease as a result of *overaging*. This eventual softening is a natural consequence of the approach of the alloy to the equilibrium condition with increasing time at temperature. In fact, a greatly overaged alloy would be essentially identical with an *annealed* alloy, that is, one in which the equilibrium structure is produced by slow cooling from the solution-treating temperature.

In addition to composition and heat-treating conditions, the properties obtained in certain alloys are greatly affected by cold-working after solution treatment. In some instances cold-working interferes with the development of maximum property values during the following precipitation treatment, but frequently the important design factor, yield strength, can be significantly increased by this means.

**Applications.** Precipitation hardening is the most important method of strengthening nonferrous metals by solid-state reaction. It is especially useful for aluminum, the principal metal of this class, and both cast and wrought aluminum alloys are precipitation hardened.

Since detailed information on the wide variety of aluminum alloys available commercially is given in the *Metals Handbook, Vol. 1*, only a brief survey of representative alloys will be presented here. The alloy designations used for the wrought alloys in Table 10.4 developed from an earlier system in which only one to three numbers were used. For example, alloy 5052 was previously 52S, where the S identified it as a wrought alloy. The first number in the present designation specifies the alloy group according to the following system.

| Alloy number | Major alloying element |
|---|---|
| 1xxx | Commercially pure aluminum (99+ % Al) |
| 2xxx | Copper |
| 3xxx | Manganese |
| 4xxx | Silicon |
| 5xxx | Magnesium |
| 6xxx | Magnesium + silicon |
| 7xxx | Zinc |
| 8xxx | Other element |

**TABLE 10.4**

Some Characteristics of Several Types of Aluminum Alloys

| Alloy designation | Principal alloying elements | Hardening process* | Range of properties (soft to hard conditions) | | | | |
|---|---|---|---|---|---|---|---|
| | | | Tensile strength, $10^7$ N/m² | Yield strength, $10^7$ N/m² | Elongation in 50 mm, % | Endurance strength ($5 \times 10^8$ cycles), $10^7$ N/m² | Typical applications |
| | | | *Wrought alloys* | | | | |
| 1100 | Commercial purity | Cold-working | 9–16 | 3–15 | 45–15 | 3– 6 | Cooking utensils |
| 5052 | 2.5% Mg | Cold-working | 19–29 | 9–26 | 30– 8 | 11–14 | Bus and truck bodies |
| Alclad 2024 | 4.5% Cu, 1.5% Mg (with protective sheet of pure aluminum) | Precipitation | 19–47 | 8–32 | 22–19 | 9–14 | Aircraft |
| 6061 | 1.5% Mg$_2$Si | Precipitation | 12–31 | 6–28 | 30–17 | 6–10 | General structural |
| 7075 | 5.6% Zn, 2.5% Mg, 1.6% Cu | Precipitation | 23–57 | 10–50 | 16–11 | 12–16 | Aircraft |
| | | | *Cast alloys* | | | | |
| 195 | 4.5% Cu | Precipitation | 22–28 | 11–22 | 8.5–2 | 5– 6 | Sand castings |
| 319 | 3.5% Cu, 6.3% Si | Precipitation | 19–25 | 12–18 | 2–1.5 | 7– 8 | Sand castings |
| 356 | 7% Si, 0.3% Mg | Precipitation | 23–26 | 15–19 | 5–4 | 8– 9 | Permanent mold castings |

* In addition to alloy hardening.

To convert from N/m² to lb/in², multiply by $1.450 \times 10^{-4}$

Table 10.4 shows the range in mechanical properties of each alloy from the soft to the hard condition. The condition or *temper* of an alloy can be indicated by adding a symbol to the alloy designation. The principal part of this symbol is a letter having the following significance.

| Letter | Condition of alloy |
|--------|--------------------|
| *F* | As fabricated |
| *O* | In soft condition as a result of annealing (recrystallizing) |
| *H* | Strain hardened by a cold-working process |
| *T* | Heat treated |

The letters *H* and *T* are usually followed by numbers indicating in more detail the treatment that the alloy received. For example, *H*1 designates an alloy that has been strain hardened only, *H*2 designates one that has been strain hardened and partially annealed, and *H*3 designates one that has been strain hardened and stabilized by suitable annealing. A second number, 2, 4, 6, 8, or 9, is used to indicate increasing amounts of strain hardening. For example, the second alloy in Table 10.4 is designated 5052-*O* in its soft condition, while its hardest commercial temper is 5052-*H*18.

The various tempers produced by heat treatment are indicated by *T* combined with the following numbers:

*T*2   Annealed (applies only to castings, annealed to improve ductility for example)

*T*3   Solution heat-treated and then cold-worked

*T*4   Solution heat-treated and naturally aged

*T*5   Artificially aged only

*T*6   Solution heat-treated and artificially aged

*T*7   Solution heat-treated and stabilized (by an overaging heat treatment)

*T*8   Solution heat-treated, cold-worked, and then artificially aged

For example, the fifth alloy in Table 10.4 is designated 7075-*O* in its soft condition, which is produced by annealing for a few hours at 410°C (770°F). The hard temper, 7075-*T*6, is produced by solution heat-treating at 465°C (870°F) and aging (precipitation heat-treating) at 120°C (250°F) for about 25 hours. Figure 10.30 shows the property changes that occur during this precipitation treatment. Since alloying and the precipitation heat treatment decrease the corrosion resistance of aluminum, certain of the high-strength alloys are protected by a layer of pure aluminum firmly bonded to the surface by a hot-rolling process. Alclad 2024 alloy is an example of this type of product.

The heat treatment of aluminum alloys illustrates the relation between phase diagrams and the structure of alloys. For example, the structure of alloy 195 in

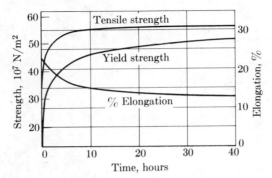

**Fig. 10.30** Property changes during the precipitation heat treatment of 7075 aluminum alloy sheet at 120°C (250°F). (After Dix)

the as-cast condition, Fig. 10.31(b), shows coarse particles of the $\theta$-phase, $CuAl_2$, that formed as a divorced eutectic structure during solidification. The matrix is $\alpha$ solid solution, in agreement with the phase diagram, Fig. 10.31(a). However, there is another phase in this structure, the dark (Al, Fe, Si) constituent that is formed by the iron and silicon impurities in the alloy. Solution heat treatment dissolves both the $\theta$ phase and the impurity phase, leaving only the supersaturated $\alpha$ solid solution when the alloy is quenched, Fig. 10.31(c). A subsequent aging treatment leads to the formation of GP zones and of various modifications of the $\theta$ phase, described previously in connection with Fig. 9.15.

Nickel is another metal whose alloys are hardened principally by precipitation. Nickel is much like iron in its mechanical properties, but its corrosion properties are so much better than those of iron that it is used for many purposes despite the fact that it costs about ten times as much. Typical nickel alloys are surveyed in Table 10.5. Other nickel alloys with special electrical and magnetic properties will be considered in Chapter 12. These include Alnico 5B and Cunife, which are hardened by precipitation.

Precipitation hardening in steels is of minor interest compared with eutectoid decomposition hardening, but there are several aspects of precipitation in iron-base alloys that deserve mention. One of these is the decreased ductility of low-carbon steels resulting from the precipitation of carbon or nitrogen during *quench aging* or *strain aging*. Quench aging is the usual type of precipitation hardening, while strain aging is precipitation resulting from a cold-working operation. Both these aging effects are undesirable in mild steel, and various means, such as fixing of the carbon and nitrogen, are used to minimize them.

The following three examples of beneficial precipitation hardening in steel are of considerable industrial importance. A class of low-alloy steels, known generically as *Ni-Cu-age* steels, employs copper as the precipitation hardener. They typically contain 0.5–3.0% Ni, 0.5–2.0% Cu, plus minor additions of such elements as molybdenum, chromium, titanium, or niobium. The conditions of processing can usually be controlled so that the copper remains in solution after

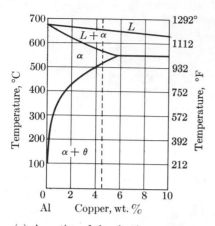

(a)  A portion of the aluminum-copper phase diagram showing the position of alloy 195.

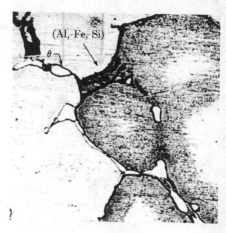

(b)  As-cast alloy with particles of $\theta$ and $\alpha$ (Al, Fe, Si) in dendrites of the $\alpha$-solid solution.

50 $\mu$m

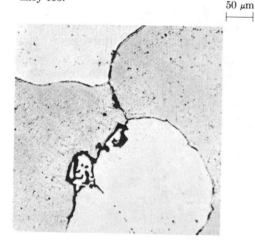

(c)  The alloy after solution heat treatment at 515°C [960°F] for 12 hours; only the solid solution remains.

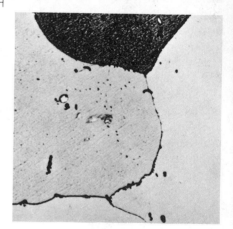

(d)  After aging for 4 hours at 155°C [310°F] no change is seen in the microstructure.

**Fig. 10.31**  The microstructure of aluminum alloy 195 (4.5% Cu, 0.8% Si, 0.5% Fe) after various heat treatments.  (Courtesy Alcoa Research Laboratories)

the hot-rolling operation.  Heat treatment then consists merely in an aging treatment to precipitate the copper-rich phase.  Because carbon is held below 0.1%, toughness, formability, and weldability are far superior to the usual alloy steels at similar yield strengths, 35–85 × $10^7$ N/m² (50–125 × $10^3$ lb/in²).  An additional advantage of these copper-bearing steels is their significantly greater resistance to atmospheric corrosion.  The precipitation hardening of *maraging* steels permits

**TABLE 10.5**
Some Characteristics of Several Nickel Alloys

| Alloy | Principal alloying elements | Mechanical properties | | | | Typical applications |
|---|---|---|---|---|---|---|
| | | Tensile strength, $10^7$ N/m² | Yield strength, $10^7$ N/m² | Elongation in 50 mm, % | Endurance strength ($10^8$ cycles), $10^7$ N/m² | |
| Nickel 200, hot-rolled | None | 45 | 21 | 50 | 21 | Chemical industry |
| Duranickel, Alloy 301, cold-drawn and precipitation hardened | 4.4% aluminum 0.6% titanium | 130 | 105 | 20 | 41 | Springs, plastics, extrusion equipment |
| Monel, Alloy 400, hot-rolled | 30% copper | 62 | 38 | 35 | 29 | Oil refinery parts |
| Monel, Alloy K500, cold-drawn and precipitation hardened | 29% copper 3% aluminum | 125 | 110 | 5 | 32 | Pump rods, valve stems |
| Inconel, Alloy 600, hot-rolled | 15% chromium 8% iron | 83 | 45 | 35 | 31 | Gas turbine parts |
| Inconel, Alloy X–750, hot-rolled and aged at 700°C (1300°F) | 15% chromium 7% iron 2.5% titanium | 125 | 90 | 25 | 41 | Springs and bolts subjected to corrosion |

To convert from N/m² to lb/in², multiply by $1.450 \times 10^{-4}$

the attainment of yield strengths as high as $210 \times 10^7 \text{ N/m}^2$ ($300 \times 10^3 \text{ lb/in}^2$). A typical composition is 18% Ni, 5% Mo, 0.5% Ti, and 8% Co. Martensite is first produced by cooling the steel in air from 815°C (1500°F). This martensite is relatively soft because carbon is essentially absent; therefore, the steel can be given a final machining operation or can even be cold-worked. A precipitation heat treatment, typically for three hours at 480°C (900°F), hardens the steel through the formation of complex Ni-Mo-Ti compounds. The third example is the precipitation hardening of stainless steel by the use of aluminum (see Chapter 11). Prior to this development the popular 18 chromium-8 nickel-type stainless steels could be hardened only by cold-working.

**Diffusion-reaction hardening.** Precipitation and resulting hardening can be produced in some systems by changing the composition of the alloy; that is, by adding a suitable element by means of a diffusion treatment. For example, if metal $A$ dissolves metal $B$ but has limited solubility for the compound $B_xC_y$, it is sometimes possible to obtain considerable hardening by diffusing metal $C$ into the solid solution, thereby tending to precipitate $B_xC_y$. The hardenable alloy is initially the alpha solid solution (metal $B$ dissolved in metal $A$). As metal $C$ diffuses into this solid solution, the overall composition gradually shifts into the alpha-plus-beta region of the diagram, and the beta phase ($B_xC_y$) tends to precipitate. Presumably, hardening is the result of lattice coherency which exists in the early stages of the precipitation of $B_xC_y$ in the solid solution. Only the surface layer is hardened by this process, since only a small depth is normally reached by the diffusing element.

The nitriding process for the surface hardening of steels, briefly described in Table 4.5, is an example of industrial use of diffusion-reaction hardening. The elements aluminum, chromium, and vanadium have a strong tendency to form nitrides. Therefore, when nitrogen diffuses into a steel containing one or more of these elements, particles of the nitride phases form within the steel. The extreme surface hardness that results from the nitriding treatment is attributable to the state of fine dispersion of the nitride particles rather than to the inherent hardness of large nitride grains. Softening that occurs when a nitrided steel is subjected to prolonged heating above 550°C (1000°F) may be caused by loss of lattice coherency as the nitride particles become larger.

## FIBER STRENGTHENING

Since nondeforming particles give the maximum strengthening possible from a dispersed phase, various methods have been used to obtain this type of structure. Incorporation of a fiber into a supporting matrix, especially by the *composite technique*, Fig. 10.9(b), affords excellent control of the important variables (such as the configurations, amounts, and properties of the matrix and of the dispersed phase). The significant properties of the reinforcing fibers, Table 10.6, are high

**TABLE 10.6**

Properties of Typical Fibers*

| Material | Density, g/cm$^3$ | Tensile strength, $10^7$ N/m$^2$ | Modulus of elasticity, $10^{10}$ N/m$^2$ |
|---|---|---|---|
| Tungsten | 19.4 | 400 | 41 |
| Molybdenum | 10.2 | 220 | 36 |
| Beryllium | 1.84 | 125 | 24 |
| Boron | 2.36 | 280 | 40 |
| Borsic | 2.5 | 280 | 38 |
| Silicon carbide | 3.5 | 250 | 44 |
| Glass | 2.5 | 410 | 9 |
| Graphite | 1.63 | 200 | 35 |
| Alumina whiskers | 3.96 | 1000 | 45 |

* Adapted from compilation of J. N. Fleck and E. J. Jablonski.
To convert from N/m$^2$ to lb/in$^2$, multiply by $1.450 \times 10^{-4}$.

strength and high modulus of elasticity. For applications in which a high strength-to-weight ratio is important, nonmetallic fibers such as boron have a distinct advantage because of their low density. The metal or alloy chosen for the matrix must combine adequate strength and ductility with other properties that make it compatible with the fiber in question. Often the fiber is given a special coating (for example, SiC on boron filaments) to help achieve the dual purpose of bonding the fiber/matrix interface while avoiding deleterious chemical reactions. If the fiber is relatively short, it requires aligning by a special (electrical, magnetic, etc.) technique. Incorporation of the matrix may be accomplished by casting, powder-metallurgical procedures, pressure bonding or other techniques.

Consider first the behavior of a composite composed of long filaments of the reinforcing phase, Fig. 10.32. Provided that the filaments are adequately small for uniformity, typically 100 μm in diameter, the strength properties depend primarily on the volume fraction, $V_f$, of fiber. Deformation proceeds in four stages. (1) Since the matrix is well bonded to the fibers, elastic deformation occurs equally in both phases until just beyond the elastic limit of the matrix. (2) The matrix then begins to experience plastic deformation although the fibers are still in their elastic range. (3) The fibers may begin to deform plastically, though fibers generally have little ductility. (4) Finally, the fibers begin to fracture and cause the eventual fracture of the composite. For well bonded fibers, the tensile strength $\sigma_c$ of a composite can be approximated by the relation,

$$\sigma_c = \sigma_f V_f + \sigma_m(1 - V_f).  \tag{10.7}$$

$\sigma_f$ is the tensile strength of the fiber and $\sigma_m$ is a value somewhat greater than the

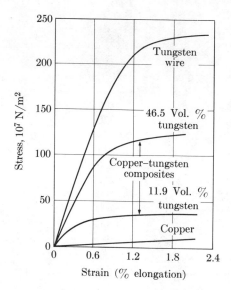

**Fig. 10.32** Initial portion of stress-strain curves for copper, tungsten, and composites made by imbedding continuous tungsten wire in a copper matrix. (Data of D. L. McDaniels *et al.*)

yield strength of the matrix. Since $\sigma_f$ is much larger than $\sigma_m$, the strength of a composite increases almost directly with the volume fraction of fiber, $V_f$.

In the case of relatively short fibers, a new feature is the behavior of a small length $\delta$ at each end of a section of fiber. This is the length over which the transfer of stress from the matrix can be pictured as building up from zero (at the end of the fiber) to $\tau_m$, the shear stress of the matrix. Consequently, when a fiber of diameter $d$ is stressed to almost its tensile strength $\sigma_f$ in a composite, the force transferred over a total length of $2\delta$ is $\pi d\tau_m\delta$. If the fiber is not to be pulled out of the matrix, this force must be at least equal to the tensile force on the fiber, $(\pi d^2/4)\sigma_f$. Therefore,

$$\frac{2\delta}{d} = \frac{l_c}{d} = \frac{\sigma_f}{2\tau_m}. \tag{10.8}$$

The *critical aspect ratio*, $l_c/d$, is the minimum ratio of length to diameter of discontinuous fibers for which adequate transfer of stress from the matrix occurs. Each matrix/fiber combination has a characteristic critical aspect ratio, typically about 10. These values can usually be exceeded in practice without difficulty; directional solidification of eutectics, for example, gives values of 100–500, Fig. 10.9(a).

### PROBLEMS

**1.** List the processes of strengthening that can be used with

a) pure metals,

b) alloys.

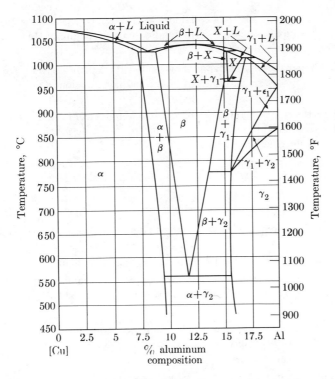

**Fig. 10.33** A portion of the copper-aluminum equilibrium diagram.

**2.** The strength of attraction between a solute atom and a dislocation can be expressed as a binding energy $E_b$. By analogy with Eq. (6.43), the concentration $c$ of solute atoms at the dislocation (that is, within about $3|\mathbf{b}|$ of the core) is given by

$$c = c_0 \exp(E_b/kT),  \tag{10.9}$$

where $c_0$ is the average solute concentration. Using the value $E_b = 0.5$ eV, calculate the fraction of carbon atoms at the dislocations in a steel containing 0.05 a/o carbon if the density of dislocations is $10^{14}/\mathrm{m}^2$.

**3.** Two common forms of $AB$ ordering are CuAuI and CuZn (beta brass). CuAuI consists of alternate sheets of copper and of gold atoms on {200} planes of an FCC distribution of lattice points. In CuZn, the copper atoms occupy the corner points in a BCC array and the zinc atoms the center points. Determine for each of these two structures the energetically most favorable superdislocation. See Eq. (3.14).

**4.** Derive Eqs. (10.4) and (10.5) by analogy with the derivation of Eq. (3.20). Recall that $F_x = \tau_{yx}b_x$ per unit length.

**5.** Make phase analyses for

a) 0.4% C,

b) 0.8% C, and

c) 1.2% C steels under equilibrium conditions at (1) a temperature in the austenite region, (2) 725°C, (3) 720°C, and sketch the corresponding microstructures. How much pearlite is present at 720°C in each of these steels?

**6.** a) In the portion of the Cu-Al diagram shown in Fig. 10.33 there are two eutectoid reactions. Specify their locations by giving the temperature and composition of each of the eutectoid points.

b) In Table 7.6 the composition of the hardenable aluminum bronze is given as 10.5% Al. Give a possible reason for the choice of this composition. (Note the elongation values of the 5% and 10.5% Al alloys in Table 7.6.)

c) Make phase analyses for the 10.5% Al alloy under equilibrium conditions at the following temperatures: (1) 900°C, (2) 700°C, (3) 570°C, (4) 560°C, and sketch the corresponding microstructures.

d) How much eutectoid microconstituent (analogous to pearlite) is present at 560°C?

**7.** a) In a steel to be austenitized, is it preferable that the initial structure be coarse spheroidite or fine pearlite?

b) Explain your answer in terms of the mechanism of isothermal austenite formation.

c) Specify the heat-treating conditions (time, temperature, cooling rate, etc.) suitable for (1) annealing, (2) normalizing, (3) spheroidizing a bar of hot-rolled 0.8% C steel one inch in diameter.

**8.** a) Describe a practical procedure for determining experimentally the isothermal reaction curve for the decomposition of austenite in a given steel at 400°C.

b) Sketch a typical reaction curve.

c) Show that most of the information given by this curve can be recorded in the TTT diagram for the steel.

**9.** a) If a hot-rolled 0.8% C steel bar one inch·in diameter is heated to 650°C, held for 15 seconds, and quenched in water, can the TTT diagram of Fig. 10.16 be used to predict the structure produced?

b) Explain.

**10.** Since the austenite grain size influences such properties as impact strength and hardenability, it is often necessary to determine the grain size of the austenite in a given heat of steel at the austenitizing temperature. Making use of the fact that pearlite, proeutectoid ferrite, and proeutectoid cementite form preferentially at the austenite grain boundaries, suggest possible procedures for determining the austenite grain size in steels containing

a) 0.4% C,

b) 0.8% C, and

c) 1.2% C.

[Note: A completely pearlitic structure is not convenient for determining grain size.]

**11.** Compare the usefulness of

a) the equilibrium diagram,

b) the TTT diagram, and

c) the CCT diagram in predicting the effect of normalizing a bar of eutectoid steel.

**12.** "Steel is made hard by quenching." List at least three requirements that must be met to justify this statement.

**13.** Determine the ideal critical diameter ($D_I$) of AISI 6145 steel that has a No. 8 grain size. An average composition is

$$
\begin{array}{ll}
\text{C—0.45\%} & \text{Cr—0.95} \\
\text{Mn—0.8} & \text{V—0.2} \\
\text{Si—0.3} &
\end{array}
$$

[*Note*: The effect of vanadium is variable and may be neglected here.]

**14.** What diameter bars of the steel of Problem 13 could be fully hardened (50% martensite at the center) using each of the following quenching mediums: (a) still air, (b) still oil, (c) still water, and (d) agitated brine?

**15.** What condition is necessary to render an alloy system capable of precipitation hardening?

**16.** Criticize the statement, "A precipitation hardening alloy can be annealed by water quenching from a suitable high temperature."

**17.** Use the copper-beryllium equilibrium diagram, Fig. 10.34, to choose a suitable composition for a wrought, hardenable alloy in this system.

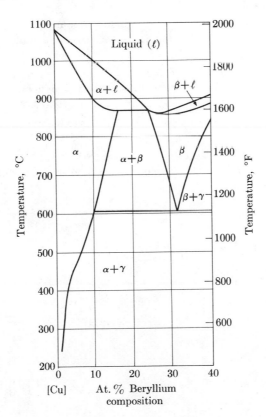

Figure 10.34

a) What type of solid-state reaction is the basis of hardening in this case? [Note that eutectoid decomposition hardening is not feasible in this system because of the brittleness of alloys in the corresponding range of compositions.]

b) Approximately what composition is suitable for a wrought alloy?

c) Describe the procedure for the solution heat treatment of this alloy.

d) Estimate the heat-treating temperature and time that would be suitable for the precipitation heat treatment in this case.

**18.** Assuming that the precipitation reaction is controlled by diffusion, use the data given in Chapter 8 to estimate the time to reach maximum hardness at 100°C in an aluminum-4% copper alloy if the time at 150°C is 10 hours.

# REFERENCES

Argon, A. S. (Ed.), *Physics of Strength and Plasticity*, Cambridge, Mass.: MIT Press, 1969. An advanced treatment of the theory of the strength of solids.

Broutman, L. J. and R. H. Krock (Eds.), *Modern Composite Materials*, Reading, Mass.: Addison-Wesley, 1967. A comprehensive treatment of the production, properties, and theory of strengthening of composites.

Cottrell, A. H., *Introduction to Metallurgy*, London: Edward Arnold, 1967. Chapters 20, 21, and 25 are good summaries of strengthening mechanisms and of heat-treating methods and their applications.

Honeycombe, R. W. K., *Plastic Deformation of Metals*, London: Edward Arnold, 1969. A clear, comprehensive treatment of strengthening mechanisms at an intermediate level.

Reed-Hill, R. E., *Physical Metallurgy Principles*, 2nd edn., New York: Van Nostrand and Co., 1973. Chapters 16, 17, 18, and 19 are good introductions to the strengthening of steel and the martensite reaction.

Tegart, W. J. M., *Elements of Mechanical Metallurgy*, New York: Macmillan, 1966. Chapter 6 is a good summary of data on strengthening effects in polycrystalline metals and alloys.

CHAPTER 11

# CORROSION AND OXIDATION

## INTRODUCTION

Corrosion, the attack of metals by their environment, is a primary engineering problem for two reasons. First, it is an economic matter involving billions of dollars annually, and each design must be decided on a practical cost basis rather than through the selection of an ideal, but expensive material. Second, corrosion of a typical industrial structure is a complex phenomenon, electrochemical in nature, but generally requiring an engineering approach to the many variables involved. These two aspects of corrosion, its great financial waste and its complexity, have produced an unusual situation. Extensive research has resulted in a good understanding of many phases of corrosion, and thousands of new investigations are conducted each year. Nevertheless, in many important applications engineers must rely on the empirical results of service tests, since actual performance may differ markedly from the predictions of laboratory investigations.

Although corrosion theory cannot be applied rigorously to all practical problems, in many cases its basic principles offer a guide to corrosion control. Most of the subsequent discussion is devoted to these principles. Chemical behavior is fundamentally important, but other factors also influence the choice of corrosion-resistant materials for engineering applications. For example, although tantalum has excellent corrosion resistance, it is expensive and often difficult to fabricate in the required form. The usual criteria for selecting an alloy for a given application are adequate strength, ease of fabrication, low initial and maintenance costs, and corrosion resistance in a specific environment.

The prime cause of all reactions between metals and their environments is a decrease in the free energy of the system as a result of the reaction. In only a few instances is it economical to reverse this driving force for corrosive action by the use of noble metals such as gold or platinum. In most practical situations corrosion is controlled by reducing the *rate* at which the corrosion reaction proceeds. As a background for studying this control of corrosion rates, we shall consider here the main types of corrosion, *electrochemical corrosion*, which usually involves a corroding liquid, and *high-temperature oxidation*, which generally occurs in the absence of moisture.

## ELECTROCHEMICAL CORROSION

The type of corrosion most frequently encountered takes place at or near room temperature as a result of the reaction of metals with water or with aqueous solutions of salts, acids, or bases. These reactions are part of the broad field of *electrochemical corrosion*. Although there are many variations of this type of corrosion, the action of an aerated salt solution at the junction between a piece of iron and a piece of copper* illustrates the essential characteristics (Fig. 11.1).

---

* Strictly speaking, this is an example of *galvanic corrosion*, but other types of electrochemical corrosion involving only a single metal occur in a similar manner.

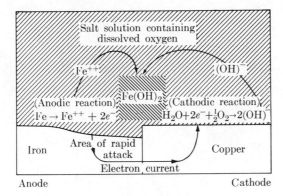

Salt solution containing dissolved oxygen

$Fe^{++}$

$(OH)^-$

(Anodic reaction)  $Fe(OH)_2$  (Cathodic reaction)

$Fe \rightarrow Fe^{++} + 2e^-$    $H_2O + 2e^- + \frac{1}{2}O_2 \rightarrow 2(OH)^-$

Iron    Area of rapid attack    Copper

Electron current

Anode    Cathode

**Fig. 11.1** Schematic illustration of typical electrochemical corrosion.

Two principal reactions occur, one at the anode and another at the cathode. Anodic reactions are always oxidation reactions and therefore tend to destroy the anode metal by causing it to dissolve as an ion or to revert to a combined state such as an oxide. Cathodic reactions are always reduction reactions and usually do not affect the cathode metal, since most metals cannot be further reduced. The electrons produced by the anodic reaction flow through the metal and are used up in the cathodic reaction. The disposition of the reaction products is often decisive in controlling the rate of corrosion. They may go into solution or be evolved as a gas and thus not inhibit further reaction. In other cases an insoluble compound is formed that may cover the metal surface and be effective in reducing the rate of additional corrosion. However, the insoluble compound in Fig. 11.1 is formed at a distance from the corroding area and has little protective value.

**Electrode potential.** The essence of electrochemical corrosion is the occurrence of an anodic reaction, which involves the giving up of electrons by the corroding metal. An electrical measurement of the tendency of the metal to give up electrons can therefore serve as a basic criterion of corrodibility, expressed as the electrode potential of the corroding system. In the sketch of Fig. 11.2, for example, when an iron atom goes into solution as $Fe^{++}$ ion, two electrons are given up to the iron specimen and contribute to the creation of an electrical potential. Two points must be borne in mind in connection with quantitative values of these potentials. First, the metal specimen and its solution constitute only half of a complete cell, and therefore a second half-cell must be chosen to form an electrochemical system on which measurements can be made. The usual choice is the standard hydrogen half-cell consisting of gaseous hydrogen, $H_2$, at one atmosphere pressure and hydrogen ions, $H^+$, at unit activity (concentration) in contact with a specially prepared platinum electrode. In the *electromotive series* (Table 11.1) which is based on hydrogen as a reference, it can be seen that the (Fe, $Fe^{++}$) half-cell has a standard equilibrium value, $E^0 = -0.44$ volt.

Electrode potential, $E$

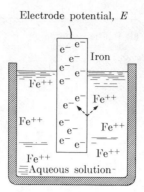

**Fig. 11.2** Schematic illustration of the anodic (corrosion) reaction, $Fe = Fe^{++} + 2e^-$, showing the important role of the liberated electrons in producing an electrode potential.

The second point concerning electrode potentials is already involved in fixing the value $-0.44$ volt. Namely, the nature of the corroding solution in contact with the iron has been given a convenient, standard value of unit activity with respect to $Fe^{++}$ ions. If the ion concentration is reduced below this standard value, there will be a greater tendency for the iron to go into solution, and consequently the electrode potential will change from the standard value, $E^0$, to a new value, $E$, given by the relation

$$E = E^0 + \frac{0.059}{n} \log c, \tag{11.1}$$

where $c$ is the concentration* of $Fe^{++}$ ions in units of moles per liter, and $n$ is the valence of the ion in question ($n = 2$ in this case).

**Pourbaix diagrams.**  An important instance of the use of Eq. (11.1) in understanding corrosion is given by the diagram of electrode potential versus hydrogen-ion concentration ($E$ versus pH)—the Pourbaix diagram—that summarizes the chemical equilibria underlying the corrosion behavior of a given metal.  In the case of the corrosion of iron by water, for example, the diagram has the form shown in Fig. 11.3 and delineates areas corresponding to three different types of behavior: *corrosion*; *immunity* to corrosion; and *passivation*.  The significance of each of these areas will now be considered.

It is easy to understand the existence of the large region labeled "corrosion" in terms of Fig. 11.2 and the known value $E^0 = -0.44$ volt for iron.  When iron is in contact with an aqueous solution, electrode potentials near $E^0$ (or algebraically larger) correspond to large equilibrium concentrations of iron ions in solution. Therefore, it can be concluded that iron will tend to go rapidly into solution

---

* At the low concentrations of interest for corrosion problems, it is permissible to employ concentration values rather than the more exact chemical activity.

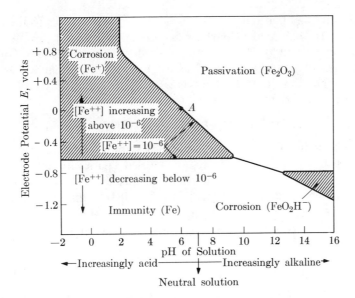

**Fig. 11.3** Simplified Pourbaix diagram for the iron-water system showing the domains of corrosion behavior. The symbol $[Fe^{++}]$ represents the equilibrium concentration of $Fe^{++}$ ions in units of moles per liter of solution.

(corrode) when it is in contact with water in this range of electrode potentials. The potentials in question need not be artificially applied since they develop naturally as a result of corrosion reactions. For example, when a piece of iron is placed in a beaker of water, the electrode potential is found to be about $-0.45$ volt.

If the natural electrode potential is suitably modified through an external electrical source, it is possible to cause the iron specimen to pass into the region labeled "immunity" in Fig. 11.3 and to be essentially free of corrosive attack. The change from corrosion to immunity is not a sharp one, the dividing line at $E = -0.62$ volt representing an arbitrary but useful value based on the following reasoning. If the *equilibrium* concentration of $Fe^{++}$ ions is sufficiently small, $c = 10^{-6}$, then the amount of iron that will dissolve will be negligibly small, and the iron can be said to be immune from corrosion. Now, the value of electrode potential corresponding to $c = 10^{-6}$ can be calculated easily using Eq. (11.1):

$$E = E^0 + \frac{0.059}{2}\log 10^{-6} = -0.44 + (0.0295)(-6) = -0.62 \text{ volt.}$$

Thus, at electrode potentials of $-0.62$ volt or greater, iron is effectively immune from corrosion. This phenomenon is the basis of cathodic protection, a method of corrosion control discussed in a later section.

In the same way that the reaction $Fe = Fe^{++} + 2e^-$ leads to the horizontal line at $E = -0.62$ volt on the Pourbaix diagram, other possible reactions between iron and water account for the remaining boundary lines in Fig. 11.3. The passivation reaction, leading to the sloping line at the center of the diagram, is of special importance. This is a reaction involving $Fe^{++}$ ions and a layer of $Fe_2O_3$ that forms on the surface of the iron:

$$2Fe^{++} + 3H_2O = Fe_2O_3 + 6H^+ + 2e^-. \tag{11.2}$$

It will now be shown how this reaction can maintain the concentration of $Fe^{++}$ ions below $[Fe^{++}] = 10^{-6}$ in the region labeled "passivation." By use of the known thermodynamic values of the chemical species involved, a relation can be obtained between the concentrations $[Fe^{++}]$ and $[H^+]$ that exist at equilibrium with a solid layer of $Fe_2O_3$. An example for the special case $E = 0$ will illustrate the method of calculation, the numerical relation then being

$$\frac{[H^+]^6}{[Fe^{++}]^2} = 10^{-24}. \tag{11.3}$$

Recalling that the value $[Fe^{++}] = 10^{-6}$ is taken by definition as the boundary of the region of corrosion, it follows that the concentration of $H^+$ ions required to maintain this value in equilibrium with a layer of solid $Fe_2O_3$ on the surface of the iron is

$$[H^+] = (10^{-24} \cdot 10^{-12})^{1/6} = 10^{-6},$$

corresponding to pH = 6, the value at $E = 0$. This point is plotted at $A$ in Fig. 11.3. The degree of corrosion protection afforded a metal by the formation of a layer of solid reaction product on its surface is widely variable, and therefore the practical significance of passivation as a means of corrosion control correspondingly differs from one metal to another.

It will be seen in Fig. 11.3 that there is a second region of corrosion at the right of the diagram. This is a result of the occurrence of such reactions as

$$Fe + 2H_2O = FeO_2H^- + 3H^+ + 2e^-, \tag{11.4}$$

with the production of ions that are stable in alkaline solutions. Although this type of corrosion occurs at high temperatures, it is fortunate that the rate becomes negligibly small at normal temperatures so that iron vessels can be used industrially for handling alkaline solutions. Under special conditions the corrosion problem known as *caustic embrittlement* occurs. For example, a high-temperature steam boiler may be operated with water specially treated to maintain a pH value in the safe range from about 10 to 13. In the vicinity of crevices at rivets, however, the alkali can become concentrated, and if the metal at this point contains nonuniform stresses, greatly accelerated corrosion can occur. A further complication is the decomposition of the corrosion product, $Na_2FeO_2$ in the case of a sodium hydroxide solution, with the regeneration of NaOH and the liberation of nascent hydrogen. The hydrogen may then be absorbed by the steel to produce the low-ductility condition known as hydrogen embrittlement.

**Factors determining corrosion rates.** As shown by the Pourbaix diagram in the case of iron, the typical engineering use of metals is under conditions where corrosion can occur. The practical problem is then to control the rate of corrosion so that a satisfactory service life is obtained, a topic discussed from several viewpoints at the end of this chapter. However, it will be useful here to consider briefly the principles that govern rates of corrosion.

Essentially, corrosion occurs at the rate at which an electrical current, the corrosion current, can be driven through the corroding system by the existing electrode potentials. This idea can be seen especially clearly in the case of galvanic corrosion (Fig. 11.1). The actual potential for copper-iron corrosion is discussed later, but as a very crude starting value, the standard electrical potential can be borne in mind. This value is the algebraic difference between the single electrode potentials of the two metals (Table 11.1) and has the value

$$E^0_{Cu} - E^0_{Fe} = 0.34 - (-0.44) = 0.78 \text{ volt.} \tag{11.5}$$

This electrical driving force can be pictured as causing the following series of steps.

1. The anodic (corrosion) reaction on the iron
2. The current of electrons in the metallic path from iron to copper
3. The cathodic reaction on the copper, which uses up the electrons
4. The ionic current of $Fe^{++}$ and $(OH)^-$ within the aqueous solution

These steps form a series in the electrical sense also, in that corrosion current, $i$, must pass through each step in turn. This fact is the basis of several possibilities for decreasing the rate of corrosion as the following discussion will show.

The corrosion current can be expressed in terms of Ohm's law as

$$i = \frac{E_3 - E_1}{R_2 + R_4}, \tag{11.6}$$

where the numbers refer to the steps listed above. Consider first the possibility of decreasing the corrosion current by increasing the electrical resistance. The metallic resistance, $R_2$, is small under the conditions for galvanic corrosion being discussed here, but it can evidently be made very large by electrically insulating the iron from the copper. The cell resistance of electrochemistry, $R_4$, has a high value in dilute solutions but decreases as the solution becomes more concentrated. The electric-potential terms in Eq. (11.6) are generally of decisive importance. First, it should be noted that the relation of $E_{Fe}$ to $E_{Cu}$ determines which of the two metals is corroded; namely, the one that stands higher in the electromotive (galvanic) series. Thus, although iron is the corroding member in a galvanic couple with copper, the direction of the corrosion current is reversed in the couple iron-zinc, and the iron is protected (see Problem 2). The industrial process of *galvanizing*, in which steel is covered with a thin layer of zinc, makes use of this principle of corrosion protection. Consequently, a zinc coating on steel gives both mechanical and chemical protection (Fig. 11.4a), whereas a tin coating can cause severe corrosion under a scratch (Fig. 11.4b). In this connection, a *galvanic series* is useful in practical corrosion design. Such a series is simply the relative order of chemical reactivity of metals and alloys in a given corroding medium, sea water in the series listed in Table 11.1.

**TABLE 11.1***

Comparison of the Galvanic Series in Sea Water and the Electromotive Series

| Galvanic series in sea water | Electromotive series† | | |
|---|---|---|---|
| Anodic (corroded) end | | | |
| Magnesium | Li, | $Li^+$ | $-3.02$ volts |
| Magnesium alloys | K, | $K^+$ | $-2.92$ |
| Zinc | Na, | $Na^+$ | $-2.71$ |
| Galvanized steel | Mg, | $Mg^{++}$ | $-2.34$ |
| Aluminum | Al, | $Al^{+++}$ | $-1.67$ |
| Cadmium | Zn, | $Zn^{++}$ | $-0.76$ |
| Aluminum alloys | Cr, | $Cr^{++}$ | $-0.71$ |
| Steel | Fe, | $Fe^{++}$ | $-0.44$ |
| Wrought iron | Cd, | $Cd^{++}$ | $-0.40$ |
| Cast iron | Co, | $Co^{++}$ | $-0.28$ |
| 50-50 Solder | Ni, | $Ni^{++}$ | $-0.25$ |
| 18-8 Stainless steel (active) | Sn, | $Sn^{++}$ | $-0.14$ |
| Lead | Pb, | $Pb^{++}$ | $-0.13$ |
| Tin | $H_2$, | $H^+$ | 0.00 (Reference) |
| Muntz metal | Bi, | $Bi^{+++}$ | 0.23 |
| Nickel | Cu, | $Cu^{++}$ | 0.34 |
| Yellow brass | Hg, | $Hg^{++}$ | 0.80 |
| Red brass | Ag, | $Ag^+$ | 0.80 |
| Copper | Pt, | $Pt^{++}$ | 1.2 |
| 70-30 Cupronickel | Au, | $Au^+$ | 1.7 |
| 18-8 Stainless steel (passive) | | | |
| Cathodic (protected) end | | | |

* Adapted from the *Corrosion Handbook*.      † Voltages are for oxidation reaction.

The notation $E$, rather than $E^0$, was employed in the paragraph above to indicate that the actual electrode potentials generally have values significantly different from the electromotive series (determined under specified standard

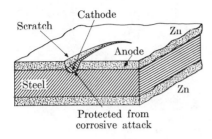

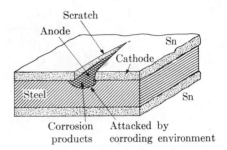

(a) In galvanized steel the zinc coating is sacrificially corroded, thus protecting the iron.

(b) In tinplate the tin layer is cathodic and causes severe local corrosion of the iron.

**Fig. 11.4** Difference in corrosion behavior of two types of coated steel when the mechanical protection is locally removed by a scratch.  (After Van Vlack)

conditions).  Polarization is another aspect of the electric-potential term $E_3 - E_1$ in Eq. (11.6) that has great significance for actual corrosion processes.  Polarization refers to the composite of various changes at either electrode causing the difference in potential to be only a small fraction of the equilibrium value.  Typical polarization effects are the accumulation of reaction products in the solution near the electrode in question or the formation of a protective layer on the electrode surface.  Since the anodic and cathodic reactions are linked in an electrical series, it follows that the corrosion rate is decreased by either cathodic or anodic polarization or by a combination of the two.  The passivation reaction in a Pourbaix diagram is an example of anodic polarization, and such a protective film can be an effective means of decreasing corrosion rates.  However, anodic polarization has the danger that any accidental breaks in the film can lead to intensive localized corrosion, such as the pitting discussed below.

Cathodic reactions, while not causing direct corrosive destruction of the type produced by the anodic reactions so far considered, are of basic importance for the control of corrosion.  This fact can be illustrated for the case of galvanic corrosion of Fig. 11.1, where the cathodic reaction occurs on the copper.  Actually, the copper is not directly involved in the cathodic reaction, since the electrons flowing to the copper surface are discharged by reaction with hydrogen ions,

$$H^+ + e^- = H. \tag{11.7}$$

Figure 11.5(a) shows how the layer of atomic hydrogen so produced can cover the cathode surface, thus polarizing it and preventing further reaction.  Problem 3 is an example of the effectiveness of this phenomenon for corrosion control.  In sufficiently acid solutions (see Fig. 11.20) the hydrogen atoms can combine to form $H_2$ gas and escape from the cathode surface, but usually a reaction involving oxygen comes into play (Fig. 11.5b).  Oxygen dissolving at the surface of the cor-

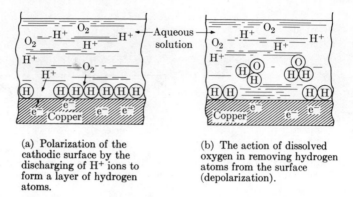

(a) Polarization of the cathodic surface by the discharging of $H^+$ ions to form a layer of hydrogen atoms.

(b) The action of dissolved oxygen in removing hydrogen atoms from the surface (depolarization).

**Fig. 11.5** Schematic illustrations of two important cathodic reactions.

roding solution makes its way to the cathode surface and removes the protective layer by the reaction

$$\tfrac{1}{2}O_2 + 2H = H_2O, \tag{11.8}$$

thus leading to the overall cathodic reaction shown in Fig. 11.1 (see Problem 4). This discussion explains a remarkable aspect of the corrosion of iron; namely, that the corrosion rate is often determined by the rate at which oxygen can diffuse to the reacting metal. Application of this fact for control of corrosion is discussed later.

It has been convenient to use galvanic corrosion as an example in the above discussion, because the anodic and cathodic reactions can then be discussed especially simply since they occur separately on the two metals in question. The corrosion mechanism is fundamentally the same, however, when only a single metal is involved in the corrosion process. Taking the corrosion of iron as an example, it is clear that the anodic reaction, $Fe = Fe^{++} + 2e^-$, will tend to occur, but at first it might seem that there is no possibility for the corresponding cathodic reaction to take place in the absence of a second metal such as copper. It will be recalled, however, that the copper was not directly involved in the two cathodic reactions, Eqs. (11.7) and (11.8), and, in fact, these reactions can occur on portions of a steel specimen. Furthermore, since the essential cathodic reaction is the discharging of hydrogen ions, the electrical driving force is effectively the same whether the reaction occurs on copper or on iron. The standard value can be determined from the electromotive series of Table 11.1 as

$$E_H^0 - E_{Fe}^0 = 0.00 - (-0.44) = 0.44 \text{ volt}, \tag{11.9}$$

although the effective value is very much smaller, as was discussed above.

The distribution of cathodic areas is often an important factor in the corrosion of a given metal. Normally myriad microscopic anodic and cathodic areas exist adjacent to one another, but their positions fluctuate. Thus a uniform corrosion is produced even though at any instant the cathodic areas are protected regions

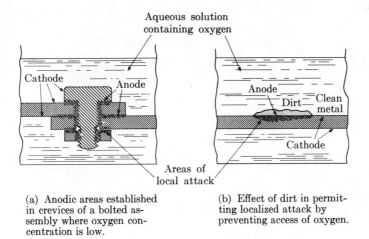

(a) Anodic areas established
in crevices of a bolted assembly where oxygen con-
centration is low.

(b) Effect of dirt in permitting localized attack by
preventing access of oxygen.

**Fig. 11.6** Examples of localized corrosion caused by differential aeration. (After Van Vlack)

of the specimen. Under certain conditions, however, the location of the anodic or
the cathodic regions may be stabilized and cause severe local corrosion. A commonly occurring instance of this kind is illustrated in Fig. 11.6, the cathodic areas
being established where oxygen is available for the reaction of Eq. (11.8) and the
anodic reaction being localized in areas relatively inaccessible to oxygen but in
contact with the corroding solution. For example, in the case of an ordinary spot
of dirt on a metal surface (Fig. 11.6b), the destructive anodic reaction occurs
beneath the dirt in the presence of a corroding solution, the extent of corrosion
progressively increasing with increasing time of exposure. Eventually a deep pit
will have formed at this location, although the surrounding metal may be almost
free of corrosive attack. The corroding liquid may also stabilize the cathodic and
anodic areas by other mechanisms; for example, through differences in solute
concentration, local velocity, temperature, bacteria count, or intensity of illumination.

Distinct anodic and cathodic areas can also be caused by inhomogeneities
within the metal or alloy. Figure 11.7 illustrates two instances in which local
anodes are produced at cold-worked areas in a specimen, thus causing pronounced
attacks at these points. Particles of a second phase in an alloy evidently represent
regions with an electrode potential different from that of the bulk of the alloy and
therefore can localize corrosive action. Grain boundaries, segregation of impurities, and nonuniformly heat-treated areas also represent sources of electrochemical
differences in alloys.

Another aspect of the electrochemical behavior of certain alloys is their conversion to the *passive** state under the action of oxygen or oxidizing media, a state

---

* The characteristics ascribed to the *passive* condition are somewhat different from those of the
*passivated* state in the Pourbaix diagram.

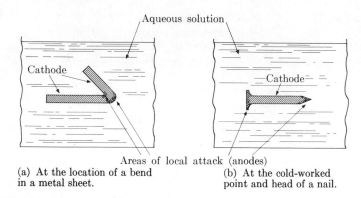

Aqueous solution

Cathode

Cathode

Areas of local attack (anodes)

(a) At the location of a bend
in a metal sheet.

(b) At the cold-worked
point and head of a nail.

**Fig. 11.7** Examples of local anodes produced by cold working. (After Van Vlack)

in which they are less subject to corrosion than when they are in the normal or *active* condition. For example, when normal iron is exposed to concentrated nitric acid, it becomes passive and is then virtually unattacked by strong solutions of this acid. Passivity is a relative term, however, and a metal may be passive under one set of conditions, but corrode rapidly under another. This behavior is shown for stainless steel in sea water in Table 11.1. Under most conditions stainless steel is passivated by sea water and acts like a more noble metal, but if a pit develops on the surface of the steel, the metal in that local area becomes active, and corrosion is accelerated. Passivity is believed to be caused by the presence of special oxide films at the metal surface.

**Specific corrosion types.** To complement the broad theoretical discussion of electrochemical corrosion presented in the previous section, it is desirable to consider now the more specific classifications that are widely used for certain types of industrially important corrosion. The broadest of these descriptions is the term *uniform* attack. This term is applied to any form of corrosion in which the whole surface of the metal is corroded to the same degree. One type of uniform attack is shown in Fig. 11.10. Under these conditions the useful life of a given material is easily estimated, and unexpected failure need not be feared. *Pitting* and *intergranular corrosion*, on the other hand, are types of nonuniform corrosion that can proceed in otherwise undamaged metal. Both are initiated by inhomogeneities in the metal or on its surface, but may be accelerated by such additional factors as a localized change from the passive to the active state after corrosion has begun. The deep, isolated holes caused by pitting (Fig. 11.8) are especially difficult to avoid, but improving the homogeneity of the metal and of the corroding medium is often helpful.

Intergranular corrosion occurs when a pronounced difference in reactivity exists between the grain boundaries and the remainder of the alloy. In stainless

**Fig. 11.8** Examples of pitting in a copper pipe used to handle drinking water.

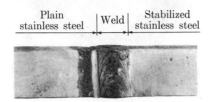

**Fig. 11.9** Intergranular corrosion of welded stainless steel subjected to a corrosive environment. (Courtesy M. G. Fontana)

steel such a difference is set up when chromium carbides form at the grain boundaries during heating of the steel in the range 480–760°C (900–1400°F). The grain-boundary region is thereby depleted in chromium and becomes anodic with respect to the surrounding alloy. Corrosion can then occur along the grain boundaries and can produce serious damage like that shown in Fig. 11.9 for welded stainless steel subjected to a corrosive environment. Usually corrosion in welded stainless steels is not a serious problem, since corrosion resistance can be restored by a heat treatment that dissolves the chromium carbides precipitated during welding. However, when stainless steel assemblies are too large to be heat treated or are to be subjected in service to temperatures in the 480–760°C range, it is then necessary to prevent the depletion of chromium by special means. The use of very low carbon contents (about 0.03%) is one remedy, but generally the customary 0.08% carbon is *stabilized* by the addition of titanium or columbium as in type 321 stainless steel, Table 11.3. In Fig. 11.9 note that the stainless steel stabilized by the addition of 0.5% titanium has not been attacked. Titanium and columbium are extremely strong carbide formers, and they leave almost no carbon available for combination with chromium.

Single-phase alloys usually have better corrosion resistance than similar two-phase alloys. Electrochemical corrosion is encouraged by the presence of two dissimilar constituents, and one of them tends to be preferentially attacked. Moreover, even in certain single-phase solid-solution alloys, preferential corrosion of one of the component metals may occur. The best-known example of this type of behavior is the *dezincification* that takes place in brasses containing more than 15% zinc. Figure 11.10 shows this kind of corrosion in a brass pipe. Ordinary water is capable of dissolving the zinc and leaving the copper, perhaps by redeposition, as a spongy mass that has almost no strength. Plug-type (localized) dezincifi-

**Fig. 11.10**  Uniform dezincification that has proceeded halfway through a brass water pipe. (Courtesy M. G. Fontana)

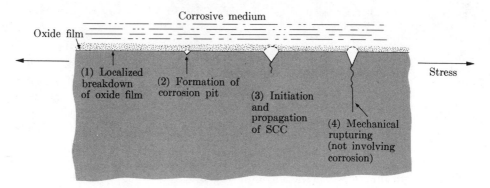

**Fig. 11.11**  Schematic illustration of the relation of stress-corrosion cracking (SCC) to preparatory processes and to final rupture.  (After B. F. Brown)

cation is especially severe.  The addition of small amounts of alloying elements, such as tin or arsenic, to high-zinc alloys effectively controls this type of corrosion in some cases.  A second type of preferential loss of zinc (that is also termed dezincification) occurs when brasses are heated to high temperatures at which the zinc vaporizes rapidly.

*Stress corrosion* refers to greatly accelerated corrosion that takes place in certain environments when metals contain internal tensile stresses.  Well-known examples are *season cracking* that occurs in brasses, especially in the presence of moisture and traces of ammonia, and *caustic embrittlement* of steel exposed to solutions containing sodium hydroxide.  The most effective control is by elimination of the tensile stresses.  Usually this is accomplished by a stress-relief anneal that need not greatly decrease the strengthening effect of a previous cold-working operation.  Occasionally it is possible to produce unobjectionable compressive stress by a means such as shot-peening.  Since a specific environment is often

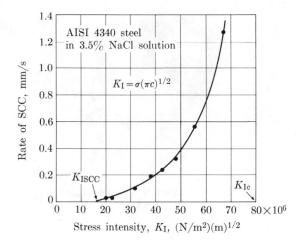

**Fig. 11.12** Data on rate of stress-corrosion propagation of an existing crack as a function of stress intensity $K_I$. The high-strength alloy steel has a yield strength of $155 \times 10^7$ N/m$^2$ ($225 \times 10^3$ lb/in$^2$). (Data of M. H. Peterson *et al.*) To convert from (N/m$^2$)(m)$^{1/2}$ to (lb/in$^2$) (in.)$^{1/2}$, multiply by $0.910 \times 10^{-3}$.

necessary for rapid stress corrosion, adequate control may sometimes be obtained by treating the corroding medium. High-strength alloys, including alloy steels and special titanium and aluminum alloys, may fail in service under a relatively low applied stress as a result of stress-corrosion cracking (SCC). Various corroding media can cause SCC; even ordinary humid air is sometimes responsible for it. The study of SCC is complicated by the fact that other processes may be required to produce the initial defect, such as a pit, Fig. 11.11, at which the stress-corrosion crack initiates. Also, the process of cracking terminates by mechanical rupturing, for which no corrosion is required. Laboratory studies of SCC are made on specimens containing an initial crack, and therefore the stress intensity $K_I = \sigma(\pi c)^{1/2}$ [see Eq. (4.30)] is the pertinent quantity, rather than simply the applied stress $\sigma$. As shown by the data in Fig. 11.12, if a high-strength material is in an appropriately corrosive medium, SCC occurs at stresses far below the ordinary fracture stress $\sigma_f$; that is, at values of $K_I$ far below $K_{Ic}$. As $K_I$ is progressively decreased, the rate of growth of the initial crack also decreases. Finally, below a critical value $K_{ISCC}$, an existing crack does not grow. Since the numerical value of $K_{ISCC}$ depends on the length $c$ of the crack in question, use of $K_{ISCC}$ data in design calculations requires information on the length of the longest crack. This information can be obtained by a suitable inspection procedure.

*Corrosion fatigue* is the combined action of corrosion and repeated stresses and is far more serious than the sum of these two factors acting individually. To express the influence of corrosion on fatigue strength, the *damage ratio* is frequently

**Fig. 11.13** The erosion corrosion of an elbow in a steam-condensate return line produced by impingement attack. (Courtesy M. G. Fontana)

used:

$$\text{Damage ratio} = \frac{\text{corrosion fatigue strength}}{\text{normal fatigue strength}}.$$

This ratio for salt water as a corroding medium is about 0.2 for carbon steels, 0.5 for stainless steels, 0.4 for aluminum alloys, and 1.0 for copper. Suitable protective measures against corrosion fatigue include treatment of the corroding medium and surface protection of the metal. Nitriding of steels is often useful for this purpose.

Because most corrosion testing is done in liquids and gases at rest, the data are usually not applicable to service conditions with rapid movement past the metal surface. Such *erosion corrosion* causes accelerated attack, because it mechanically removes the protective layer that normally builds up on the corroding surface. Impingement attack of elbows in pipelines (Fig. 11.13) is typical of this type of corrosion. Improved mechanical design is often a sufficient remedy for erosion corrosion, but treatment of the corroding medium or choice of a different alloy may be necessary.

Corrosion processes are frequently classified according to the type of environment in which they occur, and data on various alloys can generally be found listed under several such general headings. Exposure to the atmosphere (weathering) is an important classification of this kind for which the relative corrosion resistances of several alloys are listed in Table 11.2. It should be noted that atmospheric corrosion involves exposure to varying conditions such as rain, humid air, sun, dust particles, and chemical contaminants. A somewhat less complex environment is produced when the service conditions involve immersion in water or in a solution of acids, alkalies, or salts. Much more extensive data than those listed in Table 11.2 will be found on this topic in the *Corrosion Guide* and in the *Corrosion Handbook*, which are listed among the references at the end of the chapter. Another general classification is underground corrosion, such as that experienced by buried pipe lines. Under most soil conditions this type of corrosion occurs by one of the usual electrochemical mechanisms, but in water-logged soils micro-biological corrosion can occur. Certain anaerobic bacteria may then find suitable growth conditions and cause severe local corrosion of iron and steel by depolarizing the layer of atomic hydrogen.

## OXIDATION

This type of corrosion involves the reaction of metals with oxygen at high tem-peratures, usually in the absence of moisture. (It is noteworthy that a type of oxide formation, such as the rusting of iron, that occurs in the presence of moisture is not considered to belong to this category, since a different mechanism is involved.) The nature of the oxide that is formed plays an important part in the oxidation process.

1. The oxide may be unstable, as in the case of gold oxide, and oxidation does not occur.

2. The oxide may be volatile, as in the case of molybdenum oxide, and oxidation occurs at a constant, relatively high rate.

3. One or more oxides may form a layer or layers at the metal surface.

The third occurrence is the most common, and various aspects of its behavior are considered here.

**Oxide films.** Surface oxide layers are called *films* when their thickness is less than about 3000 Å, and they are called *scales* when their thickness exceeds this value and is more easily measured. Since films are so thin, their destructive effect on most metal parts can be neglected. However, under certain conditions of wear, the rate of metal loss is greatly increased as a result of the removal by abrasion of the thin protective film. Films usually decrease the rate of additional oxidation and may, as in the case of the invisible film on aluminum, be almost completely protective. An example of film formation in steel heat treatment is the formation of *temper colors* during the heating of steels in the range 230 to 320°C (450 to 600°F). These colors range from light straw to dark blue and are the interference effects produced as the oxide film increases in thickness as a function of time and tem-perature. The laws governing the growth of films are similar to those for scale formation.

**Scale formation.** Thick oxide layers or scales are divided into two categories, protective and nonprotective, on the basis of the *Pilling–Bedworth rule*. According to this rule an oxide is protective if the volume of the oxide is at least as great as the volume of metal from which it formed. If the volume of oxide is less than this amount, the scale is not continuous and hence is comparatively ineffective in preventing the access of oxygen to the metal surface. Although there are many exceptions to the Pilling–Bedworth rule, it is a useful guide when the specific

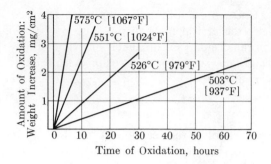

**Fig. 11.14**  The oxidation of pure magnesium in oxygen. (After T. E. Leontis and F. N. Rhines)

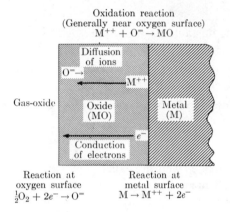

**Fig. 11.15**  The course of oxidation of a metal through a protective oxide layer.

oxidation characteristics of a metal are unknown.  Metals that have these *non-protective* oxides tend to increase the weight, $W$, of their scale at a *linear* rate according to the equation

$$W = At, \qquad (11.10)$$

where $A$ is a constant that depends on the temperature and $t$ is the time.  Data on the oxidation of magnesium are given in Fig. 11.14.  Some of the alkali and alkaline earth metals also oxidize in accordance with Eq. (11.10).  Usually the mechanism of growth of nonprotective films involves the passage of gaseous oxygen through pores or fissures in the oxide film.

When a *protective* oxide forms on the exposed metallic surface, diffusion must occur through the scale for additional growth to take place.  Figure 11.15 is a schematic representation of a process of oxidation under these conditions.  This figure shows that the metal ionizes at the oxide-metal surface, and then both the metal ion $M^{++}$ and the electrons diffuse through the oxide layer to the oxygen

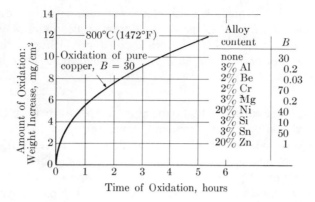

**Fig. 11.16** The parabolic oxidation of copper in air and the effect of various alloying elements on the rate constant, $B$.

surface. Here the electron aids in forming the oxygen ion. This ion reacts near the oxygen surface with the metal ion to form the oxide, represented by MO. As shown in Chapter 8, the rate of oxidation in this case should follow the *parabolic law*,

$$W^2 = Bt, \qquad (11.11)$$

where $B$ is a constant that depends on the temperature. This behavior is observed experimentally in many instances, and Fig. 11.16 shows typical data for copper and several binary alloys.

The oxidation of iron is an example of the simultaneous formation of two or more oxide layers. At high temperatures the scale is composed largely of FeO, which forms an inner layer; $Fe_3O_4$ forms a middle layer, and $Fe_2O_3$ is at the oxygen surface. The first two oxides grow by the mechanism described above, but the oxidation of iron to $Fe_2O_3$ takes place in a different manner. In this case the oxygen is more mobile and diffuses inward from the oxygen surface. At temperatures below about 570°C (1060°F) the scale formed on iron has a different character since the FeO phase is absent. This scale is more adherent and is harder to remove by the acid pickling operation customarily used to clean steel surfaces.* The iron oxides illustrate an important characteristic of oxides in general; namely, that the crystal lattice of the oxide must contain defects (vacancies or interstitial ions) in order to be able to form. The most common defect structure, characteristic of FeO and $Fe_3O_4$, contains cation (metal ion) vacancies. $Fe_2O_3$ is an example of an oxide with anion (oxide ion) vacancies, while ZnO depends on interstitial ions for its process of growth.

---

* To remove surface oxides, steels are immersed for a suitable period in a pickling solution (5–10% $H_2SO_4$) at about 70°C (160°F).

Some metals with protective oxides have reaction rates that diminish more rapidly with time than Eq. (11.11) predicts. For example, under many conditions the oxide layer on aluminum approaches a constant thickness with increasing time and so obeys an *asymptotic law*. A *logarithmic* increase in the weight of oxide,

$$W = C \log (Dt + E), \tag{11.12}$$

where $C$, $D$, and $E$ are constants that depend on temperature, is observed in other cases, such as for iron or nickel at moderate temperatures.

When the oxidation of an alloy (rather than a pure metal) is considered, several new features may appear. The separate oxides may form a solid solution; a multiphase scale may appear; or only a single component in the alloy may be oxidized, a process called selective oxidation. When the oxides which form are highly soluble in each other, certain generalizations regarding the oxidation behavior of the alloy system can be made. For example, NiO is basically a cation-deficient structure and has high mutual solubility with $Cr_2O_3$. It is observed that small additions of Cr to Ni increase the oxidation rate, while additions of Ni to pure Cr decrease the oxidation rate. The general rule for this type of behavior is that additions of a higher valence species tend to be deleterious, while additions of a lower valence species are usually beneficial. On the other hand, in oxides which have excess cations and electrons in interstitial positions, such as ZnO, the reverse is true. Aluminum additions to zinc lower the oxidation rate of zinc, while lithium additions have been shown to increase the rate.

When the oxide scale consists of more than a single phase, the oxidation characteristics of the alloy are usually governed by the properties of that oxide phase which is continuous. For example, in the oxidation of some Cu-Zn alloys, islands of ZnO are found in a matrix of $Cu_2O$, and the oxidation process is governed by diffusion through $Cu_2O$. A notable exception to this rule is the case of dilute additions of vanadium and molybdenum to steels. The oxides of these components form a low-melting eutectic with the oxides of iron, and catastrophic oxidation results. In the case of selective oxidation of one component at the surface of an alloy, great benefit can be derived from certain surface films if they are adherent, have poor electrical conductivity, and possess a complex crystal structure that hinders diffusion through them. A special case of selective oxidation is discussed below.

**Internal oxidation.**   Under special conditions, oxidation of an alloying element may occur below the surface of the base metal. This *internal oxidation* which may take place in the absence of appreciable film or scale formation at the surface usually requires the following conditions.

1. The alloying element must have a greater affinity for oxygen than does the base metal.

2. Oxygen must diffuse rapidly in the base metal, compared with the diffusion of the alloying element.

3. Scaling of the base metal must not be so rapid as to destroy the surface region in which internal oxidation is occurring.

In some cases internal oxidation is useful as a method of surface hardening. More often it is the cause of difficulties, for example when certain copper or silver alloys are given high-temperature treatments in mildly oxidizing atmospheres. The internal oxidation produced under these conditions may cause processing problems or mar the surface of the metal. Compared with ordinary oxide layers these subscales are difficult to remove by the usual cleaning methods.

Oxidation rarely occurs in only one of the ways described here. Combinations of two or three types of reaction may occur in turn or simultaneously in different parts of the metal. The stress condition and the orientation of the oxide layer may vary with time or thickness, and discontinuous cracking or spalling of the layer may cause sudden changes in the reaction rate. If more than one oxide is stable under the conditions existing during oxidation, a series of oxide layers is formed in the manner discussed for the commercially important case of the scaling of iron. The book *Oxidation of Metals*, listed among the references to this chapter, is an excellent source of additional information on these more advanced topics.

**Liquid-metal corrosion.** Corrosion of metals and alloys at high temperatures by liquid metals flowing past them is important because of its occurrence in devices for nuclear power. The corrosion reaction is essentially a process of mass transfer and is not dependent upon local cell potentials for its driving force. Usually the driving force for this form of corrosion is the tendency of the solid to dissolve in the liquid up to the solubility limit at the given temperature. The specific types of liquid-metal attack include the following: simple solution of the solid metal, formation of a chemical compound, intergranular attack, and selective extraction of one of the component metals in a solid alloy. The usual causes of these phenomena is the existence of a temperature gradient or a concentration gradient within the solid-liquid system.

The most serious cases of damage by liquid-metal attack occur in heat exchangers carrying liquid-metal coolants. As the solid container, usually iron or copper tubing, approaches equilibrium with the liquid-metal coolant in the hot zone of the heat exchanger, a portion of the solid goes into solution in the liquid. When the liquid moves to a cooler part of the heat exchanger, the solubility limit decreases and deposition of solids takes place, usually on the walls of the exchanger tubes. In a *thermal loop* of this kind, the hot zone is continually corroded, and the cold zone becomes plugged with the deposited corrosion products. Shutdown of liquid-metal heat exchangers may be necessary after less than one month of continuous operation unless specific preventive measures are taken. The most promising technique for prolonging the life of exchangers is the addition of certain inhibitors to the liquid alloy. It is believed that the inhibitors adsorb on the walls of the solid vessel or form protective films that hinder the corrosion processes at elevated temperatures.

## CONTROL OF CORROSION AND OXIDATION

Since the types of corrosion are so numerous and the conditions under which corrosion occurs are so extremely varied, it is not surprising that many methods are used to deal with corrosion problems. The method or combination of methods used in a given instance is determined largely by economic considerations. In some cases it is cheaper to replace a low-cost metal at more frequent intervals than to use a somewhat superior, higher-cost alloy. However, when corrosion resistance is vitally important, a larger investment in corrosion control is usually advisable. Corrosion control measures can be considered under four general headings: (1) designing against corrosion, (2) improving the characteristics of the metal, (3) protecting the metal with a coating substance, and (4) treating the corroding medium to reduce its corrosive action.

**Designing against corrosion.** Several principles of good design, based on the electrochemical nature of corrosion processes, merit primary consideration for control of corrosion. While the choice of corrosion-resistant materials or the use of corrosion-retarding treatments also enter in the over-all design, attention will be devoted in the present section to a different aspect of design. This is the problem of ensuring that, whatever rate of corrosive attack is operative, it occurs in as uniform a manner as possible and does not cause premature failure because of intense, localized corrosion.

By far the most important design principle in this respect is the avoidance of dissimilar metals in contact in the presence of a corroding solution. Such a galvanic couple localizes corrosion on the electrochemically more active metal, the other metal being substantially protected. Furthermore, the corrosion is often further localized to the immediate vicinity of the dissimilar metal contact. An extreme example of poor design in this respect is the use of a small part made of an anodic (active) metal in a much larger expanse of cathodic metal; e.g., a steel pipe in a copper tank. The large cathodic area encourages a large corrosion current, which is then localized in the iron and causes rapid deterioration and eventual failure. Clearly, a copper pipe in a steel tank offers far less danger, although localized attack may still be a problem. A much better design, of course, is the use of the same metal for both the pipe and the tank. Whenever the direct joining of dissimilar metals is unavoidable, insulation to prevent electrical contact between them will often avoid intensified corrosion.

Good design can help prevent the occurrence of inhomogeneities (both in the metal and in the corrosive environment) which are potential causes of localized corrosion. Crevices are to be avoided since they permit concentration differences to arise in the manner previously described in connection with Fig. 11.6. Bolts and rivets are undesirable for this reason (see Fig. 11.6a) and are preferably replaced by a butt weld. When crevices are unavoidable in a given design, their deleterious effect can be minimized if the corroding medium is denied access to them. Effective measures for this purpose are improvement of the fit, filling the crevice with an impervious material, or painting the matching parts before assembly. The accumulation of dirt or deposits of various kinds can also cause localized corrosion, so the design should allow for adequate cleaning and flushing of critical parts of the equipment. Uniform flow of a corrosive liquid is also a design consideration since both stagnant areas on the one hand and impingement conditions on the other hand can cause accelerated corrosion.

When assigning working stresses to machine components that must operate in a corrosive environment, consideration should be given to the possible occurrence of corrosion fatigue or stress corrosion. The previous discussion of corrosion fatigue pointed out that a metal with adequate corrosion resistance should give satisfactory performance, but generally a conservative working stress should be used when fatigue conditions are involved. Stress corrosion occurs under static stress conditions for a specific combination of metal and corroding medium; the examples of steel (sodium hydroxide solution) and brass (ammonia solution) have previously been considered. Other important combinations are: austenitic stainless steel (chloride solution), hardened alloy steels (hydrogen sulfide solution), and aluminum alloys (chloride solution). Use of these dangerous combinations should be avoided, but in the case of necessity, stress-relief treatments of the alloy or modification of the corroding environment may give adequate corrosion control.

**Characteristics of the metal.** It is obviously impossible to describe here the corrosion tendencies of hundreds of different alloys under thousands of different service conditions. An excellent summary is contained in the *Corrosion Handbook*, and the answers to many corrosion problems can be found there. To give a brief, over-all view of metal characteristics, qualitative information on the corrosion of representative materials is presented in Table 11.2,* but even these qualitative relations may change as conditions of service, such as the operating temperature, are altered.

The corrosion resistance of a given metal may be improved by increasing its purity, but the low strength of pure metals is often a disadvantage. It is more common to increase both strength and corrosion resistance by the use of suitable alloying elements. Examples of such alloying can be found for almost all metals, but iron-base alloys are especially significant because of the wide use of stainless steel. Many alloying elements are useful in giving iron limited corrosion resistance. For example, small amounts of phosphorus and copper improve the resistance of structural steels to atmospheric corrosion, high silicon contents are useful in acid-resistant cast alloys, and about 10% aluminum renders iron extremely resistant to high-temperature oxidation but also makes it brittle. However, chromium dwarfs all other alloying elements in importance. The reasons are many, but the primary one is the effect of chromium in increasing corrosion

---

* Adapted from a *Metal Progress* Data Sheet.

**TABLE 11.2**    Relative Corrosion Resistance of Several Metals and Alloys

| Material | Atmosphere Sea shore | Atmosphere Indus- trial | Water Domes- tic | Water Sea | Water Wet steam | Oxidizing gases °C | Oxidizing gases °F | Reducing fuel gas °C | Reducing fuel gas °F | Food products | HCl | H₂SO₄ | Acetic | 1–20 % Solutions of alkalies | NH₄Cl | MgSO₄ |
|---|---|---|---|---|---|---|---|---|---|---|---|---|---|---|---|---|
| Ingot iron or wrought iron | P | P | F | F | FG | 540 | 1000 | 540 | 1000 | P | P | P | P | G | P | FG |
| Low-carbon steel | P | P | F | F | FG | | | | | P | P | P | P | G | P | FG |
| Low-alloy, high-strength structural steel | F | F | F | F | FG | | | | | | P | P | P | | P | FG |
| Galvanized steel | G | G | FG | FG | FG | | | | | P | P | P | P | E | P | FG |
| Gray cast iron | F | FG | G | F | F | | | | | P | P | P | P | P | P | FG |
| 3½ % Nickel cast iron | FG | FG | F | FG | F | 650 | 1200 | 650 | 1200 | P | P | P | P | GE | P | FG |
| 17 % Chromium-iron | G | G | G | PG | G | 840 | 1550 | 840 | 1550 | GE | P | P | E* | G | FG | G |
| 18 % Cr-8 % Ni stainless steel | E | G | E | FG | E | 900 | 1650 | 900 | 1650 | E | P | PG | E | E | G | G |
| 25 % Cr-20 % Ni stainless steel | E | G | E | G | E | 1150 | 2100 | 1090 | 2000 | | P | F | E | E | G | G |
| Nickel (99.2 %) | E | G | E | FE | E | 1040 | 1900 | 1260 | 2300 | GE | FG | FG | FG | E | G | G |
| Inconel (80 % Ni, 14 % Cr, 6 % Fe) | E | G | E | FE | E | 1090 | 2000 | 1150 | 2100 | E | F | F | G | E | G | G |
| Hastelloy C (58 % Ni, 17 % Mo, 14 % Cr, 5 % W, 6 % Fe) | E | E | E | E | E | 1150 | 2100 | 1150 | 2100 | E | G | E | E | E | E | E |
| Copper (99.9 %) | G | G | G | G | G | | | | | FG | PG | FG | FG | G | FG | G |
| 5 % Tin-bronze | G | FG | G | G | G | | | | | F | P | FG | FG | G | FG | G |
| Aluminum (99.2 %) | GE | E | FG | G | G | 430 | 800 | 430 | 800 | G | P | F | G | P | F | G |
| 202476 Aluminum alloy (4.5 % Cu, 1.5 % Mg, 0.6 % Mn) | PG | FG | PF | P | F | | | | | F | P | P | F | P | P | P |
| Magnesium | G | G | F | P | P | 200 | 400 | 200 | 400 | PE | P | P | P | E | P | P |
| Wrought magnesium alloys | G | G | F | P | P | | | | | PE | P | P | P | E | P | P |
| Tin (99.9 %) | G | G | E | G | G | | | | | G | P | P | P | F | FG | G |
| Lead (99.9 %) | G | G | PG | G | G | | | | | P | F | E | F | P | | |

* Service temperature is always an important variable, and in this instance the rating applies only for temperatures below 50°C (120°F).

**TABLE 11.3** Some Properties of Typical Stainless Steels

| Description of steel | | | Composition, % | | | | Yield strength, $10^7$ N/m² | Tensile strength,* $10^7$ N/m² | % Elongation in 50 mm | Applications |
|---|---|---|---|---|---|---|---|---|---|---|
| Class | Type | Condition | C | Cr | Ni | Other elements | | | | |
| Martensitic | 420 | Annealed | 0.15+ | 13 | | | 45 | 70 | 30 | Cutlery, surgical instruments, springs |
| | 420 | Heat-treated | | | | | 85–140 | 105–210 | 12–2 | |
| | 501 | Annealed | 0.10+ | 5 | | 0.5 Mo | 20 | 50 | 28 | Petroleum industry |
| Ferritic | 405 | Annealed | 0.08 – | 13 | | 0.2 Al | 25 | 40 | 20 | Turbine buckets |
| | 446 | Annealed | 0.35 – | 25 | | | 40 | 60 | 25 | Heat-resisting parts |
| Austenitic | 301 | Annealed | 0.15 | 17 | 7 | | 30 | 85 | 80 | Transportation industries |
| | 301 | Work-hardened | | | | | 35–120 | 90–1 0 | 60–10 | |
| | 304 | Annealed | 0.08 – | 18+ | 8+ | | 30 | 65 | 70 | Standard 18–8 for general purpose applications |
| | 304 | Work-hardened | | | | | 35–105 | 70 120 | 50–10 | |
| | 310 | Annealed | 0.25 – | 25 | 20 | | 35 | 70 | 50 | For severe oxidation and corrosion conditions |
| | 321 | Annealed | 0.08 – | 17+ | 8+ | 0.5 Ti | 30 | 60 | 60 | For use when welding or heating in service might induce intergranular corrosion |
| | AM-355 | Double-aged | 0.13 | 16 | 4 | 2.75 Mo | 105 | 135 | 10 | Useful where strengthening cannot be obtained by cold work |
| | AM-355 | Subzero cooling | | | | 0.12 N | 130 | 150 | 13 | |
| | 201 | Annealed | 0.15 | 17 | 4.5 | 6.5 Mn | 40 | 80 | 55 | Low-nickel alternative to 300 series |
| | 202 | Annealed | 0.15 | 18 | 5 | 8.5 Mn | 40 | 70 | 55 | |

* The endurance strength is usually about half of the tensile strength, corresponding to an endurance ratio of 0.5.
To convert from N/m² to lb/in², multiply by $1.450 \times 10^{-4}$.

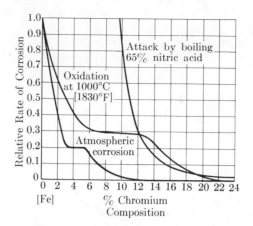

**Fig. 11.17** The effect of chromium in decreasing the rate of corrosion of iron-base alloys under three different conditions of corrosion.

resistance in almost all environments. Figure 11.17 shows the relation between chromium content and the corrosion rates under three different conditions.

The subject of steels based on chromium—the stainless steels and related oxidation-resistant alloys—is a most extensive one. Thus in the *Metals Handbook, Vol. I,* the space devoted to the engineering aspects of this topic is about equal to the whole of this textbook. The purpose of the present treatment, therefore, is to furnish a useful orientation in the field and to provide sufficient theoretical background for effective use of such handbook material. The summary of typical stainless steels in Table 11.3 shows that some of the topics of special importance are: wrought versus cast alloys, effects of chemical composition, strengthening through cold-working, possibility of precipitation hardening, and the classification of alloys under the groupings *martensitic, ferritic,* and *austenitic.* This classification of alloys has basic significance, as the following discussion of the appropriate phase diagrams will show.

The *martensitic* stainless steels are those that have the possibility of forming this phase during rapid cooling from the austenitic condition. The iron-chromium phase diagram (Fig. 11.18) shows that the martensite reaction is restricted to steels with limited chromium contents since 100% austenite phase can be produced only for chromium contents up to 12%. In alloys containing about 0.6% carbon, the limit of the austenite region (called the gamma loop) extends to 18% chromium. Therefore *cutlery steels* containing 12 to 18% chromium can be made hard by martensite formation, and yet they have relatively good corrosion resistance. The type 501 stainless steel containing about 5% chromium has only limited corrosion and oxidation resistance as can be seen from the curves of Fig. 11.17. However, this low-cost alloy is the proper engineering selection for many applications in the petroleum industry.

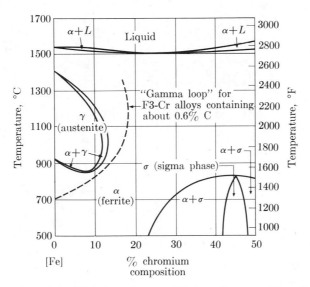

**Fig. 11.18** A portion of the iron-chromium equilibrium diagram. The effect of carbon in expanding the austenite region is indicated.

When substantial amounts of chromium are present in iron-chromium alloys, so that the corrosion resistance is greatly improved, the alloys fall in the ferrite area of the phase diagram (Fig. 11.18) and are classed as *ferritic* stainless steels. Commercial alloys of this class contain 13–25% chromium. Since these alloys cannot be hardened by heat treatment, they are generally used in the annealed condition, but sometimes cold-working is employed as a means of increasing their strength. The usual cold-forming operations can be employed in fabrication, but welding tends to produce an embrittlement that may not be completely removed by subsequent heat treatment. The relatively low cost of these steels accounts for their use in automobile trim as well as in applications where resistance to high-temperature oxidation is important. The *sigma phase* that appears at very high chromium contents is almost always undesirable in stainless steels because it has an adverse effect on corrosion resistance and impact strength. However, the hardness of this phase has been found useful in applications such as automobile engine valves, where both wear and moderate corrosion problems are encountered.

The iron-chromium phase diagram is adequate for explaining the nature of martensitic and ferritic stainless steels, but to understand the most important and varied class, the *austenitic* stainless steels, it is necessary to consider the ternary iron-chromium-nickel diagram. Since the standard stainless steel, called 18–8, contains about 18% chromium and 8% nickel, it is possible to understand the essential role of nickel by considering a vertical section at 18% chromium

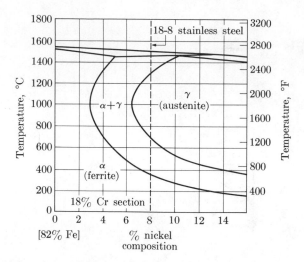

**Fig. 11.19** Vertical section at 18% chromium through a portion of the iron-chromium-nickel equilibrium diagram. (After E. C. Bain and R. H. Aborn)

through the ternary diagram (Fig. 11.19). At elevated temperatures the addition of nickel to an 18% chromium alloy causes a gradual change from the ferrite phase, first into the two-phase region, and finally to the completely austenitic condition somewhat above 6% nickel. One aspect of these equilibrium relationships is significantly modified by the occurrence of metastability at higher nickel contents. For example, although Fig. 11.19 indicates that 18–8 stainless steel should be completely ferritic at room temperature, in practice this steel remains austenitic on being cooled from high temperatures. Consequently, austenitic stainless steels cannot be hardened by quenching since the austenite does not transform to martensite.

Austenitic stainless steels have an outstanding combination of properties. Of foremost importance is the enhancement of both corrosion and oxidation resistances by the presence of nickel, as indicated in Table 11.2. Fabrication by welding is also generally superior compared to the other two classes, and the availability of the stabilized grades such as type 321 (Table 11.3) permits application of this method of joining even when subsequent heat treatment is impractical. Austenitic stainless steels have a special advantage with regard to cold-forming operations since it is possible to promote either: (1) ease of forming; or (2) rapid work hardening (i.e., strengthening), through appropriate modification of the chemical composition. The explanation lies in the fact that, while metastable austenite tends to decompose into a martensite-like structure during cold-working, the degree of metastability in this regard depends on the composition. Elements such as manganese, carbon, and nitrogen act with nickel to stabilize austenite, while molybdenum aids chromium in producing the ferritic condition. In many commercial stainless steels the effects of only nickel and chromium need be considered. At a composition of 17% chromium and 7% nickel an exceptionally high degree of work hardening is present, but the amount of decomposition of austenite produced by cold-working decreases with increasing nickel content and is almost zero in the 18% chromium-12% nickel alloy.

A further versatility of chromium-nickel stainless steels is the availability of certain compositions, such as AM–355 in Table 11.3, that are capable of being strengthened by age hardening. The precipitation reaction can be made to occur in the metastable austenite by an aging treatment (double-aging), or the austenite can be transformed to martensite by cooling to $-75°C$ ($-100°F$) (subzero cooling) followed by a tempering treatment. Alloys of this type can also be strengthened by cold-working. The corrosion resistance of hardened AM–355 is somewhat inferior to that of standard 18–8, but it is adequate for many applications.

The examples of stainless steels listed in Table 11.3 and discussed above were chosen from among the *wrought* stainless steels, available in the usual standard shapes such as sheet and bar stock. For designs in which a casting is needed, alloys having similar chemical compositions are available in cast form, and information on them is given in two sections of the *Metals Handbook*. The section, "Corrosion-Resistant Steel Castings," deals with alloys suitable for applications involving ordinary corrosion or oxidation below 650°C (1200°F). The identifying initial letter "C" is used for these alloys, some examples being: CA–40, CC–50, and CK–20, corresponding essentially to types 420, 446, and 310, respectively, in Table 11.3. The second section, "Heat-Resistant Alloy Castings," considers alloys for use at temperatures above 650°C (1200°F). These alloys, identified by the initial letter "H", run to higher chromium and nickel contents; for example, type HL contains 30% chromium and 20% nickel, while type HU contains 19% chromium and 39% nickel. The group of heat-resistant alloys includes nickel-base and cobalt-base compositions as well as those based on iron.

**Protective coatings.** The choice of an engineering material involves many factors, including cost. In the majority of applications of metals the importance of the corrosion factor does not warrant the use of expensive corrosion-resistant alloys. However, in general, when an easily corroded alloy like steel is used, some provision is made to prolong its life or improve its appearance by protecting it from its surroundings. By far the most important protective measure is some kind of coating.

*Nonmetallic coatings.* Paint is the most widely used protection against corrosion. For adequate protection of steel surfaces, paint must provide not only a continuous barrier against the corroding environment but also a rust-inhibiting action. A pigment such as red lead ($Pb_3O_4$) satisfies the second requirement. It tends to maintain an alkaline, corrosion-retarding environment at the metal surface, and it also acts to make the metal passive. Other nonmetallic coatings include porcelain, plastics, and rubber. Bituminous coatings such as pitch are especially useful for protecting underground tanks and pipes.

A second type of nonmetallic coating is sometimes called a *chemical coating*, since its formation involves chemical reaction with the metal on which it is formed. *Anodic coatings* on aluminum consist of a layer of aluminum oxide about 10 mμ thick. They are formed when the metal is made the anode in an acid electrolyte. Because these coatings are about 1000 times thicker than the natural oxide film, they improve resistance to corrosion as well as to mechanical injury. *Chromate coatings* are widely used on magnesium alloys and to a smaller extent on zinc. These thin coatings are produced by many different procedures, but one of the simplest merely requires dipping the metal in a chromate solution. The protective value of these coatings is a result of some mechanical protection plus the action of the chromate ion in decreasing the anodic reaction in electrochemical corrosion. *Phosphate coatings* (parkerizing, bonderizing) are produced on steels by dipping them in suitable phosphate solutions. They are used principally as an adherent base for paint or a protective oil.

*Metallic coatings.* Two factors are involved in the protection of the underlying metal by a metallic coating. One is mechanical isolation of the metal from the corroding environment; the other is the galvanic relation of the coating metal and the base metal. As was discussed previously, if the coating metal is higher in the galvanic series than the base metal, discontinuities in the coating are not a serious problem, since the base metal is cathodic and is protected. Zinc and cadmium are examples of metals that afford galvanic protection to steel. The superior properties of cadmium coatings under such conditions as attack by salt spray or alkalies may justify their higher cost. Tin, nickel, and chromium are below steel in the galvanic series for most service conditions. Consequently, corrosion of the underlying steel occurs at small discontinuities that may be present in the protective coating. This disadvantage is outweighed for some applications by the attractive appearance of nickel and chromium plating, and by the excellent corrosion and fabricating qualities of tin plate for use in food containers.

A number of methods are used to apply metallic coatings. Zinc coatings on steel (galvanized coatings) and tin plate are usually produced by dipping clean sheet steel into a molten zinc or tin bath. Thinner, uniform coatings of zinc and tin can be obtained by electroplating, and this process is becoming increasingly important for these metals. Cadmium coatings and decorative nickel and chromium plates are almost always produced by electroplating. *Diffusion coatings* can be formed on steel by high-temperature treatment in the presence of powdered aluminum (calorizing), chromium (chromizing), ferrosilicon (ihrigizing), zinc (sherardizing), and other metals. These coatings are relatively thick and are usually applied to cast or machined parts such as bolts, valve guides, and turbine buckets. Thick coatings of almost any metal can be produced by *metal spraying*. In this process a special gun melts and atomizes the coating metal and then uses compressed air to drive the small particles against a prepared surface. Since the metal particles solidify in the air, the bond they form is primarily mechanical,

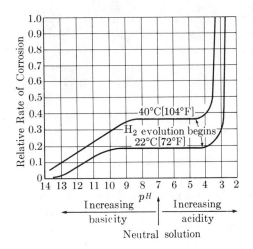

**Fig. 11.20** Effect of hydrogen ion concentration on the corrosion of steel. (After Whitman *et al.*)

and the coating is somewhat porous. However, the coating is sufficient for many purposes in addition to corrosion protection; for example, thick sprayed metal coatings can be used to build up worn shafts and other damaged parts.

*Metal cladding* is the joining of two different metals or alloys by forming a strong alloy bond between plates of the two materials. The bond may be formed in many ways (by casting one alloy against a solid plate of the other, for example). This thick, composite ingot is then hot-worked to reduce it to a convenient form such as clad sheet. The production of a corrosion-resistant surface is the principal objective for cladding; for example, *Alclad* aluminum alloys (see Chapter 8) are used when superior corrosion resistance is required. Cladding is also used for such purposes as making bimetal strips for temperature-control devices.

**Environment control.** Sometimes the rate of corrosion can be greatly reduced by small changes in the corroding environment. Since the rate of corrosion is usually an exponential function of temperature, a slight decrease in the temperature of the corroding medium may cause a pronounced decrease in the amount of corrosion. Also, if the velocity of the corroding medium can be diminished by a change in operating conditions or in design, the rate of corrosion is usually lowered. This factor is especially important when erosion corrosion is occurring.

*Treatment of liquid media.* Changes in the chemical composition of the corroding medium may have a great effect on corrosion behavior. Figure 11.20, for example, shows that in distinctly acid solutions steel corrodes rapidly with the evolution of hydrogen gas. In mildly acid or basic solutions, on the other hand, the cathodic reaction involves the formation of water rather than of hydrogen, and the reaction is relatively slow. Since oxygen is necessary for this second process, the corrosive action of ordinary water in boilers can be greatly decreased by removing its dissolved oxygen. Although lowering the hydrogen ion or oxygen concentration of solutions in contact with steel effectively reduces corrosion, such radical changes in composition are frequently impractical. More useful in most instances are small additions of *inhibitors* to the corroding medium. Inhibitors are organic or inorganic substances that dissolve in the corroding medium but are capable of forming a protective layer of some kind at either the anodic or cathodic areas. Anodic inhibitors, such as chromates and phosphates, are those that stifle the destructive anodic reaction. Although this type of control is effective, it may be dangerous, since severe local attack can occur if certain areas are left unprotected by depletion of the inhibitor. Magnesium and calcium salts are examples of cathodic inhibitors that act by forming a deposit on iron through which oxygen must diffuse to take part in the cathodic reaction. The more effective anodic inhibitors are most often used to treat boiler water, and care is taken to ensure that an adequate amount of inhibitor is always present.

Electric currents exist between the (dissolving) anodic regions and the (protected) cathodic regions during the course of corrosion. One method of stopping the flow of these local currents and thereby controlling corrosion is the imposition of opposing potentials, a procedure known as *cathodic protection*. The metal to be protected, such as a buried pipeline, is connected to external anodes that are in the same corroding environment. The anodes may be a reactive metal like zinc, which is replaced at intervals, or they may be essentially unreactive like graphite, and the protective potential is then supplied by a suitable direct-current source such as a rectifier. To neutralize adequately the corrosion currents in iron, a system of anodes must supply on the order of $0.1 \text{ A/m}^2$. Cathodic protection has been applied extensively to buried tanks and pipelines, and to water-storage tanks.

*Protective Atmospheres.* The use of protective atmospheres in heat-treating furnaces is an important example of industrial control of corrosive environments. Two defects may be introduced when alloys are heated at high temperatures. The first is scaling or discoloration of the surface as a result of reaction of the base metal with the surrounding gases, especially oxygen, water vapor, or carbon dioxide. A second, less easily observed defect is a change in the chemical composition of the alloy (for example, the loss of carbon from the surface of steel, a process known as *decarburization*). Since the reactivity of alloying elements is different from that of the base metal, a given atmosphere may protect metal parts from only one of these two types of damage. Many unfinished metals, such as steel ingots, are heated in ordinary air or in the oxidizing combustion products of the gases which heat the furnace. Under these conditions the surface layer that is changed in alloy content is effectively removed by scale formation. The poor surface appearance and the dimensional changes produced by these untreated atmospheres are often objectionable in finished metal parts. By heating such parts in a controlled-

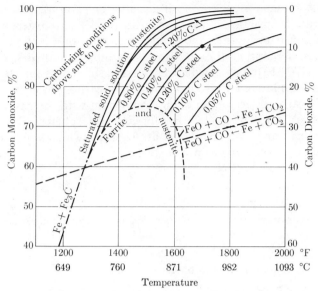

**Fig. 11.21** Curves showing the equilibrium between $CO$-$CO_2$ gas mixtures and steels of various carbon contents at heat-treating temperatures. The total pressure of $CO + CO_2$ is at one atmosphere. (After Becker)

atmosphere chamber (called a *muffle* or *retort*) that is isolated from the air and from the combustion products of the heating gases, it is possible to surround them with a protective atmosphere that leaves the parts *clean* (unscaled) or even *bright* after a high-temperature treatment. Changes in alloy composition can also be minimized.

Many different types of protective atmospheres are used. An inert gas like helium is ideal, but its high cost limits its use to special applications such as research projects. Hydrogen and the mixture of hydrogen and nitrogen produced by dissociating ammonia ($NH_3$) are moderate-cost atmospheres that are successfully used in the heat treatment of stainless steel and in sintering many types of powder-metal compacts. Low-cost atmospheres for heat treating steel, brass, and other alloys are prepared from the products of combustion of fuel gases. By removing water vapor and carbon dioxide from partly or completely burned natural gas, an atmosphere can be obtained that prevents both decarburization and discoloration during the hardening of plain carbon and alloy steels. Liquid salt or lead baths also are frequently used as inert atmospheres in the heat treatment of steel.

An example of a gaseous controlled atmosphere, employed to prevent both oxidation of steel and change in its surface carbon content, is a mixture of carbon monoxide and carbon dioxide. The use of this atmosphere is governed by the curves of Fig. 11.21. Two reactions are considered in this figure: (1) the oxidation of iron by $CO_2$ occurs for gas compositions and temperatures below the sloping line labeled with this reaction; (2) the carburization of a given steel occurs for gas compositions and temperatures above and to the left of the curve labeled with the composition in question. For example, an atmosphere of 90% CO and 10% $CO_2$ at 927°C (1700°F) (point $A$) would carburize a steel containing 0.10% carbon. Under the same conditions a steel containing 0.40% carbon would be decarburized, while a 0.20% carbon steel would be in equilibrium with the gas and would be unaffected. The data of Fig. 11.21 are for *dry* atmospheres. The presence of appreciable water vapor can change an atmosphere (which would otherwise be protective) into one that is oxidizing for steels.

## PROBLEMS

**1.** Compare the corrosion that occurs (a) when iron and copper are in electrical contact in a dilute sodium chloride solution containing dissolved oxygen; (b) when isolated pieces of iron and copper are in the sodium chloride solution containing oxygen.

**2.** a) Use the electromotive series to obtain the value $E_{Fe}^0 - E_{Zn}^0$ for the galvanic couple iron-zinc.

b) Use a sketch similar to Fig. 11.1 to describe the anodic and cathodic reactions in this case.

c) If the positions of the metals are reversed so that the galvanic couple can be called zinc-iron rather than iron-zinc, show that the electrical potential changes sign and the direction of current flow is reversed.

**3.** Explain the following phenomena: (a) a piece of iron sealed in a small glass tube filled with water showed no signs of rusting over a long period, but it began to rust almost immediately when air was allowed to come in contact with the water; (b) a specimen of an iron pillar that was almost unaffected by weathering in an arid climate for hundreds of years was found to corrode rapidly in England.

**4.** Show that the over-all cathodic reaction in Fig. 11.1 is the sum of three simpler reactions. Recall that the $H^+$ ions needed for the reaction of Eq. (11.7) are obtained from the decomposition of water, a process that yields $OH^-$ ions as well. To obtain the over-all reaction, make use of the usual procedure of merely adding the three simpler reactions.

**5.** Explain why an oxide that is unprotective according to the Pilling-Bedworth rule should increase in amount at a *linear* rate.

**6.** Test the following data for "fit" to the parabolic, linear, or cubic rate laws for the oxidation of a metal.

| Oxide thickness, Å | Time, minutes |
|:---:|:---:|
| 0 | 0 |
| 500 | 10 |
| 1000 | 40 |
| 2000 | 160 |

**7.** If a first-order interference color of yellow is seen on a slightly oxidized tool steel, estimate the oxide thickness $t$. Assume a representative blue radiation of 4700 Å, and that first-order extinction of blue occurs when $t = \lambda/4\eta$, where $\lambda$ is the wavelength being destroyed and $\eta$ is the index of refraction of the oxide film ($\eta \approx 3$).

**8.** Is internal oxidation more likely to occur in aluminum- or in copper-base alloys? Why?

**9.** A widely used measure of corrosion is the weight loss per unit area per unit time, for example mdd (the loss in *milligrams* per square *decimeter* per *day*). Contrast the usefulness of this measure when applied to uniform corrosion and when applied to pitting.

**10.** Often a second phase forming within a solid solution tends to be localized at the grain boundaries. Explain whether this effect can cause intergranular corrosion.

**11.** a) Name three applications in which special corrosion-resistant alloys are used.
 b) Name three applications in which the corrodible alloys used must be protected from the corroding environment.

**12.** Compare the three general types of stainless steels (martensitic, ferritic, and austenitic) with respect to (a) composition, (b) possible phases, and (c) typical uses.

## REFERENCES

Evans, U. R., *An Introduction to Metallic Corrosion*, 2nd ed., London: Edward Arnold, 1963. Gives a good elementary treatment of the principles of corrosion.

Fontana, M. G. and N. D. Greene, *Corrosion Engineering*, New York: McGraw-Hill, 1967. Discusses practical corrosion problems; also contains a section on the theory of corrosion.

Godard, H. P., *et al.*, *The Corrosion of Light Metals*, New York: John Wiley, 1967. Detailed, specialized treatment of the corrosion of such metals as aluminum, magnesium, and titanium.

Hauffe, Karl, *Oxidation of Metals*, New York: Plenum Press, 1965. Advanced treatment of especially the theoretical aspects of oxidation.

Logan, H. L., *The Stress Corrosion of Metals*, New York: John Wiley, 1967. Gives detailed treatments of both the theoretical and practical aspects of this special topic.

Pourbaix, M., *Atlas of Electrochemical Equilibria in Aqueous Solutions*, New York: Pergamon Press; and Brussels: Cebelcor, 1966. Lists the Pourbaix diagrams for almost all the metallic elements.

Rabald, E., *Corrosion Guide*, Amsterdam: Elsevier, 1968. Contains detailed tabulated data on metallic and other materials suitable for use in a given corrosive environment.

Scully, J. C., *The Fundamentals of Corrosion*, Oxford: Pergamon Press, 1966. Gives an elementary introduction to the broad subject of corrosion.

Shreir, L. L. (Ed.), *Corrosion* (2 vols.), New York: John Wiley, 1963. Advanced treatments of virtually every topic in the field of corrosion.

Uhlig, H. H., *Corrosion and Corrosion Control*, 2nd ed., New York: John Wiley, 1971. Gives an introduction to the principal topics in the theory and practice of controlling corrosion.

Uhlig, H. H. (Ed.), *Corrosion Handbook*, New York: John Wiley & Sons, 1948. A compilation of extensive data on the corrosion of metals and alloys in various environments.

CHAPTER 12

# ELECTRONIC STRUCTURE AND PHYSICAL PROPERTIES

The study of electronic structure divides itself naturally into two parts: the electronic structure of isolated atoms (discussed in Chapter 1), and that of an aggregate of atoms; for example, a solid alloy. For topics such as plastic deformation, phase transformations, and heat treatments for strengthening, only the more elementary part of electronic theory is usually employed at the present state of development of theory and practice. An adequate treatment of many physical properties, on the other hand, must take account of the mutual electronic interactions of the multitude of atoms composing a piece of solid metal. The theory is helpful in understanding many aspects of electrical, magnetic, thermal, and optical behavior. In a few instances even the goal of quantitative prediction is being approached. The following brief treatment can, of course, be only an introduction to the large field of the physics of metals.

## ELECTRON ENERGIES IN SOLIDS

The theory discussed previously in Chapter 1 was concerned with "free" atoms, that is, atoms sufficiently isolated from one another that their interactions can be ignored. Most metal vapors, which are typically monatomic, are composed of free atoms of this kind. Brief consideration of the properties of solid (or of liquid) metals, on the other hand, suggests that the constituent atoms must be exerting profound influences on one another. In principle it should be possible, as in the case of the hydrogen atom, to apply wave mechanics to the assembly of atoms in a solid metal to determine the characteristic energy values; in practice, mathematical difficulties have prevented the use of this rigorous method. Consequently, it has been necessary to introduce various approximations in solving the problem. Three solutions that form a connected sequence of increasing rigor will now be considered.

**Drude-Lorentz theory.** Not all the electrons in the atoms that compose a solid metal take an equal part in the interactions that result in the characteristic properties of strength, hardness, and so on. Evidence from x-ray spectra indicates that the inner (core) electrons, which form closed shells in an atom, are essentially unaffected and that only the outer (valence) electrons play an active role in metallic bonding. The Drude-Lorentz theory made an extreme distinction between these two types of electrons; it completely neglected all of the atom except the valence electrons, and it pictured the aggregate of valence electrons from all the atoms in a given piece of metal as forming an "electron gas" that was free to move throughout the volume of the metal. The total energy of the electrons was determined as the sum of the kinetic energy (taken to be that of an ideal gas, $\frac{3}{2}RT$) and the potential energy $-W$, relative to a free electron in the space outside the metal. This total energy led to a gross overestimation of the electronic specific heat and to other conflicts with experimental data, so that the principal legacy of this theory is the concept of an "electron cloud" within a metal.

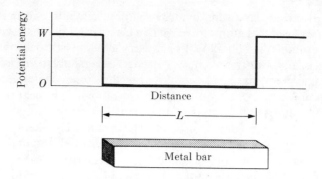

(a) Decreased potential energy within a metal bar of length $L$.

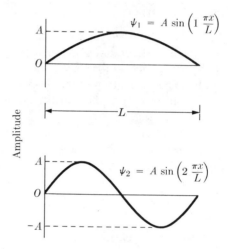

(b) The first two wave functions that can exist in a bar length $L$.

**Fig. 12.1**   Schematic illustrations of important concepts in the free-electron theory of metals.

**Free-electron theory.** This theory retains the concept of a uniform potential within the metal, but it uses the Schroedinger equation to calculate the total energies $E$ in a manner analogous to that described in Chapter 1 for the hydrogen atom. If the potential energy is set equal to zero inside a metal bar of length $L$, Fig. 12.1(a), the solution of the Schroedinger equation is especially convenient, since Eq. (1.13) then has the form of Eq. (1.8), with

$$k^2 = \frac{8\pi^2 \, mE}{h^2}. \tag{12.1}$$

From the discussion of Eq. (1.8) it follows that the solutions of this equation are sine waves, $\psi = A \sin kx$. By analogy with the standing waves in an organ pipe it can be shown that a condition for these wave functions to exist in the metal is that the amplitude be zero at the two ends. Two wave functions that satisfy this condition are shown in Fig. 12.1(b) and, in general,

$$\psi_n = A \sin \frac{n\pi x}{L}, \tag{12.2}$$

so that $k_n = n\pi/L$. When this expression for $k$ is substituted in Eq. (12.1), the corresponding permissible values of total energy are found to be

$$E_n = n^2 \frac{h^2}{8mL^2} \tag{12.3}$$

for the one-dimensional problem considered here. When the same problem is solved for a three-dimensional block of metal with sides $L_x, L_y$, and $L_z$, the following analog of Eq. (12.3) is obtained,

$$E_n = \frac{h^2}{8\pi^2 m}\left(k_x^2 + k_y^2 + k_z^2\right) = \frac{h^2}{8\pi^2 m}\left[\left(\frac{n_x\pi}{L_x}\right)^2 + \left(\frac{n_y\pi}{L_y}\right)^2 + \left(\frac{n_z\pi}{L_z}\right)^2\right]. \tag{12.4}$$

A more convenient form (see Problem 1) is

$$E_n = \frac{h^2}{8mV^{2/3}} n^2, \tag{12.5}$$

where $V = L_x L_y L_z$, the volume of the block, and $n^2 = (n_x^2 + n_y^2 + n_z^2)$. The integers $n_x, n_y$, and $n_z$ can independently have positive values 0, 1, 2, and so on to arbitrarily large numbers (as needed to accommodate all of the valence electrons in the block of metal). A certain value of $n^2$ (say, $n^2 = 1$) is given by a number, $g$, of different combinations of the integers $n_x, n_y$, and $n_z$ ($g = 3$ for $n^2 = 1$). Since $g$ different quantum states correspond to the same energy $E_n$, this energy level is said to be $g$-fold degenerate.

    For calculating the distribution of energies of the valence electrons in a given metal, use is made of the *density of states*, $\rho(E)$, which is the number of states per unit interval of energy. This function can be easily calculated for the free-electron theory because the electrons are assumed to occupy the quantum states of lowest energy. However, because of the mutual interactions among all the electrons that form the "electron cloud," it is necessary to consider that all the electrons are in a single system and that the Pauli exclusion principle applies. This means that a given state (specified by the three quantum numbers $n_x, n_y$, and $n_z$) can be occupied by only two electrons and that these must differ in the fourth quantum number, $(+)$ or $(-)$ spin. Since there are about $10^{24}$ valence electrons in a piece of metal, an enormous number of states of continuously increasing energy are occupied.

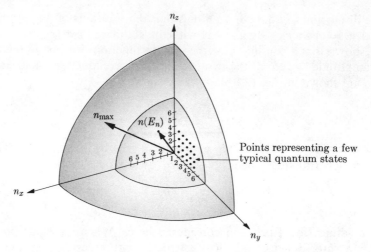

**Fig. 12.2**  Method of plotting the integral values $n_x$, $n_y$, and $n_z$ so that all possible quantum states can be represented.  Since $n_{max}$ is about $10^8$, the portion near the center has been greatly expanded to show the details of plotting.

In the construction shown in Fig. 12.2, the occupied states occur in a spherical region out to some radius $n_{max}$ corresponding to the maximum energy necessary to accommodate all the electrons.  The number of states out to any given value $n$ (corresponding to energy $E_n$) is then simply

$$N = \frac{1}{8}\left(\frac{4\pi}{3}n^3\right). \tag{12.6}$$

The factor $\frac{1}{8}$ appears because positive values of $n_x, n_y$, and $n_z$ occur only in one quadrant of the sphere.  Use of Eq. (12.5) permits the following reformulation of Eq. (12.6) in terms of the energy $E_n$,

$$N = \frac{\pi V}{6}\left(\frac{8m}{h^2}\right)^{3/2} E_n^{3/2}. \tag{12.7}$$

In view of the definition of $\rho(E)$, the desired expression can now be obtained as follows:

$$\rho(E) = \frac{dN}{dE_n} = \frac{\pi V}{4}\left(\frac{8m}{h^2}\right)^{3/2} E_n^{1/2}. \tag{12.8}$$

A plot of $\rho(E)$ versus energy is therefore a parabola, Fig. 12.3(a), according to this simple argument.

**Energies near the Fermi level.**  The total energy of a system can be distributed among the constituent particles in various ways; that is, according to various types of statistics.  One type, Boltzmann statistics, is represented by Eq. (6.45) and

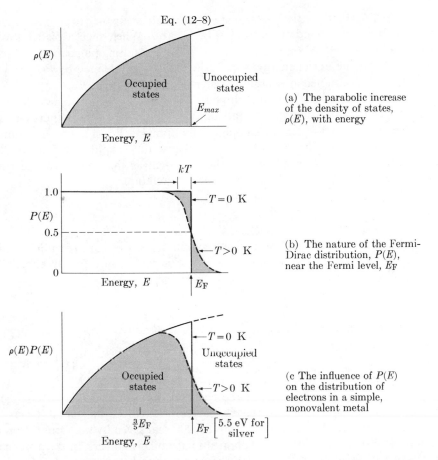

Eq. (12–8)

$\rho(E)$

Occupied states

Unoccupied states

$E_{max}$

Energy, $E$

(a) The parabolic increase of the density of states, $\rho(E)$, with energy

$kT$

1.0

$P(E)$

0.5

0

$T=0$ K

$T>0$ K

Energy, $E$

$E_F$

(b) The nature of the Fermi-Dirac distribution, $P(E)$, near the Fermi level, $E_F$

$\rho(E)P(E)$

$T=0$ K

Unoccupied states

Occupied states

$T>0$ K

$\frac{3}{5}E_F$

Energy, $E$

$E_F$ $\begin{bmatrix} 5.5 \text{ eV for} \\ \text{silver} \end{bmatrix}$

(c The influence of $P(E)$ on the distribution of electrons in a simple, monovalent metal

**Fig. 12.3** The distribution of electrons as a function of energy, including the special behavior near the Fermi level.

applies to such classical systems as the atoms composing a gas. Another principal type, Fermi-Dirac statistics, applies to quantum-mechanical systems in which allowed energies must be compatible with the Schroedinger equation and with the Pauli exclusion principle. In this case, the probability $P(E)$ that a state with energy $E$ is occupied by an electron is given by the Fermi-Dirac distribution,

$$P(E) = \frac{1}{\exp\left[(E - E_F)/kT\right] + 1}. \tag{12.9}$$

This equation shows that the *Fermi energy* $E_F$ is the energy at which $P(E)$ is $\frac{1}{2}$. For $T = 0$ K, $P(E)$ is unity below $E_F$ and zero above $E_F$, Fig. 12.3(b). This distribution corresponds to the simple density-of-states curve, Fig. 12.3(a), in which all states are occupied below $E_{max}$ and are empty above $E_{max}$. For any temperature above zero, Eq. (12.9) predicts the transition from full occupancy to zero occupancy shown by the corresponding curve in Fig. 12.3(b). When such a distribution is applied to the density-of-states curve, Eq. (12.8), the resulting plot shows a gradual transition near the Fermi energy, Fig. 12.3(c). $E_F$ is still defined by $P(E) = \frac{1}{2}$ but some states are occupied at energies greater by about $kT$ than $E_F$.

Energies of electrons are much higher than those of particles, such as atoms of a gas, governed by Boltzmann statistics (for which the average energy is only $kT$). The Fermi energy can be calculated by applying Eq. (12.7) to the number $N_e$ of electrons in a volume $V$ of metal. Because of the spin quantum number, the number of states $N$ need only be half as great as the number of electrons. Rearrangement of Eq. (12.7) leads to the desired relation,

$$E_F = \frac{h^2}{8m} \left( \frac{3}{\pi} \frac{N_e}{V} \right)^{2/3}. \tag{12.10}$$

Thus, the Fermi energy is a function of the number of electrons per unit volume, $N_e/V$. The average energy $\bar{E}$ of the valence electrons is not greatly affected by temperature since $kT \ll E_F$. As shown in Problem 2,

$$\bar{E} = \frac{\displaystyle\int_0^{E_F} E\, \rho(E)\, dE}{\displaystyle\int_0^{E_F} \rho(E)\, dE} = \frac{3E_F}{5}. \tag{12.11}$$

The total energy of all $N_e$ electrons is therefore simply $3N_e E_F/5$.

Since only a small fraction of the electrons are influenced by a change in temperature, the free-electron theory correctly predicts the small electronic specific heat of metals, a value only about 1/200 that given by the Drude-Lorentz theory. Electrical conduction is explained in a similar manner. In the absence of an applied voltage the electrons are moving randomly, and the effect of those traveling in a positive direction is balanced by the effect of an equal number moving in a negative direction, so that there is no net current in the metal. An applied voltage does not produce a net effect for the vast majority of the electrons because the energy states into which they might be promoted by additional electrical energy are already occupied. Again, only those electrons near the Fermi level experience a net change in energy and contribute to the motion in a preferred direction that constitutes a current. While the free-electron theory accounts satisfactorily for electrical conductivity in most metals, it fails to explain why other substances that also contain mobile electrons have virtually no conductivity and are considered to be excellent insulators. A solution to this problem is given by the zone theory.

**Zone theory.** The two previous theories have neglected the effect of the ion cores and have assumed that a uniform potential exists within the metal. Evidently this is a poor approximation to the actual environment of an electron moving

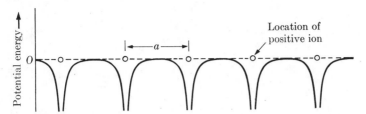

**Fig. 12.4** The variation in potential energy of an electron moving along a one-dimensional array of positive ions.

through the uniform array of positively charged atomic cores that compose a metal crystal. The potential field experienced by the electron must vary periodically with the repeated atomic pattern in the crystal. Figure 12.4 is a schematic illustration of the potential energy for a one-dimensional array of positive ions. In the spaces between ions the potential energy is zero, as in the free-electron theory, but in the vicinity of an ion the potential energy has a large negative value. It is possible to determine the energies of the electronic states by using an appropriate expression for the periodic potential energy in solving the Schroedinger equation. The essential results of this analysis will be discussed in connection with the following, simpler treatment of this problem.

Since electrons have associated wave properties, they can be diffracted by the periodic "grating" formed by the atoms of the crystal. The important technique of electron diffraction, which is based on this fact, is discussed in Chapter 5. Electrons of low energy have very long wavelengths compared with the atomic spacings and move through the crystal undisturbed. More energetic electrons have wavelengths $\lambda$ nearly equal to the atomic spacing $d_{hkl}$ of the plane in question, identified by its Miller indices $(hkl)$. Such electrons can be diffracted according to Bragg's law, $\lambda = 2d_{hkl} \sin \theta$ (Eq. (5.8)]. This equation can be put into the more convenient form,

$$|\mathbf{k}| \sin \theta = \pi/d_{hkl} \tag{12.12}$$

if the *wave vector*, $|\mathbf{k}| = 2\pi/\lambda$, is used instead of $\lambda$ as a measure of the energy. Thus, diffraction occurs when the component of the wave vector normal to the diffracting planes reaches a critical value, $\pi/d_{hkl}$.

A graphical representation of this condition for diffraction, Fig. 12.5, helps explain its significance. Consider four electrons with various energies and directions, represented by the four wave vectors $\mathbf{k}_1$ $\mathbf{k}_2$, $\mathbf{k}_3$, and $\mathbf{k}_4$. These vectors are plotted in Fig. 12.5(a) according to the values of their $k_x$ and $k_y$ components (in two dimensions). In three dimensions this procedure is known as plotting in *k-space*. We analyze first the case of a simple-cubic lattice and begin with the {100} crystal planes. Although the four electrons have various energies, they have the common characteristic of being diffracted by a {100} plane. The graphical construction that insures this characteristic can be explained using $\mathbf{k}_1$. A vertical line is constructed through the point $k_x = \pi/d_{100}$. Since $\mathbf{k}_1$ is drawn so that it

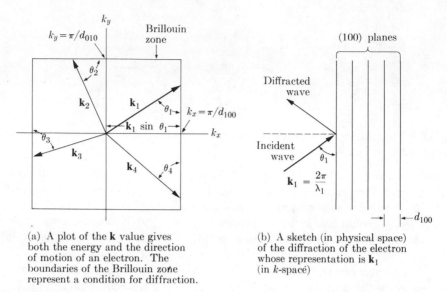

(a) A plot of the **k** value gives both the energy and the direction of motion of an electron. The boundaries of the Brillouin zone represent a condition for diffraction.

(b) A sketch (in physical space) of the diffraction of the electron whose representation is **k**$_1$ (in $k$-space)

**Fig. 12.5**  Explanation of the manner in which the diffraction condition $|\mathbf{k}| \sin \theta = \pi/d_{hkl}$ leads to the concept of Brillouin zone.

ends on this vertical line, $|\mathbf{k}_1| \sin \theta_1$ must equal $\pi/d_{100}$. But this is exactly the condition for diffraction, Eq. (12.12), expressed in terms of the wave vector. The corresponding physical picture of the diffraction at angle $\theta_1$ of an electron beam of wavelength $\lambda_1$ is sketched in Fig. 12.5(b). $\mathbf{k}_4$ represents an electron of higher energy than $\mathbf{k}_1$, but a similar argument shows that it, too, is diffracted by the (100) planes. Problem 4 analyzes the corresponding diffraction of $\mathbf{k}_2$ by the (010) planes. Thus the square (or cube in three dimensions) labeled *Brillouin zone* in Fig. 12.5(a) concisely describes the condition of diffraction from {100} planes in a simple-cubic lattice. A similar analysis leads to the Brillouin zones for other types of planes (see Problem 5).

The relation of the energies of electrons to the Brillouin zones is of crucial importance. The case of an electrical insulator is sketched in Fig. 12.6. A point in this diagram, such as the one indicated in the figure, is to be interpreted as follows. It represents an electron having an energy described by the wave vector whose components are $k_x$ and $k_y$. The energy of the electrons, as given by the free-electron theory, depends only on the wave vector. Increasing energy values plot as circles such as those shown near the center of the Brillouin zone. However, an important result of the treatment of this problem by zone theory is the prediction that the energy contours become distorted in the vicinity of the boundaries of the Brillouin zones; also, the energy suddenly increases as a boundary is crossed in any direction. In Fig. 12.6 the (first) Brillouin zone is just filled by the valence electrons, and therefore this material is a nonconducting substance (an electrical

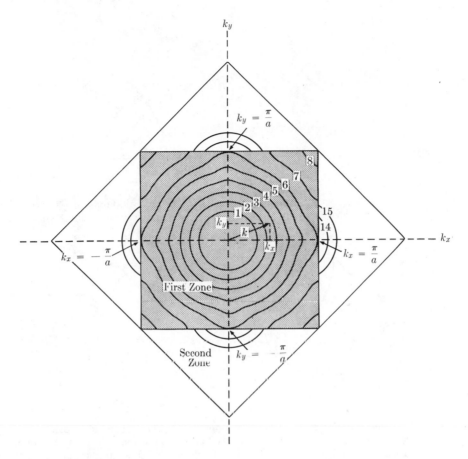

**Fig. 12.6** The energy levels in the first Brillouin zone for a square lattice. Two energy levels are shown in the empty second zone.

insulator). The reason for this behavior is more easily seen when the data of Fig. 12.6 are presented in the form of a $\rho(E)$ curve.

The energy levels predicted by the zone theory are conveniently shown by $\rho(E)$ curves that are an elaboration of the parabolic curve of Fig. 12.3. Figure 12.7 is an example of this type of representation of the energy levels sketched in Fig. 12.6. The $\rho(E)$ curve is parabolic out to 3 eV. Beyond this value the energy contours in Fig. 12.6 begin to be seriously distorted; there is an increased number of electronic states in a given energy interval, and consequently the $\rho(E)$ curve rises. At 6 eV the energy contours first touch the zone boundaries; therefore, for energies higher than this, the $\rho(E)$ value falls because the area available for additional energy states is limited to the corners of the zone. The $\rho(E)$ value falls to zero at the energy for the corners of the zone, which is somewhat below 9 eV. The second zone in Fig. 12.6 extends from the first zone to a second boundary

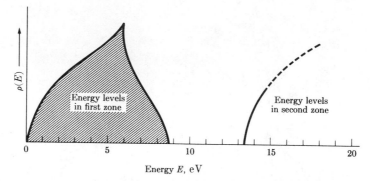

**Fig. 12.7**  $\rho(E)$ curves for an insulator corresponding to the Brillouin zones of Fig. 12.6. The shaded area indicates that there are just sufficient valence electrons to fill the first zone.

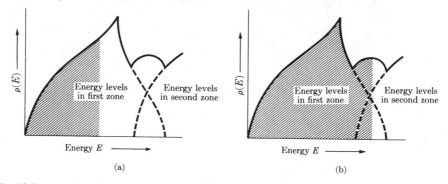

(a)

(b)

**Fig. 12.8**  Possible $\rho(E)$ curves for electrical conductors. (a) The valence electrons in a monovalent metal fill only half the energy levels of the first Brillouin zone. (b) In a divalent metal the first zone is nearly full and the lower energy levels of the second zone are occupied.

that is determined by Bragg reflection from planes at 45° in Fig. 12.5 (see Problem 5). For an insulator the energies of the lowest levels in the second zone are considerably higher than the energies of the upper levels in the first zone, Fig. 12.7.

The $\rho(E)$ curve of Fig. 12.7 represents a nonconductor because it fulfills two conditions. First, it has just enough valence electrons to completely fill one (or more) zones. The quantitative theory shows that this condition is easily fulfilled, since the number of states in each zone corresponds to two electrons per atom;* therefore any substance with an even number of valence electrons might be an insulator. The four valence electrons in diamond, for example, just fill the first two zones in this excellent insulator. The second condition requires that the higher, unoccupied zone be separated from the lower, filled zone by a large energy

---

* Strictly speaking, the quantity involved is the number of electrons per primitive cell in the crystal structure (see Chapter 1). The primitive cell contains one atom except in special cases such as "defect crystals" in which some lattice sites are not occupied.

gap. This gap in diamond is 7 eV. Under these two conditions it is clear that an applied voltage cannot cause a current. Because there are no unoccupied states in the filled zone, the electrons cannot increase their velocity in the direction of the voltage since this would require an increase in energy. Also, ordinary voltages are not sufficient to cause an electron to cross the large energy gap separating the lower and upper zones.

In good electrical conductors at least one, and usually both, of the above conditions are lacking. Figure 12.8 is a schematic $\rho(E)$ curve for a monovalent metal in which conduction occurs within the energy levels of the lower unfilled zone. In this case it is unimportant whether the second zone overlaps the first, as shown here, or is separated from it as in Fig. 12.7. Substances having an even number of valence electrons, such as magnesium, which has two, can be conductors of electricity only if the lower and higher energy bands overlap in the manner shown schematically in Fig. 12.8(b). The first Brillouin zone can still accommodate two electrons per atom. However, in this case there is an overlap of the energies in the upper part of the first zone with those in the lower part of the second zone (see Problem 6). Consequently the total energy is lower when the electrons only partially fill the energy levels in the first zone and occupy the lower portion of the second zone. Unlike the situation in Fig. 12.7, unfilled energy states are readily available in this arrangement and therefore electrical conduction is possible.

In addition to the concepts of Bragg reflection and $\rho(E)$ curves, there is one other useful tool in this area of electronic energies—the so-called *effective mass* of an electron, $m^*$. Using this quantity it is possible to retain the same type of equation to describe the behavior of electrons that interact strongly with the periodic structure of the lattice as is used for the electrons of the free-electron theory. For example, Eq. (12.5) for the permissible energy states can be written

$$E_n = \frac{h^2}{8m^* V^{2/3}} n^2, \tag{12.13}$$

by merely substituting $m^*$ for $m$. Essentially, by attributing the mass $m^*$ to an electron that is influenced by a periodic potential, its behavior can be described by the simple theory derived for a uniform potential. Thus, for free electrons $m^* = m$, but for other conditions $m^*$ can be larger or smaller than $m$, or it can even be negative.

The zone theory is often referred to as the *band theory* because it can be considered as describing a band (range) of energies occupied by the valence electrons of a metal. It may be desirable to distinguish between energy bands that arise from different types of electrons, especially as in the case of semiconductors where the bands are separated by an energy gap. This distinction between energy bands is quite clear in the approach to band theory that is based on the "tight-binding approximation" in which it is assumed that a given valence electron

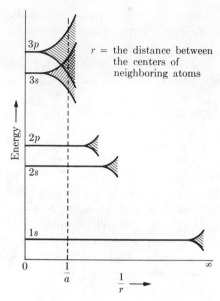

**Fig. 12.9** The broadening of the energy levels in magnesium as the free atoms are brought closer together. The dashed line represents the actual spacing of the atoms in metallic magnesium.

is bound to a given ion.* In the case of magnesium, for example, the energy levels are sharp in the free atom, as shown in Fig. 1.2(a). However, it can be shown that the broadening illustrated in Fig. 12.9 must occur as an assembly of magnesium atoms is (imagined to be) brought closer and closer together. At the actual spacing of atoms in metallic magnesium the $3s$ and $3p$ levels have broadened so that they overlap. The inner or core electron levels are unaffected at this actual inter-atomic distance and would broaden only at some unattainable compression of the atoms.

## ELECTRICAL PROPERTIES

Many properties of metallic (and also of nonmetallic) materials can be understood in terms of electronic energies and configurations. An example of a structural characteristic, the adoption by an alloy of particular crystal structures, is discussed later. Physical properties, however, are much more closely linked to electronic structure and form the main subject of this and the following sections.

Since electricity is carried in a metal by electrons, problems in electrical conductivity involve the motion of these electrons. Thus, if there are $N$ free electrons per cubic meter and if their average net (drift) velocity under the action

---

* In the zone theory discussed earlier in this section the "free-electron approximation" was made; i.e., it was assumed that the electrons were free of any particular ion.

of an electric field is $v_D$, then the electric current per square meter is

$$j = Nev_D,\qquad(12.14)$$

where $e$ is the charge on the electron. It is characteristic of an electric current that the electric field does not continuously accelerate the electrons as it would in the absence of friction, but rather, the current quickly reaches a constant value. This behavior is explained by the phenomena of random collisions of the electrons with their surroundings, where the average distance $l$ between collisions is called the mean free path. Alternatively, the corresponding time $\tau$ between collisions can be used,

$$l = \tau v,\qquad(12.15)$$

where $v$, the velocity of an electron, is much greater than $v_D$, the drift velocity. The *relaxation time* $\tau$ is the more generally useful concept, and it can be shown (see Problem 8) that $v_D$ depends on the value of $\tau$ according to the equation

$$v_D = \frac{Ee}{m}\tau,\qquad(12.16)$$

where $E$ is the electric field and $m$ is the mass of an electron. The conductivity $\sigma$ is defined as $j = \sigma E$, and therefore when Eq. (12.16) is substituted in Eq. (12.14) it follows that

$$\sigma = \frac{Ne^2\tau}{m^*}.\qquad(12.17)$$

By using the effective mass $m^*$ in this final equation, the equation correctly describes the results given by wave mechanics.

**Semiconductors.** We begin our discussion of electrical behavior with those materials (such as silicon and germanium) that exhibit semiconductivity. This type of conductivity is characterized by a very small number of electronic *carriers*, which may be electrons, or holes, or both. In the latter case Eq. (12.17) must be written,

$$\sigma = \left(\frac{N_e\tau_e}{m_e^*} + \frac{N_h\tau_h}{m_h^*}\right)e^2,\qquad(12.18)$$

where the subscripts refer to the two types of carriers. The nature of the holes (and also of the conduction electrons) can be understood in terms of the initial filled valence band of typical semiconductors, Fig. 12.10, both *intrinsic semiconductors* and *doped semiconductors* (n-type and p-type). Intrinsic semiconductors such as pure silicon have a relatively narrow energy gap $E_g$ (about 1 eV) between the top of the valence band and the bottom of the conduction band, Fig. 12.10(a). At temperatures near 0 K an intrinsic semiconductor will have a completely filled valence band and a completely empty conduction band, and therefore it behaves like an insulator. However, at higher temperatures some of the more

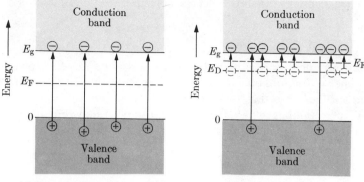

(a) In an intrinsic semiconductor, electrons are excited into the conduction band from the valence band; positive holes are produced in the valence band.

(b) In an extrinsic, $n$-type semiconductor, the major source of conduction electrons is excitation from the donor level, $E_D$.

**Fig. 12.10**  Schematic illustrations of the relation of semiconductivity to the electronic band structure for (a) an intrinsic semiconductor, and (b) an $n$-type semiconductor.

energetic valence electrons are raised into the conduction band and are available to help conduct electricity. The number $N_e$ of conduction electrons per cubic meter is given approximately by the equation (see Problem 9),

$$N_e = \frac{2(2\pi m^* kT)^{3/2}}{h^3} \exp\left(-\frac{E_g}{2kT}\right). \tag{12.19}$$

For each electron that is raised to the conduction band, an unoccupied state (a hole) is left in the valence band. The holes are equivalent to positive charges and move in the opposite direction to the conduction electrons. Equation (12.19) explains why the conductivity of intrinsic semiconductors increases with an increase in temperature, while metals characteristically show the opposite behavior.

A more versatile material for use in transistors and similar devices is an extrinsic (doped) semiconductor. We consider first an $n$-type semiconductor, in which the major carriers are ($n$, negative) electrons, Fig. 12.10(b). If a small quantity of an impurity of valence $v + 1$ is added to a crystal of an element whose valence is $v$, there will be an extra electron per impurity atom beyond the number required to fill the valence band. At low temperatures each of these extra electrons will move about one of the donor impurity ions in response to the additional positive charge on these ions. This is a localized donor state and has an energy $E_D$ within the forbidden range of the pure semiconductor, just below the conduction band. The donated electrons are easily raised (ionized) into the conduction band by small amounts of thermal energy; almost all are ionized at room temperature. The Fermi level varies with the degree of ionization, but lies between $E_D$ and $E_g$. The concentration of the majority carriers, electrons in this case, depends on the amount of dopant added.  Usually the concentration is many orders of magnitude greater than that of the minority carrier (holes, in this case), which results from the mechanism shown in Fig. 12.10(a).

Doping with an impurity having a valence of $v - 1$ leads to the production of a $p$-type extrinsic semiconductor, which has ($p$, positive) holes as the majority carriers.  In this case the dopant atoms have a deficiency of one electron relative to the pure semiconductor.  The corresponding electronic structure is conveniently represented as a full valence band (at 0 K) plus holes at an acceptor level $E_A$ (see Problem 10).  Ionization, analogous to that in $n$-type semiconductors, occurs when an electron moves from the valence band up to the acceptor level.  These electrons are localized and cannot conduct an electric current.  The resulting hole in the valence band is then the effective electronic carrier, analogous to the electron in the conduction band in Fig. 12.10(a).

**Metallic conduction.**  In contrast to conduction in insulators and in semiconductors, which involves filled energy bands, metallic conduction depends on partially filled bands.  Several deductions can be drawn from this theoretical basis.  Consider first the generally superior conductivity of monovalent metals compared with divalent metals.  Since the single valence electron in a monovalent metal only half fills the first energy band, Fig. 12.8(a), two conditions favorable for conduction are created.

1. When a band is half filled, the density of electron-energy states $\rho(E)$ is high. Consequently, when an electric field is applied, a large number of electrons are raised above the existing level.

2. These most energetic electrons have an effective mass approximately equal to $m$, the mass of a free electron.

The corresponding conditions in a divalent metal are the following.

1. The two valence electrons nearly fill the first band and only slightly fill the second, overlapping band.  The $\rho(E)$ value is low at these positions in both bands.

2. In the upper portion of the first band $m^*$ is considerably larger than $m$.

Thus both these factors tend to produce lower conductivity in divalent metals.

The anomalous *increase* in conductivity of bismuth upon melting is another phenomenon attributable to Brillouin zones.  Most metals experience a decrease in conductivity upon melting because of the resulting disruption of the orderly atomic arrangement.  This disruption also occurs in bismuth, but the predominant factor determining the change in conductivity is the breaking-down of the Brillouin zones that accompanies the disintegration of the crystalline structure of solid bismuth.  Conduction is so severely restricted in solid bismuth by the zone structure that a net increase in conductivity occurs when this restriction is substantially removed during melting.

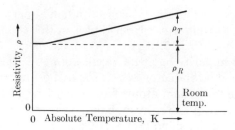

**Fig. 12.11** Comparison of the two components of the resistivity of a typical solid-solution alloy.

Engineering problems concerning electrical conductivity often involve only a given base metal, perhaps with small amounts of alloying elements. Under these conditions, variations in $\sigma$ (Eq. 12.17) are attributable almost entirely to the relaxation time $\tau$. The study of the various factors that influence $\tau$ is facilitated by *Mathiessen's rule*, which states that electrical resistivity (the reciprocal of conductivity) is the sum of a thermal term $\rho_T$ and a residual resistivity $\rho_R$; that is,

$$\rho = \rho_T + \rho_R \qquad (12.20)$$

Figure 12.11 is a schematic representation of the character of the terms in Eq. (12.20). We see that $\rho_R$ is assumed to be independent of temperature, while $\rho_T$ decreases with decreasing temperature. At 0 K, $\rho_T$ becomes negligibly small and the only remaining (residual) resistance is $\rho_R$. This term arises because of disturbances in the crystal lattice, as described below.

The variation of $\rho_T$ with temperature reveals several significant features of electronic conduction in metals. In suitably pure crystals for which $\rho_R$ is zero, the resistivity approaches zero at temperatures near 0 K. This means that the electrons move freely through the lattice without experiencing collisions with the ion cores that constitute the crystal lattice. Although this behavior cannot be explained from the classical viewpoint, it is entirely consistent with wave mechanics, since the periodic electron waves have a perfectly periodic lattice in which to move. Under these conditions, when collisions of electrons with the ion cores are virtually eliminated, the shape of the specimen affects the observed resistivity value. The electrons are reflected more often from the walls of a thin specimen, and therefore the observed resistivity is higher than in a thicker specimen. At higher temperatures the thermal disturbance of the lattice can be described in terms of quantized elastic waves or *phonons*, and the increase in $\rho_T$ can be visualized as resulting from collisions between the electrons and phonons. For most metals $\rho_T$ increases linearly with temperature near room temperature and above, although the behavior at very low temperatures is more complex.

The residual resistivity $\rho_R$ results from disturbances in the lattice other than those caused by thermal vibration. By far the most drastic increases in $\rho_R$ are

**TABLE 12.1**

Electrical Properties of Solids

| Material | Resistivity at 20°C, $\Omega \cdot m$ | Temperature coefficient $\alpha$, per °C |
|---|---|---|
| *Metals and alloys* | | |
| Copper, annealed | $1.67 \times 10^{-8}$ | $4.29 \times 10^{-3}$ (0–100°C) |
| Copper, reduced 75% by cold drawing | $1.71 \times 10^{-8}$ | — |
| Cartridge brass, annealed | | |
|    70% Cu, 30% Zn | $6.2 \times 10^{-8}$ | $1.48 \times 10^{-3}$ (20°C) |
| Aluminum, annealed | $2.65 \times 10^{-8}$ | $4.29 \times 10^{-3}$ (20°C) |
| Iron, annealed | $9.71 \times 10^{-8}$ | $6.57 \times 10^{-3}$ (0–100°C) |
| Constantan | | |
|    55% Cu, 45% Ni | $49 \times 10^{-8}$ | $0.02 \times 10^{-3}$ (25°C) |
| Manganin | | |
|    84% Cu, 12% Mn, 4% Ni | $44 \times 10^{-8}$ | $0.000 \times 10^{-3}$ (25°C) |
| | | $-0.042 \times 10^{-3}$ (100°C) |
| Nichrome | | |
|    80% Ni, 20% Cr | $108 \times 10^{-8}$ | $0.14 \times 10^{-3}$ (0–500°C) |
| *Semiconductors* | | |
| Germanium | $10^{-5}$ to 0.6 | — |
| Silicon | $10^{-5}$ to $2.3 \times 10^3$ | — |
| *Insulators* | | |
| Alumina | $10^{11}$ | — |
| Diamond | $10^{12}$ | — |

To convert from $\Omega \cdot m$ to $\Omega \cdot cm$, multiply by $10^2$.

caused by foreign atoms in solid solution in the matrix metal. Table 12.1 shows, for example, that the resistivity of copper is increased almost fourfold by the addition of 30% zinc to make cartridge brass. Cold-working, on the contrary, increases this resistivity term only slightly. For this reason cold-working is a suitable means of strengthening alloys to be used as electrical conductors.

Metals with high electrical resistance are needed for some applications, as in a toaster heating element, while low electrical resistance is essential for such uses as long-distance transmission lines. In any event, it is necessary to have a quantitative measure of this property for design purposes. The actual resistance depends on the shape of the metal as well as on the *resistivity*. Values of the latter property are given in Table 12.1. For a conductor of constant cross section the following equation is valid:

$$R = \rho \frac{l}{A},$$ (12.21)

where $R$ is the resistance in ohms, $l$ is the length in meters, $A$ is the area in square meters, and $\rho$ is the resistivity in ohm meter ($\Omega \cdot m$). Since the actual resistivity value for a given metal is strongly influenced by many factors, it is necessary to consider the nature of these influences.

The effect of temperature on resistivity is usually determined with sufficient accuracy by means of the equation

$$\rho_t = \rho_{20}[1 + \alpha(t - 20)],$$ (12.22)

where $\rho_t$ is the resistivity at $t°C$, $\rho_{20}$ is the resistivity at 20°C, and $\alpha$ is the temperature coefficient of resistivity per °C. Values of $\alpha$ are given in Table 12.1. In conductors that operate at high temperatures, such as heating elements, the temperature correction term has a large value and is an important design factor. In critical electrical devices the variation of resistance with temperature can be made negligibly small by the use of *manganin* (84% Cu, 12% Mn, 4% Ni) or similar alloys that have low-temperature coefficients.

An interesting aspect of resistivity is its dependence on the direction of the electrical current in single crystals of noncubic metals. The resistivities parallel to the $c$-axis and perpendicular to this axis for several metals are tabulated here.

| | Electrical resistivity at 20°C, $\Omega \cdot m$ | |
|---|---|---|
| Metal | Parallel | Perpendicular |
| Magnesium | $3.78 \times 10^{-8}$ | $4.53 \times 10^{-8}$ |
| Zinc | $6.13 \times 10^{-8}$ | $5.91 \times 10^{-8}$ |
| Cadmium | $8.30 \times 10^{-8}$ | $6.80 \times 10^{-8}$ |
| Tin | $14.3 \times 10^{-8}$ | $9.9 \times 10^{-8}$ |
| Tellurium | $56\,000 \times 10^{-8}$ | $154\,000 \times 10^{-8}$ |

This *anisotropy** of resistivity has potential uses in special electrical devices. Of considerable industrial interest is the variation of resistivity produced by alloying.

---

*An *isotropic* substance has the same value of a given property in every direction of testing. For example, the resistivity of an aluminum single crystal, which is face-centered cubic, is $2.65 \times 10^{-8}\,\Omega \cdot m$ in any arbitrary direction of testing. An *anisotropic* substance, such as a magnesium single crystal, shows a variation in the value of a given property as the direction of testing is varied. The values given above are only the maximum and minimum values. The value of resistivity $\rho_x$ in a direction that makes an angle $\alpha$ with the $c$-axis in hexagonal or tetragonal crystals is

$$\rho_\alpha = (\rho_{\text{parallel}}) \cos^2 \alpha + (\rho_{\text{perpendicular}}) \sin^2 \alpha.$$ (12.23)

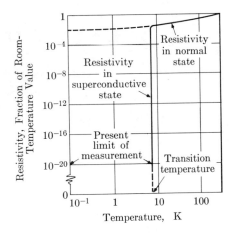

**Fig. 12.12** Schematic illustration of the transition to the superconductive state in the metal niobium. Theoretical treatments indicate that the resistivity in the superconductive state is actually zero.

Solid-solution formation causes a pronounced increase; this behavior was shown in Fig. 7.8 for the copper-nickel system. In two-phase alloys the resistivity tends to change proportionally with composition between the resistivities of the two phases.

**Superconductivity.** As a result of recent developments, interesting engineering applications are being made of the electrical materials known as *superconductors*. These are substances that lose their resistance to the flow of electric current when they are cooled below a critical transition temperature, which is near absolute zero. When electric currents are induced in a ring specimen of a superconductor, it has been impossible to detect any decrease in the currents in experiments ranging up to a year in duration. Thus, the resistivity values are in a completely different range from those met under ordinary conditions (Fig. 12.12). It is significant that the residual resistivity due both to $\rho_T$ and to $\rho_R$, Eq. (12.20), are eliminated when the transition to the superconductive state occurs. A superconducting material is also perfectly diamagnetic. Quantum-mechanical theory attributes

---

It should be noted that a given crystal may be *isotropic* with respect to one property and *anisotropic* with respect to a different property. For example, an aluminum single crystal exhibits anisotropy with respect to Young's modulus (see Chapter 2), and isotropy with respect to resistivity.

The concepts of *anisotropy* and *inhomogeneity* are sometimes confused. A homogeneous substance is defined as one that has the same properties at every *point*. At every point in a homogeneous magnesium single crystal, electrical resistivity has the same dependence on direction, that is, the same anisotropy. Thus, a homogeneous substance can be anisotropic. Inhomogeneous substances are often complex, but such substances could conceivably exhibit either isotropic or anisotropic properties.

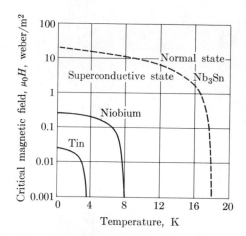

**Fig. 12.13**  Variation of the critical magnetic field with temperature for the superconducting metals, tin and columbium.  A curve for the $Nb_3Sn$ intermetallic compound is also shown. [To obtain the equivalent magnetic field in oersteds, multiply weber/m$^2$ by $10^4$.]

superconductivity to a long-range ordering of the electrons.  The presence of a small energy gap of about $3.5\,kT_c$ ($\sim 10^{-3}$ eV) above the Fermi level restricts the electrons to the ordered (superconducting) state near 0 K.  As the temperature is raised, the electrons are excited across the gap and are progressively disordered until the superconducting state completely disappears at the critical temperature $T_c$.

Two major obstacles to the engineering application of superconductivity have relegated it to the class of a laboratory curiosity until very recently.  The difficulty of maintaining the extremely low temperatures involved is being overcome by constant improvements in *cryogenics*, the field of low-temperature studies.  A second problem is the fact that the superconductive state is destroyed if the metal is subjected to a magnetic field higher than a definite critical value.  The critical field strength, $H_c$, varies with temperature, rising from zero at the transition temperature to a maximum value as absolute zero is approached (Fig. 12.13).  Since a magnetic field is associated with a flowing current, the low values of $H_c$ characteristic of pure metals had been a serious obstacle to the application of superconductivity for electromagnets and other devices involving large electrical currents.  Fortunately, many alloys and compounds (such as $Nb_3Sn$) develop a so-called *vortex phase* in response to a magnetic field and consequently exhibit a much higher critical field.  Such type-II superconductors are characterized by a two-dimensional lattice of rods of magnetic flux in a superconducting matrix, Fig. 12.14(a).  When the flux-line lattice is decorated by the use of fine ferromagnetic particles, Fig. 12.14(b), it is found to reveal the analog of such properties of atomic lattices as dislocations, stacking faults, and grain boundaries.  A present application

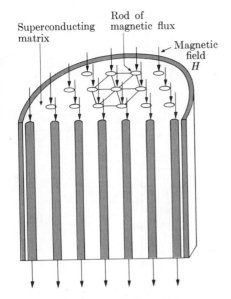

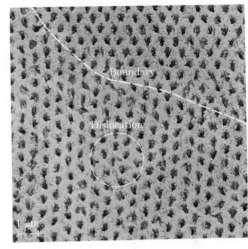

(a)  Rods of magnetic flux within a superconducting matrix take the form of a flux-line lattice. (Adapted from A. Seeger)

(b)  Micrograph of the flux-line lattice in a type-II superconductor. The analog of a grain boundary and an edge dislocation are indicated. (Courtesy of H. Trӓuble and U. Essmann)

**Fig. 12.14**  Illustrations of the vortex phase in type-II superconductors.

of superconductors is in the design of powerful electromagnets. Future possibilities include noiseless electric amplifiers, frictionless bearings, and ultra-fast switches for computers.

**Thermoelectric effects.** The term *thermoelectric* refers to the interdependence of certain thermal and electrical effects in metals. Two thermoelectric phenomena, the Peltier and Thomson effects, are of special interest because they are basic to the operation of thermocouples. Two additional quantities, Joule heat and contact potentials, are often confused with these more significant thermoelectric effects, and therefore all four of these phenomena are shown in Fig. 12.15.

The temperature of any conductor tends to rise as a result of *Joule* or *resistance heating* when there is a current in the conductor, Fig. 12.15(a). Since this factor is easily taken into account, it will not be discussed further here. The *Peltier effect* is also a heat effect associated with current. It differs from Joule heating in that heat may be *absorbed* or evolved depending on the direction of the current, Fig. 12.15(b). The Peltier effect occurs at the junction between two different metals, and its magnitude depends on the nature of the metals, the temperature, and the

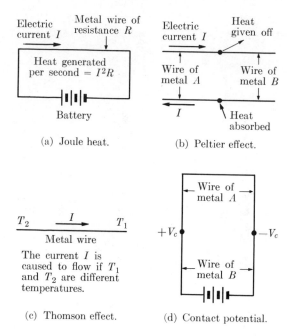

(a) Joule heat.

(b) Peltier effect.

(c) Thomson effect.

(d) Contact potential.

**Fig. 12.15**  The essential nature of the thermoelectric and related effects in metals.

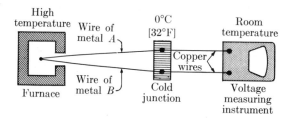

**Fig. 12.16**  One method of using a thermocouple of metals $A$ and $B$.

first power of the current. The *Thomson effect*, Fig. 12.15(c), is the production of a current in a homogeneous metal (e.g., copper) when a temperature gradient exists along its length.

An important practical application of the Peltier and Thomson effects is in *thermocouples*, which are widely used for temperature measurement and control. Figure 12.16 shows the essential features of a thermocouple and its use. The junction of two dissimilar metals is placed in the region whose temperature is to be measured. The voltage at the ends of the two wires indicates this temperature with respect to the temperature of the cold junction. This voltage is a measure of

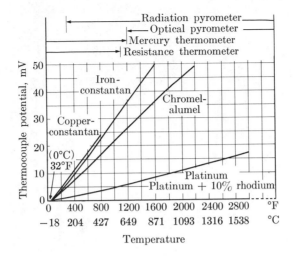

**Fig. 12.17** The voltages developed by several types of commercial thermocouples. Useful ranges of other temperature-measuring devices are shown for comparison.

the *Seebeck effect*, which is the existence of an electric current in a closed circuit composed of two different metals when the metallic junctions are at different temperatures. This current is developed as the result of (1) heat flow at the junctions of the thermocouple (Peltier effect), and (2) temperature gradients along each of the two wires (Thomson effect). When the cold junction of a thermocouple is fixed at a standard temperature like 0°C (32°F), its voltage is determined completely by the temperature of the hot junction. Figure 12.17 shows the magnitudes of the voltages developed by several of the commonly used pairs of metals. When precision greater than 0.1°C is required, a *resistance thermometer* is preferable to a thermocouple, while *radiation* and *optical pyrometers* permit the measurement of higher temperatures and do not depend on actual contact with the hot body.

It is to be noted that *contact potentials* are *not* the cause of thermoelectric effects. These potentials, $V_c$, are developed at the junction of two dissimilar metals, Fig. 12.15(d), but they cancel in a completely metallic circuit like the one shown in this figure. Contact potentials are commonly a few tenths of a volt and are essentially independent of temperature.

## MAGNETIC PROPERTIES

The engineer has a dual interest in magnetic phenomena: first, because the commercial magnetic materials, such as transformer iron, are widely used in modern industry, and second, because the science of magnetism explains many aspects of the structure and behavior of matter and is the source of many engineering developments. Since these two aspects of magnetic phenomena have much in

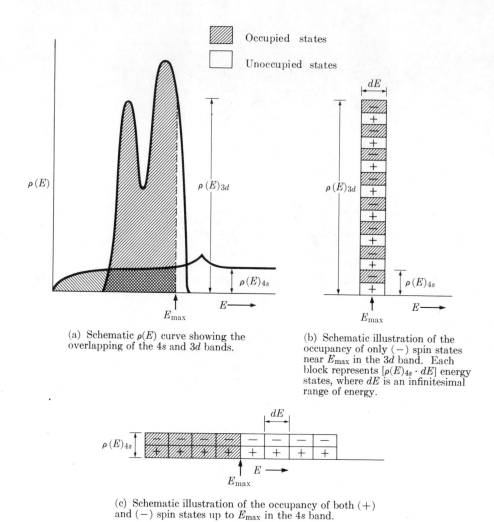

Occupied states

Unoccupied states

(a) Schematic $\rho(E)$ curve showing the overlapping of the 4s and 3d bands.

(b) Schematic illustration of the occupancy of only $(-)$ spin states near $E_{max}$ in the 3d band. Each block represents $[\rho(E)_{4s} \cdot dE]$ energy states, where $dE$ is an infinitesimal range of energy.

(c) Schematic illustration of the occupancy of both $(+)$ and $(-)$ spin states up to $E_{max}$ in the 4s band.

**Fig. 12.18**  The occupancy of energy states near $E_{max}$ in the 4s and 3d energy bands in nickel.

common, it is desirable to begin with a general treatment of magnetism and then to consider the special alloys of commercial interest.

**Atomic magnetism.**  Every electron in both inner and outer orbits of the various elements has a quantity of magnetism or magnetic moment (defined as one Bohr *magneton*) as a result of "spin" of the electron.  The sign of this magnetism may be positive or negative, depending on the direction of spin of the electron, $(+)$ or $(-)$.  It will be recalled from the discussion of quantum numbers in Chapter 1 that a given energy level can accommodate an equal number of electrons with

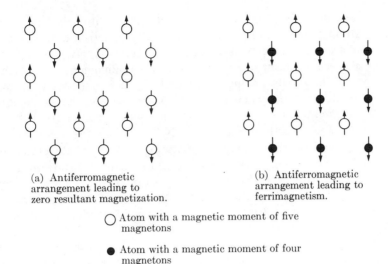

(a) Antiferromagnetic
arrangement leading to
zero resultant magnetization.

(b) Antiferromagnetic
arrangement leading to
ferrimagnetism.

○ Atom with a magnetic moment of five
   magnetons

● Atom with a magnetic moment of four
   magnetons

**Fig. 12.19** Antiparallel arrangements of resultant electron spins that characterize antiferromagnetic substances.

(+) and (−) spins.   Thus all filled energy levels have essentially no resultant magnetic moment because of cancellation.   When an energy level is only partly filled with electrons, two opposing factors determine the spin of these electrons. The dominant factor in most cases is the lower energy that results when the available quantum states are occupied in regular succession starting with the state of lowest energy.   The outer, valence electrons normally behave in this manner for the reason shown schematically for the 4s electrons of nickel in Fig. 12.18.   The 4s band in Fig. 12.18(a) covers a broad range of energies, and consequently there are relatively few electron states for each increment of energy $dE$. Figure 12.18(c) is a schematic representation of the electron states near the energy level $E_{max}$ to which the 4s band is filled.   States that differ only in spin are shown as having the same energy, but the electron energy increases toward the right as shown.   Therefore, the energy is a minimum when states of both (+) and (−) spin are occupied up to $E_{max}$.

The 3d band can accommodate a maximum of ten electrons, compared with a maximum of two for the 4s band.   Also, the 3d band shown in Fig. 12.18(a) is typical in being narrower than the 4s band.   As a result, the number of electron states at a given energy value is relatively large in the 3d band, as shown in Fig. 12.18(b). There is less advantage in having states of both (+) and (−) spin occupied in this case, and a second factor determining spin comes into play.   This is an electrostatic energy that favors the existence of electrons having the same spin. It is not shown in Fig. 12.18(b), but as a result of its action the electrons preferentially occupy the states of negative spin.   Consequently, there is an excess of electrons with negative spin and the atoms can be considered to have a net magnetic moment.

Although many of the elements satisfy the above condition, only a few metals and alloys exhibit *ferromagnetism*, the pronounced magnetism that is of primary interest in engineering. The explanation of this anomaly involves an electrostatic interaction between adjacent atoms that affects the alignment of the resultant electron spins of the atoms. If the interaction is positive, the spins are aligned parallel to one another, their magnetic moments are additive, and ferromagnetism results. If the interaction is negative, the net spins of neighboring atoms are aligned in opposite directions (antiparallel), producing the phenomenon of *antiferromagnetism*. In the latter case it might be expected that the overall magnetic moment would be essentially zero, and Fig. 12.19(a) shows an arrangement for which this is true. However, if the material is composed of two atoms having different degrees of magnetism, then in an ordered crystal structure there can be overall magnetism, Fig. 12.19(b). Although the actual behavior is more complex than suggested by the figure, the distribution of $Mn^{++}$ ions in solid MnO is similar to Fig. 12.19(a), so that MnO is nonmagnetic. On the other hand, the $Fe^{++}$ and $Fe^{+++}$ ions in magnetite, $FeO \cdot Fe_2O_3$, occupy two types of sites as in Fig. 12.19(b). This configuration causes the pronounced magnetic properties suggested by the name of this substance, "magnetic iron oxide." The term *ferrimagnetism* is used to describe this type of magnetization, which approaches ferromagnetism in intensity, and the corresponding substances are called *ferrites*.

The interaction between atoms having net spins can be nearly zero, and in this case the spins of the various atoms in a substance are randomly oriented. The magnetism is then only about one-millionth that of ferromagnetic substances. This kind of magnetic behavior is called *paramagnetism* and is characteristic of such nonmagnetic metals as tungsten.

The positive interaction that characterizes ferromagnetic materials is gradually decreased by thermal energy as the temperature is increased, and as the *Curie temperature* is approached the interaction falls toward zero. The corresponding decrease in the intensity of magnetization of iron is shown in Fig. 12.20, and it can be seen that above 770°C (1418°F) iron is paramagnetic, that is, essentially nonmagnetic. Nickel and cobalt, the other two important ferromagnetic metals, show a similar behavior.

**Magnetic domains.** In view of the alignment of spins that is characteristic of ferromagnetism, it is puzzling that a ferromagnetic body, such as the steel shaft of a screwdriver, is not actively magnetic under all circumstances. A screwdriver that initially does not attract small iron screws frequently takes on magnetism after it has been used near electrical equipment or stored in contact with a horseshoe magnet. That is, a *magnetic* material may or may not exhibit *overall magnetization*. Also, while all ferromagnetic substances become magnetized in a magnetic field, such as that generated by an electric current in a coil, there are important differences in the degree to which magnetization is retained. Many materials are easily demagnetized after the magnetizing field is removed; these are called magnetically

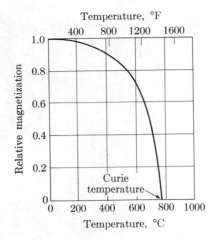

**Fig. 12.20** A plot of the decrease of magnetization of iron with temperature. Nickel and cobalt behave in a similar fashion, with Curie temperatures of 354°C (670°F) and 1115°C (2039°F), respectively.

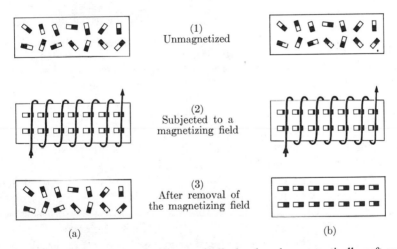

**Fig. 12.21** The difference between a magnetically hard and a magnetically soft material. The small magnets represent domains in the bars.

*soft*. Materials that strongly retain appreciable magnetization when the field is removed are called magnetically *hard*, and are used for permanent magnets. Figure 12.21 shows how these phenomena are explained by the domain structure of ferromagnetic materials.

A *domain* is a region in which the magnetic moments of the atoms are aligned and point in the same direction. The interaction energy discussed above would tend to make an entire metal crystal one large domain, but this would create magnetic poles on the surface of the metal and the associated magnetostatic

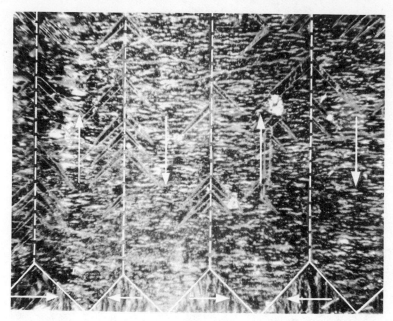

**Fig. 12.22**  Domain structure in a single crystal of iron, revealed by a colloidal suspension of fine magnetic particles.  Arrows show the direction of magnetization within each domain. Photograph by H. J. Williams.  (By permission from *Ferromagnetism* by R. M. Bozorth; copyright Van Nostrand, 1951)

energy would be very high.  A compromise between these two energies is reached when the metal assumes a domain structure, Fig. 12.22.  Here, regions with various directions of magnetization form closed loops, so that there are few magnetic poles at the surface or in the interior of the metal.

It will be noticed in Fig. 12.22 that the directions of magnetization in all the domains are either vertical or horizontal; these are [100]-type directions in the crystal.  This behavior is favored because the ease of magnetization varies with crystallographic direction within a single crystal, and the [100] direction in an iron single crystal is the direction of easiest magnetization, Fig. 12.23.  In nickel this direction is the [111] direction, and in cobalt it is the hexagonal axis, [0001]. If magnetization is to be maintained in a direction other than the direction of easiest magnetization, an additional energy, the *anisotropy energy*, is required.

An especially important feature of domain structure is the boundary between domains, called the *Bloch wall*.  Consideration of two adjacent vertical domains in Fig. 12.22 shows that the direction of magnetization must turn out of the direction of easiest magnetization in making the transition from one domain to another. The anisotropy energy increases the energy of such domain boundaries, but the increase is a minimum if the transition between domains is gradual and occurs over a region about 1000 atoms wide.  This region is the Bloch wall.  Because of the

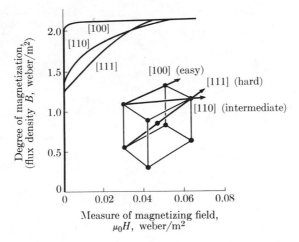

**Fig. 12.23** Effect of applying a magnetizing field in each of three different crystallographic directions in a single crystal of iron. Full magnetization is easily obtained in the [100] direction. [To convert to the equivalent plot in c.g.s. units, $B$ (in gauss) versus $H$ (in oersteds), multiply weber/m² by $10^4$ for both scales.]

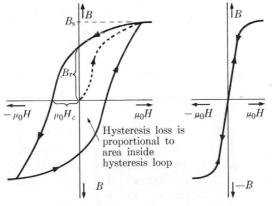

(a) Generalized $B$-$\mu_0 H$ diagram.  (b) Ideal $B$-$\mu_0 H$ diagram for magnetically soft material.

**Fig. 12.24** $B$-$\mu_0 H$ curves, illustrating some of the important properties of magnetic materials.

relatively large extent and high energy of domain boundaries, if the dimensions of a ferromagnetic substance are sufficiently small, each particle will exist as a single domain and will have exceptional magnetic properties. These fine-particle magnets are discussed below.

**Magnetically soft materials.** In electrical devices such as motors, transformers, and relays it is desirable that the magnetization of the alloy follow very closely

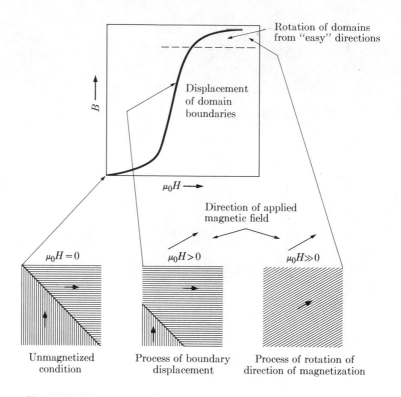

**Fig. 12.25**  The process of **magnetization** in a ferromagnetic material.

any change in the magnetizing field.  This magnetic behavior is conveniently described using a general $B$-$\mu_0 H$ diagram, Fig. 12.24(a).  If a specimen is initially unmagnetized, then its magnetization $B$ will increase along the dashed curve as the magnetizing force $H$ increases in magnitude.  The degree of magnetization eventually reaches an essentially constant value $B_s$, the *saturation magnetization*. When the magnetizing force is reduced toward zero, $B$ decreases along the solid curve and reaches the value $B_r$, the *residual magnetization*, when $\mu_0 H = 0$.  As the magnetizing force is increased *in the opposite direction*, $B$ continues to decrease and becomes zero when the reverse magnetizing force is equal to $\mu_0 H_c$, the *coercive force*.  As the reverse magnetizing force becomes even greater, the magnetization becomes reversed also and eventually reaches $(-)B_s$.  The remaining half of the solid curve is obtained when the magnetizing force is again reversed and increases in the positive sense.

Figure 12.25 shows schematically how the domain structure changes during initial magnetization, the dashed curve in Fig. 12.24(a).  In the unmagnetized specimen the magnetic effects of the domains cancel out.  When a small magnetic field, $\mu_0 H > 0$, is applied in the direction indicated by the arrow, domains with orientations close to this direction have lower energies and are more stable. There is then a tendency for a more stable domain to increase in size by motion of the Bloch wall separating it from a less stable domain. The final stage of the magnetization process involves the rotation of the direction of magnetization of the stable domains from the original easy direction toward the direction of the applied field. This process requires a relatively large magnetic field, $\mu_0 H \gg 0$.

The ideal behavior of a magnetically soft material is shown in Fig. 12.24(b) in comparison with the general $B$-$\mu_0 H$ diagram. In particular, the following property values may be of special interest, depending on the exact application.

1. *Saturation flux density, $B_s$.* This quantity* determines the degree of magnetization that can be developed in a magnetic material. A high value of $B_s$ is almost always desirable.

2. *Relative permeability, $\mu_r$.* The *permeability* is defined as $\mu = B/H$, but a more convenient quantity is $\mu_r = \mu/\mu_0$, which has the same value in both SI and c.g.s. units. The larger the value of $\mu_r$, the smaller is the magnetizing force $\mu_0 H$ necessary to produce a given magnetization $B$. A large initial relative permeability is especially significant for applications in which only a weak magnetizing force is available.

3. *Coercive force, $(\mu_0 H)_c$.* This quantity is the *reverse* magnetizing force necessary to eliminate the *residual* flux density $B_r$. The coercive force should be small in magnetically soft materials.

4. *Total core loss.* This quantity is the total power loss (watts per kilogram of alloy) under given conditions. It is the sum of *hysteresis loss* [which depends mainly on $(\mu_0 H)_c$] and *eddy current loss*.

Table 12.2 lists these data and typical applications for several magnetically soft alloys. It is seen that chemical composition has an important influence on magnetic properties. For example, the total core loss of silicon steel is reduced 50% by increasing the silicon content from 1.0 to 4.25 percent. Suitable heat treatments and the prevention of mechanical strains also contribute to desirable properties in magnetically soft alloys. Iron-silicon alloys are far easier to magnetize if the grains composing a sheet of alloy are oriented so that a [100] direction is in the direction of the field, Fig. 12.23. Table 12.2 shows the improved permeability of such "grain-oriented" steels compared with regular silicon steels. This is an example of the general phenomenon of *preferred orientation* shown in Fig. 2.9 in connection with its effect on elastic properties.

**Permanent (hard) magnets.** Although the most powerful magnetic fields are obtained by means of electromagnets, in which magnetically *soft* materials are

---

* Usually the *saturation magnetization, $M = (B - \mu_0 H)_s$* is reported in the literature; see Table 12.2. Since $\mu_0 H$ is generally small at $B_s$, $(B - \mu_0 H)_s$ is approximately equal to $B_s$.

**TABLE 12.2** Properties of Several Magnetically Soft Metals and Alloys

| Material | $(B - \mu_0 H)_s$, saturation magnetization, weber/m² | $\mu_r$, relative permeability initial | $\mu_r$, relative permeability maximum | $B_r$, residual flux density*, weber/m² | $(\mu_0 H)_c$ coercive force*, weber/m² | Hysteresis loss, J/m³ per cycle* | Total core loss, watts/kg | Resistivity, Ω·m | Typical applications |
|---|---|---|---|---|---|---|---|---|---|
| Puron (99.99% Fe) | 2.16 | 4 000 | $10^5$ | 0.90 | $0.05 \times 10^{-4}$ | 15 | — | $10 \times 10^{-8}$ | — |
| Ingot iron (99.9% Fe) | 2.16 | 250 | 7 000 | 1.10 | $1.0 \times 10^{-4}$ | 500 | — | $11 \times 10^{-8}$ | |
| 1% silicon steel | 2.16 | 500† | 6 000 | 0.90 | $0.9 \times 10^{-4}$ | — | 2.6 | $30 \times 10^{-8}$ | Small ac motors |
| 2.5% silicon steel | 2.05 | 900† | 6 000 | 0.80 | $0.8 \times 10^{-4}$ | — | 2.2 | $40 \times 10^{-8}$ | Induction motors |
| 4.25% silicon steel | 1.95 | 1 500† | 9 000 | 0.70 | $0.4 \times 10^{-4}$ | 350 | 1.3 | $60 \times 10^{-8}$ | Power transformers |
| 3% silicon steel "grain oriented" | — | 7 500† | 40 000 | — | $0.15 \times 10^{-4}$ | 75 | 0.9 | $47 \times 10^{-8}$ | Power transformers |
| Supermalloy 79% Ni, 16% Fe, 5% Mo | 0.79 | $10^5$ | $10^6$ | 0.50 | $0.2 \times 10^{-6}$ | 0.8 | — | $60 \times 10^{-8}$ | Telephone transformers |
| Hypernik 50% Ni, 50% Fe | 1.60 | 4 500 | $10^5$ | 0.80 | $0.05 \times 10^{-4}$ | 10 | — | $50 \times 10^{-8}$ | Radio transformers |
| Perminvar 45% Ni, 25% Co, 30% Fe | 1.55 | 400 | 2 000 | 0.60 | $1.2 \times 10^{-4}$ | 250 | — | $19 \times 10^{-8}$ | (Constant permeability at low fields) |

* Determined for saturation, $(B - \mu_0 H)_s$.  † Determined for $B = 0.01$ rather than for $B = 0$.  To convert weber/m² to gauss (or to oersted) multiply by $10^4$.

**TABLE 12.3**
Properties of Typical Permanent (Hard) Magnetic Alloys

| Material (Balance of composition, if any, is iron) | $B_r$, residual flux density, weber/m² | $(\mu_0 H)_c$, coercive force, weber/m² | $(BH)_{max}$, maximum magnetic energy, J/m³ | $B_d$, flux density at $(BH)_{max}$ weber/m² | Mechanical properties | Commercial methods of fabrication |
|---|---|---|---|---|---|---|
| Carbon steel (0.9% C, 1% Mn) | 1.00 | 0.005 | $0.20 \times 10^4$ | 0.52 | These magnetic steels are similar to tool steels and are usually formed by hot rolling or forging. They are quenched in oil or water to harden them and therefore have high hardness and low ductility. | |
| Tungsten steel (5% W, 0.7% C, 0.3% Mn) | 1.03 | 0.007 | $0.30 \times 10^4$ | 0.69 | | |
| Cobalt steel (36% Co, 4% Cr, 5% W, 0.7% C) | 1.00 | 0.024 | $1.0 \times 10^4$ | 0.63 | | |
| Alnico 5B (8% Al, 14% Ni, 24% Co, 3% Cu) | 1.27 | 0.060 | $4.4 \times 10^4$ | 1.00 | Hard, brittle | Casting |
| Cunife (60% Cu, 20% Ni) | 0.54 | 0.050 | $1.3 \times 10^4$ | 0.40 | Ductile, relatively soft | Cold rolling, punching, machining |
| Ferroxdur (BaFe$_{12}$O$_{19}$) | 0.34 | 0.180 | $2.5 \times 10^4$ | 0.20 | Hard, brittle | Powder metallurgy |
| Fine powder (30% Co) | 0.90 | 0.100 | $5.0 \times 10^4$ | — | Weak | Compacted by pressing |
| Bismanol (MnBi) | 0.43 | 0.340 | $4.3 \times 10^4$ | — | Hard, brittle | Powder metallurgy |

To convert weber/m² to gauss (or to oersted) multiply by $10^4$.

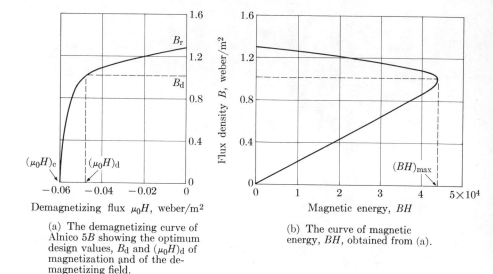

(a) The demagnetizing curve of Alnico 5$B$ showing the optimum design values, $B_d$ and $(\mu_0 H)_d$ of magnetization and of the demagnetizing field.

(b) The curve of magnetic energy, $BH$, obtained from (a).

**Fig. 12.26** A typical curve of demagnetization and its use to determine the maximum external energy, $(BH)_{max}$.

used, it is frequently desirable to have a constant magnetic field that does not depend on an electric current. Permanent magnets made of magnetically *hard* materials satisfy this need. Some of the basic uses of permanent magnets are: (1) to convert electrical energy to mechanical motion, as in electric clocks and meters; (2) to convert mechanical motion to electrical energy, as in a magneto or a phonograph pickup; (3) for mechanical holding as in magnetic separation or as a latch for a refrigerator door.

Properties of typical permanent magnet alloys are listed in Table 12.3. Two of these materials, Ferroxdur and Bismanol, are examples of the strong magnetism that can be developed by *ferri*magnetic substances, which were discussed above in connection with antiferromagnetism. The remaining alloys in the table are *ferro*magnetic. The fine powder magnet was developed after it had been shown theoretically that a particle of iron about 300 Å in size would be a single domain and would have remarkable magnetic properties, since the direction of magnetization could change only by rotation. An interesting magnet of this class, known as a "samarium-type" magnet, employs fine powder of a rare-earth compound such as $SmCo_5$. After alignment of the particles in a magnetic field during the process of producing a sintered magnet, the remarkably high $(BH)_{max}$ value of $20 \times 10^4$ is obtained. Single domains also account for the excellent properties of Alnico 5B. It has been shown that this alloy consists of fine, platelike particles, and that these single-domain particles orient themselves in the direction of a strong magnetic field that is applied during heat treatment.

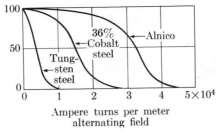

(a) Effect of a demagnetizing field

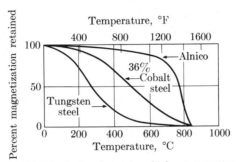

(b) Effect of exposure to a higher temperature

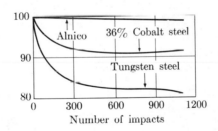

(c) Effect of mechanical impacts

**Fig. 12.27** Relative resistances of three permanent-magnet alloys to demagnetization. (Courtesy General Electric Company)

Permanent-magnet materials have the characteristic of strongly retaining magnetization when the magnetizing field is removed. The residual flux density $B_r$ is the degree of magnetization that remains when the (saturation) magnetizing force is reduced to zero. The coercive force $(\mu_0 H)_c$ is the *reverse* magnetizing force required to decrease magnetization to zero; $(\mu_0 H)_c$ is thus a measure of the resistance of the permanent magnet to demagnetizing forces, such as stray magnetic fields. A more generally useful design quantity than either $B_r$ or $(\mu_0 H)_c$ alone is the *maximum external energy*, $(BH)_{max}$. This quantity is derived from the demagnetization portion of a $B$-$\mu_0 H$ curve, Fig. 12.26(a), by finding the value of magnetization

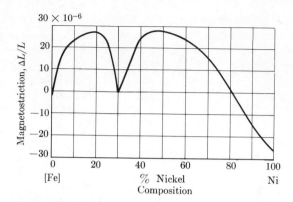

**Fig. 12.28** The longitudinal magnetostriction produced in iron-nickel alloys by a magnetizing field corresponding to $\mu_0 H = 0.025$ weber/m$^2$.

$B_d$ at which the magnetic energy required for demagnetization reaches a maximum value, Fig. 12.26(b). The product of $B_d$ and $H_d$, the corresponding demagnetizing field, is $(BH)_{max}$. When a permanent magnet is used most efficiently, its magnetization is thus $B_d$ rather than $B_r$.

Other factors besides those given by the demagnetization curve are important in the practical application of permanent magnets. In addition to cost, these factors include the forming methods that can be used, the mechanical properties of the final product, the density, and the resistance of the magnetized alloy to various demagnetizing influences. Figure 12.27 shows the ability of several permanent magnet materials to resist demagnetization when subjected to stray magnetic fields, high temperatures, or mechanical impact.

**Magnetostriction.** When the degree of magnetization of a ferromagnetic material is changed, a small corresponding change in length results. This phenomenon is called *magnetostriction*. Conversely, an externally produced strain in a magnetized material causes a change in the amount of magnetization. The actual magnitude of magnetostriction in iron-nickel alloys is shown in Fig. 12.28. In spite of the small changes of length involved, this phenomenon is the operating principle of *Sonar*, a type of underwater radar, and of certain supersonic generators for metal testing and treatment. Magnetostriction is also the explanation of the "60-cycle hum" of ordinary transformers.

## PREDICTION OF ALLOY PHASES

In view of the strong influence of crystal structure on mechanical and physical properties of alloys, metallurgical science has sought relations between electronic configuration and crystal structure. The *Engel–Brewer theory* is a significant step in this direction because of its applicability to complex alloys such as superalloys.

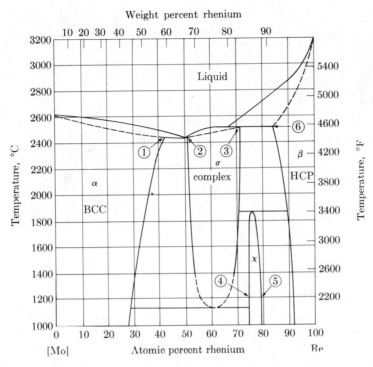

**Fig. 12.29** A typical example of the occurrence of intermetallic phases, the molybdenum-rhenium system. (Diagram adapted from R. P. Elliott, *Constitution of Binary Alloys*, McGraw-Hill, New York, 1965)

We start with a typical phase (molybdenum and its alpha solid solution in Fig. 12.29) and try to determine why it has a definite crystal structure, body-centered cubic in this case. For a few normal metals like potassium, electron theory can supply the answer by calculations of the energies of alternative crystal structures. Similar calculations are not yet feasible for transition metals or alloys. In the most general case, however, a correlation exists between the stability of the BCC structure and a hybrid $s^1 p^x$ electronic structure in which the $p$-type contribution is in the range $x = 0$ to $0.5$. In pure molybdenum the electrons beyond the filled core are $(4d)^5$ and $(5s)^1$, the latter corresponding to $s^1 p^0$. (The role of the $d$-type electrons is discussed later.) In the BCC solid solution of 40 a/o Re in molybdenum, a new feature enters—the promotion of electrons. The electron configuration of a free atom of rhenium (in the vapor phase, for example) is $(5d)^5 (6s)^2$. The paired $6s$ electrons are ineffective in bonding, but each can participate if one of them acquires the energy required to raise (promote) it to the $6p$ level. The energy for promotion is less than the energy gained from the two bonds formed, one by the $6s$ electron and another by the $6p$ electron. The electron configuration of the

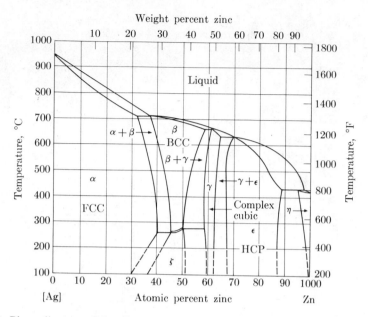

Weight percent zinc

**Fig. 12.30** Phase diagram of the silver-zinc system, showing the range of existence of the two primary solid solutions, $\alpha$ and $\eta$, and of the intermediate phases, $\beta$, $\gamma$, and $\varepsilon$. The latter are called *electron phases* because their structures depend on electron configurations. (Diagram adapted from M. Hansen and K. Anderko, *Constitution of Binary Alloys*, McGraw-Hill, New York, 1958)

solid solution of molybdenum containing 40 a/o Re is $s^1 p^{0.4}$, near the end of the range characteristic of the BCC structure.

At the rhenium side of the diagram the correlation rule that applies is $s^1 p^{0.7-1.1}$, which describes the range of $p$-character associated with the HCP structure. For pure rhenium the configuration is $s^1 p^1$, while the limit of the rhenium-rich solid solution is at $s^1 p^{0.85}$. An unfavorable size effect accounts in part for the absence of the full extent of solubility permitted by the correlation rule. The two remaining phases in the diagram, $\sigma$ and $\chi$, have complex structures (about 30 atoms in the unit cell) and undesirable properties such as brittleness. A method for designing high-temperature alloys to avoid the presence of $\sigma$-type and similar objectionable phases is based on the $sp$ correlation. To discuss this method, we need first to consider the third common metallic structure, the FCC.

The silver-zinc system, Fig. 12.30, is interesting for several reasons, but we begin with the silver-rich solid solution. The free silver atom has the electronic configuration $(4d)^{10}(5s)^1$, with only one electron available for bonding. Through promotion of two of the $4d$ electrons to the $5p$ level, five electrons are available for bonding in pure metallic silver (see Problem 19). The appropriate correlation rule associates a configuration in the range $s^1 p^{1.5-2.0}$ with the FCC structure.

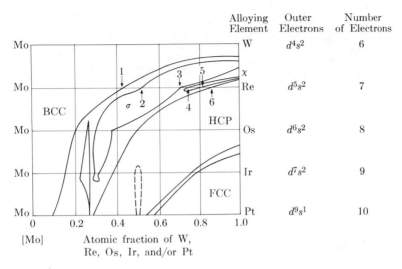

| Alloying Element | Outer Electrons | Number of Electrons |
|---|---|---|
| W | $d^4s^2$ | 6 |
| Re | $d^5s^2$ | 7 |
| Os | $d^6s^2$ | 8 |
| Ir | $d^7s^2$ | 9 |
| Pt | $d^9s^1$ | 10 |

**Fig. 12.31** Plot based on the region of existence of various phases in the binary diagrams of molybdenum with the series of transition elements from tungsten to platinum. (Adapted from Leo Brewer)

Since zinc after promotion has the $s^1p^1$ configuration, the maximum solubility of zinc in silver, 40 at. %, produces a configuration of $s^1p^{1.6}$ in the solid solution, a value still within the range for the FCC structure. The $\beta$ phase is not merely the $\alpha$ (silver) phase with some additional zinc; it is a much different material and requires a separate calculation. Because of lack of sufficient opportunities for $d$-type bonding, negligible promotion of electrons occurs in the silver atoms, leaving the electron configuration $s^1$. The zinc atoms continue to use the promoted configuration, $s^1p^1$, and therefore the $\beta$ phase at 50 at. % Zn has the configuration, $s^1p^{0.5}$, that correlates with the BCC structure. A range in $p$-character from 0.37 to 0.59 occurs over the complete range of composition of the $\beta$ phase. The remaining phases in the Ag-Zn system are the $\gamma$ phase, which is complex cubic with 52 atoms in the unit cell, and the $\varepsilon$ and $\eta$ phases, both of which are hexagonal (see Problem 20). The $\zeta$ phase (hexagonal) is related structurally to the $\beta$ phase.

A useful method for designing multicomponent alloys involves Brewer diagrams, an example of which is shown in Fig. 12.31. This particular diagram is a concise representation of phase formation in the binary systems Mo-W, Mo-Re, Mo-Os, Mo-Ir, and Mo-Pt. For example, in the Mo-Re phase diagram, Fig. 12.29, the maximum range of the BCC phase is out to 43 at. % Re, point 1. This point is plotted on the Mo-Re line in Fig. 12.31, and together with similar points from the other four binary diagrams is the basis for plotting the boundary line for the region of existence of the BCC phase. Similarly, points 2 and 3 describe the maximum range of the $\sigma$ phase, points 4 and 5 of the $\chi$ phase, and point 6 the limit of the HCP phase. Once the Brewer diagram has been constructed, the

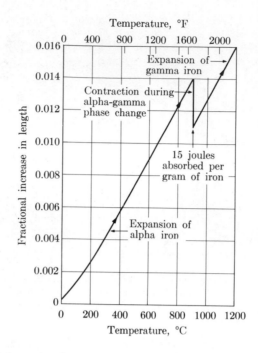

**Fig. 12.32** Idealized thermal expansion curve for iron during slow heating. The discontinuous contraction and heat absorption during the allotropic change are indicated.

correlation rules predict that it applies for a multicomponent alloy and depends only on the average number of electrons. For example, although tungsten is BCC and iridium is FCC, an alloy of 50 at. % W in iridium (7.5 electrons per atom) is HCP, as predicted by the diagram. Furthermore, when molybdenum is added to such a tungsten-iridium alloy, the succession of phases is predicted by the corresponding horizontal line in the Brewer diagram. Although the predictions are not always accurate, this type of diagram is used successfully in designing alloys containing several components, and is especially useful in pointing out those compositions that might form objectionable $\sigma$ type phases.

## THERMAL PROPERTIES

A number of properties determine the total effect of adding heat to a metal or alloy. For example, the *specific heat* $c_p$ controls the temperature increase $\Delta T$ produced by the addition of a given quantity of heat $Q$ to one kilogram of metal:

$$Q = c_p \Delta T. \qquad (12.24)$$

TABLE 12.4

Thermal Properties of Several Metals and Alloys

| Material | Temperature, °C | Specific heat $c_p$, J/(kg·K) | Thermal conductivity $k$, W/(m·K) | Coefficient of thermal expansion $\alpha$, per °C |
|---|---|---|---|---|
| Aluminum | 0 | $0.93 \times 10^3$ | $0.24 \times 10^3$ | $22 \times 10^{-6}$ |
| | 100 | $0.96 \times 10^3$ | $0.24 \times 10^3$ | |
| | 400 | $1.06 \times 10^3$ | $0.22 \times 10^3$ | |
| | 600 | $1.12 \times 10^3$ | $0.21 \times 10^3$ | |
| Copper | 0 | $0.38 \times 10^3$ | $0.40 \times 10^3$ | $16 \times 10^{-6}$ |
| | 100 | $0.39 \times 10^3$ | $0.40 \times 10^3$ | |
| | 500 | $0.43 \times 10^3$ | $0.38 \times 10^3$ | |
| | 1 000 | $0.48 \times 10^3$ | $0.34 \times 10^3$ | |
| Iron | 0 | $0.43 \times 10^3$ | $0.84 \times 10^3$ | $11.5 \times 10^{-6}$ |
| | 100 | $0.48 \times 10^3$ | $0.74 \times 10^3$ | $12.6 \times 10^{-6}$ |
| | 500 | $0.68 \times 10^3$ | $0.46 \times 10^3$ | $16.5 \times 10^{-6}$ |
| | 700 | $0.96 \times 10^3$ | $0.35 \times 10^3$ | |
| Invar (36% Ni, 64% Fe) | 0 | | | $1.4 \times 10^{-6}$ |
| | 100 | $0.51 \times 10^3$ | $0.11 \times 10^3$ | $1.5 \times 10^{-6}$ |
| | 200 | | | $6.3 \times 10^{-6}$ |
| | 300 | | | $13.4 \times 10^{-6}$ |
| Kovar (29% Ni, 17% Co, 54% Fe) | 0–450 | | | $5.0 \times 10^{-6}$ (Similar to hard glass expansion; used in glass-to-metal seals) |

The rate at which heat can flow through a material under the influence of a given temperature gradient is determined by the *thermal conductivity k*:

$$\frac{dQ}{dt} = k(T_2 - T_1)\frac{A}{l} \quad \text{J/s,} \quad (12.25)$$

where $A$ is the area through which flow occurs, and $l$ is the distance separating the surfaces that are at temperatures $T_2$ and $T_1$. A third property that is frequently of interest is the coefficient of thermal expansion, $\alpha$, defined so that the length at temperature $T$ is given by

$$l_T = l_{20}[1 + \alpha(T - 20)], \tag{12.26}$$

where $l_{20}$ is the length at 20°C. Table 12.4 lists these thermal properties for several metals and alloys.

The magnitudes of thermal properties depend on the nature of the solid phase and on its temperature. Moreover, these properties show abrupt variations when a phase change takes place. Figure 12.32 shows the sharp decrease in length and the absorption of heat that occur when alpha iron changes to gamma iron at 910°C. Sudden changes of this kind are especially troublesome in commercial heat-treating operations.

## OPTICAL PROPERTIES

We usually think of transparent substances like glass or quartz when *optical properties* are mentioned, but the data of Table 12.5 show that such quantities as refractive index $n$ and absorption coefficient $\kappa$ can be measured for metals. The absorption coefficient is nearly zero in transparent substances, so that only the refractive index is required for a description of their optical properties. For metals, on the other hand, both optical constants are significant. For example, the most important optical property, reflectivity, is given by the following equation for normal incidence:

$$\text{Reflectivity} = \frac{(n - 1)^2 + \kappa^2}{(n + 1)^2 + \kappa^2}. \tag{12.27}$$

Reflectivity influences the choice of metals for mirrors and also determines the color exhibited by an alloy in ordinary light. The exact value of reflectivity depends on such factors as the condition of the metal surface and the angle of incidence, but its dependence on wavelength is of principal interest. Figure 5.4, giving the reflectivity curves for aluminum, copper, and iron, shows that copper reflects the red end of the visible spectrum much more strongly than it does the violet. As a result, copper appears reddish when seen in ordinary white light. Aluminum reflects red light even more strongly than copper does, and it reflects almost equally well light of other wavelengths. Consequently aluminum appears metallic gray in ordinary light. Iron has only moderate reflecting power for all wavelengths and therefore shows a duller metallic gray color.

The actual appearance of a metal is greatly influenced by a relatively thin oxide film that forms on the surface with exposure to air. For example, unoxidized iron is much brighter than the pieces ordinarily seen. This surface film has been put to good use in aluminum by dyeing the artificially thickened (anodized) film and thus producing a variety of bright surface colors.

All metals absorb light strongly, and ordinary pieces of metal are opaque to visible light. However, if an appreciably transmitted beam is obtained by the use

**TABLE 12.5**

Optical Properties of Several Metals

| Metal | Wavelength of the incident light, $\mu m$ | Refractive index $n$ | Absorption coeff. $\kappa$ | Reflectivity $R$, % |
|---|---|---|---|---|
| Silver | 0.400 | 0.075 | 1.93 | 94 |
| | 0.600 | 0.060 | 3.75 | 99 |
| | 0.750 | 0.080 | 5.05 | 99 |
| | 0.950 | 0.110 | 6.56 | 99 |
| Gold | 0.450 | 1.40 | 1.88 | 40 |
| | 0.600 | 0.23 | 2.97 | 91 |
| | 0.750 | 0.16 | 4.42 | 97 |
| | 0.950 | 0.19 | 6.10 | 98 |
| Copper | 0.450 | 0.87 | 2.20 | 58 |
| | 0.600 | 0.17 | 3.07 | 94 |
| | 0.750 | 0.12 | 4.62 | 98 |
| | 0.950 | 0.13 | 6.22 | 99 |
| Aluminum | 0.400 | 0.40 | 3.92 | 91 |
| | 0.600 | 0.97 | 6.00 | 90 |
| | 0.750 | 1.80 | 7.12 | 88 |
| | 0.850 | 2.08 | 7.15 | 86 |
| | 0.950 | 1.75 | 8.50 | 91 |
| Steel | 0.589 | 2.49 | 3.44 | 58 |
| Sodium | 0.589 | 0.005 | 2.61 | 99.7 |

of extremely thin metal foil, the transmitted color is complementary to the color seen by reflected light. This phenomenon is explained by the fact that absorption and reflection vary in a parallel fashion with change in wavelength. A wavelength that is strongly reflected and contributes significantly to a reflected beam is also strongly absorbed and tends to be absent from a transmitted beam.

## NUCLEAR THEORY

By far the largest number of reactions and properties of general interest involve principally the valence electrons. Since these electrons are relatively loosely bound to the atom, the energies needed to influence them are low, on the order of one electron volt per atom. A second type of reaction, such as that involved in the production of x-rays, requires the removal of electrons from the inner orbits of the atom. The energies necessary for this purpose range up to thousands of electron volts. Finally, changes that occur in the nucleus of the atom involve energies on the order of a million electron volts. These nuclear reactions are the basis of nuclear power, a field in which metallurgy is vitally important.

**Structure of the nucleus.** Two kinds of particles, protons and neutrons, can be considered to make up the atomic nucleus. Attractive forces, the natures of which are not completely understood, overcome the force of repulsion between the positively charged protons and cause the volume of the nucleus to be very small compared with that of the entire atom. The constitution of a given nucleus is described by two integers: $Z$, its charge, which is equal to the number of protons, and $M$, its mass number, which is equal to the sum of its protons and neutrons. The magnesium nucleus, for example, is represented by the symbol $^M_Z\text{Mg}$. The charge number $Z$ can be omitted since it duplicates the information given by the chemical symbol; that is, the number of protons, $Z$, equals the number of electrons characteristic of the atom. Three *stable* isotopes of magnesium exist, $^{24}\text{Mg}$, $^{25}\text{Mg}$, and $^{26}\text{Mg}$, which differ in the number of neutrons present in the nucleus. There is only a limited range of masses over which a given element is stable. For example, the magnesium isotopes $^{23}\text{Mg}$ and $^{27}\text{Mg}$ decompose within a few minutes. Much attention is devoted to the production and disintegration of these *unstable isotopes*, and the following law governing their useful life is widely used:

$$N = N_0 \exp\left(-\lambda t\right), \tag{12.28}$$

where $N$ is the number of unstable atoms remaining after the time $t$ has elapsed, $N_0$ is the number of unstable atoms present initially, and $\lambda$ is the decay constant for a particular isotope. The *half-life*, $T$, is defined as the time during which half of a given number of unstable atoms decompose, and it is related to $\lambda$ by the equation

$$T = \frac{\log_e 2}{\lambda}. \tag{12.29}$$

Various time units, from seconds to years, are used for $T$ and $\lambda$, depending on the stability of the isotope.

**Nuclear reactions.** The principal energy-producing reactions are of course the *fusion* of two light nuclei to form a heavier nucleus and the *fission* of a very heavy nucleus into fission products. An example of the latter reaction is the fission of uranium-235 by a neutron,

$$^{235}\text{U} + \text{n} = \begin{pmatrix} \text{fission products} \\ \text{such as } ^{140}\text{Ba} \end{pmatrix} + (1 \text{ to } 3)\text{n},$$

in which 200 MeV (million electron volts) of energy are released. It will be shown later that the metallurgical problems in nuclear engineering largely arise from two features of this reaction: (1) the energetic fission products and radiation, and (2) the necessity for conserving as many as possible of the neutrons produced (about 2.5 on the average) both for causing additional fission reactions and for breeding new fissionable material.

**TABLE 12.6**

List of Atomic Entities

| Name | Symbol | Mass | Charge |
|------|--------|------|--------|
| Electron | $\beta^-$ | $9.11 \times 10^{-28}$ gram | $-1.602 \times 10^{-19}$ C |
| Positron | $\beta^+$ | $9.11 \times 10^{-28}$ gram | $+1.602 \times 10^{-19}$ C |
| Proton | p | $1.67 \times 10^{-24}$ gram | $+1.602 \times 10^{-19}$ C |
| Neutron | n | $1.68 \times 10^{-24}$ gram | 0 |
| Deuteron | d | (proton + neutron) | |
| Alpha particle | $\alpha$ | (2 protons + 2 neutrons) | |
| Beta particle | $\beta^-$ | (electron) | |
| Gamma ray | $\gamma$ | (electromagnetic radiation of short wavelength) | |

Less violent nuclear reactions lead to the production of unstable nuclei. These reactions are produced by bombardment of the various elements with alpha particles, protons, deuterons, neutrons, and gamma rays (see Table 12.6). Neutron bombardment may result in the important *neutron capture* type of reaction, which is illustrated by the following example:

$$^{107}\text{Ag} + \text{n} = {}^{108}\text{Ag} + \gamma.$$

The stable nucleus $^{107}\text{Ag}$ absorbs a neutron and releases energy in the form of gamma radiation. The silver isotope $^{108}\text{Ag}$ thus produced is unstable and decomposes with emission of an electron:

$$^{108}\text{Ag} = {}^{108}\text{Cd} + \beta^-.$$

The loss of a unit of negative electricity coincides with an equal gain in positive charge by the nucleus. A second example of nuclear reactions is the following:

$$^{52}\text{Cr} + \text{n} = {}^{52}\text{V} + \text{p}.$$

The chromium nucleus captures a neutron and changes into a vanadium isotope with emission of a proton. The unstable vanadium atom decomposes in accordance with Eq. (12.28), with a half-life of four minutes. The decomposition reaction is

$$^{52}\text{V} = {}^{52}\text{Cr} + \beta^-.$$

In this instance the electron is emitted with an energy of 2 MeV.

A convenient measure of the likelihood that a metal will absorb a neutron is given by its nuclear *cross section*. This method pictures the probability of interaction of a neutron with a given nucleus as depending on the "size" of the nucleus expressed in units of barns, defined as $10^{-24}$ cm$^2$. Table 12.7 shows the great range in the cross sections of various metals for thermal (low-energy) neutrons. Rare metals, such as zirconium and boron, occupy a prominent position in nuclear engineering because of their unusual cross sections.

TABLE 12.7

Absorption Cross Sections of Several Metal Atoms for Thermal Neutrons

| Atom | Cross section, barns |
|------|---------------------|
| Beryllium | 0.009 |
| Magnesium | 0.059 |
| Zirconium | 0.18 |
| Aluminum | 0.22 |
| Iron | 2.4 |
| Copper | 3.6 |
| Nickel | 4.5 |
| Manganese | 12.6 |
| Hafnium | 115 |
| Boron | 750 |
| Cadmium | 2 400 |
| Gadolinium | 44 000 |

**Metals in nuclear engineering.**   Nuclear reactors of various types are employed in the technical applications of nuclear energy.   Metallic materials are applied in several distinctly different ways in nuclear reactors: as fuels, as structural members, as control materials, etc.   Let us now consider the special properties required for these applications.

*Fuels.*   The principal nuclear fuels are the fissionable isotopes of uranium, plutonium, and thorium.   These elements are occasionally used in the metallic state in alloys, but usually as oxides.   The nuclear fuel is subjected to a damaging environment due to high thermal stresses and bombardment by fission fragments. Damage by fission fragments occurs in three forms.   First, a significant fraction of the ultimate fission products are gas atoms, such as xenon and krypton.   These gas atoms collect as bubbles at discontinuities in the fuel structure (such as grain boundaries and dislocations) and cause the fuel to expand.   Second, high-energy fission fragments displace groups of atoms from their normal lattice positions, creating a *displacement spike* (Fig. 12.33).   Because the displaced atoms occupy interstitial lattice positions, the resulting vacancies can cluster and cause further

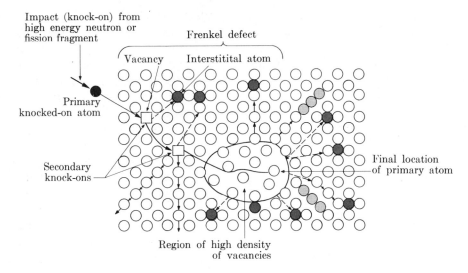

Impact (knock-on) from
high energy neutron or
fission fragment

Frenkel defect

Vacancy    Interstitital atom

Primary
knocked-on atom

Secondary
knock-ons

Final location
of primary atom

Region of high density
of vacancies

**Fig. 12.33** Schematic illustration of the type of complex atomic displacements denoted by the term *displacement spike*. (After A. Seeger)

expansion of the fuel. Third, some of the fission products are additional metal atoms and also tend to increase the dimensions of the fuel. The resulting distortion and growth of the fuel may be so severe as to limit its useful life to a time much shorter than that required for optimum "burnup" of its fissionable material. The swelling of uranium fuel elements can be reduced by alloying with aluminum or zirconium. The resulting alloy provides a more stable matrix and also facilitates the removal of heat.

*Cladding and structural materials.* The cladding of the nuclear fuel serves the important function of protecting the chemically reactive fuel material from the coolant (water, air, or liquid metal). It also prevents radioactive fission products from entering the coolant stream and being distributed throughout the reactor system. The principal cladding materials used at present are aluminum, Zircalloy (a zirconium alloy), and stainless steel. Aluminum is suitable for low-temperature reactors, which are cooled by air or water. Zircalloy is widely used in water-cooled reactors for generating electric power. Stainless steel can operate at higher temperatures than can Zircalloy, but its cross section for neutron capture is higher. Stainless steel is currently used in high-temperature (fast-neutron, breeder) reactors which employ liquid sodium as the coolant.

Alloys used for structural components of reactors (and for cladding, also) must have adequate engineering strength and ductility and must maintain dimensional stability during service. The properties of structural alloys change significantly during service in a reactor because of *radiation damage*, which results from bombardment of the metal by neutrons escaping from the fissioning fuel. If the energy of the neutron is above a minimum value, called the *displacement energy* (usually about 25 eV), the atom may be knocked off its lattice site. A billiard-ball-type reaction can then occur, with this atom knocking other atoms off their lattice sites, the end result being the displacement spike mentioned above. The interstitial atoms and vacancies then proceed to cluster in various ways, depending on the temperature and the character of the neutron flux. At low temperatures (less than about $0.3T_f$, where $T_f$ is the melting temperature of the alloy) the clusters are small and numerous, appearing as tiny black spots in electron micrographs. This *black-spot damage* increases the strength, but also causes a loss in ductility. At higher temperatures (between 0.3 and $0.5T_f$), low neutron fluxes produce little damage because the interstitial atoms and the vacancies largely annihilate each other. If the neutron flux is high,* however, the vacancies cluster into voids and the interstitial atoms produce dislocation loops. This structure causes pronounced increase in strength and decrease in ductility. The formation of voids decreases the density of the alloy and therefore increases the dimensions of structural members. Density changes as great as 10% occur in type 304 stainless steel during fast-neutron irradiation. At temperatures above $0.5T_f$, the main source of radiation damage is the formation of helium bubbles, the helium being derived from alpha particles produced during irradiation. The bubbles form preferentially at grain boundaries and along dislocations, and cause serious embrittlement of alloys such as stainless steel.

*Reactor-control materials.* Nuclear reactors are regulated and shut down through the action of control rods that can absorb the neutrons required to maintain a chain reaction. The principal requirement of a control-rod material is a high cross section for neutrons; the materials used in reactor control rods include gadolinium, hafnium, cadmium, boron, and boron alloys and compounds, and tantalum. Gadolinium is extremely rare, and even hafnium, which is obtained as a by-product in the refining of zirconium, is too expensive for most applications. The use of cadmium metal has been restricted because of its low melting point. Metallic boron does not have desirable mechanical properties, but various boron compounds and alloys are widely used. A commercial control-rod alloy is a 2%-boron steel that can be forged or rolled.

Other aspects of a reactor environment include the creation of induced radiation in a metal, thus creating a problem in maintenance and in disposal of used components. Induced activity occurs by the formation of unstable long-lived isotopes. The best-known example is the creation of $^{60}$Co in cobalt, but manganese has a similar objectionable behavior and consequently is maintained at a low level in steels for nuclear applications. Another possible action of radiation is to cause unusual chemical reactions. For example, some types of corrosion that are not observed under normal conditions can occur in the presence of radiation.

---

* A fast-neutron breeder reactor, for example, has a flux of about $10^{19}$ neutrons/(m$^2$ · s), the neutrons having an average energy of 7 MeV.

**Radioactive tracers.** Unstable (radioactive) atoms behave like stable atoms of the same element in most respects, but the location of the unstable atoms is easily determined because of their occasional decomposition. It is possible to add a small number of unstable atoms to an ordinary piece of metal and to follow certain changes in the metal by means of these "tagged" atoms. This procedure is called a radioactive tracer technique. It is used, for example, in studies of self diffusion. Other applications include the investigation of steelmaking reactions, wear resistance, heat treatments, and oxidation.

## PROBLEMS

**1.** For the special case of a cubical block of metal ($L_x = L_y = L_z = L$), show that Eq. (12.4) can be put in the form of Eq. (12.5).

**2.** a) Using suitable sketches, show that the integral in the numerator of Eq. (12.11) is the total energy of the electrons and that the denominator is the total number of electrons. Therefore $\bar{E}$ is the average energy. More generally, Eq. (12.11) defines the *first moment* of the distribution $\rho(E)$.

b) Using Eq. (12.8), evaluate the two integrals and verify the value of $\bar{E}$ given in Eq. (12.11).

**3.** An electron beam of which of the following velocities would be most strongly diffracted by a crystal having periodic spacings of 3 Å separation: a) $2 \times 10^8$, b) $2 \times 10^6$, c) $2 \times 10^4$ m/s?

(Recall that appreciable diffraction occurs only when the wavelength of the beam and the spacing of the grating are of the same order of magnitude.)

**4.** Using sketches similar to those in Fig. 12.5, explain why the electron represented by $\mathbf{k}_2$ in Fig. 12.5(a) is diffracted by the (010) plane.

**5.** Prove that the boundaries of the second Brillouin zone shown in Fig. 12.6 are correct. Since these boundaries are determined by reflections from planes at 45° in Fig. 12.5, it is convenient to use coordinates $x'$ and $y'$ that are rotated 45° relative to the $x$- and $y$-axes. The same method can then be used as for the first zone except that the spacing of the planes is $a/\sqrt{2}$ rather than $a$.

**6.** Determine approximate values for the two energy levels in the second zone of Fig. 12.6 that would give an $N(E)$ curve similar to Fig. 12.8(b).

**7.** Why is it unsatisfactory to explain electrical conduction entirely on the basis of free electrons?

**8.** The drift velocity $v_D$ is the velocity that an average electron has when a conducting substance is acted on by an electric field. Consider how $v_D$ can be expressed as a function of the relaxation time $\tau$ (Eq. 12.16), starting with the defining equation

$$v_D(t) = v_D(0)e^{-t/\tau}. \tag{12.30}$$

An example of the application of this equation is the decrease in $v_D$ when the electric field is removed at time $t = 0$; $v_D(0)$ is then the value of $v_D$ in the presence of the field $E$, and $v_D(t)$ is the value at any later time $t$. After a sufficiently long time, $v_D = 0$.

a) Using Newton's second law and assuming a frictional force proportional to velocity, show that the equation

$$Ee = m\left(\frac{dv_D}{dt} + \frac{1}{\tau}v_D\right)$$
(12.31)

describes the motion of an electron of charge $e$ and mass $m$ in an electric field $E$.

b) If $v_D$ is the steady-state value in Eq. (12.31), show that $v_D(t)$ at time $t$ after removal of the electric field is given by Eq. (12.30). This proves that $(1/\tau)v_D$ is a suitable frictional term in Eq. (12.31).

c) At the steady-state condition $v_D$ is constant. Show that Eq. (12.31) then reduces to the desired expression, Eq. (12.16).

9. The methods used in the free-electron theory can be applied to determine the number $N_e$ of electrons in the conduction band in an intrinsic semiconductor.

a) Explain why

$$N_e = \int_{E_g}^{\infty} \rho(E)\,P(E)\,dE$$
(12.32)

is an appropriate equation for evaluating $N_e$.

b) Using Eqs. (12.8) and (12.9), perform the integration of Eq. (12.32).

c) Explain why the Fermi level, $E_F$, is at $E_g/2$ in this case.

d) Therefore, show that your result in (b) is identical with Eq. (12.19).

10. Make two sketches analogous to Fig. 12.10(b) explaining the relation of $p$-type semi-conductivity to the band structure. The localized acceptor level, $E_A$, lies about 0.1 eV above the top of the valence band.

a) In the first sketch, show the condition at 0 K (in the absence of ionization).

b) Explain how the concept of holes at the $E_A$ level permits the valence level to be full in this sketch.

c) Make a second sketch showing the condition at 50% ionization.

11. Assume that a rough calculation has shown that a small electric furnace can be heated to 500°C if 1000 watts are supplied. How many meters of nichrome wire 1 mm in diameter are needed to give this wattage at 115 volts, based on the room temperature resistance value? Assuming that the wire will operate at 500°C, show how this will affect the length of wire needed.

12. You are asked for an opinion on the following suggestion. Since silver has excellent conductivity and since several percent of it can be dissolved in solid aluminum, why not strengthen aluminum for use in high-voltage transmission lines by solid-solution hardening using silver?

a) Is this idea basically sound?

b) Can you suggest a better strengthening method for the purpose?

c) Explain the advantages of the method you suggest.

13. A resistor of exactly 5 ohms resistance is needed for a routine laboratory test. Contrast the errors introduced by normal temperature fluctuations if (a) copper or (b) constantan wire is used in winding this resistor.

**14.** How may (a) Joule heating and (b) contact potential effects be reduced to a negligible value in thermocouple circuits?

**15.** Explain what commercial thermocouple is most suitable for measurements at 200°C.

**16.** Contrast the meanings of "magnetic alloy" and "magnetized alloy."

**17.** Give reasons why pure iron is not used in transformers.

**18.** Would it be feasible to use the magnetostriction effect in pure nickel to operate a valve requiring a linear displacement of 5 mm to operate it?

**19.** Pauling's theory of valence pictures a filled shell containing $N$ electrons as consisting of $N/2$ pairs of electrons of opposite spin. The removal of $n$ electrons from a filled shell leaves $n$ unpaired electrons, which are then available for bonding in the previously filled shell. Explain why the electronic configuration $(4d)^8(5s)^1(5p)^2$ gives silver a metallic valence of five.

**20.** Apply the correlation rules to the $\varepsilon$- and $\eta$-phases of the Ag-Zn system, Fig. 12.30, and show that both should be hexagonal-close-packed.

**21.** It is impossible to cool a piece of metal *instantly* from a high temperature. Consider a large plate of iron 0.1 m thick initially at 700°C. Make an approximate analysis to estimate the time necessary to cool the center of the plate to 350°C, assuming that the two *surfaces* of the plate can be cooled instantly to 0°C. The density of iron is about $8 \times 10^{-3}$ kg/m³.

**22.** In order to make readings of the relative displacement of two arms in a creep-testing device, a dial gage is to be equipped with extension bars that have a total length of 0.5 m. Assuming a maximum variation in room temperature of 10°C:

a) What error would be introduced by expansion if the extension bars were made of iron?

b) If they were made of invar? The dial gage can be read to $\pm 10\,\mu$m.

**23.** From the data given for the vanadium isotope:

a) determine the time constant $\lambda$,

b) calculate the fraction of a sample of this isotope that would remain after 24 hours.

## REFERENCES

Cullity, B. D., *Introduction to Magnetic Materials*, Reading, Mass.: Addison-Wesley, 1972. A comprehensive treatment of magnetic phenomena and materials at an intermediate level.

Hume-Rothery, W., *Atomic Theory for Students of Metallurgy*, 3rd ed., London: Institute of Metals, 1962. The essential ideas of wave mechanics are explained and then applied to metals.

Hutchison, T. S. and D. C. Baird, *Physics of Engineering Solids*, 2nd ed., New York: John Wiley, 1968. A relatively elementary treatment with applications to many aspects of metallic behavior.

Kittel, C., *Introduction to Solid State Physics*, 4th ed., New York: John Wiley, 1971. The standard, advanced treatment of the physical properties of metals and other materials.

Smith, C. O., *Nuclear Reactor Materials*, Reading, Mass.: Addison-Wesley, 1967. A thorough treatment of materials for nuclear reactors in terms of the design and operation of various reactors.

Stringer, J., *Introduction to the Electron Theory of Solids*, New York: Pergamon Press, 1967. A treatment of quantum mechanics at an intermediate level; metals are included among other materials.

Wert, C. A. and R. M. Thomson, *Physics of Solids*, 2nd ed., New York: McGraw-Hill, 1970. A broad treatment of the physical properties of metals and other materials at an intermediate level.

# APPENDIX


# SI UNITS

The main purpose of SI (Système International) units is to facilitate interdisciplinary exchange of information. Physical metallurgy is itself an interdisciplinary subject matter related principally to the fields of physics, chemistry, and engineering. As an example of the advantage of a unified system of units, consider the measures of energy encountered by the metallurgist in the typical literature of these three fields. They are the *erg* (physics), the *calorie* (chemistry), and the *BTU* (engineering). Interdisciplinary cooperation would evidently be facilitated if a single unit were universally employed, such as the *joule* in the international system. Another aspect of SI units, which evolved during extensive experience with the m.k.s. (meter-kilogram-second) system, is an integration of the units for all areas of science and technology such as mechanics, electricity, magnetism, and thermodynamics. In particular, the practical engineering units in electricity (volt, ampere, and watt) are part of the international system.

**TABLE A.1**

SI Base Units

| Physical quantity | Name of unit | Symbol for unit |
| --- | --- | --- |
| Length | meter | m |
| Mass | kilogram | kg |
| Time | second | s |
| Electric current | ampere | A |
| Thermodynamic temperature | kelvin | K |
| Amount of substance | mole | mol |
| Luminous intensity | candela | cd |

The international system employs a few *basic units*, Table A.1, and many *derived units*, Table A.2, obtained from the basic units.* An important derived unit is the *newton*, used to measure force. From Newton's second law of motion,

$$\text{force} = \text{mass} \times \text{acceleration}, \tag{A.1}$$

it follows that in SI units the newton is the force that causes a mass of 1 kg to accelerate at 1 m/s$^2$. This unit of force is $10^5$ larger than the dyne (c.g.s. units) and is smaller than the pound (force) by a factor of 4.45. Coulomb's law of electrical attraction has the following form for use with SI units:

$$\text{force} = -\frac{1}{4\pi\varepsilon_0} \frac{(\text{charge})(\text{charge})}{(\text{distance})^2}, \tag{A.2}$$

where $\varepsilon_0$ is the permittivity of free space (see Table A.3). The pertinent relation

---

* A more complete list of units can be found in *Metallurgical Transactions*, Vol. 3, pp. 355–358 (1972).

**TABLE A.2**
Representative SI Derived Units

| Physical quantity | Name of unit | Symbol for unit |
|---|---|---|
| Velocity | — | m/s |
| Acceleration | — | $m/s^2$ |
| Force | newton | N |
| Pressure, stress | pascal | $Pa \, (=N/m^2)^*$ |
| Energy, work, quantity of heat | joule | $J \, (=N \cdot m)$ |
| Power | watt | $W \, (=J/s)$ |
| Electric charge | coulomb | $C \, (=A \cdot s)$ |
| Potential difference | volt | $V \, (=W/A)$ |
| Electric resistance | ohm | $\Omega \, (=V/A)$ |
| Capacitance | farad | $F \, (=C/V)$ |
| Magnetic flux | weber | $Wb \, (=V \cdot s)$ |
| Inductance | henry | $H \, (=Wb/A)$ |
| Surface tension | — | $N/m$ or $J/m^2$ |
| Permittivity | — | F/m |
| Permeability | — | H/m |

* In this book, only the units $N/m^2$ (but not Pa) are used for stress.

**TABLE A.3**
Values of Fundamental Constants in SI Units

| Name of quantity | Symbol | Value |
|---|---|---|
| Velocity of light | $c$ | $2.998 \times 10^8$ m/s |
| Charge of electron | $e$ | $1.602 \times 10^{-19}$ C |
| Mass of electron | $m$ | $9.109 \times 10^{-31}$ kg |
| Planck's constant | $h$ | $6.626 \times 10^{-34}$ J $\cdot$ s |
| Boltzmann's constant | $k$ | $1.380 \times 10^{-23}$ J/K |
| | | $(8.616 \times 10^{-5}$ eV/K) |
| Avogadro's number | $N$ | $6.023 \times 10^{23}$ (g mol)$^{-1}$ |
| | | $6.023 \times 10^{26}$ (kg mol)$^{-1}$ |
| Gas constant | $R$ | 8.314 J/(g mol $\cdot$ K) |
| | | $8.314 \times 10^3$ J/(kg mol $\cdot$ K) |
| Volume of ideal gas at 0°C and atmospheric pressure | $V^0$ | $22.41 \times 10^{-3}$ m$^3$/g mol |
| | | 22.41 m$^3$/kg mol |
| Acceleration of gravity | $g$ | 9.781 m/s$^2$ |
| Faraday constant | $F$ | $9.649 \times 10^4$ C/g equivalent |
| | | $9.649 \times 10^7$ C/kg equivalent |
| | | $[2.306 \times 10^4$ cal/(V $\cdot$ g equivalent)] |
| Permeability of vacuum | $\mu_0$ | $4\pi \times 10^{-7}$ H/m |
| Permittivity of vacuum | $\varepsilon_0$ | $8.854 \times 10^{-12}$ F/m |

**TABLE A.4**

Useful Conversion Factors

| | |
|---|---|
| *Length* | *Energy* |
| 1 meter = 39.37 in. = 3.281 ft | 1 joule = $10^7$ erg = 0.2390 cal |
| 1 inch = 25.40 mm | $= 0.625 \times 10^{19}$ eV |
| 1 Å = $10^{-10}$ m | $= 9.485 \times 10^{-4}$ BTU |
| | 1 erg = 1 dyn $\cdot$ cm = $10^{-7}$ J |
| *Area* | $= 0.625 \times 10^{12}$ eV |
| 1 $m^2$ = $1.550 \times 10^3$ $in^2$ = 10.76 $ft^2$ | 1 electron volt = $1.602 \times 10^{-19}$ J |
| 1 $in^2$ = $6.452 \times 10^{-4}$ $m^2$ | $= 1.602 \times 10^{-12}$ erg |
| 1 $ft^2$ = $9.29 \times 10^{-2}$ $m^2$ | 1 eV/atom = $2.306 \times 10^4$ cal/g mol |
| | 1 calorie = 4.184 J = $3.968 \times 10^{-3}$ BTU |
| *Volume* | 1 BTU = $1.054 \times 10^3$ J = $2.520 \times 10^2$ cal |
| 1 $m^3$ = $10^3$ liter = $6.102 \times 10^4$ $in^3$ | |
| $= 35.1$ $ft^3$ | |
| 1 liter = $10^3$ $cm^3$ = 61.02 $in^3$ | *Miscellaneous* |
| $= 3.51 \times 10^{-2}$ $ft^3$ | 1 atmosphere = $1.013 \times 10^5$ N/$m^2$ |
| 1 $in^3$ = $1.639 \times 10^{-5}$ $m^3$ | $= 760$ mm Hg (torr) |
| 1 $ft^3$ = $2.832 \times 10^{-2}$ $m^3$ | $= 1.013 \times 10^6$ dyn/$cm^2$ |
| | $= 14.7$ lb/$in^2$ |
| *Velocity* | 1 lb/$in^2$ = $6.90 \times 10^3$ N/$m^2$ |
| 1 m/s = 3.281 ft/s | 1 N/$m^2$ = $1.450 \times 10^{-4}$ lb/$in^2$ |
| 1 ft/s = 0.3048 m/s | $= 0.981 \times 10^7$ kg/$mm^2$ |
| | 1 oersted = 79.6 ampere-turns/m |
| *Force* | 1 gauss = $10^{-4}$ weber/$m^2$ |
| 1 newton = $10^5$ dyne = 0.2247 lb (force) | |
| 1 lb (force) = 4.45 N | |

among magnetic quantities for the purpose of this book is

$$\text{magnetic flux density} = \mu_0 \times \text{magnetic field,} \tag{A.3}$$

where $\mu_0$ is the permeability of free space and the magnetic field has units of ampere-turns/meter. Two units are in common use, although not an integral part of the international system. These are the angstrom unit (1 Å $= 10^{-10}$ m) and the electron volt (1 eV $= 1.602 \times 10^{-19}$ J). An electron volt is the energy gained by an electron in falling through a potential difference of one volt.

For the convenience of the reader whose previous experience has been principally with the c.g.s. and English units, conversions to these units are given frequently in this book. Table A.4 lists conversion factors to meet other needs of the reader.

# INDEXES

# AUTHOR INDEX

# SUBJECT INDEX